Rural Change and Sustainability
Agriculture, the Environment and Communities

RURAL CHANGE AND SUSTAINABILITY
Agriculture, the Environment and Communities

Edited by
Stephen Essex
University of Plymouth, UK
Andrew Gilg
University of Exeter, UK
Richard Yarwood
University of Plymouth, UK
with John Smithers
University of Guelph, Canada
and Randall Wilson
Gettysburg College, US

CABI Publishing

CABI Publishing is a division of CAB International

CABI Publishing
CAB International
Wallingford
Oxfordshire OX10 8DE
UK
Tel: +44 (0) 1491 832111
Fax: +44 (0) 1491 833508
E-mail: cabi@cabi.org
Web site: www.cabi-publishing.org

CABI Publishing
875 Massachusetts Avenue
7th Floor
Cambridge, MA 02139
USA
Tel: +1 617 395 4056
Fax: +1 617 354 6875
E-mail: cabi-nao@cabi.org

A catalogue record for this book is available from the British Library, London, UK.
A catalogue record for this book is available from the Library of Congress, Washington, DC, USA.

Library of Congress Cataloging-in-Publication Data

Rural Change and sustainability: agriculture, the environment and communities / edited by Stephen Essex … [et al.].

p. cm
includes biographical references and index.
ISBN-13: 978-0-85199-082-8 (alk. paper)
ISBN-10: 0-85199-082-7 (alk. paper)
1. Rural development—Environmental aspects—Congresses. 2. Agriculture—Economic aspects—Congresses. 3. Rural Geography—Congresses. I. Essex, Stephen.
HN49.C6R84 2005-05-21
307.1'412'0917521—dc22

2005009234

ISBN 0 85199 082 7

Printed and bound in the UK by Cromwell Press, Trowbridge from copy supplied by the editors.

Contents

Communities

Governance

List of contributors

Katie Algeo, Department of Geography and Geology, Western Kentucky University, 1 Big Red Way, Bowling Green, KY 42101, United States. (katie.algeo@wku.edu).

Holly Barcus, Institute for Regional Analysis and Public Policy, Morehead State University, 114 Rader Hall, Morehead, Morehead, KY 40351, United States. (h.barcus@morehead-st.edu).

Lyndenn Behm, Department of Rural Development, Brandon University, Brandon, Manitoba, Canada, R7A 6A9.

David Bruce, Rural and Small Town Programme, Mount Allison University, Avard-Dixon Building, 320A-144 Main Street, Sackville, New Brunswick, Canada, E4L 1A7. (dwbruce@mta.ca).

Lisa Butler Harrington, Department of Geography, Kansas State University, Seaton Hall, Manhattan, KS 66506-2904, United States. (lharrin@k-state.edu).

Kirsten Valentine Cadieux, Department of Geography and Planning, University of Toronto, Toronto, Ontario, Canada, M5S 3G3. (kvcad@yahoo.com).

Deborah Che, Department of Geography, Western Michigan University, Wood Hall, Kalamazoo, MI 49008, United States. (deborah.che@wmich.edu).

Christine Corcoran, Research Fellow, School of Education, University of Birmingham, Birmingham, B15 2TT, UK. (c.e.corcoran@bham.ac.uk).

Ann K. Deakin, Department of Geosciences, State University of New York Fredonia, 114, Houghton Hall, Fredonia, New York, United States. (Ann.Deakin@Fredonia.edu).

Stephen Essex, School of Geography, University of Plymouth, Drake Circus, Plymouth, PL4 8AA, UK. (sessex@plymouth.ac.uk).

John Everitt, Department of Geography, Brandon University, Brandon, Manitoba, Canada, R7A 6A9. (everitt@brandonu.ca).

Owen J. Furuseth, Department of Geography and Earth Sciences, University of North Carolina at Charlotte, UNC Charlotte, 9201 University City Boulevard, Charlotte, NC 28223-0001, United States. (ojfuruse@email.uncc.edu).

Hugh J. Gayler, Brock University, St. Catherines, Ontario, Canada, L2S 3A1. (hjgayler@spartan.ac.brocku.ca).

Andrew Gilg, Department of Geography, University of Exeter Amory Building, Rennes Drive, Exeter, EX4 4RJ, UK. (a.w.gilg@exeter.ac.uk).

Alison M. Gill, Department of Geography, Simon Fraser University, Burnaby, British Columbia, Canada, V7N 3K7. (agill@sfu.ca).

Greg Halseth, Department of Geography, University of Northern British Columbia, Prince George, British Columbia, Canada. (halseth@unbc.ca).

Lewis Holloway, Department of Geography, University of Hull, Hull, HU6 7RX, UK. (L.Holloway@hull.ac.uk).

Brian Ilbery, Geography Subject Area, Coventry University, Priory Street, Coventry, CV1 5FB, UK. (b.ilbery@coventry.ac.uk).

Ian MacLachlan, Department of Geography, University of Lethbridge, Lethbridge, Alberta, Canada, T1K 3M4. (maclachlan@uleth.ca).

Joan Marshall, McGill School of Environment, Raymond Building, R3-021 21,111 Lakeshore, Ste. Anne-de-Bellevue, Quebec, H9X 3V9, Canada. (marshallj@macdonald.mcgill.ca).

Chris Mayda, Department of Geography and Geology, Eastern Michigan University, Strong 217 Ypsilanti, MI 48197, United States. (cmayda@emich.edu).

Damian Maye, Geography Subject Area, Coventry University, Priory Street, Coventry, CV1 5FB, UK. (d.maye@coventry.ac.uk).

Douglas Ramsey, Department of Rural Development, Brandon University, Brandon, Manitoba, Canada, R7A 6A9. (ramsey@brandonu.ca).

Guy M. Robinson, Department of Geography, Kingston University, Surrey, UK. (G.Robinson@kingston.ac.uk).

John Smithers, Department of Land Resource Science, University of Guelph, Guelph, Ontario, Canada, N1G 2W1. (jsmither@uoguelph.ca).

Michael Troughton, Department of Geography, University of Western Ontario, London, Ontario, Canada, N6A 5C2. (mjtrough@uwo.ca).

Gregory Veeck, Department of Geography, Western Michigan University, Wood Hall, Kalamazoo, MI 49008, United States. (gregory.veeck@wmich.edu).

Ann Veeck, Department of Marketing, Western Michigan University, Haworth College of Business, Kalamazoo, MI 49008, United States. (ann.veeck@wmich.edu).

Peter W. Williams, Department of Geography, Simon Fraser University, Burnaby, British Columbia, Canada, V7N 3K7. (agill@sfu.ca).

Randall K. Wilson, Department of Environmental Studies, Campus Box 2455, Gettysburg College, Gettysburg, PA 17325, United States. (rwilson@Gettysburg.edu).

Richard Yarwood, School of Geography, University of Plymouth, Drake Circus, Plymouth, PL4 8AA, UK. (ryarwood@plymouth.ac.uk).

Preface

The chapters in this book have been selected from the papers given at the fifth British-American Rural Geography Symposium held at the Universities of Exeter and Plymouth, in the County of Devon in the UK during July 2003. The Symposium has been held roughly every four years since the first in Guelph, Ontario in 1985. Since then there have been other meetings in the English Midlands (1991), North Carolina (1995) and Newfoundland (1999). CAB International published selected papers from the 1991 and 1995 symposia and is proud to be publishing papers from this fifth conference.

The longevity of the Symposia series is a reflection of the ties that bind the Anglo-Saxon school of Geography based on the English language, and indeed some of the meetings have seen contributions from our cousins from the Antipodes and Africa. However, for British delegates, who also collaborate with many colleagues in continental Europe through the successful and long-standing Rural Geography Study Group links with notably, France, Spain and Germany, the North American links provide a different perspective on Rural Geography, as Chapter 1 highlights. For example, there is a striking difference between the ways in which agriculture is perceived and planned either side of the Atlantic Ocean.

Thus, the theme of the conference, 'Contrasting Ruralities', sought to stress the different responses to changing rural circumstances between Europe and North America in the core areas of economics, society and landscapes. However, the contents of the delivered papers encouraged a revision of the main themes of this book which emphasizes that key differences and similarities are best grouped under the overarching concept of 'Rural Sustainability' with four key sub-issues: agricultural responses, environmental issues, communities and finally, governance.

Accordingly, this book is much more than a collection of conference papers since only 21 papers have been selected from the 40 odd papers given at the Symposium. A few authors decided not to seek publication since their papers were for example, work in progress. Selection by the editors, CABI Publishing, their anonymous referees and selected delegates from the USA and Canada was based on the contribution that the papers offered to the four main issues outlined above. Once selected, authors were asked to embark on rewrites to emphasize the four key sub-issues and to fit the parameters of this book.

The Symposium could not have gone ahead without a great deal of hard work by the organizers, myself, Stephen Essex and Richard Yarwood, but I must thank Stephen Essex in particular who was a tower of strength throughout the Symposium and was the major architect of its success. Well done, Stephen. Other people who made a great contribution were Charles Watkins (Chair of the Rural Geography Study Group of the Royal Geographical Society and Institute of British Geographers), from across the Atlantic, Leslie Duram (Contemporary Agriculture and Land Use Speciality Group of the Association of American Geographers) and Doug Ramsey (Rural Study Group, Canadian Association of Geographers) played excellent roles in marshalling their delegates and papers from their respective nations. The Symposium was superbly organized by the Conference and Hospitality Units at the Universities of Plymouth and Exeter under the extremely efficient hands of Denise Horne and Jayne Garcia from Plymouth. Thanks are also due to the Royal Geographical Society and the Institute of British Geographers for subsidizing the costs of

UK postgraduate delegates. Finally, thanks are due to Michael Winter (Centre for Rural Research, University of Exeter) for organizing a very useful field day in Devon, and Stephen Essex, Richard Yarwood, Clive Charlton and Mark Brayshay (University of Plymouth) for the multifunctional day out in Cornwall.

Andrew Gilg,
Exeter,
January 2005.

Chapter 1

Rural change and sustainability: key themes

Stephen Essex, Andrew Gilg and Richard Yarwood

Introduction

Rural areas in Britain and North America are facing complex and conflicting sets of demands, pressures and challenges, which have far-reaching implications for rural spaces. These issues were the subject of the Fifth British-American-Canadian Conference on rural geography, which sought to make sense of these changes by considering the implications on the sustainability of rural places. Emphasis was placed on understanding the importance of local actions and how these effect, and are affected by, global processes of change. Locality and community have been enduring themes in rural geography, despite falling in and out of academic favour. Currently, however, policy responses to rural change have given increased emphasis to local action in rural places. It is argued that local action offers a more participatory, sustainable and sensitive way of addressing rural needs and issues (Moseley, 2003). For example, alternative food networks have been promoted to connect consumers more closely with local farmers; local distinctiveness is emphasized in the promotion and reimaging of rural places for tourism; community-based strategies have been used to develop housing and services and, more generally, partnership working has been heralded as the best way of empowering and informing local issues. Although these actions are aimed at sustaining particular local places, such actions may also benefit spaces beyond the immediate locality. Indeed, they will only succeed if they are placed in a global, rather than parochial, context.

Geographers are, of course, well-positioned to understand the (re)positionings of local spaces in a global context. The sub-discipline of rural geography has witnessed many theoretical and methodological advances over the past few years, yet has retained a focus on spatial relationships. This focus has leant a vibrancy and diversity to rural geography, witnessed by a selection of titles from the 'Perspectives on Rural Policy Planning' series that have edited and written by leading rural geographers in the UK and North America:

- *Geographies of Rural Cultures and Societies* (Holloway and Kneafsey, 2004)
- *Big Places, Big Plans* (Lapping and Furuseth, 2004)
- *Young People in Rural Areas of Europe* (Jentsch and Shucksmith, 2004)
- *Power and Gender in European Rural Development* (Goverde *et al.*, 2004)
- *Equity, Diversity and Interdependence: Reconnecting Governance and People through Authentic Dialogue* (Murray and Murtagh, 2004)

- *Multifunctional Agriculture* (van Huylenbroeck and Durand, 2003)
- *The Reform of the CAP and Rural Development* (Kasimis and Stathakis, 2003)
- *Women in the European Countryside* (Buller and Hoggart, 2004)
- *Leadership and Local Power in European Rural Development* (Halfacree *et al.*, 2002)

It has now been over 12 years since Chris Philo (1992) suggested stirring additional 'ingredients into the mix of rural studies', notably by listening to 'other voices' and embracing the wider 'cultural turn' then influencing geography. This call has been widely heeded by British and, to a degree, by North American rural geographers. This activity has placed the sub-discipline into an expansive mode, not just in activity (Cloke, 2000), but also in its subject areas. Whether or not rural geographers have fully embraced the 'cultural' turn and adopted a 'post-rural' paradigm as Murdoch and Pratt (1993) advocated, the influence of cultural approaches on British rural geography has been significant. This influence has led, among other things, to new conceptualizations of people, animals and nature; the significance of the body; and how people understand and interact with rural landscapes in emotional, sometimes spiritual, ways (see Cloke, 2004 for a good collection of essays on these topics). Further, familiar topics, such as agriculture and community, have been re-examined with cultural lenses to produce new perspectives on the practices of living, working and visiting the countryside (Holloway and Kneafsey, 2004).

If cultural geography has sometimes been accused of either being so theoretically obscure that it serves no practical purpose (Miller, 1996), or, at the other extreme, rather 'common sense' and descriptive (Hoggart, 1998), then it is important to remember, as Philo originally advocated, that it is but one ingredient in rural geography's rich kitchen: an exiting spice for particular dishes rather than a ubiquitous dollop of tomato ketchup to be applied liberally, and often inappropriately, to every recipe. Indeed, it is important to remind those who enjoy a traditional rural fare that significant research progress continues to be made in arenas more traditionally associated with the political economy of the countryside. For example, widely held notions of a 'post-productive' transition have been questioned (Evans *et al.*, 2003), at least in the UK, and changes in structures of governance have led to wide ranging debate and research on the changing nature of power and politics in rural places (Goodwin, 1998). In both Britain and North America, empirical investigation remains on rural geography's menu and remains a popular serving that continues to fulfill the hunger to understand rural change.

If such diversity presents a difficulty in comprehending whether there is, or even should be, a coherent discipline of rural geography, then it is a focus on place that gives geography its identity and lends a perspective that is important in the understanding of rural sustainability. Whether geographers are mapping physical space with increasingly sophisticated GIS packages or conceptualizing how it is a hybrid of imagined identities, spatial perspectives allow connections to be made between the local and the global, aiding understanding of and, hopefully, contributing to the sustainability of the countryside in different parts of the world.

This book examines rural places in three countries: Canada, the USA and the UK. This geographical focus not only allows contrasts and connections to be made between the different rural spaces of three First World countries, but also allows the opportunity to pursue the menus of rural geography being offered in them. Thus, while the 'cultural turn' has made its contribution to UK geography, it has been less evident in North America. Here, the region remains a strong focus, characterized by the number of thorough and detailed investigations in this book that draw upon a rich vein of empirical data. Despite their local grounding, however, these case studies are guided by theoretical understandings of rural change and draw important conclusions for rural spaces elsewhere and, indeed, the practice of rural geography.

Another difference is that rural land uses are still seen as 'productivist' in North America, whereas in Britain, rural land uses are increasingly seen as an adjunct to other uses, like recreation or nature conservation in a multifunctional countryside. This contrast is, of course, a key difference: Britain has countryside while North America has rural space in which one type of land use dominates. Indeed, in Britain, there is a fierce debate about whether productivism has been replaced with multifunctionality. In Britain, environmental issues are focused on quality of life and managing farm habitats back towards some supposed elysian field, but in North America environmental issues are more focused on resource depletion and management. Likewise, most North American communities do not have the long history found in their British counterparts, and single purpose settlements, like those illustrated in several of the North American chapters, are not commonly found in Britain. The differences and similarities between rural geography in Britain and North American can be summarized in four key areas:

- Agriculture: North America remains focused on productivism whereas in Britain there is a fierce debate about whether productivism has been replaced with multifunctionalty.
- Environmental issues: in Britain, environmental issues are focused on quality of life and managing farm habitats back towards some supposed elysian field, but in North America environmental issues are more focused on resource depletion and management.
- Communities: in Britain, recent research has focused on issues of gentrification and recomposition (often caused by population growth), whereas in North America, decline and transition of single purpose towns are still important themes.
- Governance: in Britain and North America, work has focused on whether governance works, how decisions that are made work and why minorities are still excluded from power.

Conferences and symposia, with their eclectic range of paper topics, can highlight a failure 'to see the wood for the trees'. Conversely, by forcing people to sit through papers which they would never read in their normal daily round, conferences reunite the subject and give people different perspectives to view their own narrow topic.

This book, which has reduced eclectic disparity by selecting papers that embrace four key sub-issues, thus represents a very useful mid-point on the publishing spectrum, between the eclectic mix of papers published in research journals and presented as conference papers and the somewhat artificial rationalization of rural geography in text books.

The structure of the book has been organized around four broad headings (agriculture, environment, communities and governance), although there is inevitable blurring between the sections. While themes of place and locality are strong in all four sections, it is not the intention to present a unified or holistic vision of sustainable development. Such a quest is exclusive and, often, unhelpful. Rather, these different sections seek to demonstrate how particular aspects of sustainable development of particular importance for particular people in particular places. Geography, as Williams and Millington (2004) emphasize, is at the heart of understanding and promoting sustainable rural futures.

Agricultural Responses

The chapters in the agriculture section, while dominated by contributions about the North American industry, illustrate how productivism has responded to and survived changes in world markets and pressures for greater environmental sustainability. Such processes continue to disconnect farmers from the local environment and cultures. Even alternative modes of production associated with post-productivist agriculture in the UK, such as local specialized livestock production, often/inevitably remains intermeshed with conventional food systems. Similarly, while agri-tourism and the changing function of agricultural shows provide some apparent reconnection to rural cultures and environments for the general public, there is recognition of the partial images and distancing from the realities of modern agriculture.

Troughton (Chapter 2) argues that, in contrast to assertions of both a post-Fordist industrial era and post-productivist agriculture, the agro-food system in North America continues to evolve based on overt productivist policies, with continued industrialization employing 'neo-Fordist', rather than post-Fordist, models and practices. Industrialization of the agro-food system, which pre-dates Fordism itself, remains strongly wedded to Fordist ideals of scale, mass production and the assembly (or dis-assembly) line. Continued industrialization of farm-level production and its consolidation into the input supply, output processing and distribution sectors of the system, is being facilitated through a combination of corporate Fordism and productivist, government support.

Thus, Troughton demonstrates that he is resolutely researching in the big bad corporate world and making the sort of holistic points made by traditional agricultural geography. The message may be rather bleak for the anti-McDonalds camp, but it is one rooted in practical reality and reasserts focus on mainstream issues and productivism.

MacLachlan (Chapter 3) takes a similar approach based on Fordism, but modifies it in a neo-Fordist interpretation. He shows that in just 20 years, the feedlot finishing of beef cattle has transformed the landscape of southern Alberta. In many ways, feedlots are exemplars of Fordist agroindustrialization, yet they are also flexible. Trends in the distribution of slaughter cattle production demonstrate increasing levels of concentration as a key dimension of agroindustrialization. Demand pull and supply push account for the rapid growth in intensive cattle feeding. The flexibility of feedlot producers is manifest in finding alternative markets when Canadian beef consumers turn fickle, new sources of grain supply when weather conditions and input costs are adverse, and rapid reductions in scale when export cattle markets collapsed due to the discovery of BSE in Alberta.

In other words, MacLachlan demonstrates that productivism on the industrial model is alive and well, because it is capable of continual transformation as the world about it changes. Most of these responses are due to economic changes, but some are due to cultural changes, for example, a reaction against meat, notably red meat in recent years.

Ramsey, Everitt and Behm (Chapter 4) examine the dramatic rise of hog (pig) farming and attempts to impose environmental controls on its development in Manitoba. Hog farming in North America is an interesting mix between corporate capital that provides the hogs and feedstuffs in many cases, while farmers provide the land and the labour. In other areas, where hog farming has expanded rapidly, strict controls have been imposed, for example, in North Carolina and Quebec. However, in a Province with a strong farming history and a relatively small urban population, controls have been resisted and the governance of the new industry has developed patchily with tensions between national, provincial and local legislatures. A number of case studies show that productivism does not always have its sway as Manitoba moves from grain to meat production in response to changing world agricultural markets.

Marshall (Chapter 5) appears to leave behind the world of corporate finance in her work on local foods. But, in fact, her study of a small traditional fishing community in New Brunswick demonstrates that international organizations and methods have used Fordist principles to develop industrial aquaculture based on the mass production of farmed salmon. The impacts of this development have been the decline of the traditional fishery, and environmental degradation in spite of increasingly restrictive regulatory structures. The native community has found it hard to adapt with growing social problems, youth alienation and income disparities between those still trying to fish traditionally and those employed in the globalized fish farms. In particular, locals have lost control of their own destiny, will not buy the farmed fish and the fish farming systems are unlikely to become embedded in local culture.

Thus, another chapter has focused on productivism as its main theme, but this time there is a discussion of how the local community has suffered badly from the invasion of alien economic systems based on corporate capital rather than family ownership. The community has been transformed, but not for the better, and this cultural interpretation bears similarities to similar British studies of village transformation, perhaps because Newfoundland has a long history of settlement based on fishing.

The first British chapter by Ilbery and Maye (Chapter 6) also focuses on food produced in one locality, but this time the focus is on traditional specialized livestock production. They argue that interest in traditional food production has been engendered by recent food scares and diseases in the UK (e.g. BSE, foot and mouth) that have increased public and political concern about where food comes from and how it is produced. Their chapter introduces findings from survey work conducted with producers of specialist livestock products in the Scottish-English borders. Using supply chain diagrams, it shows the personalized, embedded, customized and differentiated nature of specialist livestock supply chains. The heterogeneity of surveyed producer initiatives questions the often simplified conceptual distinctions that are often drawn between conventional and alternative food systems, not least because many 'alternative' producers are obliged/choose to 'dip in and out' of different conventional nodes (e.g. abattoirs, wholesalers). The key finding thus is that local speciality food is not independent of the mainstream industrial food system, but is intermeshed with it at various stages of the food production process.

The chapter by Che, Veeck and Veeck (Chapter 7) also undermines the potential divide between corporate America and individual Britain with its study of how small farmers use agritourism to sell their produce to visitors in Michigan. The chapter illustrates how agritourism successfully sells American tradition. First, agritourism sells the tradition of farming and local food production in its provision of much-needed education on where food comes from for those removed from a farming background. Second, agritourism sells traditional rural American landscapes and values. Agritourism's strength is in its marketing rural or small town virtues such as simplicity, service and a personal touch, since its prices for comparable items are higher than those at supermarkets. Finally, agritourism taps into family traditions, such as parents bringing children back to the same farms they visited as children and/or reliving their own farming background. According to its proponents, agritourism promotes family values and creates new family-oriented farm traditions by providing a safe, intergenerational, nature-based setting for family outings.

The next chapter (Chapter 8), Holloway's exploration of agricultural shows in Britain continues this trend. For example, shows once represented mainstream agriculture when it had unquestioned hegemony as the dominant land use in Britain, but in order to survive they have had to reach out to new markets while focusing around their core purpose to show off farming. Thus agricultural shows have become part of a struggle by some sections of the farming industry to legitimize its status and practices against widespread criticism. The way in which agricultural societies fulfill their objective of 'promoting' agriculture has thus partly changed, from an investment in the development of agriculture, to engaging in public relations for farming. Shows present a particular vision of farming, emphasizing the successes of expertise in food production and conservation. Although agricultural societies stress the importance of establishing reconnection between farmers and non-farmers, a farming/non-farming distinction is reproduced in showing farming to the public. In particular, non-farmers are represented as lacking knowledge of farming and food, and as inhabiting lifeworlds distinct from farmers. While strategies of 'educating' non-farmers seek to inform, they are implicated in presenting a particular and partial image of farming which maintains the distancing of non-farmers from many of the 'realities' of agriculture.

The tone of the chapter is thus implicitly cultural and based on a study of 'others' and the dwindling power of the farming community in Britain. The fact that farmers see themselves as endangered and marginalized is a cultural construction that contrasts strongly with the productivist agenda of corporate food companies examined in Chapters 2, 3 and 5.

Environmental Issues

In the environment section, two themes emerge. The first is the emergence and evolution of public policies to encourage environmental protection. Contrasts between the agri-environment schemes formulated in Canada and Europe reflect the context of the changing external frameworks within which farming is operating and emphasize the significance of different production, environmental and consumption agendas. Similarly, while the effectiveness of urban containment policies may have been hindered in the past by the realities of farm economies and middle-class demand for rural life, the protection and sustainability of both the environment and the rural economy has gained renewed significance. The second theme in this section is the vulnerability and sustainability of rural communities in light of emerging environmental hazards and changes. The potential for change in response to environmental change in traditional agricultural communities is enormous, as areas adjust to new levels of water availability and weather systems, although various environmental crises from a variety of causes have already created a series of 'environmental ghost towns'.

The contrast between the dominance of agricultural productivism in North America and the transition towards post-productivist agriculture in the UK and Europe, highlighted in the first section of this book, is also reflected in the formulation and implementation of agri-environment schemes. These have been introduced as a means of reducing food production as well as a response to the adverse effects of modern farming on the environment. In Chapter 9, Robinson contrasts the different approaches taken in Canada and Europe, and particularly focuses upon Ontario's Environmental Farm Plan (EFP) scheme, first introduced in 1992 and to be extended to all provinces throughout Canada. Unlike many agri-environmental schemes operating in Europe, which are top-down and agency-led, the EFP scheme owes its origins to a mixture of farmers' genuine concerns for environmental well-being and a fear of being forced by government-introduced regulation into different types of farm management practices to achieve greater environmental sustainability.

Farmers have been encouraged to develop and implement an environmental farm plan tailored to their farm, covering issues such as water quality, soil quality, air quality, agricultural inputs and the conservation of natural areas. Subsequently, through a province-wide system of workshops, over half of Ontario's farmers have participated voluntarily in the programme, though with considerable sectoral and spatial variation. Robinson suggests that the Canadian scheme represents a 'soft' approach to the agri-environmental policy, which contrasts with the emergence of 'harder edge' implementation within the EU involving decoupling, cross-compliance and modulation. This contrasting approach reflects the context of the changing external frameworks within which farming is operating and emphasizes the significance of different production, environmental and consumption agendas.

Both North America and the UK continue to share the planning goal of urban encroachment as a means of environmental protection of rural areas. Urban expansion, involving low-density suburbanization and urban sprawl, are recognized as unsustainable. Recent policy initiatives on both sides of the Atlantic have emphasized so-called 'smart growth' policies, which involve the use of brownfield land and the creation of more inclusive urban environments. Gayler (Chapter 10) provides a historical perspective on the implementation of planning policies to control urban expansion in the 'Golden Horseshoe' region of southern Ontario from 1946. Urban expansion has become critical in this region because of the projected rapid growth of the Greater Toronto area and the threats that are placed on the different and highly valued agricultural and natural resources. Planning policies to control urban expansion have not been effective in this region because of the continued strong public demand for low density residential development and because of loopholes that allow farmers to sell land for residential development as an economic survival strategy. Ironically, Ontario's Niagara Wine Region, which has recently acquired a strong reputation as a quality product and introduced new prosperity to the area, now faces severe repercussions if the threats and realities of urban development are not curtailed by more draconian measures. The case illustrates the renewed significance of seeking effective urban containment policies for the protection and sustainability of both the environment and the rural economy.

Wider issues relating to environmental sustainability are also discussed. Resource-based environmental issues, especially related to the concepts of vulnerability and sustainability science, are of much greater concern in North America than in the UK at present. In Chapter 11, Harrington reviews the vulnerability and sustainability of the High Plains of southwest Kansas, part of the US Great Plains sub-region, where an economy largely based on agriculture is vulnerable to various hazards and environmental stresses. Natural stressors include drought, ground-water decline, high temperatures and humidity, tornadoes, storms, floods and climate change. Socio-economic conditions, such as low levels of economic diversification, the dependency on a limited number of employers and fluctuations in market prices, also create vulnerability for the agricultural sector. Short-term vulnerability can be overcome by adaptability and resiliency, but the sustainability of agriculture in the long term is questionable.

Harrington's suggestion that environmental factors will play an increasingly significant role in the sustainability of some rural communities in North America is illustrated in Mayda's chapter on environmental ghost towns (Chapter 12). Mayda identifies the emergence of a new type of abandoned town where environmental factors have caused the resettlement of whole communities. Three case studies involving different environmental causes are discussed, namely the loss of land and resources to coastal erosion and storms linked to: global warming at Shishmaref, Alaska; pollution from an energy plant at Cheshire, Ohio; and the collapse of cod fisheries at Great Harbour Deep, Newfoundland. These examples indicate the potential role of environmental change in the disruption and upheaval of rural settlements and communities.

Communities

Community has been an enduring theme in rural geography. Although the geographical significance of 'community studies' has waxed and waned, current interests in regulation, governance and developmental networks have prompted reconsiderations of the ways in which local actors and local cultural glues serve to regulate rural change in different localities. Focuses on locality allows consideration of the various ways that local rural spaces effect, and are affected by, global restructuring and change. Key to these understandings is an appreciation that 'the local' is culturally constructed and contested by different actors for different ends in different places. The chapters in this section examine local and community actions and consider how imaginations of places and communities are used as a response to, and regulator of, rural change.

In Chapter 13, Smithers makes an important link between the productivist approach of North American agriculture and the role of the farming family within their local rural communities. As a result of the transformations in agriculture, involving mechanization, intensification and amalgamation, there has been a growing belief that farming and rural communities have become largely disconnected from one another, both economically and socially. Smithers, using a survey of family farms in north Huron County, southern Ontario, examines this 'decoupling thesis' with particular attention to the increasingly differentiated nature of family farms and the evolving social and economic character of rural communities. He identifies six farm business development trajectories with the sample of family farms, with those displaying expansionist trajectories demonstrating less commitment to both involving themselves in local social activities and purchasing farm inputs from local sources. Non-expansionist farming families gave a stronger recognition to the local community, mainly because of the necessity for alternative employment locally. Smithers uses the concepts of integrative and divisive change to explain the processes involved in the integration or separation of family farms within rural communities.

Valentine Cadieux's study of rural migration (Chapter 14) examines a very different set of patterns, processes and impacts. Here, emphasis is given to the environmental and social dilemmas faced by people moving from Canadian cities to their rural hinterlands. In a wide-ranging paper, she notes that many rural in-migrants are interested in processes of agriculture, fishing, and forestry as well rural land management for open-space or ecological conservation, yet access to these productive processes is obstructed by cultural norms and narratives of consumption and preservation. By engaging in rural living as a form of residential tourism, and by succumbing to consumerized forms of leisure environmentalism, many genuinely well-intentioned migrants end up damaging both environmental quality and the prospect of a future rural landscape of the sort in which they wish to live. She questions whether we can disrupt destructive countryside consumption through facilitation of a more rigorous engagement with the *processes* of rural landscapes.

While the designation of National Parks have an important role to play in the protection and conservation of the rural environment, they also influence the type of economic development within the affected rural communities through the creation of tourist facilities, services and supplementary recreational experiences. National Parks assist in the achievement of the goals of environmental sustainability, but the effect of the designation of economic sustainability of rural communities can be detrimental. Algeo in Chapter 15 examines the role of National Parks in altering the trajectory of the economic development within the rural settlements of the Mammouth Cave National Park in Kentucky, first designated in 1941. Early national park boosters made extravagant predictions of the economic prosperity that would follow federal investment in infrastructure, federal management, and the cache of national park designation. This chapter critically assesses that promise. The results of the study indicate that tourism has failed to create a robust economy in the Kentucky cave region. The tourism economy is distributed unevenly and excludes some sections of the community. The presence of the National Park has stifled development in some parts of the designation, such as around Park City, and allowed growth in other locations, especially in gateway locations at the fringe next to major road intersections (e.g. Cave City). Socio-economic indicators for the national park suggest high rates of poverty and cross-county commuting, and limited manufacturing and retailing activity. While gateway tourism appears to be a preferred type of economic development, it is often at the expense of landscape aesthetics.

Housing, and access to housing, is a crucial component of social recomposition in rural communities: providing a gateway to rural settlements for some and a barrier to others. In Canada, the UK and the USA, the gentrification of rural properties and the exclusion of lower income groups from rural homes have been in sharp contrast, highlighting social disparities in rural places. Affordable housing is needed in all three countries for the development of socially balanced, sustainable communities. If greater political will is needed to provide these homes, then a basis for this action is a better understanding of the needs of local income households in rural places. Chapters 16 and 17 provide a thorough analysis of housing need for low income groups in the USA and Canada respectively. Holly Barcus' chapter explores the sub-county distribution of vulnerable populations in six rural counties in the Great Plains, Upper Great Lakes, and Appalachian regions of the USA. She emphasizes that local housing quality and access to housing is in part determined by wider processes of demographic change and economic restructuring. Using data at the county and census-tract level, she evaluates the spatial distributions of people and housing stocks in a range of communities, revealing how growing and declining communities have significant, but often ignored, differences in their housing needs. Similarly, David Bruce's chapter provides a thorough and informative analysis of housing need in rural Canada. Far from being homogenous, a typology of low income groups is used to demonstrate how housing need and circumstances vary between different groups in different rural communities. Based on this analysis, Bruce highlights the local barriers and opportunities to the development of affordable housing. Both Barcus and Bruce conclude that locally-sensitive policies and strategies are crucial to meeting geographically specific social housing needs in different communities.

These analyses of housing situations link the economic growth or decline of communities to changing social circumstances, a theme also developed by Owen Furuseth in Chapter 18. Emphasis is placed on the ways in which rural economic change in and outside the rural USA has influenced the migration of Latino groups, particularly from Mexico, to North Carolina. While many British accounts of minority groups in rural areas have favoured qualitative approaches, Furuseth makes effective use of US census data to examine how, where and why these demographic changes are occurring. Interviews with a local parish priest supplement this analysis and begin to describe the 'stories' of Latino migrants, foregrounding the attractiveness of rural places and employment, and allowing some initial consideration of the social impacts of migration on both migrants and rural communities.

As all the chapters in this section show, rural communities are continuing to experience profound social, economic and demographic change. The challenge for policy makers and planners is to develop policies that are effective at a range of scales: locally sensitive, yet appreciative of the broader changes impacting on rural places. The following section examines how government and governance are changing in light of these challenges.

Governance

Many commentators have suggested that rural areas are becoming subjected to new forms of governance that blend and empower actors from the public, private and voluntary sectors. Although the significance and form of this apparent shift from government to governance is still debated, it is clear that rhetoric of community and locality-based participation is emerging in the way that decisions are made about rural places in Canada, the USA and the UK.

However, as Ann Deakin demonstrates (Chapter 19), advocates of regional governance in the USA have failed to recognize that 'region' is not necessarily synonymous with administrative units, such as counties. To potentially inform the movement towards rural regional governance, her chapter makes a preliminary exploration of the extent to which the area covered by an administrative rural county coincides with local understandings of region or regions. This exploration draws on theories of the 'new' regional geography and advances in Geographical Information Science (GIS) to inform an analysis of environmental and social data in a particular county of western New York State. In doing so, it reveals that a range of real and imagined geographical regions intersect the legislative space of the county.

In Chapter 20, Gill and Williams use stakeholder theory to examine how a corporation works with a community to shape the locality within which it is located. Corporate-community relations in Whistler, British Columbia are explored, where the Intrawest ski resort company (and its predecessors) have been instrumental in the town's development and economic fortunes as a mountain ski resort. The successful integration of corporate and community stakeholders have become essential to the sustainability of both groups. For the company, intangible assets, such as good community relationships and environmental protection, can produce competitive commercial advantages, while for the community, employment opportunities and measures to avoid the 'corporatization of place' are important for future generations. After an initial phase of uncontested growth within the resort (1975 to 1990), in which the needs of the resident population were overlooked in favour of maintaining economic investment, a period of controlled growth management was evident in which social and community values were considered within a more participatory and accountable form of governance. A convergence of vision between the community and the corporation, especially with respect to environmental quality and sustainability, began to emerge in this phase. A key factor in this change appeared to be a localized commitment from the fact that many of the senior managers of the corporation were also residents within the community. Nevertheless, the authors suggest that in response to the changed power relations that have resulted from the forces of globalization, the necessity of working with local actors in participatory processes is now a requirement in many rural business sectors. The relationships are complex and the drivers are challenging.

Likewise, Greg Halseth (Chapter 21) explores how contrasting understandings of rural Canada frame the starting points for community development and revitalization debates, but also constitute a barrier to change. In British Columbia, rounds of social, political and economic restructuring are bringing disjunctures between contrasting understandings of rurality to the fore. His chapter evaluates two resource-dependent towns that have faced the closure of a major industry and the responses of three key participants in debate about their revitalization. He notes how the provincial government remains dependent upon traditional resource revenue streams and is reluctant to change; that local governments also remain committed to a vision of resource extraction vision; and others see mill/mine closures as an opportunity for change.

Wilson (Chapter 22) critically evaluates collaborative community-based approaches to public land management and planning, which have proliferated rapidly in the American West during the 1990s. Dominant explanations posit collaboration as a response to the political gridlock produced by the failure of top-down environmental regulation to balance the ecological health of public lands with the changing socio-economic realities of rural western communities. Relying upon ideas of civic republicanism, these accounts view the structural contradictions of federal governance as leading to an eroded sense of community belonging and responsibility in many rural communities. Collaboration is thus presented as a means of revitalizing these communities, resolving land use conflicts and improving management decisions. This chapter draws upon a case study of collaborative forest planning and argues that these narratives set up a number of problematic assumptions and expectations that hinder the potential of collaborative projects. Such explanations tend to simplify socio-environmental and political contexts, which mean that collaboration may exacerbate rather than ameliorate local land use conflicts. These explanations underplay the essential role that Federal government has to play in local environmental action, but also that access to the collaborative process is not always equal or open, with some social and economic groups likely to be excluded, while others legitimatized by the state. Different stakeholder groups have different agendas and fail to establish common objectives. Wilson suggests that such perspectives raise questions about the future sustainability of collaboration unless greater attention is given to these considerations.

In the final chapter of this sector, Christine Corcoran (Chapter 23) questions the role of the geographer and geographical research in new structures of governance. Her paper draws on her doctoral research on a partnership that was formed to develop anti-poverty policies in an English county. The chapter addresses the tensions between partnerships and young people living in rural areas and, specifically, the problem of combining 'top-down' and 'bottom-up' voices. Despite rhetoric and intent to represent and integrate young people, the partnership under scrutiny had no direct input from the young. This essential but missing element questioned the integrity of the partnership philosophy and added to concerns identified by those involved in youth studies that the young are indeed powerless in an adult world. She outlines how 'action research' and her own positioning as a geographer in the partnership enabled her to bring the voices of young people into the policy arena.

Conclusions

These chapters illustrate how contemporary rural change is experienced and managed both with and between regions. Contrasts and connections can be made between rural spaces in the UK and North America. The final chapter reflects on these contrasts and connections from a North American perspective and highlights key themes, policy implications and research avenues. While many authors in these chapters have emphasized sustainability, the range of issues and approaches studied in this book confirms that rural geography itself is engages with an impressively broad and continually evolving research agenda.

References

Buller, H and Hoggart, K (eds.) (2004) *Women in the European Countryside* Ashgate, London.

Cloke, P (1997) Country backwater to virtual village? Rural studies and the 'cultural turn', *Journal of Rural Studies* 13, 367-375.

Cloke, P. (ed) (2004) *Country Visions* Pearson, Harlow.

Evans, N., Morris, C., and Winter, M. (2002) Conceptualising agriculture: a critique of post-productivism *Progress in Human Geography* 26, 313-332.

Goodwin, M. (1998) The governance of rural areas: some emerging research issues and agendas. *Journal of Rural Studies* 14, 5-12.

Goverde, H., de Hann, H. and Baylina, M. (eds) (2004) *Power and Gender in European Rural Development* Ashgate, Aldershot.

Halfacree, K., Kovach, I. and Woodward, R. (eds) (2002) *Leadership and Local Power in European Rural Development* Ashgate, Aldershot.

Hoggart, K. (1998) Rural cannot equal middle class because middle class does not exist. *Journal of Rural Studies* 14, 381-386.

Holloway, L. and Kneafsey, M. (eds) (2004) *Geographies of Rural Cultures and Societies* Ashgate, Aldershot.

Kasimis, C and Stathakis, G. (2003) *The Reform of the CAP and Rural Development in Southern Europe* Ashgate, Aldershot.

Jentsch, B and Shucksmith, M. (eds) (2004) *Young People in Rural Areas of Europe* Ashgate, Aldershot.

Lapping, M. and Furuseth, O. (eds) (2004) *Big Places, Big Plans* Ashgate, Aldershot.

Miller, S. (1996) Class, power and social construction: issues of theory and application in thirty years of rural studies. *Sociologia Ruralis, 36*, 93- 116.

Moseley, M. (2003) *Rural Development: Principles and Practice* Sage, London.

Murdoch, J. and Pratt, A. (1993) Rural studies: modernism, post-modernism and post-rural. *Journal of Rural Studies,* 9, 411-428.

Murray, M. and Murtagh, B. (2004) *Equity, Diversity and Interdependence: Reconnecting Governance and People through Authentic Dialogue* Ashgate, London.

Philo, C. (1992) Neglected rural geographies: a review. *Journal of Rural Studies* 8, 193-207.

Van Huylenbroeck, G. and Durrand, G. (eds) (2003) *Multifunctional Agriculture A New Paradigm for European Agriculture and Rural Development* Ashgate, Aldershot.

Williams, C and Millington, A (2004) The diverse and contested meanings of sustainable development. *The Geographical Journal* 170, 99-104.

Chapter 2

Fordism rampant: the model and reality, as applied to production, processing and distribution in the North American agro-food system

Michael Troughton

Introduction: objectives, models and arguments

The objective in this chapter, in line with directions which have underlain previous British-American-Canadian (BAC) Rural Conferences, is to review the situation vis-à-vis a key aspect of our rural geography; in this case the fundamental and ongoing industrial restructuring of agriculture and food in North America (i.e. the USA and Canada). Earlier BAC reviews centred on the restructuring of farming in Canada (Troughton, 1992 and 1997) and on sector change in US livestock production (Furuseth, 1997; Hart and Mayda 1998). Here, I seek to examine in broader terms the industrialization and restructuring of an increasingly integrated, continental agro-food system.

The North American agro-food system, both in model terms and in reality, seeks to function as an economically efficient, input-to-output assembly line or commodity chain, the components of which include the range of agribusiness inputs to food production, the farm production level, and agribusiness food processing and distribution beyond the farm gate. The agro-food system is marked by adoption of a common agro-industrial technology and is further articulated by a common ideology of economic efficiency and rationalization, promoted by various levels of government and supported by public and private research (Friedland *et al.* 1991; Troughton, 1986, 2002; Bowler, 1992). In what is perhaps the clearest exposition of the context of change, Whatmore (1995) suggests that the agro-food system is built upon capital accumulation that is the product of uniting the various agribusiness sectors, which in turn facilitates a tighter regulatory framework.

In seeking to describe the evolution of the system, according to an overarching model, I take a cue from Fordism and its assembly line, which dominated agribusiness until the 1970s, and is now the dominant corporate industrial model applied to the total agro-food system.

According to Dicken (1992, p.116),

> 'The Fordist system is characterized by very large scale production units using assembly-line manufacturing techniques and producing large volumes of standardized products for mass market consumption.'

My contention is that, upon examination, one finds the Fordist model and its elements recurring throughout the evolution of the North American agro-food system, including its present configuration, wherein each part of the current system, including farm-based restructuring, is seeking even greater integration within a corporate, agribusiness-determined, (neo-)Fordist framework.

This view is somewhat at odds with the post-Fordist viewpoints that dominated economic and industrial geography in the late twentieth century. However, North American agriculture was largely absent from these discussions. I believe that this reflects not only a concentration of attention on other sectors, such as the auto and electronic industries, but also that the Fordist model remains most attractive to those who have been most influential in the industrialization of the North American agro-food system, over the last 30 to 50 years. In addition, while rural-agricultural geographers, notably in the UK, have described the political economy of agriculture as moving from a productivist to a post-productivist mode (Marsden, 1988; Marsden *et al.*, 1996; Ilbery and Bowler, 1998), I feel that this shift has not been the dominant trend in North America.

Rather, I suggest that corporate integration of the total agro-food system is proceeding apace, following a predominantly productivist direction, determined by a powerful combination of economic, technological and political forces.

Consequently, three premises support this attempt to review the evolution of and the current situation in the North American agro-food system:

1. That the North American agro-food system has been dominated by the Fordist industrial model from early in its European development, and that its evolution can be best understood based on the persistence of that model.

2. That the dominant operational model, underlying greater contemporary corporate integration of the various sectors of the agro-food system, and its continental configuration, has a strongly (neo-)Fordist (rather than a post-Fordist) orientation.

3. That the underlying institutional frames of reference, namely, government policies and public and private research agendas, for both the farm sector and throughout agribusiness in North America, have always been, and remain strongly productivist.

A brief history: industry and agriculture in North America

The objective in this initial section of the paper is to examine the agriculture-industry linkages and the strong Fordist lineage of agribusiness in North America.

Fordism in the first century of agro-industrial development

The interconnections between agriculture and industry have been both direct and critical to the development of western industrial economies. Yet recent post-Fordist analyses have tended to ignore these inherent linkages, in what Page (1996) calls 'the great divide between agriculture and the new industrial geography'; notwithstanding that food and beverage production and textile manufacture are all basic agriculture-based industries, which were established before, and contributed to, the Industrial Revolution.

Grigg (1992) has described the linkages with industry as the catalyst to the nineteenth-century transformation of agriculture in the west, through application of manufactured inputs such as the steel plough, rail transportation and the growth of an urban-industrial market.

Page and Walker (1991) have described the transformation of the North American frontier to a capitalist, commercial farm economy, centred on maize and livestock production and processing in the Mid-West Corn Belt. From the 1830s on, agro-industrialization was the product of a set of production, transportation and processing technologies, pioneered by aggressive individual farmers and capitalist entrepreneurs.

Farmers sought a range of inputs to improve productivity, and the market in which to sell. Industrialists responded, and moved rapidly from local to urban centres of farm input manufacturing and food processing. Using railways to assemble grain and livestock, (Raup, 1973) processing took place using large, often unskilled, immigrant labour and mass production techniques, including the *dis-assembly line*. The latter, pioneered by pork packers in Cincinnati in the 1860s, was applied to beef slaughter and packing by the 1880s in other centres, notably Chicago, Omaha and Kansas City. The *dis-assembly* line was the direct precursor to the assembly line developed by Ford in his auto plant in 1913, and thereafter proclaimed as the critical Fordist innovation (Rifkin, 1992; Schoenberger, 1994 and MacLachlan, 2001). Rifkin (1992) stresses the early associations between meat packing and mass production based on rail-based assembly of raw materials (cattle), and the use of a large, predominantly immigrant labour force.

These characteristics, and the bitter fights to gain union recognition from anti-union companies, were all later features of 'classic' Fordism in the auto plants.

By the 1880s similar increases in scale and mass production were also occurring in the manufacture of farm machinery (Winder, 2002). Thus agriculture, both at the farm level and beyond the farm gate was critical to the development of mass production industry in the USA (Hounshell, 1984), and also in Canada (Walder and Hiebert, 1993).

Pre-Fordist mass production in meat packing and farm machinery industries also contributed to the oligopolistic consolidation of those sectors, again similar to the auto industry. Thus, although the Fordist auto-assembly-line methods were seen as 'revolutionary', and expanded across a range of non-agricultural industries, they were actually 'borrowed' from the agricultural input supply and output processing sectors.

Fordist principles, including large scale, mass production, a task segmented assembly line, and the use of a semi-skilled, non-union labour force, were also part of the break with craft unions, which began in the meat packing plants. As MacLachlan (2001) has demonstrated, these 'principles' have reappeared and are operating in today's North American beef industry.

Other sectors of agricultural processing and distribution were already industrializing by the late nineteenth century. These included the mass manufacture of the first convenience foods, notably, breakfast cereals and canned soups, vegetables and ketchup, and the first national grocery chains (Pillsbury, 1998 and Connor, 1999). The range of products proliferated in the twentieth century, both in the US and Canada. In Canada, a smaller, scattered market contributed to major amalgamations to achieve economies of scale, such that by the 1920s, one conglomerate (Canadian Canners) dominated fruit and vegetable processing, and just two companies (Canada Packers and Swift Canadian) controlled 85% of meat packing (Winson, 1992). Already, by the 1930s, national oligopolies in Canada and regional oligopolies in the US dominated virtually all agricultural-industry sectors. Self-service grocery stores appeared as early as 1916 and evolved into the first supermarkets by 1932, also characterized by consolidation into a small number of companies (Pillsbury, 1998). Although, these food distribution elements were somewhat different from the meat packing or auto plants, they too sought to achieve the economies of scale and efficiencies inherent in an assembly-line approach to securing product and developing commodity chains to service mass consumer demands.

The crisis in Fordism and the post-Fordist debate

The period between 1913 and 1970 has been loosely labelled the era of Fordist production, stemming from Ford's 'revolution' in automobile manufacture, and subsequent adoption of Fordist methods, including variants of the assembly line, across a wide array of mass-production industries, including agribusiness (Friedmann, 1982).

However, by the 1970s many economists, sociologists and economic and industrial geographers identified 'a crisis in Fordism'. The 'crisis' was linked partly to problems of mass-market saturation, and partly on the inappropriateness of the rigid Fordist model (including its post-1930s linkages to strong labour unions) vis-à-vis new technologies and management concepts, particularly the impact of computers and IT. Post-Fordism was the term used to describe a break from mass production to new industrial models, based on ideals of 'flexibility' of production and labour-management relations. Given the potential for re-thinking the whole structure of industrial production and society, the 'post-Fordist debate' has produced a wide-ranging set of interpretations (Amin, 1994; Clarke, 1992; Gertler, 1988; Jessop, 1992; Mathews, 1996; Scott, 1988).

Post-Fordism was seen as offering opportunities for change to existing regimes of capital accumulation and the labour processes, through different modes of regulation and societalization within industrial sectors, and including greater institutional flexibility (Jessop, 1992). However, the so-called post-Fordist era has been associated with the social impacts of the weakening of the Keynsian 'welfare state', the rise of neo-conservative political models, and the rush to a global economy. While these changes have stimulated corporate business, they have greatly reduced security of employment, not least in agriculture. Consequently, while the post-Fordist literature generally extols the virtues of flexibility in optimistic terms, in reality, flexibility has translated into insecurity, the collapse of many traditional employers, and even several employers in the post-Fordist high-tech employment sectors.

By and large, however, the post-Fordist debate ignored agriculture. One major reason was that, until recently, agricultural geographers, economists and sociologists were still concentrating on the farm, especially its internal relationships (production functions, land use allocation, family and gender relationships, etc.), and ignoring the fundamental and growing links to agribusiness (Bowler and Ilbery, 1987). On the other hand, industrial and economic geographers tended to be obsessed with developments in the auto industry (the Japanese model) and those of the new IT sector, including its various, new 'silicon valley' locations.

More recently, debate in agricultural geography, largely by British geographers, has advocated a more overtly political economy approach, and a shift from concentration on the farm, to that of the whole agro-food system (Bowler and Ilbery, 1987; Marsden 1988 and Whatmore, 1995). However, in the theoretical debate that has focused primarily on post-Fordist interpretations of political economy, notably regulation theory, with a stress on post-productivism and the changing role of agriculture in its environmental and social (countryside) settings (Cloke and Goodwin, 1992; Marsden, 1998b; Marsden *et al.*, 1996 and Murdoch *et al.*, 2000). These discussions reflect a general Western European concern over the range of social and political aspects of rural restructuring, as well as the particular political economy of the European Union's agricultural and regional policies, which represent a partial departure from a Fordist, productivist context. On the other hand, the North American response has been to interpret opportunities to be almost exclusively market oriented (Lawrence *et al.*, 1997; Lamb and Beshear, 1998; Drabenstott, 1999). In this context, the 'crisis' is 'solved' by a return to 'classic' Fordist capitalist ideology and labour processes, facilitated by the incorporation of late twentieth-century technology and the acceptance by industry and governments of neo-liberal economic concepts and policies.

In effect, the North American agro-food system is being reconfigured, according to what can only be termed a neo-Fordist model. Paradoxically, both sides of the Atlantic shared in the major change to agriculture in the last 40 years, namely, the industrialization of farming. But the responses to the incorporation of the food production sector into a more integrated agro-food system have differed quite sharply, both in theory and practice, between a search for a more flexible post-Fordist paradigm in Western Europe and a reaffirmation of (neo-)Fordism in North America.

Agricultural industrialization and the agro-food system in North America

In this second part of the chapter, I seek to support the premise that, neo-Fordist structures have emerged as the dominant features throughout the North American agro-food system, based on the rapid industrialization of farming, and aggressive agribusiness, since World War II.

The industrialization of farming

Prior to World War II the farm sectors in both the USA and Canada were at or near their peak number of individual farms. With a few notable exceptions, such as California's corporate horticulture, and some western ranches, it was predominantly a sector of small to medium sized, mixed farms, albeit with strong regional specializations, such as corn (maize) and hogs (pigs) in the US Mid-West and commercial grain across the Canadian Prairies. Farms supplied the raw materials for agricultural industries, but were largely distinct from them in both economic and social terms.

By the 1940s, however, the pressures to produce food to fuel the war effort, allied to the rapid application of latent mechanical and chemical energy, produced significant changes in scale of output, especially in the USA. By the 1950s, the impacts of inputs of machinery, fertilizers and the new pesticides, had produced huge surpluses, especially of maize and soybeans. Fortuitously, however, the US was on the verge of a livestock 'revolution', which, together with a rapid growth of exports via Public Law 480, would quickly divert the crop surpluses and increase demand for feed grains and oilseeds. Together, the greatly increased crop capacity and the ability to raise livestock in greater concentrations than ever before, propelled North America into the industrialization of farming and its rapid subsumption into a neo-Fordist agro-food production, processing and distribution system.

The sequence of stages through which the different livestock types shifted from widely raised small herds and flocks, to large-scale production has been succinctly described by Hart and Mayda (1998). From beginnings involving battery laying hens and broiler chickens, thence to turkeys, dairy and beef cattle, and lastly, to pigs, production of each major livestock type is now carried out along industrial lines. This includes specialized production of huge numbers of birds and animals, fully housed or in large feedlots, fed high energy feed, to increase weight gain, milk or egg production, and to reduce the time until slaughter (Boyd and Watts, 1997). The great majority of the output from the large, specialist 'industrial' production-units is processed in huge capacity centralized plants, whose corporate owners control major segments of this assembly line or commodity chain system (Troughton, 2002).

The causal factors in the industrialization of farming and the greater integration of the agro-food system included both the availability of the biological and technological inputs necessary to facilitate raising large flocks and herds in confined spaces, and the effect of a farm economy serving an inelastic market, in conditions of a cost-price squeeze.

An almost immediate effect of the shift to capital intensive, specialized farming, was to begin the drastic reduction in the number of farms and the farm labour force throughout the western (OECD) countries (Troughton, 1986). Early in the process, a set of new, very large corporate farms emerged in the USA. These were the vanguard of the changing scale of the production sector, and some of which grew into processing companies (Smith, 1980 and Windhorst, 1989). Beginning in the early 1960s, a series of studies documented the growth of clusters of large enterprises, including drylot dairying and specialized horticulture in California (Gregor, 1963; 1974). Elsewhere, Gregor (1982) confirmed the presence of regions of large scale, intensive crop and livestock across the USA. More recent studies have confirmed new regional configurations of both crops and livestock, as part of the industrialization of production and its links to the processing industry (Heffernan and Constance, 1994; Boyd and Watts, 1997; Furuseth, 1997; Lawrence *et al.*, 1997; Page, 1997; Drabenstott, 1998; Hart and Mayda, 1998; Page, 1997).

While the initial focus of industrialized farming was in the USA, Canadian agriculture followed similar trends (Troughton, 1986 and 1992). In both countries overall numbers for farms, farm population and labour force have contracted to less than one-third of their peak totals. Canada's narrower production base has meant that specialization has been in fewer sectors. However, like the USA, a minority of Canadian farms (less than 30%) contribute over 80% of output and gross farm receipts (Troughton, 1997). Canadian production of all livestock and poultry has followed a similar trajectory to the USA; i.e. large scale, specialized and capital intensive. The only exception is dairying, where supply management has operated to maintain smaller herds. On the other hand, supply management of broilers, turkeys and egg production has not stopped control of over 90% of production by less than 5% of producers in each sector (Troughton, 1989). Pigs and beef cattle are becoming more and more concentrated in large housed or feedlot premises, respectively (MacLachlan, 2001; Ramsey and Everitt, 2001).

Production on both sides of the border emphasizes rapidly increasing large scales of enterprize in both crop and livestock production. Specialized grain and cash crop operations in the maize and wheat belts average over 500 hectares. Thresholds for economically viable hog operations are in the 5,000+ sow range; beef cow calf operations average over 500 head, and feedlots of over 50,000 are becoming the norm, and major broiler chicken producers raise millions of birds per year (Hart, 2003). The reasons include the cost-price squeeze; the need to raise production against falling output prices. This situation is part of the set of cause and effect relationships within which even greater concentration and consolidation, by agribusiness elements beyond the farm gate, has taken place (Drabenstott, 1999).

If one looks again at the principles of a Fordist approach, three elements dominated; mass production, adoption of the assembly line as the vehicle, and a compliant, often semi- or unskilled labour force. Industrial agriculture, both at the farm level, and beyond the farm gate, in North America, now exhibits these dominant Fordist characteristics.

Farm production, including both crop and livestock elements, is increasingly based on large scale, mass production. The assembly line is both an overarching model and, in the case of livestock, a reality. The large commercial farm is increasingly integrated into an input-output, 'assembly line' that includes the key secondary inputs of machinery, fertilizers, pesticides, hybrid seeds, uniformly bred livestock, automated feed systems, etc. These inputs are geared to mass produce uniform output of meat, milk and eggs, conforming to strict standards of size, weight and consistency. Downstream, beyond the farm gate, this uniform output, is the feedstock of a specialized primary and secondary food processing sector. While some farmers have achieved a 'mega farm' scale of production that makes them somewhat independent (Hart and Mayda, 1998), many more farmers are caught in the 'cost-price squeeze' and on the 'treadmill' of production, in which costs of inputs and prices of outputs are controlled by the processing sector.

Increasingly, many farmers, especially in the US poultry and pig producer groups, through contract provisions and/or corporate ownership of livestock, are reduced to the (semi-skilled) role of growers (Furuseth, 1997). Many other farmers, who used to grow both crops and livestock, are today contract feed growers.

A neo-Fordist assembly-line model applied to the agro-food system

The latest, most aggressive stages in the industrialization of the North American agro-food system, underline its increasingly neo-Fordist direction, and include: a rapid and comprehensive integration of agribusiness commodity chains: increased consolidation at each level in the system, involving major players: and the development of a new geography of the system based in part, at least, on the labour process and regulatory frameworks (Drabenstott, 1999; Drabenstott *et al.*, 1999).

Exemplification of a neo-Fordist model is widespread across the North American agro-food system, and was already well established in the input supply, output processing and distribution sectors of agribusiness (Wallace and Smith, 1985). The agribusiness input sectors, namely, agricultural machinery, fertilizers and chemical manufacture, have long been based on an oligopolistic structure in which a handful of very large firms dominate both production and distribution. As early as the 1930s, two-thirds of tractors were produced by Ford (Winder, 2002). In the post-war era, a handful of companies dominated 'full-line' farm machinery, including Massey-Harris and Cockshutt in Canada, Deere, Case and International Harvester based in the USA. Today, despite some inroads of Japanese and European tractors, just two 'full-line' companies remain (Deere and Case-International). Fertilizer and agricultural chemical production are in turn dominated by a handful of large chemical companies for which agriculture may represent just one division (e.g. Dow Chemical, Dupont). Most recently, research, development and control of genetically modified organisms (GMOs) are being concentrated in the hands of a few international chemical-pharmaceutical corporations (e.g. Monsanto, Pharmacia and Upjohn). While the technology is innovative, the push to become even larger, in neo-Fordist terms of scale, output and market dominance, has resulted in several mergers of some of the largest players.

The process of consolidation in the food processing commodity chain has built up over the last 30 years, to a level where the system has come to be dominated by large, transnational companies. Heffernan and Constance (1994) in a widely referenced study, see the process as involving three stages; vertical integration, followed by conglomerate integration, leading to global integration. Using the USA poultry industry as a key example, they document the integration of 20 or so dominant firms, noting increased concentration ratios and contract levels that reached 90% by 1980. Building on information from food processing industry researchers, they note the emergence of a smaller set of conglomerates that by the 1990s were dominating across the range of processing. In the livestock sectors, this dominance included poultry (i.e. Tyson and ConAgra in broilers; ConAgra, Philip Morris and Cargill in turkeys), beef (IBP, ConAgra and Cargill) and pork (IBP, ConAgra and United Brands). A similar set of corporations dominated the related areas of grain and feed elevators (Cargill, Bunge, ADM and ConAgra), flour milling (ConAgra, ADM and Cargill), soybeans (ADM, Cargill and Bunge) and wet corn (maize) milling (ADM and Cargill) (Heffernan and Constance, 1994).

Since the study by Heffernan and Constance, there have been further concentrations, which Drabenstott (1999) saw as accelerating in the late 1990s. He cites the recent merger of Cargill and Continental, two of the largest grain trading companies, as an illustration of the ultimate level in consolidation within the agro-food system. Besides its dominance across the range of USA food processing sectors (above) Cargill is aggressively active in Canada (Kneen, 1990). Together with other continental players (e.g. ADM), Cargill plays a major role in challenging the Canadian Wheat Board and the Prairie cooperative grain Pools, and through its High River and Calgary beef packing operations, has become one of the two dominant livestock processors in Alberta (MacLachlan, 2001). The merger with Continental provides an ultimate example of cost-cutting via economies of scale and market dominance. Cargill also illustrates what Drabenstott (1999) sees as the other major direction being taken by the dominant firms, namely, the integrated commodity supply chain. Large companies are linking together into ever larger, consolidations of the whole array of food-chain elements.

The approach taken in virtually every case harks back to the 'classic' Fordist model, and includes regimes of regulation and labour processes that remind one of the 'robber baron' era of capitalism. Control of resources is maintained by take-over of competitors and control of the market. A key element is that of the move to new production and industrial locations, which are chosen because of lower costs. Production in traditional areas in the agricultural heartland of the Mid-West, for example, is being transferred to new locations across the southern USA and High Plains regions; processing is leaving traditional metropolitan areas and is moving to rural and small town locations in the new production regions.

The 'new geography' includes a significant shift in pig production from the Corn Belt to states such as North Carolina, where production is shared between a set of local corporations and the national players (i.e. Cargill and Tyson) (Furuseth, 1997 and Drabenstott, 1998). Meanwhile, the corn-hog association in the Corn Belt is giving way to the region as a feed production area. This feed, largely maize and soybeans is also a major input to the various segments of the poultry production and processing industries (Boyd and Watts, 1997). Hart and Mayda (1998) provide a series of maps that show the new concentrations of broiler, egg and turkey production. While broiler production has concentrated in the old Confederacy states, large egg producers are located closer to major urban centres in both the eastern and far western states, with turkeys produced in a small set of widely distributed concentrations, including Minnesota. Whereas demand for poultry meat and pork has been expanding, beef consumption has fallen, stimulating the beef industry to cost-cutting consolidation and new production and plant locations (Drabenstott *et al.*, 1999). Beef feeding is now concentrated in the southern High Plains, from Nebraska to northern Texas, with major concentrations in western Kansas (Stull and Broadway, 1990). The new region places the industry closer to supplies of feeder cattle from the western ranching areas. Cattle in the feedlots consume a mixture of locally produced, irrigated and imported feed crops (Hart and Mayda, 1998).

In each new production region, large packing plants have been constructed at strategic locations. The linkages between these huge cattle feed lots or pig and poultry barn complexes, close to their huge new plants, emulates the 'just-in-time' auto component supply system, albeit through a whole-owned supply chain (Xia and Buccola, 2002).

Location decisions, especially as regards pig and beef cattle, are often geared to lax state and/or local environmental regulations, which are less restrictive on manure and plant wastes, which present a major pollution control problem.

Other location factors have included the search for lower energy, and especially lower labour costs. The latter takes two forms: at the farm level, remaining farmer producers are squeezed by contract arrangements which operate in virtually all sectors in US production

(Heffernan and Constance, 1994), and which often leaves them bearing both low prices and the capital cost of production. Increasingly, the corporations own a larger share of production, and as a result many farmers are reduced to the role of managers on their own land. Classic Fordism involved major attempts by Ford and others to operate non-union shops and to use semi- and unskilled, often migrant labour in their plants. The new meat packers are adopting a neo-Fordist strategy; locating in 'right to work' (i.e. non-union) states, and employing large numbers of recent immigrants, notably Hispanic and Asian people. Although one might claim this to be flexibility of labour, it is the exact opposite of the flexible management-labour equality or cooperation envisaged by post-Fordist models. Rather than freeing labour, it enslaves it with low wages, and creates major physical (injury) and social problems in many of the small towns in which the new plants have been located (Stull and Broadway, 1990 and Schlosser, 2001).

Although the essence of the new agro-industry developed and is most advanced in the USA, it has spread to Canada, albeit with some modifications attendant on a smaller market. Despite the operation of supply management in the poultry and dairy sectors, less than 1% of poultry enterprises account for over 90% of output. Although dairy production remains largely in the hands of medium sized producers, the dairy industry has been systematically pared down to just three large dairies operating nationally. In the red meat sectors the cross-border linkages are most advanced. The beef industry has become concentrated in Alberta, with processing dominated by the two large US packers, Cargill (Cargill Foods and X-L) and IBP (Broadway, 1996; MacLachlan, 2001). Under their influence the majority of cattle for slaughter are fattened in feed lots adjacent to the two plants, in Brooks and High River. These two small towns are also experiencing some of the negative impacts of the influx and use of immigrant labour (Broadway 2001). As MacLachlan (2001) also notes, the restructuring of the industry has been at the expense of de-unionization and deskilling of a formerly, highly skilled, well-paid strongly unionized labour force.

Meanwhile, Canadian pig production is increasingly concentrated in a small number of very large production units, including locations in southern Saskatchewan and Manitoba (Qualman, 2001). This area was identified as the cheapest pig production location in the continent (Drabenstott, 1998), and selected by Maple Leaf for the location for Canada's newest, largest pig slaughtering plant (Ramsey and Everitt, 2001). This company consolidation was also at the expense of several plant closures in older locations and a lowering of wages in the industry.

A primary stimulus to the development of a Fordist industrial regime lies in the existence of a mass market for a mass produced product. Post-Fordist discussion suggests that this market has been saturated and/or that consumers are demanding a more varied product, the output of flexible, short-runs rather than a uniform mass product. However, despite calls for alternative agriculture and for slow food, the dominant development over the last 30 years has been in the area of mass food production, marketing and consumption, which supports the expansion of the Fordist model. As many observers have noted, the key food retailing outlets have been the large supermarket and the multiple fast food outlet system, each served by mass produced foods. Both of these systems provide evidence of Fordist or neo-Fordist structures, especially in their North American development.

In both Canada and the USA food retailing is dominated by a small set of supermarkets (Winson, 1992 and Pillsbury, 1998). In the USA the result is regional oligopolies, which are currently being refigured (Wrigley, 1997). In Canada a national oligopoly is made up of just five companies contributing to over 75% of all sales. Likewise, on both sides of the border, less than ten major fast food companies dominate a market which serves a large majority of all meals eaten outside the home. The supermarket chains are fully part of the vertically integrated agro-food system; several own and control integrated supply chains; others are owned by food processors. The fast food companies are major customers of the red meat and poultry and potato processors, which contract growers to produce raw materials to fast food company specifications. The fast food outlets, in turn, are also major employers of part-time, unskilled, non-union, low wage labour, thus completing a neo-Fordist, assembly line labour process, which dominates North American domestic food supply (Schlosser, 2001).

Productivist versus post-productivist political economy in the agro-food system

In this last section of the chapter I seek to support my contention that a productivist mode remains dominant in the North American agro-food system. While the descriptions in the preceding sections emphasize a generally productivist context, there are a number of historic and specific contemporary aspects that differentiate North America from Western Europe.

Fordism and productivism; production, research and politics

A number of writers have suggested that 'classic Fordism' only existed in North America and that the proliferation of mass production, and other features of Fordism never matched the original model anywhere else (Sayer, 1989). This may be true with respect to agriculture-industry links, although Europe witnessed the growth of consolidated agricultural machinery and fertilizer manufacturing industries and, more recently, major consolidation in the food retail sector (Wrigley, 2001). Likewise in the immediate past, farming has adopted industrial characteristics, including 'factory farming' (Bowler, 1992). Historically, however, there are some distinctions that have emphasized an almost complete adherence in North America to productivism throughout the evolution of the agro-food system, which is less the case elsewhere.

As agrarian-based settler nations, both Canada and the USA looked to a rapid shift from 'pioneer' to commercial production. Most farmers were owner-operators, and this engendered a collective, dominant economic outlook. Agribusiness capitalized on farmers' needs, and concentrated on increasing output of inputs to production, production itself and value-added farm-based manufacturing. In addition, each segment of agriculture and industry looked to government and related institutions to foster production. The result, as has been clearly presented by Vogeler (1981), was a system in which, in the USA, national (federal) and state governments and agricultural organizations (farm and non-farm) were allied in promoting productivist policies and practices, even at the expense of the family farm. For example, Goldschmidt's classic study (1947) describes the manner in which the interests of large land-holding corporations and their political supporters suppressed a major study of farm and farm-worker social conditions. Vogeler cites a range of farm legislation, even including the Land Grant Colleges' (i.e. state universities) research programmes which have served the interests of corporate agribusiness, rather than farmer needs (1981). Nevertheless, until World War II increases in population and domestic demand for food and other agricultural products supported an emphasis on increased farm-level production.

Somewhat different conditions operated in Canada, but post Confederation (1867) the National Policy (federal) and provincial agricultural departments were heavily oriented to increased farm output, especially in western Canada. Given the much smaller population and limited domestic demand in Canada, productivism was oriented more to the development of exports, both of primary production and processed agricultural products, especially of grains and dairy products (wheat, flour and cheese). Farms continued to proliferate in Canada until the 1930s and agriculturally based industries, notably farm machinery manufacture (Massey-Harris), grain milling, meat packing, fruit and vegetable canning, brewing, distilling and woollen textiles constituted a very large part of the industrial sector, prior to the post World War II energy and mineral boom.

Productivism in a neo-Fordist context

As noted, in the post-war period in both the USA and Canada, the primary changes affecting the agro-food system have been the industrialization of farming and its integration into the agribusiness sectors, and the rapid consolidation of the latter and their control of the various commodity supply chains. These linked processes have been the product of a productivist, growth oriented industrial and commercial outlook, fully supported by governments. A major input has been the enormous investment in both private (corporate) and public research. The vast majority of this investment has been aimed at the rapid adoption of innovative technologies to increase productivity and efficiency, Investments in areas such as the development of the biological, bio-medical, automated feed and related inputs to facilitate intensive, housed livestock production, are a case in point, as are the current researches into GMOs. The latter have received strong government support, which has resulted in the US and Canada accounting for about 85% of current global transgenic crop production (James, 1998). In addition, food processors and retailers, especially fast food corporations, are huge advertisers, often with budgets in excess of costs of food itself (Connor, 1999).

Despite concerns expressed in both the US and Canada over the actual and potential impacts of intensive agriculture on the physical and social environments (Berry, 1996 and Coote 1983), legislation to protect activities such as livestock effluent is poorly developed, with many states and several provinces failing to develop effective policies. On the other hand, government support for measures to increase output at the production level and to facilitate agricultural corporate activities is given more priority. In the USA this is particularly apparent in the provisions of the last two Farm Bills passed by Congress. The 1996 Farm Bill was dominated by the provisions which gave over $22.5 billion in direct subsidies to farmers, most of which went to the minority of large crop and livestock producers (STATpub.com, 1999). The latest, 2002 Farm Bill, although titled the *Farm Security and Rural Investment Act of 2002* and containing more provisions for conservation and organic agriculture, is still overwhelmingly weighted in terms of payments for production (USDA, 2002). In addition, corporate agribusiness is the recipient of major tax concessions. Again, Canada operates at a lesser level, and complains about US subsidies as they affect both global and continental markets in which Canadian producers compete. Nevertheless, the agri-food policies of the Canadian government, and the majority of provincial policy, including research area support, are all strongly oriented in a productivist direction (Agriculture Canada, 1989).

Concluding remarks

During the last decades, changes have forced re-evaluation of the farm landscape and its communities in the developed world. However, there are significant differences between Western Europe and North America. In the former, although there has been structural change and some integration within the agro-food system, this is tempered by agricultural policies that have, by and large, included the farm constituency, and which seek to maintain a balance between more concentrated, 'industrial' forms of agricultural production and a less productivist countryside. Examples of post-productivist measures in the European Union include set-aside and various schemes to support farmers, including diversification grants (Ilbery and Bowler, 1998). In general terms, these measures are not popular in North America, where more dominant productivist norms influence farm policies. To some extent, the emphasis on utilitarian, economic criteria has been embedded in the continental agrarian systems since their inception, and has been consistently reinforced by the pressures of government-supported, corporate agribusiness, and rarely more so, than in the context of the 1994 continental Free Trade Agreement (NAFTA), and the global economy at the beginning of the twenty-first century.

A recent assessment of changes taking place in the US agro-food system brings home both the dynamic nature of the neo-Fordist trajectory being pursued and the productivist context which is regarded as mainly positive. In testimony given to both the (USA) Senate and House Committees on Agriculture, Drabenstott (1999) describes the consolidation in US agriculture, based on the cost savings and supply chain consolidation, as beneficial to the agro-food system and to consumers, despite the continued rapid decline in farm numbers attendant on low commodity prices, the huge corporate mergers that threaten monopoly situations and the impacts on many rural communities. Drabenstott concludes that there has been a dramatic redrawing of the rural landscape as production has concentrated in fewer places and that far fewer farm communities will be viable in the future. His neo-Fordist conclusion, nevertheless, is that:

> 'The economic forces behind this trend are so powerful and the benefits to consumers so substantial that it is neither possible nor desirable to legislate consolidation away.' (Drabenstott, 1999, p.69)

References

Agriculture Canada (1989) *Growing Together: A vision for Canada's Agri-food Industry* Pub. 5269/E Agriculture Canada, Ottawa.

Amin, A. (1994) Post-Fordism: models, fantasies and phantoms of transition. In: Amin, A. (ed) *Post-Fordism: A Reader.* Blackwell, Oxford, pp.1-39.

Berry, W. (1996) *The Unsettling of America: Culture and Agriculture.* Avon, New York.

Bowler, I. (1992) The Industrialization of Agriculture. In: Bowler, I. (ed) *The Geography of Agriculture in Developed Market Economics*. Longman, London, pp 7-31.

Bowler, I. and Ilbery, B. (1987) Redefining agricultural geography. *Area* 19 (4), 327-332.

Boyd, W. and Watts, M. (1997) Agro-Industrial just-in-time. The chicken industry and post war American capitalism. In: Goodman, D. and Watts, M. (eds) *Globalising Food: Agrarian Questions and Global Restructuring.* Routledge, London/New York, pp.192-225.

Broadway, M. (1996) Following the leader: IBP and restructuring of Canada's meatpacking industry. *Culture and Agriculture* 18 (1), 3-8

Broadway, M. (1998) Where's the Beef? The integration of the Canadian and American beef packing industries. *Prairie Forum* 23 (1), 19-26.

Broadway, M. (2001) Here's the beef: the social costs of beef-packing's move to rural Alberta. *Alberta Views* Jan-Feb, 26-31.

Clarke, S. (1992) What in the F---'s Name is Fordism? In: Gilbert, N. Burrows, R. and Pollert, A. (eds) *Fordism and Flexibility: Divisions and Change.* St. Martin's Press, New York, pp.13-30

Cloke, P. and Goodwin, M. (1992) Conceptualizing countryside change: from post-Fordism to rural structural coherence. *Transactions Institute of British Geographers New Series* 17 (3), 321-336.

Connor, J. (1999) Breakfast cereals: the extreme food industry. *Agribusiness* 15 (2), 247-259

Coote, D. (1983) Stresses on land under intensive agricultural use. In: Simpson-Lewis, W. *et al.* (eds) *Stress on Land in Canada*. Lands Directorate Environment Canada, Ottawa, pp.227-257.

Dicken, P. (1992) *Global Shift: the Internationalization of Economic Activity*. Guilford Press, New York 2nd Edition

Drabenstott, M. (1998) This little piggy went to market: will the new pork industry call the heartland home? *Federal Reserve Bank of Kansas City, Economic Review* 83(3), 79-97.

Drabenstott, M. (1999) Consolidation in USA Agriculture: the new rural landscape and public policy *Federal Reserve Bank of Kansas City, Economic Review* 84(1) 63-71.

Drabenstott, M. Henry, M. and Mitchell, K. (1999) Where have all the packing plants gone? The new meat geography in rural America. *Federal Reserve Bank of Kansas City, Economic Review* 84(3) 65-82.

Friedland, W. Busch, L. Buttel, F. and Rudy, A. (eds) (1991) *Toward a New Political Economy of Agriculture*, Westview Press, Boulder Colorado.

Friedmann, H. (1982) The political economy of food: the rise and fall of the post war international food order. *American Journal of Sociology* 88(2), 248-286.

Furuseth, O. (1997) Sustainability issues in the industrialization of hog production in the USA. In: Ilbery, B., Chiotti, Q. and Rickard, T. (eds) *Agricultural Restructuring and Sustainability: A Geographical Perspective*. CAB International, Wallingford, pp.293-311.

Gertler, M. (1988). The limits to flexibility: comments on the post-Fordist vision of production and its geography. *Transactions Institute of British Geographers New Series* 13(4), 19-32.

Goldschmidt, W. (1947) *As You Sow*. Harcourt Brace, New York.

Gregor, H. (1963) Industrialized drylot dairying: an overview. *Economic Geography* 39(3), 299-318.

Gregor, H. (1974) *An Agricultural Typology of California Geography of World Agriculture 4* Akedemiai Kiado, Budapest.

Gregor, H. (1982) *Industrialization of USA Agriculture: An Interpretive Atlas*. Westview Press, Boulder Colorado.

Grigg, D. (1992) *The Transformation of Agriculture in the* West. Blackwell, Oxford.

Hart, J. (2003) How farming has changed. Paper given in session on 'Explorations in Regional Geography'. Association of American Geographers, Annual Meeting. New Orleans Louisiana.

Hart, J. and Mayda, C. (1998) The industrialization of livestock production in the USA. *Southeastern Geographer* 38(1) 58-78.

Heffernan, W. and Constance, D. (1994) Transnational corporations and the globalization of the food system. In: Bonanno, A., Busch, L., Friedland, W., Gouveia, L. and Mingione, E. (eds) *From Columbus to ConAgra: the globalization of agriculture and food*. University Presses of Kansas, Lawrence Kansas, pp.29-51.

Hounshell, D. (1984) *From the American System to Mass Production, 1800-1932* Johns Hopkins University Press, Baltimore.

Ilbery, B. and Bowler, I. (1998) From agricultural productivism to post-productivism. In: Ilbery, B. (ed) *The Geography of Rural Change*. Longman, Harlow, pp.57-84.

James, C. (1998) *Global Review of Commercialized Transgenic Crops: ISAAA Brief 8* International Service for the Acquisition of Agro-biotech Applications, Ithaca New York.

Jessop, B. (1992) Fordism and Post-Fordism. A critical reformulation. In: Storper, M. and Scott, A. (eds) Pathways *to Industrialization and Regional Development*. Routledge, London, pp.46-69.

Kneen, B. (1990) *Trading Up: How Cargill, The World's Largest Grain Company is Changing Canadian Agriculture*. NC Press, Toronto.

Lamb, R. and Beshear, M. (1998) From the plains to the plate: can the beef industry regain market share? *Federal Reserve Bank of Kansas City, Economic Review* 83(4), 49-66.

Lawrence, J. Rhodes, J., Grimes, G. and Hayenga, M. (1997) Vertical coordination in the USA pork industry: status, motivations and expectations. *Agribusiness* 13(1), 21-31.

MacLachlan, I. (2001) *Kill and Chill: Restructuring Canada's Beef Commodity Chain*. University of Toronto Press, Toronto.

Marsden, T. (1998) Agriculture beyond the treadmill? Issues for policy, theory and research practice. *Progress in Human Geography* 22(2), 265-275.

Marsden, T. (1988b) Exploring political economy approaches in agriculture. *Area* 20(4), 315-322.

Marsden, T., Munton, R., Ward, N. and Whatmore, S. (1996) Agricultural geography and the political economy approach: a review. *Economic Geography* 72(4), 361-375.

Mathews, R. (1996) Fordism, *Flexibility, and Regional Productivity Growth*. Garland Press, New York.
Murdoch, J., Marsden, T. and Banks, J. (2000) Quality, nature, and embeddedness: some theoretical considerations in the context of the food sector. *Economic Geography* 76(2), 107-125.
Page, B. (1996) Across the Great Divide: Agriculture and Industrial Geography. *Economic Geography* 72(4), 376-396.
Page, B. (1997) Restructuring pork production, remaking rural Iowa. In: Goodman, D. and Watts, M. (eds) *Globalising Food; Agrarian Questions and Global Restructuring*. Routledge, London, pp.135-157.
Page, B. and Walker, R. (1991) From settlement to Fordism: the agro-industrial revolution in the American mid-west. *Economic Geography* 67(4), 281-315.
Pillsbury, R. (1998) *No Foreign Food: the American Diet in Time and Place* Westview Press, Boulder Colorado.
Qualman, D. (2001) Corporate hog farming; a view from the family farm. In: Epp, R. and Whitson, D. (eds) *Writing Off the Rural West*. University of Alberta Press, Edmonton, pp.21-38.
Ramsey, D. and Everett, J. (2001) Post Crow farming in Manitoba: An analysis of the wheat and hog sectors. In: Epp, R. and Whitson, D. (eds) *Writing Off the Rural West* University of Alberta Press, Edmonton, pp.3-20.
Raup, P. (1973) Corporate farming in the USA. *The Journal of Economic History* 33(3), 274-290.
Rifkin, J. (1992) Beyond Beef: The Rise and Fall of the Cattle Culture. Penguin, New York.
Sayer, A. (1989) Post-Fordism in question. International Journal of Urban and Regional Research 13(6), 666-695.
Schlosser, E. (2001) *Fast Food Nation: The Dark Side of the All-American Meal*. Houghton Mifflin, Boston/New York.
Schoenberger, E. (1994) Competition, time, and space in industrial change. In: Gereffi, G. and Korzeniewicz, M. (eds) *Commodity Chains and Global Capitalism*. Greenwood Press, Westport Connecticut, pp.51-65.
Scott, A. (1988) *New Industrial Spaces: Flexible Production Organization and Regional Development in North America and Western Europe*. Pion, London.
Smith, E. Jr (1980) America's richest farms and ranches. *Annals Association of American Geographers* 70(4), 528-54.
STATpub.com (1999) Direct USA farm subsidy reaches record $22.5 billion (www.statoub.com/stat/open/2741.html) Accessed 21 May 2003
Stull, D. and Broadway, M. (1990) The effects of restructuring of beef packing in Kansas. *Kansas Business Review* 14(1), 10-16.
Troughton, M. (1986) Farming systems in the modern world. In: Pacione, M. (ed) *Progress in Agricultural Geography*. Croom Helm, London, pp.93-123.
Troughton, M. (1989) The role of marketing boards in the industrialization of the Canadian agricultural system. *Journal of Rural Studies* 5(4), 367-383.
Troughton, M. (1992) The restructuring of agriculture: the Canadian example. In: Bowler, I. Bryant, C. and Nellis, D. (eds) *Contemporary Rural Systems in Transition: Vol.1 Agriculture and Environment*. CAB International, Wallingford, pp.29-42.
Troughton, M. (1997) Scale change, discontinuity and polarization in Canadian farm-based rural systems. In: Ilbery, B. Chiotti, Q. and Rickard, T. (eds) *Agricultural Restructuring and Sustainability: A Geographical Perspective*. CAB International, Wallingford, pp, 279-291.
Troughton, M. (2002) Enterprises and commodity chains. In: Bowler, I. Bryant, C. and Cocklin, C. (eds) *The Sustainability of Rural Systems: Geographical Interpretations*. Kluwer Academic Publications, Dordrecht Netherlands, pp.123-145.
USDA (2002) *Farm Security and Rural Investment Act of 2002* (Farm Bill) House of Representatives/Senate 107th Congress (www.usda.gov/farmbill.html) Accessed 15 May 2003.
Vogeler, I. (1981) *The Myth of the Family Farm*. Westview Press, Boulder Colorado.
Walder, R. and Hiebert, D. (1993) The developing industrial heartland 1871-1891. In: Gentilcore, R. and Matthews, G. (eds) *Historical Atlas of Canada Volume 2: The Land Transformed 1800-1891.*, University of Toronto Press, Toronto, Plate 48
Wallace, I. and Smith, W. (1985) Agribusiness in North America. In: Healey, M. and Ilbery, B. (eds) *The Industrialization of the Countryside*. Geobooks. Norwich. pp.57-74.

Whatmore, S. (1995) From farming to agribusiness; the Global agro-food system. In: Johnson, R. *et al.* (eds) *Geographies of Global Change: Remapping the World in the Late Twentieth Century.* Blackwell, Oxford, pp.36-49.

Windhorst, H. (1989) Industrialization of agricultural production and the role of large-scale farming in US agriculture. *Tijdskrift voor Economische en Sociale Geografie* 80(5), 270-282.

Winder, G. (2002) Following America into the second industrial revolution: new rules of competition and Ontario's farm machinery industry, 1850-1930. *The Canadian Geographer* 46(4), 292-309.

Winson, A. (1992) *The Intimate Commodity: Food and the Development of the Agro-Industrial Complex in Canada.* Garamond Press, Toronto.

Wrigley, N. (1997) European retail giants and the post-LBO reconfiguration of US food retailing, paper given to Canadian Association of Geographers, St John's Newfoundland.

Wrigley, N. (2001) The consolidation wave in US food retailing: a European perspective. *Agribusiness* 17(4), 489-513.

Xia, Y. and Buccola, S. (2002) Size, cost and productivity in the meat processing industries. *Agribusiness* 18(3), 283-299.

Chapter 3

Feedlot growth in southern Alberta: a neo-Fordist interpretation

Ian MacLachlan

Introduction

According to some industrial and agricultural geographers, we are entering a post-Fordist era in which economies of scope and flexible production will triumph over the economies of scale and specialized mass production of Fordism. Others point to post-productivist trends as an alternative development path for agriculture. But from a southern Alberta perspective, the geography of the countryside is still dominated by a process of agroindustrial concentration that can be traced to the 1920s (Troughton, 1998 and 2005 and Bowler, 1992). One of the most prominent manifestations of this process in Canada is the concentration of the majority of physical farm product and cash receipts in a minority of large farms. Dating back to 1941, this process of concentration may be observed in every kind of animal agriculture by the decline in the number of farms reporting livestock and simultaneous growth in average herd or flock size. With suitable arrangement of the axes, these processes may be portrayed as an x-shaped pair of contrary trends. Figure 3.1, the 'Big X', measures and compares the remorseless decline in the number of Canadian census farms reporting cattle and the steady growth in the average number of cattle per farm.

This 60 year period of farm concentration is emblematic of a process of rural agroindustrialization with clear parallels to the Fordist industrialization of the early twentieth century. Yet there is also evidence of a striking flexibility in specialized livestock production. The concept of *neo-Fordism* captures the renewal of this Fordist-style industrialization process in a rural context (Troughton, 2004), though not without contemporary features which bear some resemblance to *post-Fordism*.

Following a discussion of the fundamentally neo-Fordist character of cattle feedlots, this chapter describes the factors accounting for the recent surge in feedlot production in southern Alberta. Five dimensions of agroindustrialization are reviewed, focusing on the measurement of three forms of concentration with results that are entirely consistent with the neo-Fordist hypothesis. The chapter then moves on to an analysis of supply-side trends (feed grain subsidies) and demand-side trends (growing export markets) to show that feedlot operators have shown a remarkable degree of resiliency in the face of adverse markets. Thus far, the feedlot sector seems to have adapted quite flexibly to Alberta's bovine spongiform encephalopathy (BSE) crisis. Average gross margin losses of about $60 per head in 2004 were little different from the losses sustained in 2002 due to the effect of drought conditions on feed grain prices (Alberta Auditor General, 2004).

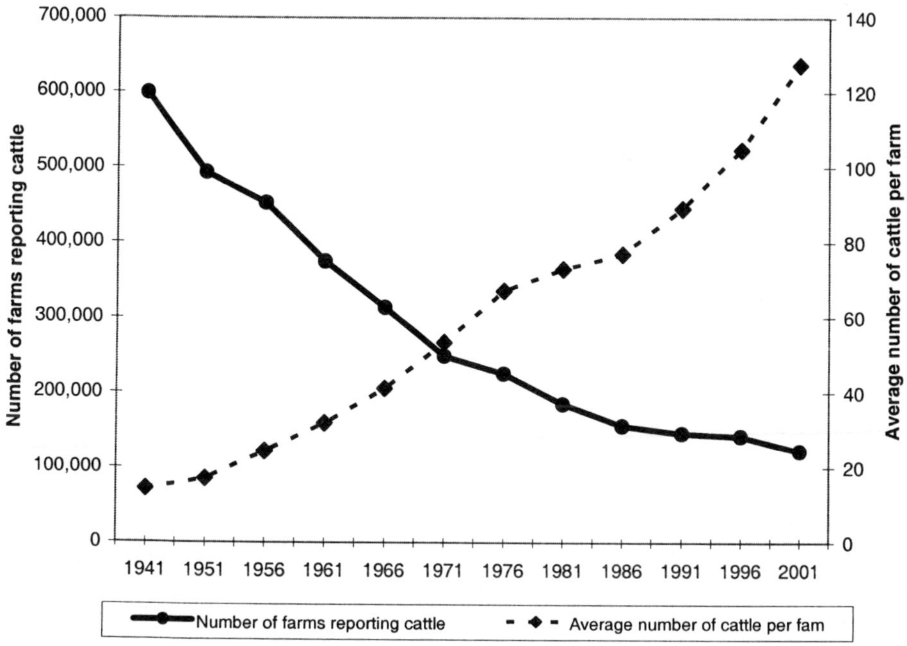

Fig. 3.1 The 'Big X': average number of cattle per farm and number of farms reporting cattle in Canada, 1941-2001. (Source: Statistics Canada, Census of Agriculture.)

Feedlots and neo-Fordism

A feedlot is an intensive livestock operation (ILO) that specializes in feeding a high-energy ration of grain and silage to young stocker cattle until they reach slaughter weight. When cattle are 'finished,' they are ready for shipment to a processing plant for slaughter. The high level of throughput in ILOs and prevailing attitudes towards them are captured in the notion of 'factory farming, redolent with prejudice' (Symes and Marsden, 1985, p.100).

Fordism is a technological paradigm (essentially mechanized Taylorism) operating in a regime of accumulation characterized by rapid growth in productivity which is shared with waged labour and matched by a massive growth in consumption. The regime is situated in a mode of regulation, a 'macroeconomic principle' to adjust individual behaviour to collective principles, balancing production and consumption, thus regulating the accumulation process (Lipietz, 1992). With the crisis of Fordism that came to a head in the 1970s and early 1980s, a post-Fordist future modelled on the 'Third Italy' with flexibility as its most enduring trait was proposed as a resolution to the crisis (Piore and Sabel, 1984). Gertler's (1988) four-level hierarchy of flexibility provides for the next section of this chapter a useful analytical schema, namely: flexible machines, flexible manufacturing systems, flexible specialization and integration and flexible accumulation. None of these notions of flexibility is associated with smallness since as Sayer (1989) notes mass production is not necessarily synonymous with inflexibility. Indeed, the adaptability of feedlot cattle producers demonstrates that mass production and flexibility are not mutually exclusive.

Flexible machines

In a juxtaposition of the multipurpose machine tools that were adapted for single-purpose use in Ford's Highland Park factory (Williams *et al.*, 1994), much of the farm machinery and material handling equipment used in feedlots (front-end loaders, feed trucks) are flexible and adaptable to many other uses. Ironically, such machines are the product of Fordist production processes. Feed mills installed for the purpose of rolling water-tempered barley into digestible form have been easily converted to handle corn (maize), a grain with very different properties. Feedlot machines are thus by necessity flexible.

Flexible manufacturing systems

Clearly feedlots are neither a manufacturing industry nor are they flexible *qua* systems. As a mass production facility, the feedlot as a whole seems hugely inflexible. The fixed capital that is sunk in feedlot infrastructure (e.g. animal pens, drainage conduits and manure lagoons) is rigidly dedicated to the confinement, care and feeding of cattle with no alternative function. It is also clear that intensive livestock agriculture does not operate with a moving assembly line in the literal sense, thus it does not share Ford's most popularly celebrated innovation, the moving assembly line which Nevins (1954) regards as the, crowning achievement in the creation of mass production techniques. However, the enormous capital investment imposes high fixed capital costs which must be spread over the greatest possible production volume. The high rate of turnover, as animals move through the growth cycle, is strongly reminiscent of the high-volume repetitive manufacturing that was the true source of Ford's success as an industrial pioneer (Williams *et al.*, 1993). In this respect feedlots exemplify 'Fordism rampant' in the North American agro-food system (Troughton, 2005).

Yet as production systems, feedlots do have some flexibility which is manifest in the type of cattle fed (steers, heifers or occasionally young cows) and in the degree of exposure to the risk of cattle ownership. For instance, the feedlot operator may elect to buy and sell cattle on the spot market, to forward contract for purchase or sale or to function as a custom feedlot, charging a daily fee per head, externalizing market risk. Notwithstanding fixed capital costs, feedlot operators often choose to leave their pens empty when stocker cattle prices are too high.

Flexible specialization and integration

The third level in Gertler's (1988, p.420) flexibility hierarchy refers to a firm's, 'ability to respond to fluctuations in market demand and to adopt new products quickly'. To this trait, I shall add, 'or to adopt new production inputs quickly'. In common with some other farming specialties, feedlot operators have shown a remarkable flexibility and adaptability. They responded to high input prices by turning to alternate sources of grain and when Canadian beef consumers turned fickle, they developed export markets. A large scale feedlot operator is perforce a commodities trader and *arbitrageur*, tracking the minute-to-minute price changes of grain, feeder cattle and slaughter cattle on futures markets. Depending on their market assessment, feedlot operators may select among a variety of hedging strategies and varying degrees of upstream integration with calf producers and downstream integration with packing plant buyers.

Flexible accumulation

The fourth level of the flexibility hierarchy refers to a form of capitalist political economy that stabilizes the technological paradigm described above with new forms of consumption and a neo-liberal stance for the state that is, in essence, the prevailing 'regime of accumulation'. Just as the Fordist regime of accumulation was associated with the $5 day and the wage-effort bargain between capital and labour in automotive production, it also coupled farm production with farm consumption and integrated farms and farm households into Fordist consumption patterns in agricultural production (Kenney *et al.*, 1991). Under Fordism, the state played a *dirigiste* role to foster infant industries with protectionist trade policies and moderated the distribution of productivity gains using permissive labour laws. In a rural context, the state also funded resource development initiatives, price supports, risk reduction and supply management institutions and in many other ways, assisted agriculture in emulating the productivity achievements in manufacturing (Kenney *et al.*, 1991). From an agricultural perspective, the current (neo-Fordist) regime of accumulation has been marked by the liberalization of trade policy, the deregulation of utilities such as transportation (notably by terminating subsidized freight rates for grain) and the encouragement of agroindustrialization, especially intensive livestock production and food processing (Alberta Agriculture, 2003; Novek, 2003).

Friedmann (1991, p.68) describes agriculture as, 'sandwiched between major corporations' (one might say 'embedded'), thus feedlot cattle production is less a distinct economic sector and more an integral component of the livestock-feed complex. Fordism has been unevenly experienced within and among commodity chains and regions (Gertler, 1988 and Sayer, 1989). Upstream of the feedlot, ranch production of calves shares few features with Fordism. Downstream from the feedlot, the carcass disassembly process is quintessentially Fordist, indeed on-the-rail dressing of beef carcasses pioneered in Chicago was the progenitor of Henry Ford's moving assembly line (Ford, 1922). This chapter shows that the feedlot revolution is a kind of rural reprise of Ford's pioneering technological paradigm of the early twentieth century. Profound changes in the broader regime of accumulation and sadly, the obsolescence of the wage-effort bargain, make the term 'neo-Fordist' a more appropriate description of agroindustrialization. But in arguing for this neo-Fordist interpretation, Gertler's hierarchy of flexibility and evidence of adaptive behaviour are presented to show that considerable scope for flexible specialization must be built into Fordist production systems if they are to persist in a dynamic economic environment.

The feedlot revolution in southern Alberta

In southern Alberta, feedlots have become the most common and conspicuous form of agroindustrial concentration, as evidenced by the number of cattle fed which grew from 1.27 million cattle in 1987, to 1.86 million in 1992 and 2.5 million by 2002, 71.5% of Canada's total fed cattle production (Anon, 1993; CanFax, 2003a). Since no direct measure of feedlot production as compared to all fed cattle is available through the census of agriculture, the distribution of steers is used as a rough proxy. In Ontario, the number of steers per farm was essentially flat between 1986 and 2001 with about 27 steers per farm while in Alberta, the average size of farms reporting steers increased abruptly from 36 to 60 between 1986 and 1991 and from 69 to 129 between 1996 and 2001. These sudden jumps in average steer herd sizes are indicators of the timing and growth of specialized feedlots and their concentration in Alberta.

Two factors account for the feedlot revolution in Alberta: demand pull and supply push. In demand pull, consumer preferences grew to favour higher quality beef and Canada developed a meat grading system based on youth and marbling as measures of quality to warrant higher prices. In supply push, the termination of both artificially low rates for the transportation of grain to tidewater for export and provincial feed grain subsidies encouraged intensive livestock production and increased the supply of grain fed cattle.

Demand pull: rising beef consumption

In the 1950s and 1960s, grain finishing of cattle was encouraged by demand from retail chain stores and the manipulation of consumer food preferences using the emergent science of marketing and mass advertising (Uvacek, 1983; Skaggs, 1986). Independent grocers and butcher shops gave way to national supermarket chains. In the battle for retail market share, there was little scope for competitive advantage in lower prices – the margins on branded groceries were already lower than any other retail category. Fresh meats constituted a significant proportion of revenues and retailers still added considerable value at the butcher counter. National chain stores used full page ads to claim that they sold only the finest federal government inspected steer beef (Toronto Daily Star, 1965). Supermarket chains such as Dominion Stores declared their competitive advantage by slogans such as: 'It's mainly because of the meat!' To justify the premium price of high quality grain fed beef, the chain stores required some authoritative means of differentiation. Canada inaugurated its beef grading system in 1929 which employed government graders to classify beef carcasses as 'Choice' (red brand) or 'Good' (blue brand) and each primal cut was stamped to indicate that it was both wholesome and of high quality. Thus the state gave its imprimatur to the Canadian beef carcass, sanctioning the consumption of youthful and well marbled beef.

But the supermarket chains could not have used meat sales to their advantage had disposable income not been rising rapidly (in a Fordist regime of accumulation) and were Canadian consumer preferences not so easily shifted in favour of red meat. Steak on the patio barbecue became symbolic of middle-class suburban affluence and beef consumption grew. Per capita domestic disappearance of beef more than doubled over a 25-year climb from 22.4 kg per capita in 1951 to peak at 50.4 kg in 1976 (Statistics Canada, 2003a).

Consumers had acquired a taste for quality cuts of beef and supermarkets featured fresh meats in their promotions. Thus the onus was on the beef industry to fulfil the demand for this high quality, high margin product. This was the 'demand pull' that encouraged grain finishing of slaughter cattle which set the stage for the feedlot revolution. Once the red brand was established as a target, cattle producers had an incentive to produce a higher quality product which could only be consistently produced from animals fed on a high energy grain ration.

Supply push: public policy incentives to cattle feeding on the Prairies

Until the 1980s, Canada was at a disadvantage in feeding cattle compared to its principal cattle trading partner: the USA. Virtually all of the cattle exported from Canada were finished for slaughter on Midwestern feedlots. To meet domestic needs, some Canadian cattle were finished locally, fed mainly on industrial by-products such as spent grain from distilleries and low quality wheat not suitable for export. Most cattle feeding took place in Ontario with a considerable proportion of stocker cattle shipped east by rail from the farms and ranches of British Columbia and the Prairies (McFall, 1927). Cattle reproduction and feeding for slaughter tended to be divided between western Canada and Ontario much as the ancient British cattle system was divided between Celtic breeders in the highlands and Saxon feeders in the lowlands.

It was commonly believed that Canada was poorly suited to finishing cattle on grain. The west had ample grassland available at low cost but it did not possess the climatic advantages of the Corn Belt states to make it competitive in feed grain production. Despite exhortations by the federal government to diversify into livestock feeding (Rutherford, 1909; Spencer, 1913; Booth, 1936) there wasn't any beef-cattle finishing tradition in Canada. Thus Prairie beef producers specialized in grass-fed stockers and feeders which were shipped to Ontario and the USA for finishing (Sinclair, 1931, p.434). As long as high wheat prices and artificially low freight rates for grain made wheat production so lucrative, cattle production was viewed as a sideline for land unsuitable for cultivation.

Grain cropping for export, on the other hand, received a massive transportation subsidy known as the 'Crow rate'. The 'Crow rate' originated with the Canadian Pacific Railway's second trans-mountain rail line which was built in 1897 through the Crowsnest Pass to consolidate Canadian control over the resources of British Columbia's Kootenay. In return for federal government grants to complete the rail line, the Canadian Pacific Railway agreed in 1897 to reduce its freight rate on grain to ½¢ per ton-mile. From its inception, the 'Crow rate' was a disincentive for raising livestock. Reduced freight rates on grain and inadequate local markets for livestock, encouraged wheat cultivation and discouraged the breeding and feeding of livestock. Pastures were broken up, barns left unused or converted into horse stables, straw piles were burned instead of being fed to cattle, money came easy and wheat was crowned king (Spencer, 1913).

Spencer's observations were as apt in the 1970s as they had been in 1913. The 'Crow rate' encouraged prairie farmers to grow wheat for export and discouraged adding any value to grain *in situ* – such as feeding it to Prairie livestock. Calves were raised on extensive rangeland not suitable for grain production and, once weaned, were shipped east for finishing on Ontario corn.

Cattle production in Alberta was at least 21% lower than would have been the case without the 'Crow rate' (MacEachern, 1978, pp.28-29). This had a retarding effect on Prairie development because cattle production and meat products have a greater growth inducing impact on the economy than does grain. The multiplier for wheat was 1.59; for every dollar of wheat output, thus $1.59 in total economic activity was created to supply the direct and indirect inputs required to grow and market the grain. The multipliers for beef cattle and for meat products were 2.16 and 2.83 respectively (Josling and Trant, 1966). By encouraging cattle feeding and cattle slaughter in Central Canada, statutory grain rates directed the growth inducing effects of the livestock and meat sectors towards the heartland and away from the Prairie West.

By 1976 the 'Crow rate' covered only about 30% of the actual cost of moving grain by rail (Snavely *et al.*, 1977). Nonetheless, in 1981 it cost only $0.22 per hundredweight to move a carload of grain from Saskatoon to Thunder Bay but $7.67 per hundredweight to move a carload of fresh meat by rail from Saskatoon to Ottawa (Wilson, 1981). The 'Crow rate' was finally abolished on 1 August 1995, 98 years after its inception, making grain feeding of western livestock a potentially more attractive proposition with indirect incentives for livestock processing and meat-packing investment.

The seeds for the demise of the 'Crow rate' had been sown in 1983, when the *Western Grain Transportation Act* (WGTA) allowed freight rates on grain to increase, gradually transferring the real costs of the 'Crow rate' from the railways (which had simply passed the costs on to other shippers) to the federal government. The WGTA was a boon to the railways but turned out be an even further disincentive to livestock producers. By subsidizing the shipment of export grain to the tune of over $700 million per year, federal payments under the WGTA caused feed grain prices to rise. To assist livestock producers in offsetting the negative impacts of the WGTA, each of the Prairie Provinces brought in incentive programmes to subsidize the cost of feed grain and foster increased livestock production (Canadian International, 1993). These provincial feed grain incentives came at an opportune time, the start of an upswing in the cattle cycle and rising cattle prices. Alberta subsidized the purchase of feed grain for ten years, longer than any other province. In its first year of operation, feed barley expenses in Alberta were reduced by $78 million (Kraft and Doiron, 2000). Alberta's subsidy was gradually reduced and finally discontinued on 1 August 1995, coincident with the elimination of subsidies under the WGTA and the inauguration of the Agricultural Agreement as part of the General Agreement on Tariffs which morphed into the World Trade Organization (Kraft and Doiron, 2000).

By subsidizing feed grain expenditures so long and so generously, the Alberta government encouraged the establishment of many new feedlots in the 1980s which prompted the expansion of beef processing (MacLachlan, 2001). These provincial pump-priming incentives encouraged the development of an agroindustrial cattle finishing industry in the mid-1980s which grew dramatically in the neo-liberal policy environment of the 1990s. Thus feedlots are at once a product of post-Fordist deregulation by the federal government (termination of the 'Crow rate' and WGTA) and neo-Fordist re-regulation by the provinces (Crow offsets).

Agroindustrialization and intensive livestock production

The industrialization of agriculture may be traced back to the 1920s but was not widespread in the USA until the midpoint of the twentieth century (Hart and Mayda, 1998). In Canada, the industrialization of livestock agriculture was not significant until the 1980s, paradoxically a time when deindustrialization of the manufacturing sector had taken hold and concerns were shifting to post-industrial society. ILOs are a conspicuous reminder that Fordism and the process of geographic industrialization (Storper and Walker, 1989) may develop in new forms from factory farms to fast food factories.

While sheer production scale is the most conspicuous feature of feedlot agriculture, its agroindustrialization may be characterized according to the three structural dimensions identified by Bowler (1992): *concentration*, *intensification* and *specialization*. Feeding operations are *concentrated* in a small number of large scale facilities instead of being dispersed among a large number of small family farms. *Intensification* refers to the application of capital to substitute for both labour and land, increasing the output per unit of land area. A typical large feedlot keeps 25,000 animals on a quarter section of land (65 hectares). Cattle feedlots tend to be highly *specialized*, handling but one phase of the production process and yielding just one product: slaughter cattle.

To Bowler's three dimensions of agroindustrialization, two more are suggested here: *turnover* and *standardization*. *Turnover* is the time required to transform raw material into a finished product, measured as the average number of times per year that an individual commodity unit is held in the value-added process before it is replaced by the next. It is closely related to the efficiency with which capital is employed and in agriculture, to the organic constraints on production. Given a 285-day gestation period and the time required for cows to come into estrus after birth, calf production is limited to one turn per year. Slaughter cattle are finished after 90 to 180 days on feed, allowing between two and three turns per year in a typical feedlot. Thus feedlot turnover is considerably higher than cow-calf production.

The finished product of feedlots is an increasingly *standardized* animal of the size, age, breed, conformation, carcass yield and grade of beef that packing plants demand.

Feedlots avoid mixing steers and heifers to reduce aggressive mounting behaviour which could interfere with weight gain. Further standardization of a live organic commodity is made possible by the ability of large feedlots to purchase uniform lots of feeder calves and a closely regulated feeding regimen ensuring that cattle are shipped promptly when they are ready for slaughter. The incentive for such precise timing is the complex pricing formulas applied by packing plants which penalize underweight and overweight cattle and rail-grade pricing based on the carcass grade assigned by federal inspectors.

Concentration

Concentration is singled out from Bowler's structural dimensions for detailed analysis because the restructuring of farm production in favour of fewer but larger establishments captures the agroindustrial and neo-Fordist character of cattle feeding especially clearly. Increasing concentration is evident as a measure of productivist agriculture and the weakest link in the reversal of agroindustrial trends that characterize post-productivism (Ilbery and Bowler, 1998). Evans *et al.* (2002) cite evidence to show that the concentration of agricultural production is in fact, continuing unabated. Finally, the required data to measure concentration are in the public domain which is not the case for some other measures of agroindustrialization. To build on the transition to fewer and larger farms diagrammed in Fig. 3.1, the concept of concentration is interpreted and measured in three different ways: establishment concentration, livestock unit concentration and spatial concentration.

Establishment concentration

In contrast to corporate concentration, establishment concentration refers to the increase in the proportion of production accounted for by a small number of large establishments irrespective of their affiliation. Establishment concentration may be measured graphically by Lorenz curves to chart the cumulative proportion of the cattle herd that is concentrated in different herd sizes (Fig. 3.2) and quantitatively using Gini's Concentration Ratio (GCR) (MacLachlan and Sawada, 1997). Census years 1976 and 1996 were selected to portray the beef cow and steer herd distribution before and after the development of large feedlots in Alberta. Focusing first on the national beef cow herd, it can be seen that the two Lorenz curves overlap and intersect suggesting little change in the level of concentration. The GCR in 1976 is 0.55 which actually declines to 0.53 in 1996. This is consistent with the traditional nature of cow-calf operations which tend to be less concentrated than any other stage in the cattle-beef commodity chain (MacLachlan, 2001). The steer herd, as a proxy for the size distribution of feedlots, was considerably more concentrated than beef cows in 1976 and its level of concentration increased quite markedly by 1996. In 1976, the Lorenz Curve is considerably further from the line of perfect proportionality over the full span of the distribution than is the case for beef cows and by 1996, the chord it subtends grows even larger. The GCRs for the steer herd distribution increase from 0.69 in 1976 to 0.83 in 1996. By 1996, the 404 largest steer farms (0.8% of all farms reporting steers) accounted for 57.9% of all the steers in the country. Farms with large numbers of steers are invariably operated in a high density feedlot configuration.

Livestock unit concentration

Livestock unit concentration is suggested here as a third measure of concentration, namely: the amount of meat output that is concentrated in each animal produced. The saleable beef produced by each animal is approximated by its carcass weight.[2] Feeders have an incentive to produce large cattle as they are sold by the pound. On the other hand, meat retailers do not want a carcass that is too heavy because the primal cuts are too large to be cut into steaks and roasts of the size and thickness that consumers prefer. Thus the pricing formulas used in packing plants may penalize finished cattle that are too heavy, with 610 kg live weight being generally accepted as the high end of the target weight range.

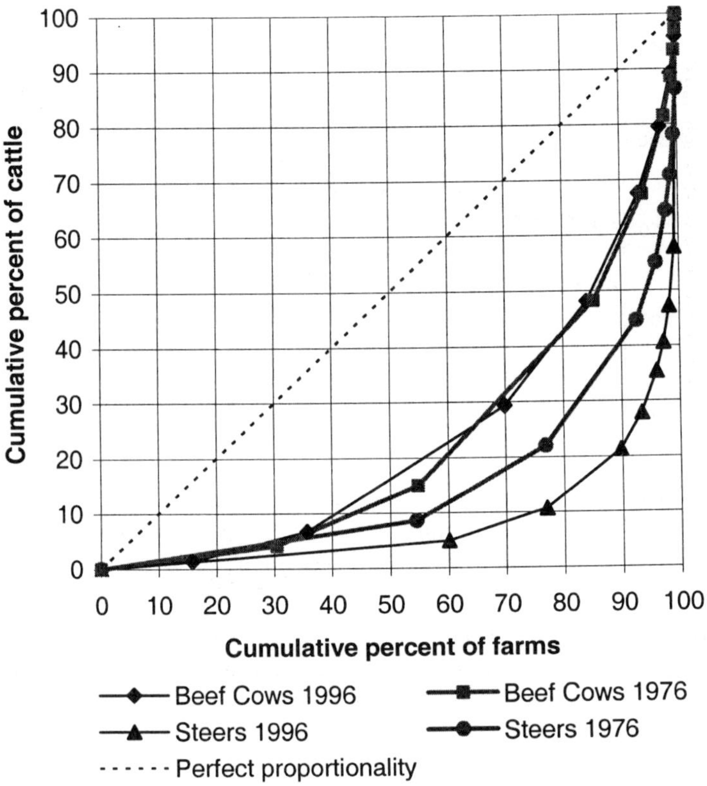

Fig. 3.2 Farm establishment concentration: Lorenz curves showing increasing concentration of Canada's cow and steer herds, 1976 and 1996. (Source: raw data from Statistics Canada, Census of Agriculture, special tabulation.)

Nevertheless, the annual average hot carcass weight of steers increased from 253 kg in 1960 to 375 kg in 2001, an increase of 47%. Heifer carcass weights increased even faster, by 60% over the same period as shown in Fig 3.3.

Several causal factors are at work here. First, the British breeds (e.g. Aberdeen Angus) have declined in favour of the heavier European breeds (e.g. Charolais) as a proportion of slaughter cattle. A second factor is that the live weight of finished cattle has increased as Alberta's grain fed and hormonally implanted cattle finish faster, heavier and more efficiently than in the past. Third, despite admonitions to the contrary, the price penalty for overweight cattle is often so small that it still pays to err on the side of being too heavy rather than too light.

Dips in the upward trend in average carcass weights coincide with peaks in the cattle cycle (1975, 1982 and 1996) while isolated peaks correspond to troughs (1970, 1979, 1987 and 1995). When the cycle peaks, farmers sell all the animals they possibly can, earlier (and lighter) than they otherwise might, to avoid potential price declines. As the cattle cycle bottoms, farmers tend to keep animals longer, hoping that prices will increase and marketing heavier animals as a consequence.

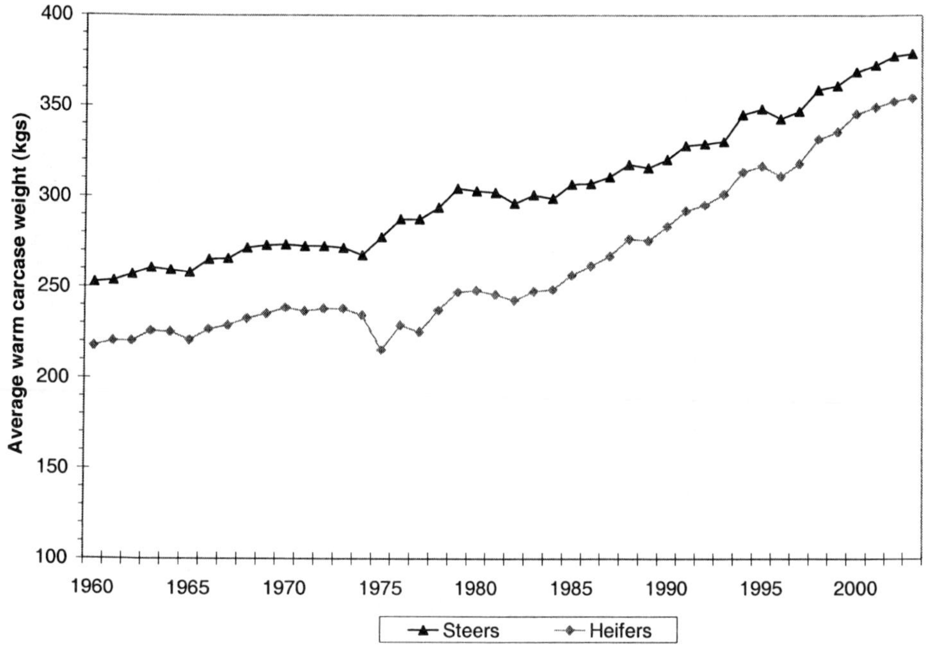

Fig. 3.3. Livestock unit concentration: Growth in average hot carcass weight of steers and heifers in Canada, 1960 - 2003. (Source: Agriculture and Agri-Food Canada Livestock Market Review, various years.)

Spatial Concentration

Feedlots are localized in five North American regions that span the Great Plains and Prairies. Southern Alberta is the smallest region, after eastern Colorado, Nebraska, western Kansas and the Texas Panhandle. Alberta feedlots are distinct from their southern neighbours because they depend on inexpensive barley which does well on the Canadian Prairies and because irrigation water is delivered by gravity from the Rocky Mountain snow-pack through superficial drainage. To the south, most feedlots rely on the Ogallala aquifer to sustain livestock and irrigate feed maize and sorghum. The rising cost of fuel to pump ground water to the surface and the gradual depletion of the Ogallala aquifer favours continued feedlot growth in Alberta, despite its shorter growing season and cooler climate.

Figure 3.4 shows that the spatial distribution of cattle feeding is highly concentrated within southern Alberta. Using bunk capacity to measure the distribution of cattle feeding, capital investment in feedlot facilities is visibly localized in just one small jurisdiction, Lethbridge County. Thus the site features which attract feedlots, (abundant irrigation water and suitable cropland to grow forage and silage), the favourable economic impacts (supplier industries and employment); and negative externalities (potential water contamination and odours), are also highly localized.

Flexible adaptation to declining beef consumption: beefing up exports

As an agricultural commodity, the cattle population is strongly cyclical with an average peak-to-peak wavelength of 12 years. At 14.7 million on 1 January 2004, the Canadian cattle population exceeded the previous all-time high of 14.3 million cattle in 1975 because of the export ban on live cattle due to the appearance of one case of BSE in May 2003. Setting aside this anomaly, the agroindustrialization of feedlot beef production was one of the primary factors in the rapid growth of the bovine population from 1987 to 1996 (Fig. 3.5). This seems all the more remarkable when the declining dairy cow population and increasing livestock unit concentration are taken into account. In short, there are proportionally more beef cattle than ever before and each animal yields more saleable beef.

Yet Fig. 3.5 also shows that per capita consumption of beef in Canada has been declining almost without interruption since it peaked in 1976. Prevailing western ideas about nutrition and epicurean veganism are sentiments promoting what some regard as a new 'quality' food regime (Evans *et al.*, 2002) that would see the gradual decline of cattle specially bred, raised and fed on grain to maximize the abundance of intramuscular fat. Could the post-modern diet of the new millennium spell the demise of feedlot beef production in Canada? Clearly not, since Fig. 3.5 also shows that the trends in per capita consumption and the cattle population have diverged ever since the cattle cycle bottomed in 1987. The uncoupling of the cattle population from domestic beef consumption can only be understood as the outcome of a rapid increase in cattle and beef exports. This uncoupling of production and consumption may be interpreted as a case of 'disembedding'. Economic activities are disembedded when they are 'lifted out' of their local social and political context and reorganized at a broader, often global scale (Novek, 2003). Until the BSE crisis of May 2003, strong global markets in the USA, Mexico and East Asia permitted Alberta's cattle and beef food chain to thrive as industry disembedded from its domestic surroundings, little affected by the slide in domestic per capita beef consumption.

Reflecting the growth of intensive livestock operations (cattle and hogs [pigs]), the structure of Prairie Canada's agricultural export trade is being transformed. Led by 'King Wheat', grain was traditionally Canada's largest agricultural export. But wheat has been diminishing in its relative importance as a staple export since the middle 1980s when the 'Crow rate' was phased out. The trend for meat and meat preparations (principally beef and pork) has been positive for many years but its growth, in structural terms, accelerated after 1995 and in 2000, its dollar value exceeded wheat for the first time (Statistics Canada, 2003b). Exports of live animals (chiefly cattle) have also grown structurally. Export markets for live cattle were favoured by the larger feedlot operators who wanted to diversify their customer base and to have some means of levering the prices offered by Canadian cattle buyers. Alberta's fed cattle production was growing faster than Alberta's cattle slaughter capacity forcing producers to turn to packing plants in Colorado and Washington. In 2001 and 2002, these trends were exacerbated by drought and high feed grain prices in Alberta which prompted the export of stockers to the USA where cheaper grain was available. Meat and meat preparations represent a higher level of processing and value added than live animals which in turn embody more value added than wheat, thus value-added agriculture and agri-food processing have assumed more prominent roles on the prairie landscape.

In 2002, Canada ranked fourth among the world's largest beef exporters (after Australia, the USA and Brazil). Traditional west-east flows of cattle have drawn to a close as slaughter cattle and beef processing are more concentrated in western Canada than ever before. West-east domestic flows of chilled meat are still significant but they are being overtaken by a new north-south trading orientation as beef exports account for a growing share of Canada's domestic beef production, reaching 39.6% in 2002 (Statistics Canada, 2003b).

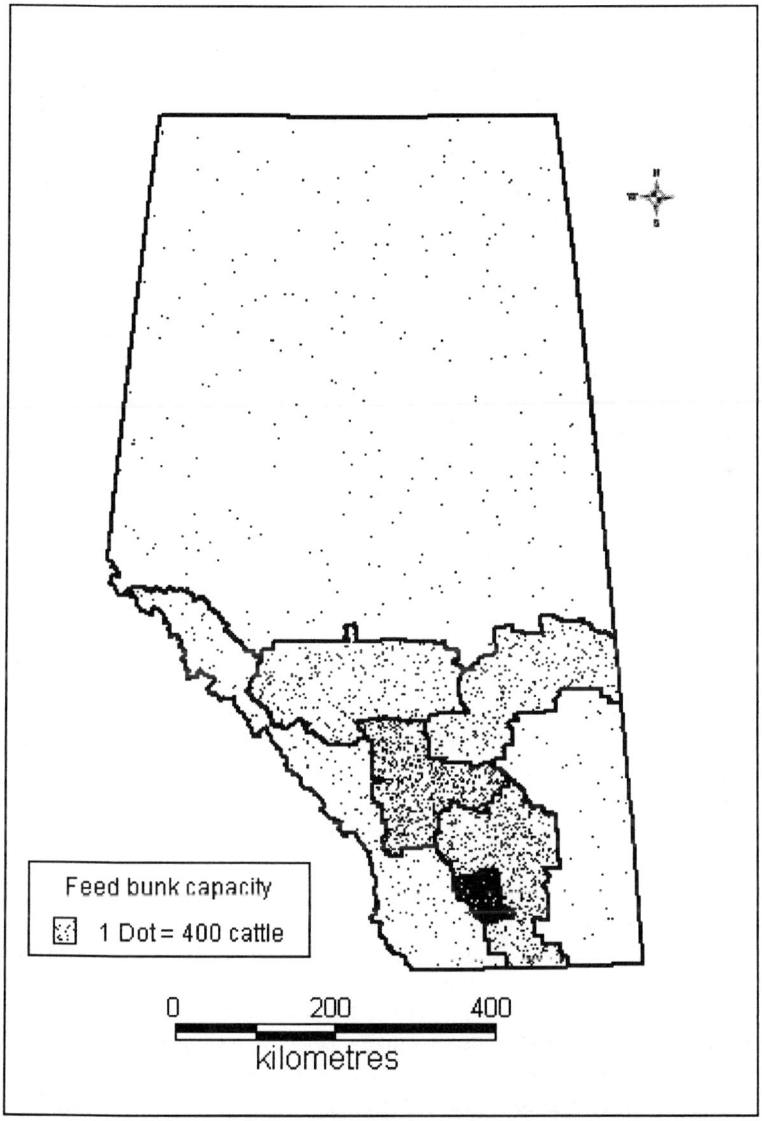

Fig. 3.4 Spatial concentration of feedlot bunk capacity in Alberta, 2003. (Source: CanFax Cattle on feed, capacity by region (http://www.canfax.ca/) Accessed 15 August 2003).

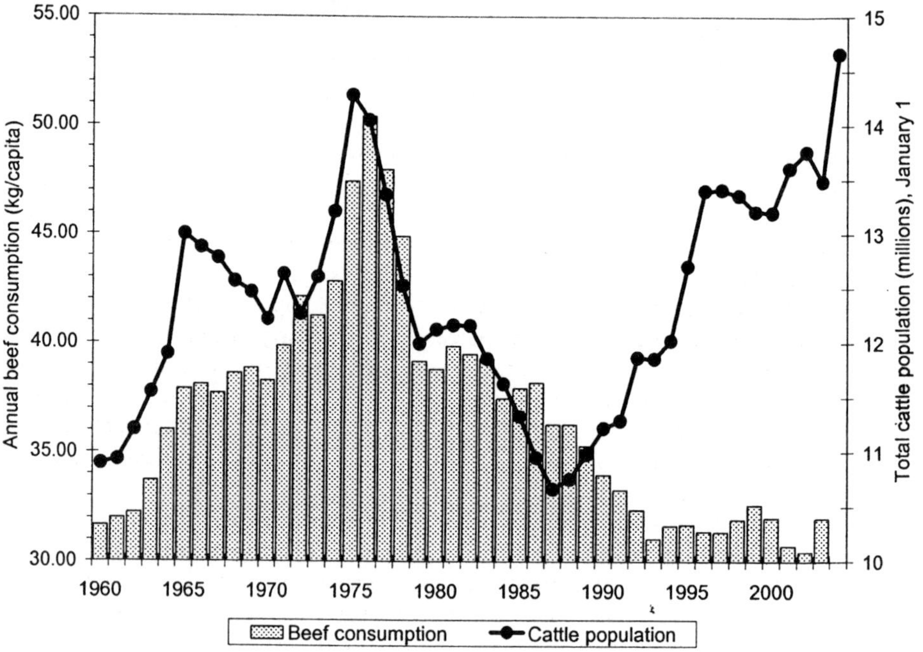

Fig. 3.5 Cattle population and beef consumption in Canada, 1960-2002. (Source: Agriculture Canada *Livestock Market Review* and Statistics Canada Cansim II series label D263633, Apparent per capita food consumption in Canada; beef, carcass weight, annual.)

Accounting for 72% of Canada's beef exports in 2001, the USA has always been the largest market due in part to its proximity and in part to the traditionally liberal terms of trade in both cattle and beef.[3] However, Canada has diversified its export markets considerably with Mexico (14%), Japan (5%), and South Korea (3%) accounting for a growing share of Canadian beef exports in 2002 (Canfax, 2003b). Thus cattle producers look to American meat packers (which sometimes offer higher prices for slaughter cattle than Canadian packer buyers) while beef processors are promoting their products in Asia and Latin America more aggressively than ever before in an effort to diversify Canada's traditional staple export dependence on the USA. This is manifest in the promotional activities of the Canada Beef Export Federation with membership from both the big beef processors (Cargill Foods, Lakeside Packers, Levinoff Meat Products and XL Foods) and affiliates from producer groups such as the Canadian Cattlemen's Association and Alberta Beef Producers.

The development of forward linkages with feed lots and food processors has added value to Canadian cattle production and enhanced beef exports, a development pattern that is consistent with the Innisian staples theory model (Innis, 1982; Watkins, 1963). For many decades cattle were produced on the extensive temperate grazing lands in which Canada has a comparative advantage, to be finished and slaughtered in the USA, or in the UK whenever American tariff policy became adverse (Foran, 1998). With the dismantling of the 'Crow rate' and the emergence of a North American free trade area, the industry had the opportunity to take the next step and begin adding more value to the grass fed stocker cattle, feeding and finishing for slaughter in the USA and, diversifying downstream to a greater extent than ever before, slaughtering and fabricating boxed beef for export.

Alberta's two largest beef processors are now foreign owned and, with exports of chilled and frozen beef on the rise, they control a growing proportion of trade. The 'shallow integration' of arm's length trade in grain differentiated only by grade and handled by the Canadian Wheat Board, has yielded to the deep integration (Dicken, 1998) of direct foreign investment, with TNCs organizing greater value added and trade in more highly differentiated meat products. This means that both transnational enterprize and rural cattle producers are globally embedded and subject to the vicissitudes of feed grain, cattle and beef commodity pricing in addition to the vagaries of consumer markets. Despite increasing agroindustrialization and diversification, prairie agriculture and trade is still mercurial.

The next step in the liberal vision of the staple model (Schedvin, 1990) is the development of vertical linkages: forward links with suppliers upstream and downstream links with value-added processors. Feedlots have clearly stimulated the expansion and specialization of cow-calf production in Alberta and Saskatchewan. Evidence of agribusiness supplier links is admittedly spotty and anecdotal. In cities such as Lethbridge one finds small-scale manufacturers of livestock handling equipment and irrigation infrastructure, suppliers of feed, pharmaceuticals and nutraceuticals, and a host of agricultural service providers which depend on feedlots as customers. Custom manure handling and livestock transportation exemplify the role of specialist sub-contractors based in the city but serving the countryside.

Downstream links create opportunities for processors to differentiate and add value to commodity beef. Put another way, these processors represent a deepening of commodity relations to produce a more durable, Fordist food product (Friedmann, 1991). Meat products cut from commodity carcass beef are processed and packaged to supply both mass retailers and purveyors, who provide portion-ready servings to restaurant and institutional markets. In many cases these value added operations are geared specifically to local markets due to the perishable nature of the product and local consumer preferences for particular cuts, seasoning, and packaging. The growing variety of 'ready meals,' supplied for both home and restaurant use, further multiplies the value of livestock production. Indeed the Alberta government has targeted meat processing as one of five key sectors in its agri-food growth strategy (Alberta Agriculture, 2003). Value-added meat processors are also moving into export markets. Products may include boneless beef for further processing, portion-controlled cuts such as steaks for the Japanese, Korean or Mexican markets, and the chilled logistics and specialized transportation services that are required (Canada Beef, 2003).

Flexible adaptation to drought: shifting feedstock

While feedlot beef production is becoming a global industry with global markets, it is still influenced by its commodity roots and is strongly cyclical. For the roughage component of the diet, Alberta feedlots use high-moisture barley silage, grown locally on irrigated land. This feed component is almost immune from the vagaries of summer precipitation; hitherto, the Rocky Mountain snow-pack has been a reliable source of irrigation water. But the barley grain which constitutes 85% of the diet by weight is another matter. It is grown on dryland in central Alberta, the region of the province most affected by drought in 2001 and 2002. Drought reduced barley yields and in 2002, Canada's barley production reached a 30-year low. Depleted barley stocks drove up prices by 54% between the 2000 and 2003 crop years, reaching an all-time high of $172 per tonne in 2002-2003 (Agriculture and Agri-Food, 2003). This prompted a radical change in trade flows of both feedstock and cattle. The industry responded in two ways. First, many Alberta feedlots switched abruptly from domestic barley to imported maize (mainly from the Midwest). Maize imports quadrupled between 1999 and 2003 with whole trainloads destined for the feedlots of southern Alberta. Feedlot operators adapted easily to this unprecedented change in their primary feedstock, making only minor technical adjustments so their feed mills could handle maize in lieu of barley.

Second, the flow of feeder cattle southward (which had slowed to a trickle in recent years) picked up dramatically. Cattle producers sold many of their feeder calves and yearlings to US cattle feeders. Canadian feedlot operators reefed their sails and purchased fewer cattle in anticipation of more favourable conditions in future. Thus the agroindustrial production of slaughter cattle retains its commodity characteristics yet producers have demonstrated a remarkable degree of flexibility.

Figure 3.6 illustrates the seasonal nature of cattle feeding and the flexible feedlot response to a shortage of feed grain. An influx of weaned calves entering feedlots in the autumn contributes to a sharp peak in the total number of cattle on feed in December of each year. The amplitude of marketings is considerably lower and its seasonal peak comes in the summer months. The net effect of the monthly fluctuations in new feedlot entrants and marketings is the number of cattle on feed, i.e. the total number of cattle in feedlots on the first day of each month. The number of cattle on feed dips sharply to a trough in August that is only 50-60% of the peak. The second feature illustrated by Fig. 3.6 is the year-to-year decline in the number of cattle on feed each month. Thus when the number of cattle on feed in 2002 peaked in December at 862,000 head, it was only 58% of the December peak in 1999. This is further evidence of flexibility as feedlot operators adapted to high feed prices by limiting their intake of feeder cattle. The corollary of lower feedlot intake is greater exports of weaned calves and stocker cattle to cattle feeders in the USA which were less affected by drought. It also indicates that if the feedlot industry had to suffer a total ban on cattle exports, the timing of Canada's BSE crisis in May 2003 was relatively good from the standpoint of cattle feeders.

The flexibility of feedlots was tested by the unprecedented crisis triggered by the discovery of a single cow with BSE in northern Alberta. Cattle exports to the US and meat exports worldwide were immediately banned. Beef exports to the US and Mexico from cattle under the age of 30 months resumed in September 2003 but the border has remained closed to live cattle. The feedlots were able to exploit their flexibility and high turnover attributes and quickly reduced their throughput to match the level of live cattle that can be processed in Canada. Figure 3.6 shows the precipitous drop in the number of cattle placed on feed in June 2003 and again in January 2004. Fed cattle prices declined, but not as precipitously as the prices for older cull cows. Most of the enormous cattle surplus is in the form of aged dairy and beef cows, which have no place on the feedlot. Feedlot producers adapted to the crisis by simply reducing the number of cattle they purchased each month. The number of cattle on feed in Alberta and Saskatchewan plummeted 55% from an annual peak of 1.5 million in December 1999 to 668,000 in December 2003. A range of provincial and federal subsidy programmes to protect cattle producers from the market downturn were developed in the summer and autumn of 2003, an initiative that was entirely consistent with the role of the state under the Fordist regime of accumulation. Of the $402 million paid out to owners of cattle under the BSE compensation programmes, 11% was received by the three large meatpacking firms, an indication of the number of cattle owned by packers as a market hedge (Alberta Auditor, 2004).

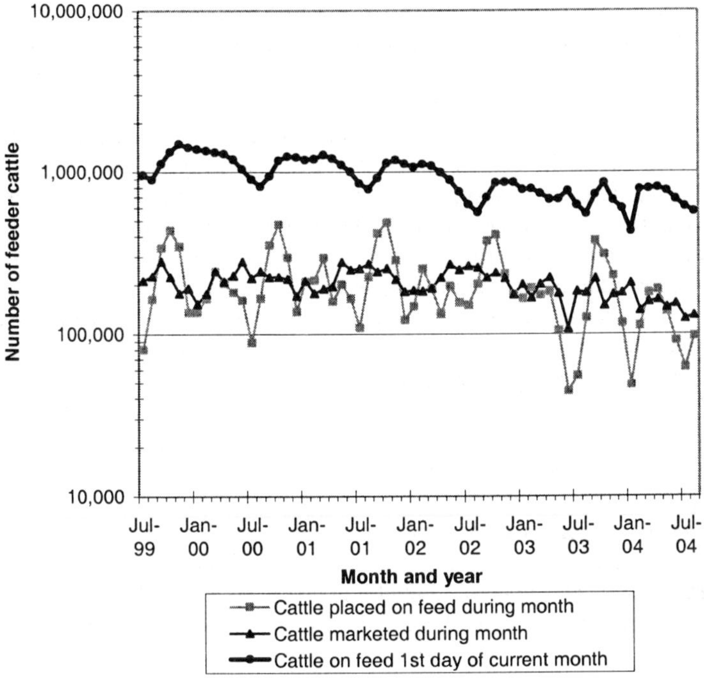

Fig.3.6 Cattle on feed, placed on feed and marketings in Alberta and Saskatchewan for feedlots in excess of 1,000 head by month, 1989-2004. (Source: CanFax, published by Agriculture and Agri-Food Canada, Cattle on feed report (http://www.agr.gc.ca/misb/aisd/redmeat/), accessed 6 November 2004)

Conclusions

Is the agroindustrialization of Alberta's feedlot cattle industry neo-Fordist? The evidence seems quite consistent with a neo-Fordist interpretation. Singling out concentration as an important measure of agroindustrialization and Fordism in the countryside, it is clear that cattle feeding has become increasingly concentrated in recent years. This is true no matter how concentration is construed; establishment, livestock unit or spatial measures of concentration all yield the same conclusions. To determine whether other dimensions of agroindustrialization are consistent with this interpretation, Census of Agriculture microdata needs to be explored in greater detail for the feedlot sector and for other types of ILOs as well.

Beginning in the 1980s, the feedlot revolution was largely the outcome of changes in the broader regime of accumulation. The industry was retarded in favour of grain production for over 80 years as a result of the 'Crow rate' as a key policy plank in the relationship of the federal government to Prairie Canada that saw most grain finishing of cattle and downstream carcass and meat processing industries localized in Ontario. The termination of the Crow was consistent with many other forms of deregulation that are characteristic of neo-liberalism in a post-Fordist regime of accumulation. However the Crow was quickly replaced by agricultural incentive programmes in each of the Prairie Provinces. These were efficacious catalysts in the development of a beef industry that was self-sustaining and without direct subsidies (until the BSE crisis) and beef exports from cattle less than 30 months of age are still competitive in the USA.

Until 1976, per capita beef consumption rose along with real disposable incomes but since that time, consumption has fallen as preferences have shifted in favour of poultry and vegetarian diet alternatives. The feedlot revolution occurred in spite of falling domestic consumption due to the buoyancy of export markets. Individually and collectively (through the Canadian Cattlemen's Association and Canada Beef Export Federation) producers and processors developed their export markets to take up the slack left by declining domestic consumption. Paradoxically in some respects, Alberta, Canada's richest province, accommodates a form of peripheral Fordism, in which mass production is embedded in global markets with the purchasing power to buy a luxury commodity and so it is partially disembedded from domestic consumption.

When the domestic feed grain supply became scarce and prices rose, producers adapted by shifting from domestic to imported grain as the feedstock to finish cattle for slaughter. This exemplifies the degree of flexibility which has always been characteristic of Fordism and mass production. However, the ability of flexible production systems to respond flexibly to exogenous shock is not infinite. Staple commodities, even those which have been processed and had some value added, have always been exposed to the vicissitudes of export market disruptions.

Eighteen months after the BSE crisis began on 20 May 2003, the American border is still closed to live cattle. Southern Alberta's countryside remains strongly productivist and dependent on an organic commodity with all the vulnerability to market disruption that animal agriculture entails. If the record of adaptation described here is any guide to future potential, it appears that Alberta's feedlot cattle production sector has the flexibility to weather this exogenous shock and retain its neo-Fordist character. But if smaller cattle finishers do not have the depth of capital required to weather the crisis, the industry may become further concentrated with the possibility of a significant increase in foreign equity participation in the agroindustrial production of fed cattle in Alberta. In Troughton's (2005) words: 'Fordism rampant'.

Notes

1. The day of record for Canada's 2001 Census of Agriculture was 14 May while 1 June 1996 was the day of record for the preceding census.
2. Hot carcass weight is measured on the rail just before the dressed and trimmed carcass goes into the cooler. As the carcass cools, it loses weight due to evaporation, a process known as carcass shrink, amounting to about 2% of hot carcass weight. Carcass yield is the hot carcass weight as a percentage of live weight and typically ranges from 55 - 60%. Annual average carcass weight is calculated as an average of all the carcasses processed through all the federally inspected plants in Canada. About 3.3 million carcasses were handled in 2002.
3. While free trade in live cattle has been a touchstone for Canadian cattle producers, several protectionist episodes have influenced continental trade patterns. Twentieth-century examples include the Fordney-McCumber tariff of 1922, the Smoot-Hawley Tariff of 1930 (Canada, 1982; Foran, 1998), an export embargo by the Canadian government during the 1940s to prevent domestic food price inflation in war-time, and a one-year ban on all meat and cattle imports from Canada, beginning in 1952 when an outbreak of foot and mouth disease was reported in Saskatchewan. On 20 May 2003, the post-mortem discovery of a BSE infected cow in Alberta prompted all importing countries to ban Canadian cattle and

beef. On 8 August 2003, the USA announced that it would start accepting low-risk boneless beef from animals under 30 months of age (effectively excluding cow meat), but not live animals. Mexico followed suit the following week.

Acknowledgements

This chapter is based in part on research which was supported by the University of Lethbridge Research Fund. Chris Birrell and Lisa Hungerford contributed as astute undergraduate research assistants. I am grateful to Diane Clark and Ivan Townshend for their generous assistance.

References

Agriculture and Agri-Food Canada (2003) *Canada: Grains and Oilseeds Supply and Disposition.* http://www.agr.gc.ca/mad-dam/e/sd1e/hsd1ez.htm. Accessed 26 September 2004.

Alberta Agriculture, Food and Rural Development (2003) *Alberta's agri-food industry growth strategy,* draft 25 March.

Alberta Auditor General (2004) *Report of the Auditor General on the Alberta government's BSE-related assistance programs*, Auditor General, Edmonton.

Anon (1993) Send us your calves and yearlings: Alberta has quietly become a leading North American feeding centre. *Alberta Report* 20(November 1), 20-21.

Booth, J. (1936) Ranching in the Prairie Provinces. Agricultural Progress on the Prairie Frontier.In: Mackintosh, W. and Joerg, W. (eds.) *Canadian Frontiers of Settlement*, Vol. 5. MacMillan of Canada, Toronto, pp.51-66.

Bowler, I. (1992) The industrialization of agriculture. In: Bowler, I. (ed.) *The Geography of Agriculture in developed market economies.* Longman Scientific and Technical, New York, pp.7-31.

Canada (1982) *Industry in turmoil: report on the long term stabilization of the beef industry in Canada.* Standing Senate Committee on Agriculture, Minister of Supply and Services, Ottawa.

Canada Beef Export Federation (2003) *Beefing up exports.* http://www.cbef.com/Pub_BEF_Beefing_Up_Exports.htm Accessed 15 August 2003.

Canadian International Trade Tribunal (1993) *An inquiry into the competitiveness of the Canadian cattle and beef industries.* Reference No. GC-92-001, Minister of Supply and Services, Ottawa.

CanFax (2003a) *Cattle on feed-feedlot capacity by region.* http://www.cattle.ca/canfax/cof_region_capacity.htm. Accessed 14 May 2003.

CanFax (2003b) *CanFax Statistical Briefer.* http://www.cattle.ca.. Accessed 14 May 2003.

Dicken, P. (1998) *Global shift,* 3rd edition. Guilford Press, New York.

Evans, N., Morris, C. and Winter, M. (2002) Coneptualizing agriculture: a critique of post-productionism as the new orthodoxy. *Progress in Human Geography* 26, 313-332.

Foran, M. (1998) Mixed blessings: the second 'golden age' of the Alberta cattle industry 1914-1920. *Alberta History* 46, 10-19.

Ford, H. (1922) *My life and work.* William Heinemann, London.

Friedmann, H. (1991) Changes in the international division of labour: agri-food complexes and export agriculture. In: Friedland, W. Busch, L. Buttel, F. and Rudy, A. (eds.) *Toward a new political economy of agriculture.* Westview Press, Boulder Colorado, pp.65-93.

Gertler, M. (1988) The limits to flexibility: comments on the post-Fordist vision of production and its geography. *Transactions of the Institute of British Geographers* New Series 13, 419-432.

Hart, J. and Mayda, C. (1998) The industrialization of livestock production in the USA. *Southeastern Geographer* 38, 58-78.

Ilbery, B. and Bowler, I. (1998) From agricultural productivism to post-productivism. In: Ilbery, B. (ed.) *The geography of rural change.* Addison Wesley Longman, Harlow, pp.57-84.

Innis, H. (1982) The importance of staple products. In: Gross, M. and Kealey, G. (eds.) *Pre-industrial Canada, 1760-1849.* McClelland and Stewart, Toronto.
Josling, J. and Trant, G. (1966) *An empirical study of interdependence among agricultural and other sectors: an input-output model.* Agricultural Economics Research Council of Canada, Ottawa.
Kenney, M. Lobao, L. Curry, J. and Goe, W. (1991) Agriculture in the USA. Fordism: the integration of the productive consumer. In: Friedland, W. Busch, L. Buttel, F. and Rudy, A. (eds.) *Toward a new political economy of agriculture.* Westview Press, Boulder Colorado, pp.173-188.
Kraft, D. and Doiron, J. (2000) *Post Crow influence on prairie feed grain prices.* Paper presented September 29 at Western Nutrition Conference, Winnipeg.
Lipietz, A. (1992) *Toward a new economic order: postfordism, ecology and democracy.* Translated by Malcolm Slater. Polity Press, Cambridge.
MacEachern, G. (1978) *Return of the Crow rate and the Alberta livestock economy.* Prepared for the Alberta Cattle Commission. Agricultural Economics Research Council of Canada, Ottawa.
MacLachlan, I. (2001) *Kill and chill: restructuring Canada's beef commodity chain.* University of Toronto Press, Toronto.
MacLachlan, I. and Sawada, R. (1997) Measures of income inequality and social polarization in Canadian metropolitan areas. *Canadian Geographer* 41, 377-397.
McFall, R. (1927) *The world's meat.* D. Appleton and Company, New York.
Nevins, A. (1954) *Ford: the times, the man, the company.* Charles Scribner, New York.
Novek, J. (2003) Intensive livestock operations, disembedding, and community polarization. *Society and Natural Resources* 16, 567-581.
Piore, M. and Sabel, C. (1984) *The new industrial divide.* Basic Books, New York.
Rutherford, J. (1909) *The cattle trade of western Canada.* Special Report, Department of Agriculture, Ottawa.
Sayer, A. (1989) Postfordism in question. *International Journal of Urban and Regional Research* 13, 666-695.
Schedvin, C. (1990) Staples and regions of Pax Britannica. *Economic History Review* 43, 533-559.
Sinclair, R. (1931) Report of the beef cattle committee. *Scientific Agriculture* 11, 433-435.
Spencer, J. (1913) *Beef raising in Canada.* Bulletin No. 13, revised edition. Minister of Agriculture, Ottawa.
Skaggs, J. (1986) *Prime cut: livestock raising and meatpacking in the USA 1607-1983.* Texas A&M Press, College Station.
Snavely, King and Associates (1977) *Costs and revenues incurred by the railways in the transportation of grain under the Statutory Rates.* Grain Transportation Branch, Transport Canada, Ottawa.
Statistics Canada (2003a) *Meat production, supply and disappearance in Canada, annual (pounds).* Cansim II, Table 003-0037.
Statistics Canada (2003b) *Merchandise imports and exports, by major groups and principal trading areas for all countries, annual (dollars).* Cansim II, Table 228-0003.
Storper, M. and Walker, R. (1989) *The capitalist imperative: territory, technology and industrial growth.* Basil Blackwell, Oxford.
Symes, D. and Marsden, T. (1985) Industrialization of agriculture: intensive livestock farming in Humberside. In: Healey, M. and Ilbery, B. *The industrialization of the countryside.* Geobooks, Norwich, pp.99-120.
Toronto Daily Star (1965) Dominion Stores advertising copy, *Toronto Daily Star*, August 25, 34
Troughton, M.J. (1986) Farming Systems in the modern world. In: Pacione, M. (ed.) *Progress in agricultural geography.* Croom Helm, London, pp.93-123.
Troughton, M. (2005) Fordism rampant: the model and reality, as applied to production, processing and distribution in the North American agro-food system. Chapter Two in this volume.
Uvacek Jr., E. (1983) Economics of feedlots and financing. In: Thompson, G. and O'Mary, C. (eds.) *The feedlot,* 3rd edition. Lea and Febiger, Philadelphia, pp.11-29.
Watkins, M. (1963) A staple theory of economic growth. *Canadian Journal of Economics and Political Science* 24, 141-158.
Williams, K., Haslam, C. and Williams, J. (1993) The myth of the line: Ford's production of the Model T at Highland Park, 1909-1916. *Business History* 35, 66-87.

Williams, K., Haslam, C., Johal, S. and Williams, J. (1994) *Cars: analysis, history, cases.* Berghahn, Providence.

Wilson, B. (1981) *Beyond the harvest.* Western Producer, Saskatoon.

Chapter 4

People and hogs: agricultural restructuring and the contested countryside in agro-Manitoba

Douglas Ramsey, John Everitt and Lyndenn Behm

Introduction

Manitoba was not able to succeed as an agricultural area until it established an export economy with eastern Canada, Europe and other parts of the world in the late nineteenth century. The export trade was tied to railway and water connections with these markets. As Morton indicated, it was not until the 'new Manitoba' of the 1880s that the development of contemporary rural society really began in this province (Everitt, 2002). In the succeeding decades a variety of changes continued to affect all aspects of prairie life, with each era differing measurably from that before it, and with each future being quite uncertain.

Restructuring in North American agriculture is not new (McMichael, 1994; Troughton, 1989, 1992 and 1997; Ramsey *et al.*, 2003). In fact, one could argue that prairie agriculture in Canada, for example, has faced continual restructuring since the first pioneer settlement - the only constant, therefore, being continued change and adaptation (Everitt, 2002). The trend towards intensive livestock (Rhodes, 1995; Hart and Mayda, 1997) and poultry (Kim and Curry, 1993) production, however, has produced new issues in agricultural communities (Furuseth, 1997; Novek, 2003; Hinrichs and Welsh, 2003) throughout North America. The Province of Manitoba, with approximately 3% of the Canadian human population, currently produces one-quarter of the hogs (pigs). While other jurisdictions such as Quebec and North Carolina have instituted moratoriums on further hog production, Manitoba is actively promoting the sector. The conflicts arising often include pitting farmer against non-farmer, but also farmer against farmer in increasingly divided agricultural communities (Ramsey, 2004). The issue of local control over the forces of restructuring is also a concern for both reasons of production response (Welsh, 1997) and community condition (Ramsey and Smit, 2002). With a large-scale hog processing plant located in the heart of southwestern Manitoba hoping to double production with a second shift, and thus processing more than 90,000 hogs per week, new facilities and operations for raising hogs are required. Thus, it is argued, the conflicts in the countryside will likely continue.

The purpose of this chapter is to discuss aspects of the various rural social and agricultural geographies in Manitoba. First, the construction of the agricultural landscapes from the 1880s up to the 1980s will be briefly discussed in order to 'set the scene' and provide the context for subsequent changes in agricultural structure. Second, some of the changes that have taken place in the past few decades will be discussed in more detail in order to clarify the contemporary agricultural situation of the prairies. Third, the hog sector in Manitoba will be discussed in order to investigate the magnitude of a selected number of these changes at a smaller regional level. In doing so, the role of the state in fostering change is outlined. Fourth, various 'expectations for the future' will be suggested, and their effects upon the cultural landscapes of the twenty-first century will be posited.

The research used to provide material for this chapter adopted a descriptive approach to analysing change. Three distinct descriptive elements are presented. First, the context of changing agricultural landscapes is described through historical data on agriculture in the Canadian prairies in general and Manitoba in particular. Second, more recent experiences with respect to hog sector restructuring and the role of the provincial government in fostering this development is described. The sources for this analysis include government and party policies, agricultural statistics, and media coverage. Third, the case studies presented in this chapter were based upon media analysis and interviews conducted with municipal leaders and administrators, as well as local citizens in three local communities in southwestern Manitoba.

The initial construction of the agricultural landscape[1]

Colonization of the prairie west in the late nineteenth century Canada was intended to lead to 'the development of one of the world's great agricultural regions'.[2] Although the transformation was not as rapid as some people expected, by 1901, 82% of the cultivated land in Manitoba was in grains, and two-thirds of this (1.9 million acres) was in wheat.[3] A number of reasons for these changes could be suggested, but the institution of the 'Crow Rate' proved to have an unexpected importance for many years to come.[4] It encouraged the wheat export trade, in some ways artificially entrenching this form of production, and leading to a series of issues that have yet to be fully resolved.

The 'Crow Rate' was a concession on freight rates made by the Canadian Pacific Railway (CPR) in 1897 (implemented in 1898-1899). The concession focused particularly on cereal crops such as wheat - in order to get a subsidy to build an expensive spur line from its mainline at Lethbridge, Alberta, to Nelson, British Columbia through the Crow's Nest Pass. This line was needed to enable the CPR to reach minerals in the Kootenays before the Americans. In later years, similar charges were applied to the Lakehead and the West Coast lines. The rates on grain and flour were given statutory status in 1925, and extended to all Canadian railways - which for the purpose of the present discussion meant the Canadian National Railway.[5] In 1930, it was extended to the shipment of grain through the northern Manitoba port of Churchill. The rates remained in place (with a few variations until 1922, and particularly during the First World War) until 1983. The 'Crow Rate' was important to Manitoba farmers as it made the shipping - and thus the sale price - of wheat relatively cheaper. In a sense the rate had the effect of shortening the distance between Manitoba's wheat farmers, and their markets. This, of course, made these farmers more competitive in the world market.[6]

In the late 1950s, the first study was produced showing that railways were losing money by shipping grain at the 'Crow Rate'. The railways became increasingly reluctant to invest in grain shipment lines and these lines deteriorated. In recent years at least, and arguably for much longer in the minds of railway company executives, the revenues raised from the rates were less than the costs of transportation.[7] Therefore, they were seen to act as a subsidy to the farmer, despite the fact that their immediate financial benefit was offset by a deterioration in rail service that was induced by the very same low freight rates. Many branch lines, in particular, were given little or no repair for many years, and services became poorer, slower, and in some instances almost non-existent. But branch lines could only be abandoned with permission from the Board of Transport Commissioners, and this would involve struggles with local farmers that were expensive and time consuming. Various agreements between the railways and the government took place over the years. In 1977, the Hall Commission report was produced, but relatively few changes resulted from this document. Farmers, however, did not feel the changes that were implemented addressed the real issues.

In 1983, came the controversial *Western Grain Transportation Act* and its passage proved to be definitive, although it was preceded and followed by much discussion and disagreement.[8] It led to a subsidy being paid to the railways, but with shippers paying a share of future rail cost increases. Over the next decade the subsidy was gradually reduced, leading to further problems with branch-line upkeep and related matters. In 1996, the subsidy was removed, but the farmers received a 'one-off' payment that was based upon their acreage and the quality of their land.[9]

In 1905, Alberta and Saskatchewan also became provinces, and the rail network in these areas increased significantly. Wheat production also increased in these areas and became subject to the 'Crow Rate'. By 1921, both Alberta and Saskatchewan had significantly more wheat acres, on larger farms, than Manitoba, and this trend has continued. By 1921, the production of wheat for export had come to characterize the Canadian prairies. By 1928, according to historian Gerald Friesen:

> 'Canadian wheat sales constituted nearly half the world export market. An entire society was organized to facilitate this activity. It was built upon rural village and transportation networks, a grain marketing system, and a family economy attuned to the rhythms of the seasons and the demands of the work itself'.[10]

Although there were some fluctuations, the wheat acreage continued to grow and between 1921 and 1931 the total rose by nearly one-third, and 'King Wheat' was unquestionably dominant. The grain economy rested upon sure foundations - although it could be argued that 'the fatal prairie habit of dependence on a single export commodity' had also been established.[11]

The early reconstruction of the agricultural landscapes

The rural aspects of western Canada have changed considerably since the 1930s. Although rural population initially outnumbered that in urban areas, by 1931 rural Canadians were in a minority. A rural majority was maintained in Manitoba until 1951, Alberta until 1956, and Saskatchewan until 1971. However, only 28% of the contemporary population of Manitoba is rural, compared to about 20% of that in Alberta, and under 37% in Saskatchewan. These demographic changes were coupled with other transformations. For instance, the number of farms decreased as changes in farming technology occurred. In 1941, there were nearly 300,000 farms on the prairies but by 1996 it had dropped to about 140,000. In Manitoba there were just over 32,000 farms in 1976, but only 24,383 in 1996. In this latter year there were 56,995 in Saskatchewan and 59,007 in Alberta, but the numbers were falling every year and faster on the prairies than in the country as a whole. From 1991 to 1996 there was a 5.3% decrease in Manitoba, in Saskatchewan it was 6.3%, and in Alberta 3.0%: the Canadian average drop was 1.8%. The 1931 Canadian census listed the prairie farming population as 1,186,000, which meant that more than one in ten

Canadians lived on a prairie farm. By 1996 the prairie farm population had dropped to 411,000, which meant that about one Canadian in 80 lived on a prairie farm. By 1996, farmers accounted for only 7.2% of Manitoba's population. Simultaneously, farm sizes increased to sizes far greater than the 160 acres decreed to be the basic unit in the 1880s. According to the agricultural census, the average prairie farm is now almost 1,000 acres in area. In Manitoba, the average is about 800 acres, compared to over 1,100 in Saskatchewan and almost 900 acres in Alberta. These changes can be clearly seen in the cultural landscapes of the prairies.

Along with the increase in farm size came a change in the way the land was worked, with horsepower declining, and mechanized power increasing. Although equipment numbers have not continued to increase as rapidly during the past 25 years as they did in earlier years, the number of tractors rose from 327,000 in 1976 (Friesen, 1984) to 358,000 just 20 years later (68,735 in Manitoba, 155,574 in Saskatchewan and 153,390 in Alberta)[12] - but only 140,000 farms remain to house this equipment.

The area given over to 'King Wheat' fluctuated over time. Overall acreage increased on the prairies until 1931, but did not exceed that level until 1981. It then increased until the early 1990s, but has since fluctuated again. At present, the future of wheat production is quite problematical - at least at the scale of the past, although over 54% of the 'census farms' are still (as of 1996) 'wheat farms' (51% in Manitoba, 75% in Saskatchewan and 34% in Alberta).

Brandon was founded in 1881, and the majority of the small towns in Manitoba date back to the late nineteenth century. The small town network peaked quite early with declines in the past few decades in terms of absolute numbers, as well as percentages of the total population. The decline reflects the drop in rural population, and the consequent decline in demand, and the landscapes of the prairies are increasingly characterized by dying settlements, or by 'ghost towns'. Agricultural settlement of the region as a whole also occurred mostly in the 1880s and 1890s and 'century farm designations' have been given to a significant number of farms in the region, indicating that lineage of those farmsteads has been intact for 100 years.

There have also been dramatic changes in the grain delivery system, including both rail lines and grain elevators. Grain elevators increased in numbers steadily until 1933, as agriculture developed, new crop varieties were adopted, and the total acreage increased. In 1933, there were 5,758 elevators with an average capacity of over 33,000 bushels. After that date numbers began to decline slowly until 1970 when there around 5,000, but then more rapidly, although average capacity sky-rocketed. By 1989, there were only 1,619 elevators in operation; by 1999 only 1,016 remained; by late 2002 only about 400 licensed elevators continued to be used - throughout all three prairie provinces and the Peace River area. This was roughly the same number as a century earlier. But the twenty-first century elevators have an average capacity of over 10,000 tonnes - or 370,000 bushels. Travelling north of Brandon on PTH 10, the first settlement reached is Forrest. In this community is an old 'Manitoba Pool' elevator that is now used by a local farmer; it holds just over 100,000 bushels. A couple of minutes further north one passes a new AgPro elevator (owned by the Saskatchewan Pool); it has a capacity of 1.5 million bushels (42,000 tonnes).

Trends in industrial agriculture

Throughout the twentieth century, the number of people involved in primary agriculture declined while production and farm size continually increased. The decline in farm populations outlined above for Manitoba, reflects trends elsewhere in North America. A second trend is related to wheat production. Although the area of spring wheat has increased since 1961, its overall percentage of land in crops - its *relative importance* - has declined from 50.2% to 38.6%. The different varieties of spring wheat (hard red spring wheat, soft white spring wheat, etc.) are addressed in detail elsewhere (Carlyle, 2002).

However, yield increases have meant that production in the prairies was much higher in 1996 than in 1961, and spring wheat remains the most dominant prairie crop. Its relative decline reflects falling farm prices for wheat, coupled with rising freight rates following the loss of the 'Crow Rate'. A third trend, that has become characteristic of the poultry and hog industries, relates to ownership. For hogs, this trend began in the eastern USA where for decades hog production has been moving towards vertically integrated systems where hog-processing companies, through contract arrangements, control decision making by hog-barn owners.

Hog production has taken place in the prairies for many years. The traditional hog-feeding operation involved relatively small numbers of animals. Recently, hog-raising has become a large scale 'productionist' operation. Many of the smaller scale operations have disappeared and, in some instances, the cultural landscape is being dramatically changed with the appearance of what John Fraser Hart has termed 'Pork Palaces'.[13] In 1996, 43% of Canada's pig farms were in the prairies, about 10% (2,064) in Manitoba, 13.5% (2,848) in Saskatchewan, and 20% (4,173) in Alberta. In Manitoba this constituted 4.2% of all farm operators, (in Saskatchewan 1% and in Alberta 2.1%). However, from 1991 to 1996 there had been a 33% drop in the number of hog farm operators in the Province of Manitoba as small producers closed down and large operators took over. There was a 38% increase in the number of pigs in this province during this time period (in Saskatchewan the number dropped by 6% and in Alberta it remained unchanged). The importance of hog production is likely to continue to increase. In 1995, for instance, Manitoba producers marketed 2.2 million hogs and in 2000, 2.9 million. Significantly, however, in 2000, 3.9 million hogs were slaughtered in Manitoba, with 1 million being brought in from Alberta, Saskatchewan and Ontario to the new industrial facilities in the province. By 2001, Manitoba accounted for 24% of total Canadian hog production.

Thus, during the 15-20 years prior to the post-Crow hog expansion in southwestern Manitoba the hog industry in the province had been undergoing a gradual change with smaller producers leaving the industry and production becoming increasingly concentrated in larger facilities. Most of these facilities, with the exception of those operated by Hutterite colonies, were in eastern and central Manitoba where a feeding and hog production industry was the most advanced. In the post-Crow era, this had political implications in that opponents of hog barns in southwestern Manitoba contended that the new, large barns were not a form of farming, but rather a type of industrial production. Proponents, on the other hand, argued that barn opponents simply were not in touch with modern agriculture.

Political structures

Canada has three levels of administrative authority: national, provincial and municipal. Both agriculture and the environment, the two sets of concerns linked with hog barns, are areas of divided responsibility between the provinces and municipalities. Municipal governments have powers and responsibilities over hog barns, although these fall under the wing of provincial governments. Technically, provinces have the ability to usurp municipal powers, or even abolish municipalities, although these actions would be subject to political considerations. For example, provincial governments have redrawn or eliminated small Rural Municipalities (RMs) but none in recent years has toyed with the idea of withdrawing recognition of a major city. Even among smaller municipalities, there is substantial opposition to reduction of RMs, despite continued population loss, except in the RMs near smaller urban centres, where urbanization is offsetting the loss of the farming population.

The federal government has only played a peripheral role in hog barn developments in southwestern Manitoba. Federal personnel sit on technical review committees, composed of federal and provincial representatives, which review proposals and make recommendations relating to such things as environmental and health concerns. The primary intersection of tensions is between the provincial government and local jurisdictions. The province has not been ambitious in its use of power, although some people fear an increase in provincial exercise of that authority. Whether those fears are grounded is debatable. Changes in the interaction between municipalities and the province, so far, have been subtle.

There have been two areas of interaction for the province. One relates to its desire to govern – or not govern – the hog industry. The other relates to its desire to have municipalities be involved in, or refrain from being involved in, decision making related to the hog industry. The province has been gradually increasing its role through direct action or through municipalities. Generally, the province has shunned being a decision maker or a legislator. There are regulations relating to environmental protection, which apply to either hog barns or the environment in general. However, its most extensive document is the *Farm Practices Guidelines*, which is simply a set of preferences, which, as hog barn opponents have noted, can be ignored by municipalities and barn proponents.

The guidelines make recommendations related to such things as minimum separation distances between barns and nearby residences. The province has also added stipulations related to the disposal of animal waste and has made technical review committee assessments, which were once an option for municipalities, mandatory. The Progressive Conservative government, the dominant party in the southern agricultural areas of Manitoba, was defeated in the 1999 provincial election and replaced by a New Democratic Party (NDP) government. A critic of industrial agriculture in opposition (Walberg and Ramsey, 2002), the NDP government announced the formation of a Livestock Stewardship Panel and a Manure Management Committee. It also provided incentives to encourage municipalities to form planning districts (Ramsey and Everitt, 2001).

Municipalities govern hog barns and other intensive livestock operations in two ways. One is through individual municipalities, acting alongside the provincial government, to create zoning and land-use bylaws. In these municipalities the RM is the decision maker on individual projects, but appeals relating to fairness of regulations can be made to the Manitoba Municipal Board (Ramsey, 2004). The other method of municipal control is through formation of planning districts. The purpose of planning districts is to provide municipalities with the personnel they need to engage in planning and zoning. Municipalities belonging to planning districts still set their own regulations and municipal councils remain the decision making body with regard to individual projects, but the planning district becomes the body of appeal when there is an objection to an individual municipality's zoning and land-use legislation (Ramsey, 2004).

There is also a third type of municipality, one that does not belong to a planning district and which does not have its own regulations. In these municipalities there are no regulations governing hog barn developments other than federal and provincial legislation, which relates primarily to such things as environmental protection. In these municipalities, the absence of municipal regulations means that hog barn developers are free to site projects where they please as long as they do not violate senior government environmental regulations. Provincial legislation does not concern itself with such things as siting, although technically complaints about siting can be referred to the Farm Practices Review Board.

In 1995, when Crow removal was announced, 19 municipalities within southwestern Manitoba were without either regulations or controls. Since then 17 municipalities have taken steps to form planning districts, although only 10 of these have completed the process. Only two municipalities in the region have no legislation relating to either the municipality or a planning district, which governs hog barn developments.

Recent actions by the Manitoba government

The provincial government has been a significant player in the development of the post-Crow hog industry in southwestern Manitoba. At the beginning of the 1990s, the province set as its goal for the decade, a doubling of hog production. The Progressive Conservatives, who governed the province from 1988 until 1999, eliminated single desk selling through the hog marketing board. The marketing board has often been regarded as an advantage for smaller hog producers. The removal of orderly marketing, which forced buyers to purchase hogs from the provincial marketing board, was cited as one of the factors that resulted in Maple Leaf choosing Manitoba as the site for its new hog-processing plant. Since the election of the NDP in 1999, the provincial government's approach could be characterized by modest steps towards increased governance of the large-scale hog industry. This has taken the form of measures involving either provincial or local authority (Walberg and Ramsey, 2002).

After the NDP government was elected it announced that it would provide financial incentives to encourage municipalities that did not have municipal hog barn regulations to join planning districts. It also formed a Livestock Stewardship Panel, which looked at a range of issues in the livestock sector. It introduced new regulations relating to the handling of hog waste and made Technical Review Committee assessments mandatory for facilities producing more than 400 Animal Units (AUs) of waste annually. In 2003, this requirement was lowered to 300 AU. Further regulations are also expected regarding how manure must be dispersed and the qualifications of personnel/firms hired to distribute waste from lagoons to farmland.

Regional Technical Review Committees are made up representatives of the provincial departments of Agriculture, Conservation and Intergovernmental Affairs as well as a representative of the Prairie Farm Rehabilitation Administration, which is under the jurisdiction of the federal agriculture department. The committee has no authority but can make recommendations to decision making bodies such as municipalities. The make-up of the committees, as well as the direction of government, may have undergone a subtle change when the newly elected government dissolved the Department of Rural Development and merged its personnel into the Department of Agriculture and Agrifoods and the newly formed Department of Intergovernmental Affairs. The mandate of Rural Development was to encourage economic development in rural areas while the mandate of Intergovernmental Affairs was more neutrally bureaucratic.

As stated earlier, Farm Practices Review Guidelines are not enforceable. Recommendations of Technical Review Committees are without legal clout, except where findings relate to other legislation such as the Environment Act. Generally speaking, the province's involvement is limited to broader issues such as the impact on subsoil and flowing water, while quality of life issues generally fall under the discretion of local decision makers. The one exception is the Farm Practices Review Board, which can make orders in response to complaints. For example, in one instance the board ordered a hog operation to cover its lagoon because of problems related to excessive odour.

Expansion of planning districts

The Manitoba government assesses hog barn development based on AUs, which are calculated by the volume and nutrient composure of manure. This is appropriate because controversies related to developments in southwestern Manitoba have primarily been related to odour drift and pollution of the soil, groundwater and surface water, all of which are a consequence of hog waste. Therefore, waste rather than hog numbers is the best measure for assessing the impacts of hog barn development. Waste volume numbers are not available for the industry in southwestern Manitoba prior to 1994, but since then 81 barns producing 46,534 AUs have been approved, while 15 projects that would have produced about 10,064 AUs have been rejected by local decision makers or cancelled by proponents for reasons that vary from local controversy to economics. In 2003, decisions were pending on four projects that would produce 1,145 AUs.

Developments in the hog industry have been accompanied by an expansion in the network of Planning Districts (PDs). Prior to 1995, 17 of 42 rural municipalities in the region belonged to planning districts. There were also six municipalities that did not belong to PDs that had municipal zoning bylaws. Six planning districts have formed since 1995: Dennis County PD, Mid-west PD, Shoal Lake PD, Tanner's Crossing PD, Carleton Trail PD, and South-west PD. The RM of Lansdowne was also added to the Neepawa and Area PD as shown in Figure 4.1. In general, the biggest change in the makeup of planning districts since 1995 has been the inclusion of purely rural RMs in PDs. Prior to 1995, with the exception of South Riding Mountain PD, PDs included only RMs that abutted, or nearly abutted villages, towns and cities. This would suggest that PDs were formed to deal with issues in semi-residential areas within sections of RMs that abutted urban boundaries. South Riding Mountain may have been an exception because much of the area consisted of trees and parkland, which contain cottages and other non-agricultural residences.

It can also be argued that prior to the arrival of hog barns; farm dwellers did not see planning and regulations as methods of dealing with issues. Rather, how neighbours handled, or managed their property was their own business and problems that might arise, such as those relating to drainage or weed control, were best dealt with through other avenues such as the Noxious Weed Act or drainage legislation. Hog barn development, with issues such as odour and intensive application of liquid manure, appears to have played a role in changing this. By 2003, there were only two RMs in the southwest region which had neither municipal legislation nor belonged to a planning district that was either established or in the process of becoming established.

Expansion of the hog industry

Opponents of large scale hog barn developments have accused the industry of targeting RMs that were without land use legislation and which did not belong to planning districts. Information here does not support this argument. It appears that there is no clear link between barn development and the existence of PDs or bylaws. This would suggest that barn construction has either been haphazard or guided by other factors such as proximity to a feed mill or prospective feed mill, or the availability of local investors. Most of the production facilities, although not always the hogs living inside, tend to be owned partially or totally by local investors. Table 4.1 lists the proposed, approved, rejected and withdrawn proposals for RMs, arranged by PD and bylaw status.

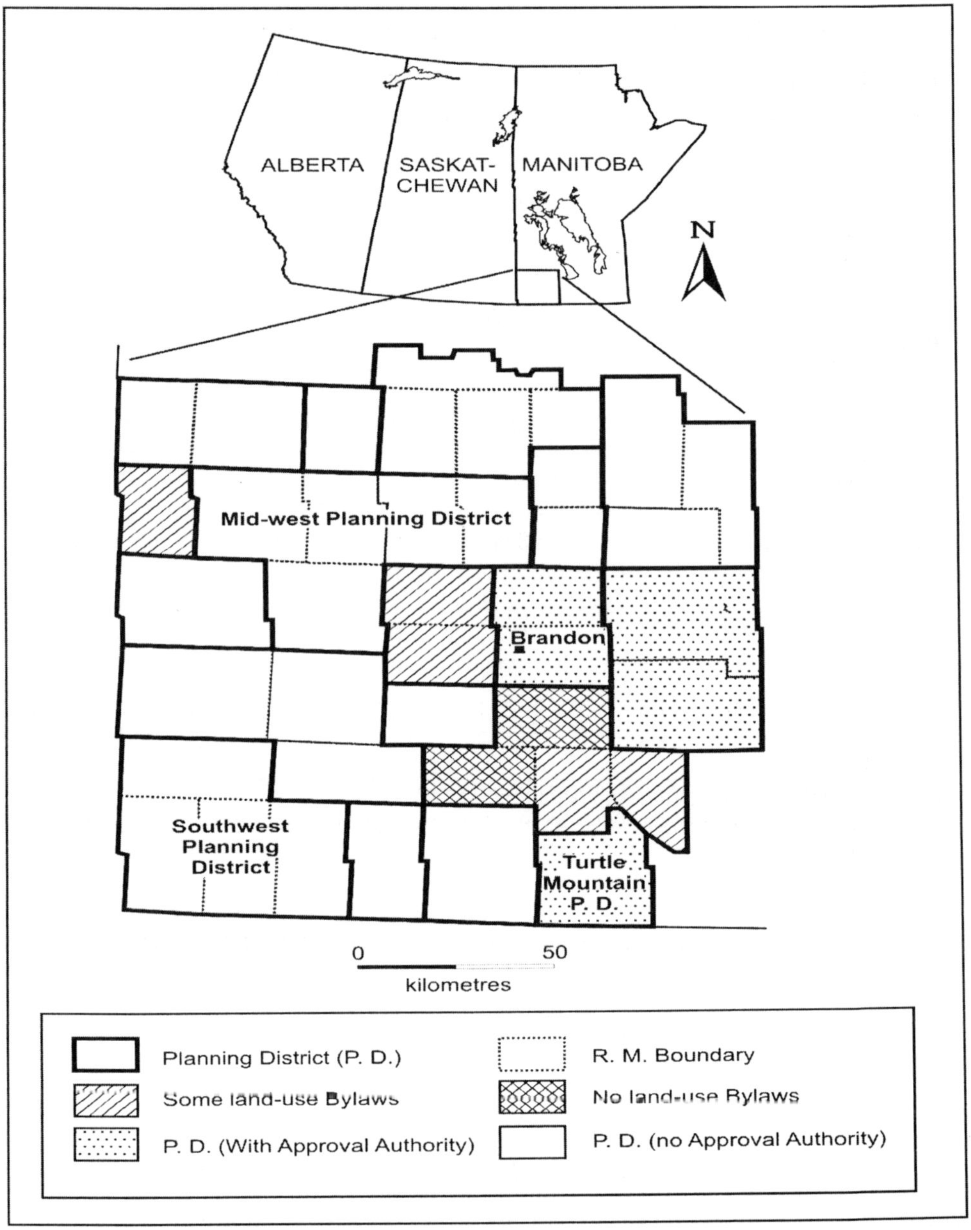

Fig. 4.1 Planning districts in southwestern Manitoba.

Table 4.1 Hog barn development in southwestern Manitoba, 1997-2002.

Municipal structure	Total RMs (with no activity)	Barns approved	AUs approved	Barns lost	AUs lost	Barns pending	AUs pending
PD completed prior to 1995	17 (6)	51	28856	10	6333	4	1145
PD completed (1995 and 2002)	10 (3)	22	12335	0	0	0	0
PD in progress (1995 and 2002)	7 (4)	3	2467	0	0	0	0
No PD (with bylaws)	6 (3)	4	2354	5	3731	0	0
No PD (without bylaws)	2 (1)	1	572	0	0	0	0
Region total (sw MB)	42 (17)	81	46534	15	10064	4	1154

Source: Manitoba Agriculture and Agrifoods (2003).
Notes: RM=Rural Municipality; AU=Animal Unit; Percentages may not add up to 100 due to rounding.

Table 4.2 Distribution of hog barns in southwestern Manitoba.

Type of municipal arrangement	% of municipalities in region	% of hog barns in region	% of AU in region	% of barns rejected in region	% of animal units rejected in region
PD prior to 1995	40.0	63.0	62.0	67.0	62.0
PD prior to 2003	23.8	27.1	26.5	0	0
PD pending	16.6	3.7	5.3	0	0
No PD, own zoning	14.2	4.9	5.0	33.0	38.0
No PD, no bylaws	4.8	1.2	1.3	0	0

Further investigation reveals that hog barn development is not distributed evenly among the categories of decision making in rural areas. As indicated in Table 4.2, planning districts that existed prior to 1995 accounted for 63% of barns and 62% of AUs, despite accounting for only 40% of total RMs in the region. This total is significant because it also represents a disproportionate allotment based on landmass. All RMs are not equal in size, but most are similar and were drawn up in the late 1800s. Most of the land in the southern region of Manitoba is cleared for agriculture.

Planning districts formed between 1995 and 2003 accounted for 23.8% of the RMs in the region, 27.1% of the hog barns and 26.5% of the AUs. The remaining three categories – planning districts agreed to but not in place, RMs with their own zoning regulations, RMs with no local control – all had disproportionately low levels of hog barn construction. In terms of barn rejections, there is a disproportionately high level of rejections among the RMs with zoning bylaws, a level of rejections that is in proportion to approvals in RMs that belonged to PDs prior to 1995, and an absence of rejections in the remaining three categories.

Given that at varying times RMs that were developing PDs between 1995 and 2003 underwent transitions from one system to another, it is necessary to separate out those affected RMs as shown in Table 4.3. Also, RMs that have begun, but not completed the process of joining PDs have been combined with RMs that still have no local control. The 23 RMs that belonged to PDs or which had municipal bylaws prior to 1995, represent jurisdictions that had local controls throughout the period of analysis, while the eight RMs where PD status is pending, or where there are no local bylaws represent the jurisdictions with the least control. Data indicates that, contrary to arguments of some opponents of the hog industry, barn developers have not been seeking out RMs where there is no local control. Concentration of hog barns is higher than normal in PDs formed prior to 1995, but below normal in areas where there are bylaws. However, barn levels are lowest in areas where there is no local control.

Table 4.3 Hog barn development in southwestern Manitoba.

Type of decision making	Approved	Animal units	Rejected	Animal units	Pending	Animal units
PD prior 1995 (17 RMs)	51	28,856	10	6333	4	1145
Municipal Bylaws (6 RMs)	4	2,354	5	2354	0	0
No controls (No PD or PD not in place yet) (9 RMs)	4	3,039	0	0	0	0

Source: Manitoba Agriculture and Agrifoods (2003).
Notes: RM=Rural Municipality; Percentages may not add up to 100 due to rounding.

All of the rejections, cancellations and withdrawals have come from either PDs formed prior to 1995 or RMs with their own zoning by-laws in place. The RMs with their own bylaws had the most substantial level of rejections. Five of the nine barn proposals never proceeded with rejected proposals accounting for 3,731 AUs, substantially more than the 2,354 AU approved. RMs with their own zoning accounted for about 5% of the regions total barns and AUs, but about a third of rejected/failed projects. It is difficult to make assumptions from the small number of applications from the nine RMs that had no local decision making authority. However, these data seem to refute charges that hog development companies are targeting those RMs.

The unequal distribution of hog barns among the types of decision making is also marked by unequal distribution within these types and categories. As a single municipality, Turtle Mountain accounts for only 2.4% of the RMs in the region, but 25.9% of the barns and 31% of the AUs. The PDs created prior to 1995 account for 63.2% of barns, but nearly half of the RMs within this decision making type have no barns. Table 4.4 ranks the five categories according to AU volumes.

Table 4.4 Ranking of hog barn development by animal units (AUs).

Animal units per RM	3000+	2000+	100+	500+	1+	0
PD prior 1995	3	1	2	2	1	8
PD prior 2003	2	1	1	3	0	3
Status pending	0	0	1	1	1	4
Municipal bylaws	0	0	1	1	1	3
No controls	0	0	0	1	0	1

Source: Manitoba Agriculture and Agrifoods (2003).
Notes: RM=Rural Municipality; Percentages may not add up to 100 due to rounding.

It should be noted that seven RMs, each of with more than 2,000 AUs, account for 34,255 AU, or 73.6% of the total AUs in the region. Interestingly, there are no municipalities with disproportionate levels of hog barn development, which is at odds with the experiences in other regions such as North Carolina (Furuseth, 1997). While the pattern in Manitoba appears less random in the sense of locating independent of the type of local decision making, other factors may be more random. One factor could be the availability of local investors, including farmers seeking to diversify from other agricultural commodities. A second factor could be that barns tend to locate in the vicinity of other barns or feed mills. Different hog companies use different management systems, but one of the more popular involves having weaners born in a farrowing barn where they are kept until about two weeks old. They are then transported to a set of barns where they are fed for about eight weeks. Finally, they are moved to another set of barns where they remain until they reach market weight. Another feature to a more patterned location is that barns need to be situated at least a few kilometres apart so that disease cannot easily spread from one facility to another. However, once this separation distance is reached it is economically advantageous to have barns as close as possible so that shipping distances can be minimized. The result is that there are often 'hot spots' within municipalities rather than among municipalities.

Two patterns appear to characterize the areas with lack of hog barn construction. First, some RMs lack hog barn construction throughout the region. However, there is also an absence of major construction near Brandon and in an area extending 100 km north to Riding Mountain National Park. Opponents of hog barns in areas near the edge of the region, as well as general anti-barn activists, have accused the people living in the Brandon area of hypocrisy, arguing that the city wants the jobs that the Maple Leaf plant creates for urban residents, but do not want to face the negative impacts of the facilities that raise the hogs. In instances where opposition to barns within 25 km of the city has arisen, opposition has tended to be stronger, better organized and better financed than in peripheral regions. There are also a number of hobby farms/acreages in the area, extending north to Riding Mountain National Park.

Case studies

In this chapter three case studies are described. The first is based on RMs in Planning Districts that are still not in place (the Southwest Planning District, consisting of the RMs of Edward, Brenda, Albert and Arthur). The remaining two case studies examine different aspects of the operations of municipalities that have belonged to planning districts since 1995. One focuses on the RM of Turtle Mountain, which is part of the Killarney and Area Planning District. Turtle Mountain has been the most intensive area of hog development, accounting for a quarter of the barns, and nearly a third of the hog waste, that will be generated by developments built or approved since 1995. The second examines the RMs of Saskatchewan and Blanshard. These two RMs are included because they illustrate the tensions that exist between the province and municipalities. These two RMs have passed restrictive legislation, which has been criticized by the provincial Agriculture Department, which is promoting the hog industry. It has also been rumoured that the province may reduce the authority of Planning District Boards as a result of the dispute with the RMs of Saskatchewan and Blanshard.

Southwest Planning District

Municipalities belonging to the Southwest Planning District, which is still in the process of being formed, have looked at ways to reverse the approach that seems characteristic throughout southwestern Manitoba in the post-Crow era. The usual approach is for a hog barn proponent to look at a prospective site and then do soil quality and water supply tests. Next, the proponent contacts the provincial government who, through the Technical Review Committee conducts an assessment relating primarily to environmental concerns. The committee is not a decision making body, but rather makes recommendations to the municipal council. Also, its findings can determine whether a project will be acceptable to the Department of Conservation which sets environmental regulations, and determines whether violators will be prosecuted by government. The final stage in the process is to submit a project proposal to the local decision making body for approval. This final stage is only necessary in RMs where there are municipal land-use bylaws or where an RM is part of a planning district.

The issue of social acceptability is either addressed at the final stage, during a municipal hearing, or else it is never addressed at all because of the absence of a local approval requirement. During the early stages of the post-Crow hog boom in southwestern Manitoba, a company that specialized in the handling of hog-barn controversies advised barn proponents to be candid and to gather support from people living in the area of a barn project at the outset of planning. Some large hog barn companies indicated that they would follow this approach, and would not build barns in areas of substantial neighbourhood opposition. Since then, there have been projects abandoned because of opposition, but there have also been proposals that have proceeded despite fierce opposition from neighbouring residents and still others that have been accused of using stealth to push projects through before opponents organized.

The southwestern approach was proposed by Harvey Dann, a Winnipeg cattle broker who wanted to encourage economic development in the area of Melita, his childhood hometown (*Brandon Sun*, 7 January 2001). He said the normal process was costly for barn developers, who sometimes spent money and time developing a proposal that was subsequently rejected because of local opposition. Also, hog barn battles were hard on the morale of communities.

Dann stated that the issue of social acceptability should be addressed first. In 2000, he announced his plan, which was to consist of mapping parts of four Melita-areas RMs – Albert, Arthur, Edward, and Brenda – to determine what locations would be suitable for intensive livestock operations. The area was to be canvassed to determine which residents were interested in investing in hogs and making land available for spreading of manure; those not interested but not opposed; and those opposed to hog developments in their area. After the RM was mapped according to landowner's attitudes, soil and environmental testing could be done. The process of site selection for barn projects could move quickly, if social and environmental issues were addressed prior to application (Dann, 2003).

The future of this plan is unclear. Dann states that he personally did not have a vested interest in it. Hog barn opponents demonstrated their opposition to the plan at public meetings, so he has decided to not pursue it any further (Dann, 2003). However, Manson Moir, a key member of the survey team says the concept is not dead. Moir, Reeve of the RM of Albert during the 2001 survey, is currently chair of the Melita and Area Economic Development Association as well as chair of the Southwest Manitoba Water Development Team, which is looking at ground and surface water supplies in 28 villages, towns and rural municipalities. He says the approach is merely on hold.

The four RMs and the town of Melita are in the process of creating a planning district and the livestock park concept cannot proceed until after the planning district is in place and the RMs have set their zoning bylaws. Mapping of the area cannot be completed until after the rules for hog barns have been determined. The planning district will likely be in place by the end of 2003 (Moir, 2003).

During the post-Crow era two barns were built in the Southwest PD area. One was a 1,500 AU barn, built in the RM of Arthur in 1999-2000. The other was a 280 AU barn in the RM of Albert in 1998. There were no proposals for the RMs of Brenda and Edward. The Albert barn encountered problems relating to water supply in 2002. More needs to be known about the water supplies in the southwestern corner of the province (Moir, 2003).

Turtle Mountain

In the post-Crow era, no community has approached the level of the RM of Turtle Mountain's hog industry development. The RM accounts for more than 25% of barns built in southwestern Manitoba and more than 30% of the animal waste. Since 1995, 24 proposals have come before the Turtle Mountain council, 21 of which have been approved, two have been cancelled or put on hold by developers, and one has been rejected. Development, it appears, is best explained in terms of positive characteristics, factors that encourage hog development, rather than the absence of negative qualities such as lack of public opposition or low levels of regulation. Reasons advanced for development include the municipal council's supports for the hog industry, the availability of local investors, and an abundance of water.

The RM council can be characterized as pro-hog development. In a September 2000 decision the reeve, who was an investor in a different hog barn said, 'I am going to stick up for farmers as long as I can' (*Brandon Sun*, 14 September 2000). Another councillor indicated that, 'I am a farmer first and I will probably be a farmer until I am shot, go broke or die' (*Brandon Sun*, 14 September 2000). During the October 2002 civic election, which saw a high turnover of municipal councillors in rural Manitoba – a development some people attributed to anti-hog barn sentiments – only one new councillor was elected to the seven-member Turtle Mountain council. Investors are allowed to keep involvement confidential, but a substantial number of farmers in the area have invested in barns. A feed mill has been built in the area and this joins good roads and water supply as part of the infrastructure support system.

A provincial agriculture official, who has served in Turtle Mountain, as well as the two RMs west of Turtle Mountain, says the abundance of water has been the key factor along with the existence of all-season roads that are not subject to truck-weight restrictions in spring. Hogs require a lot of water and this has been the main factor differentiating Turtle Mountain from the two RMs to the west, Morton, which has three post-Crow barns, and Winchester, which has none. Four locations had to be looked at before a suitable site was found for one of the Morton barns (Day, 2003). Killarney and the RM of Turtle Mountain have been part of a PD since 1983. There has only been one formal change in relation to construction of intensive livestock operations since then. In 2001, the RM and town agreed to develop a 5-km buffer zone surrounding the town. Buffers of 1.6 km were established around the villages of Ninga and Holmfield, and around the Pelican Lake resort area. Ninga, Holmfield and the southwestern shore of Pelican Lake are all part of the RM. There was also a 0.4 km buffer around any municipal water supply (Turtle mountain, 2000). The buffer zones prohibited development of large, new livestock facilities and limited expansion of existing facilities to a maximum of 230 AUs. Construction of new facilities that produced less than 230 AUs would be subject to a hearing and approval of council (Turtle Mountain, 2000). The bylaw was passed by the Turtle Mountain council and approved by the Killarney and Area Planning District.

In 2002, there was one other change, which occurred without formal measures being undertaken by the planning district. The RM council agreed that all future hog barns would have to place a straw cover on open lagoons by 15 May each year (Dowsett, 2003). Straw coverings limit the spread of odour from lagoons. The change in policy has been adopted under conditional use powers, which allow a municipality to make approvals subject to certain conditions being met. The ability to set conditional use requirements is a major difference between RMs that belong to a planning district, and those that neither belong nor have their own land-use legislation. The latter municipalities do not hold hearings and therefore cannot reject projects, or set conditions. Turtle Mountain and Killarney have a history of good working relationships. They have shared an office and administrative staff since 1983. In 1998, the two municipal councils attempted to break new ground in Manitoba with a merger plan that included a shared council and shared use of municipal equipment. In a dual plebiscite, town voters endorsed the concept but RM voters rejected it. It has been speculated that RM residents rejected the merger because the population of the town was significantly larger and felt that rural interests would lose out to the town's interests.

The only rejection of a hog barn proposal in Turtle Mountain came in June, 2001. Council voted 5:1 to reject an application by K-Line Pigs, a company based in central Manitoba. K-Line, one of three large hog companies with interests in the RM, wanted to build a 2,600-head sow barn that would have produced 55,000 weanlings and 650 AUs of animal waste annually. The barn would have been located near the eastern edge of the town buffer zone, in an area where there was already a high concentration of barns. About 60 people attended the conditional use hearing, with opposition coming primarily from residents who voiced concerns about the potential impact on soil and water. Supporters were either representatives of K-line, or investors in other hog projects, and they said concerns were exaggerated. When discussing the issue, some council members expressed concern about the level of barn concentration, which could leave some residents sandwiched between barns. The proposed K-line barn would have been the fourth in the immediate area (*Killarney Guide*, 13 June 2001). There was no petition against the project and public opposition was comparable to - or lower than - opposition the previous year to a 2,000-head feeder barn that would have produced 286 AUs.

The project that was endorsed by council, despite strong opposition, was located six kilometres northwest of Killarney and was being built by a local grain farmer who wanted to diversify his income sources. More than 100 people signed a petition opposing that project, and a similar number attended the conditional use hearing. In addition to concerns related to environment and odour, opponents also suggested the community's nickname of 'Hoglarney' was undermining efforts to attract tourists to the scenic, lakeside town (*Brandon Sun*, 14 September 2000). Supporters of the 2001 project said concerns were exaggerated. One businessperson said: 'If somebody farts on main street they blame it on the pig barns' (*Brandon Sun*, 14 September 2000).

The Killarney area is involved in the resort and recreation business. Killarney Lake borders on the town and has a beach and cabins. Pelican Lake, located 12 km northwest of town, has a golf course and cabins. The area is also attempting to attract retiring miners from the northern Manitoba mining city of Thompson (*Brandon Sun*, 26 July 1997). In recent years the Killarney area has increased its emphasis on agriculture. For example, in the mid-1990s the local economic development association was looking at construction of a resort hotel that would boost tourism (*Brandon Sun*, 30 November 1994). The resort hotel idea has been abandoned and the group is now pursuing development of an ethanol-manufacturing plant that would use farm straw to produce fuel. The group has also invested in the construction of an inland grain-handling terminal and pursued the development of a manufacturing plant that would use farm straw to produce fibre-board for building construction (*Brandon Sun*, 14 August 1996, 11 June 1997).

Midwest Planning District

Events in the Midwest PD, created in 2000, illustrate tensions between the province and local municipalities. The province has decided to continue to let municipalities remain the decision making body with regard to hog barn approvals. However, the provincial government has also been involved in promoting the hog industry. In fact, developments in the Midwest Planning District have shown that the government has been more than a disinterested bystander. The provincial Department of Agriculture and Agrifoods has appealed restrictions on hog barn developments undertaken by three of the five RMs in the PD. The PD board's rejection of appeals by the province as well as by Manitoba Pork, the provincial pork producers association, has created speculation that the province will attempt to reduce the power of local decision makers.

Members of the PD include the RM of Blanshard (6 barns, 4088 AU), RM of Hamiota (2 barns, 1596 AU), RM of Woodworth (1 barn, 655 AU), and the RMs of Miniota and Saskatchewan, which are both without hog barns, and the town of Hamiota. The town and RM of Hamiota became the Manitoba headquarters for Premium Pork, an Ontario-based corporation that expanded to western Manitoba in the post-Crow period. The company has established an office and a hog-related truck wash. The development of a feedlot has also been discussed. Hamiota was to be the focal point of southwestern Manitoba operations, which were to include a network of barns that was to create 55 - 60 jobs (*Brandon Sun*, 25 October 2000).

Developments involving Premium Pork demonstrate the controversial nature of the hog industry. On two occasions Premium barns located north of Midwest boundaries were strongly opposed by residents who signed petitions and lobbied RM councillors (*Brandon Sun*, 31 May 2001, 29 October 2000). However, in 2002 a major expansion at a cattle feedlot in the RM of Hamiota proceeded without opposition, even though it would produce 9020 AUs of waste, six times more than the largest hog barn in the southwest. Developments in jurisdictions outside the Midwest PD are worth mentioning. In the autumn of 2001, residents of the RM of Harrison, immediately north of the RM of Saskatchewan, voted 247:188 to ban hog barns that produced more than 400 AUs (Memryck, 2003). In August 2001 the provincial agriculture minister criticized the RM of Elton, immediately south of Saskatchewan, for bringing in bylaws that will make it difficult for the livestock industry to expand in the RM (*Brandon Sun*, 2 September 2001).

During 2001, the Mid-West PD was fully implemented allowing individual municipal councils to pass legislation. Initially, the province indicated it would appeal bylaws by the RMs of Hamiota, Woodworth, Blanshard and Saskatchewan, but it dropped its plan to appeal Hamiota's legislation after the RM adjusted its plans (Poppel, 2003). The Saskatchewan and Blanshard bylaws have been described as among the toughest in Manitoba in terms of minimum separation distances between Animal Production Operations (APOs) and neighbouring residences. Blanshard RM also treated liquid manure, a product of industrial sized hog farms, differently from solid manure, a feature of cattle operations. The following minimum separation distances for liquid manure are listed with solid manure separation distances in brackets: 1-100 AU: 0.8 km (0.4 km); 101-301 AU: 1.2 km (0.8 km); 301-600 AU: 2.4 km (2.4 km); 601-800 AU: 3.2 km (3.2 km); and 801+ AU: 4 km (Blanshard, 2001).

The Saskatchewan RM bylaw banned APOs producing more than 800 AU of liquid manure, if above ground tank storage was used, but banned the production of more than 250 AUs of liquid manure in earthen lagoons (the least expensive and most common form of storage). Up to 2500 AUs of non-liquid manure could be used within a 3.2 km separation distance (McGill, 2001). In its appeal, the provincial Department of Agriculture and Agrifoods accused the RM of discriminating against the hog industry by differentiating between the two types of manure and for adding 'significantly' to the cost of developing a barn because of the stipulation that operations producing between 251 and 800 AUs of liquid waste must use an above-ground storage tank (McGill, 2001).

The Midwest PD committee, consisting of representatives of the member municipalities, rejected appeals by both the Department of Agriculture and the Manitoba Pork Council. There has been speculation, but no official indication, that the province will reduce local decision making power (Poppel, 2003). Opponents of industrial-scale hog production have expressed fears that Manitoba will follow the lead of Alberta and have the province take over responsibility for approval of livestock barns. There is speculation that the Manitoba government may try to increase its control in one of two ways. One approach would be to make a provincial body, such as the Manitoba Municipal Board, the body of appeal on regulations set by the planning district. The argument would be that a planning district board, which is made up of representatives of member municipalities, should not sit in judgment of its own legislation. The other approach would be for the province to set standards for such things as separation distances from residences, which municipalities would have to adhere to in setting bylaws.

Conclusions

As throughout the first century of commercial agriculture in Canada, agriculture continues to restructure. Most visible in recent years has been the trend towards hog production. This trend comes at a time when the primary production mode has been intensive, or confinement, systems. This has resulted in controversy in many jurisdictions. The province of Quebec currently has a moratorium on further hog operations. A similar action was taken in North Carolina. In Manitoba, the reverse is the case. Faced with continued declines in the grain and uncertainties in the oilseed sectors, as well as continued depopulation of agricultural communities in the Canadian prairies, rural Manitoba has responded by greatly expanding on hog production. This direction has been fuelled by the construction of a new hog processing facility in southwestern Manitoba (Brandon) that currently processes 45,000 hogs per week, but has the potential to process 90,000 hogs per week.

However, Intensive Livestock Operations (ILOs) have not been developed without controversy. The provincial government, while in 1999 expressing concerns about such development, in recent years, has been reticent to make radical policy moves to control hog barn development. Having secured a second majority mandate in June 2003 the government does not seem to hold much promise for initiating any further controls over the sector. One thing seems to be certain: conflict in the countryside will continue.

Notes

1. The following historical discussion is developed in greater detail in Everitt, J.C. (2002) Agriculture and Rural Life in the Prairies of Canada: Evolution or Reinvention. *Zeitschrift fwr Kanada-Studien* 22. (Jahrgang, Nr. 1-2, Band 41), 100-115.
2. Friesen, G. (1984) *The Canadian Prairies: A History*. University of Toronto Press, Toronto, p.301.
3. Barley, oats, and tame hay have been the other three major 'traditional' crops on the prairies.
4. *An Historical Analysis of the Crow's Nest Pass Agreement and Grain Rates: A Study in National Transportation Policy*. A Submission of the Province of Saskatchewan to the Royal Commission on Transportation 1960 (Regina: Queen's Printer) 1961; *Keep the Crow* (Saskatoon: Saskatchewan Wheat Pool) 1980.
5. The CNR in Manitoba was made up of the lines previously owned by the Canadian Northern Railway and the Grand Trunk Pacific. The process of formation essentially took five years from 1917-1922.
6. *The Crow Rate and National Transportation Policy* (Regina, Saskatchewan: Queen's Printer) 1977.
7. This point of view has always been argued by people on the prairies, who feel that the railways' arguments are, at best, simplistic. See, for instance, note 9.
8. *Report of the Grain Handling and Transportation Commission. Grain and Rail in Western Canada*. Volume 1 (Ottawa: Government of Canada) 1977.
9. Andy Sirski (1995) Crow Benefit payout will be capital gains. *Grain News* September, 1995: 1, 4.
10. Friesen (1984) 301. See end note 2.
11. Friesen (1984) 328-329. See note 2.
12. Statistics Canada: *Agricultural Profile of Canada* Catalogue number 93-356-XPB, Table 24.1.
13. Hart, J. (2001) Pork Palaces in North Carolina. In: McKnight, T. (ed.) *Regional Geography of the USA and Canada*, Third Edition, New Jersey: Prentice Hall, pp.208-209.

References

Blanshard Municipality (2001) *Rural Municipality of Blanshard Zoning By-law, 2001-2002.*

Carlyle, W. (2002) Cropping patterns in the Canadian prairies: thirty years of change. *Geographical Journal* 168(2), 97-115.

Dann, H. (May 20th, 2003) Telephone interview

Day, S. (May 6th, 2003) Telephone interview, agricultural representative (Manitoba Agriculture and Agrifoods) for the Boissevain-Deloraine district.

Dowsett, J. (2003) Interview on May 21st at Killarney office of Dowsett, who is chief administrative officer for Killarney and the RM of Turtle Mountain.

Everitt, J. (2002) Agriculture and Rural Life in the Prairies of Canada: Evolution or Reinvention. *Zeitschrift für Kanada-Studien*22 (Jahrgang, Nr. 1-2, Band 41), 100-115.

Friesen, G. (1984). *The Canadian Prairies: A History.* University of Toronto Press, Toronto.

Furuseth, O. (1997) Restructuring of hog farming in North Carolina: Explosion and Implosion. *Professional Geographer* 49, 391-403.

Hart, J. and Mayda, C. (1997) Pork palaces on the panhandle. *Geographical Review* 87, 396-401.

Hinrichs, C. and Welsh, R. (2003) The effects of the industrialization of US livestock agriculture on promoting sustainable production practices. *Agriculture and Human Values* 20, 125-141.

Kim, C. and Curry, J. (1993) Fordism, flexible specialization and agri-industrial restructuring: the case of the US broiler industry. *Sociologia Ruralis* 33(1), 61-80.

Killarney Guide (June 13, 2001) RM council rejects hog barn application.

McGill, K. (2001) Presentation to mid-West Planning District Board Hearing, Oct. 24, 2001, by Ken McGill, manager, agriculture resources section, Manitoba Agriculture and Food.

McMichael, P. (1994) *The Global Restructuring of Agrofood Systems.* Cornell University Press, Ithaca, New York.

Memryk, D. (May 20th, 2003) Telephone interview with Memryck, chief administrative officer for the RM of Harrison.

Moir, M. (May 19th, 2003) Telephone interview.

Novek, J. (2003) Intensive livestock operations, disembedding, and community polarization in Manitoba. *Society and Natural Resources* 16, 67-581.

Poppel, W. (May 23, 2003) Interview May 23, 2003, at Miniota office of Wayne Poppel, planning officer for Mid-West Planning District.

Ramsey, D. (2004) Lack of development on the regulation of intensive livestock production in Manitoba. In: Ramsey, D. and Bryant, C. (eds) *The Structure and Dynamics of Rural Territories: Geographical Perspectives.* Brandon University, Brandon, Manitoba, pp.272-286

Ramsey, D. and Everitt, J. (2001) Post crow farming in Manitoba: an analysis of the wheat and hog sectors. In: Epp, R. and Whitson, D. (eds) *Writing Off The Rural West: Globalization, Governments, and the Transformation of Rural Life.* University of Alberta Press: Edmonton, pp.3-20.

Ramsey, D. and Smit, B. (2002) Rural community well-being: models and applications to the tobacco belt, Ontario, Canada. *Geoforum* 33, 367-384.

Ramsey, D., Stewart, C., Troughton, M. and Smit, B. (2003) Agricultural restructuring of Ontario tobacco production. *The Great Lakes Geographer* 9, 71-93.

Rhodes, V. (1995) The industrialization of hog production. *Review of Agricultural Economics* 17, 107-118.

Troughton, M. (1989) The role of marketing boards in the industrialization of the Canadian agricultural system. *Journal of Rural Studies* 5, 367-383.

Troughton, M. (1992) The restructuring of agriculture: the Canadian example. In: Bowler, I., Bryant, C. and Nellis, D. (eds) *Contemporary Rural Systems in Transition, Vol 1: Agriculture and Environment.* CAB International, Wallingford, pp.29-42.

Troughton, M. (1997) Scale change, discontinuity and polarization in Canadian farm-based rural systems. In: Ilbery, B., Chiotti, Q. and Rickard, T. (eds) *Agricultural Restructuring and Sustainability: A Geographical Perspective.* CAB International, Wallingford, pp.279-292.

Turtle Mountain (2000) *Rural Municipality of Turtle Mountain, Bylaw 6-2000.*

Walberg, B. and Ramsey, D (2002) Approving intensive livestock operations (ILOs): conflict in the countryside in southwestern Manitoba. *Prairie Perspectives* 5, 270-292.
Welsh, R. (1997) Vertical coordination, producer response and the locus of control over agricultural production decisions. *Rural Sociology* 62,491-507.

Chapter 5

Global markets, local foods: the paradoxes of aquaculture

Joan Marshall

Introduction

A casual stroll through the daily newspapers and weekly magazines confirms that our contemporary food and agriculture is very much a global reality that shapes our lives in profound cultural, ideological and economic ways (Watts and Goodman, 1997). The growth of research interest in food and agro-food networks in the context of globalization reflects a combination of factors related to concerns for the environment, the increasing importance of industrialized agriculture, counter-trends resisting genetically modified organisms, activism against liberalized trade policies and movements associated with organic farming and support for local produce. Together these many factors provide interesting problems associated with the transformation of both producer and consumer networks, that themselves are increasingly integrated and relational. The food system is squeezed into the fault line between environment and society (Atkins and Bowler, 2001), creating distinctive characteristics of supply chain integration and producer-consumer networks. Whatmore, (1995) has identified two important themes that describe the research agendas for agribusiness and food systems. As I shall show, her theoretical focus on agriculture does not negate the relevance of her arguments for the traditional fishery and aquaculture. The first theme she identified is the organization of capital accumulation in the agri-food sector, notably a focus on the social, economic and technological ties between three sets of industrial activities. The second theme is the role of the agri-food sector in the wider institutional fabric of social regulation, the focus here being the political and policy processes (Whatmore, 1995). That the traditional fishery sector has been largely absent from the food studies literature is partly related to its distinctive productive and social relations. As well, until the 1970s it was not a major part of the globalized food structures in the context of retail food chains. That situation has changed, for both the traditional fishery and for aquaculture.

The particular theme that I explore here is the significance of new globalized food structures and high technology production systems, for one small, remote, fishing community. I examine the recently established aquaculture industry on Grand Manan Island, New Brunswick, Canada, documenting the varied responses to changing social and productive relations, and to new market and industrial structures. In a strongly empirical approach, I offer only tentative formulations towards a theoretical framework drawing upon

Polanyi's (1944) notion of embeddedness. The definition and meaning of local foods is explored in the context of a long history of community reliance upon the traditional fishery, and with respect to the current debates about globalization and agrofood systems. Past research on food has focused on production and consumption, a tendency that has polarized research at two ends of the food chain, constraining integrative perspectives (Goodman, 2002; Whatmore, 2002). While aquaculture continues to realize main economic gains at the production end of the chain, as compared to the increasing proportion of value-added attributable to new technologies applied to the consumption end of the stream (Arce and Marsden, 1993), social distance in the productive system significantly impedes social embeddedness at the community level. But the reorganization of food production and the role of the state need to be seen as negotiated realms of change as Marsden *et al.* (1996) have demonstrated in their study of Chile.

Aquaculture has been part of sustainable agricultural systems since ancient times. Modern industrial aquaculture, on the other hand, is of recent origin, although its impacts can be seen to have similarities to the transformation from subsistence to commercial agriculture, affecting community and economic relations and local food choices and lifestyles. While the industry itself has made dramatic changes to the economic structures on Grand Manan Island, social and cultural adaptation to change is occurring much more slowly, erratically, and often with tragic outcomes. Despite its profound impacts on social structures, aquaculture is not truly embedded in the social relations of the island, but rather is detached from cultural traditions and patterns of community norms and values, acting *on* (and *against)* rather than *within* historical relationships and structures.

The history of aquaculture development on Grand Manan has several distinctive stages, but always moving in the direction of growth. When its impacts and networks of productive relations are understood as being among other changes associated with environmental degradation, the decline of the traditional fishery, and increasingly restrictive regulatory structures, it is apparent that many facets of globalization are extending their reach into this small community. That the community's ability to adapt is problematic is shown in the growing social problems, youth alienation and income disparities that characterize current community relations on Grand Manan.

Local foods, global markets

The globalization of agrofood systems has been defined as agriculture's continued incorporation into the general dynamics of capitalist accumulation (Le Heron, 1993). The OECD has defined the globalized food system as the set of activities and relationships that interact to determine what, how much, by what method and for whom food is produced and distributed (Whatmore, 1995, p.37).

While it has been pointed out that acceptance of an intrinsic link between forces of globalization and the agrofood system is somewhat problematic (Watts and Goodman, 1997), there are substantive factors that place aquaculture firmly within global structures, framing the production of aquaculture-produced food in a fundamentally different paradigm from the traditional fishery that has defined lives and identities on Grand Manan for over 200 years. Aquaculture is directly related to the agrarian transition, partly as being one aspect of the food production system, but also because of the new technologies that are located within the industrial complex of political economy and inter-sectoral flows.

While this discussion focuses on the changing institutional structures that characterize aquaculture, it is important to emphasize that the fundamental philosophy of globalized markets that underpins aquaculture production, had begun to evolve within the context of the traditional fishery in Canada a decade earlier. Government fishery policy through the 1970s and 1980s based its development of new policy initiatives upon the significance of the common property perspective (Matthews, 1993). Believing that the common property nature of the fishery leads to inevitable overcrowding and depletion of fish stocks, the government established a new regulatory regime that effectively moved excess labour out of the fishery, and organized a new system of limited entry. The evolution of licensing and quotas, particularly the establishment of the Individual and Individual Transferable Quotas (IQs or ITQs), together with the definition of *bona fide* fishermen (excluding part-time fishers) effectively moved policy to an emphasis on resource management that gradually marginalized small-scale and part-time fishermen. The institutionalization of the resource management model has become the authoritative account of the basic problem of the fishery. Fishermen have been recast from heroes to villains in the drama of the fisheries and must consequently be restrained rather than protected (Apostle *et al.*, 1998). The change in Canadian fisheries policy through the 1970s and 1980s fundamentally altered the relationships of fishermen to their resource by putting in place mechanisms that increasingly disembedded fishermen from local communities and re-embedded them into state and market structures (Apostle *et al.*, 1998). Insofar as government policies neglected the significance of social relations in their efforts to promulgate regulations for economic and property distribution, there was no attention given to distributional fairness, participatory access, and systemic equality (Davis and Bailey, 1996). The patterns were established for active promotion of new productive capacity that would alleviate problems associated with unemployment and a degraded marine environment that was producing less fish with fewer fishermen. The rational solution was the encouragement of aquaculture development.

Not only in Canada, but also in the wider global context, the changing patterns of institutional relations within the fishery and food production systems has also had important implications for the development of aquaculture. Four factors have been especially important in the globalization of fish markets (Arbo and Hersoug, 1997). The first factor, the establishment of the new ocean law regime of the 1970s and the delimitation of the 200-mile exclusive economic zone, radically altered the distribution of rights to fish resources. There was a *de facto* realignment of control of the ocean commons in favour of nation-states. The *nationalization of resources* through the establishment of the 200-mile limit meant that trade in fish became crucial, especially for nation-states such as Spain and Japan who had lost access to waters previously relied upon for the resource. The second factor was the development and diffusion of new *technology*. Efficient technologies for catching and processing fish, and the revolution in aquaculture, have been complemented by new information and communications technologies that have all created global links and new conditions for organizing international operations. Third, there has been the rise of *international corporations* whose growing power through markets and pricing mechanism, effectively use strategies such as global sourcing to control seafood markets. Finally, there has been a general *liberalization of trade* and movement of capital. Ultimately, a global phenomenon in the shift of regulatory focus of governments from market management to resource management has shifted fisheries policy to a greater concern with what goes on *before* the fish is landed than *after*. This means that *the closure of the resource system has coincided with an opening of the trading system* (Arbo and Hersoug, 1997). In their analysis of the significance of globalization for the fishing industry, Arbo and Hersoug (1997) argue that in the proliferation of multilateral agreements (such as through GATT and NAFTA), there has been a reduction in barriers to trade that has required new rules related to structural protection for national fishing interests.

While the former system allowed for government subsidies and support for fishers and their communities, increasingly competitive world markets are focused on the retail level represented by large corporate interests. The possibility for planning production in relation to expected market prices, and the ability to enter contracts on the sale of fish before slaughtering, increases predictability and the potential for expanded vertical integration with the food industry. The growing links into food retailing in turn create new demands for government monitored health standards in buildings, machinery and procedures that demand additional capital inputs.

Even as food production has become increasingly integrated into global networks, there has been a growing research interest in local agro-food systems, a trend that, it has been suggested, is largely related to a perceived need to balance the dominance of the globalization rhetoric, engendering an emergence and revitalization of local food studies (Hinrichs, 2003). The acknowledged importance of heterogeneity at the local level of agro-food systems and the increasing interest in a cultural economy approach to rural development are evidence that globalization as a defining paradigm for structuring our understanding of food systems is being challenged (Ward and Almas, 1997; Bessiere, 1998; Ilbery and Kneafsy, 1999). Localization has been seen as a process that reverses the trend of globalization by conscious discrimination in favour of local products. The idea of local itself is ambiguous, needing explicit definition with respect to significance and meanings for rural communities.

For Grand Manan, rather than adopting a framework of localization, it is more useful to explore the economic and social structures that are changing with the shifting emphasis from local to global food production systems. More relevant to the experience of the Grand Manan community is the partial displacement of the traditional fishery as an important source of local foods, and as a socially embedded productive activity, by the increasing importance of salmon aquaculture. The concept of local foods is frequently linked to traditional foods or to organic foods, both being a proxy for distance between production and consumption, and both loaded with normative values of goodness with respect to environmental and social outcomes. The particular meaning of local for this paper is defined as traditional foods that are historically linked to livelihoods and cultural identities. The local foods of the title include the produce of individual gardens, the wild game that is a significant source of winter protein for many families, the wild berries and greens harvested by the women through the summer months, and, of course, the rich and diverse wild fishery that has been the basis of livelihoods for over 200 years. The distinction is less about distance as a criterion of social meaning than about history and cultural identity. Food usage practices are not simply tied to physical location, but to food cultures (Tregear, 2003). It is this context within which aquaculture has been introduced that situates it as a disjuncture in the community, transforming productive relations and the local economy.

Aquaculture, ingenuity and globalization

Aquaculture has frequently been described as the Blue Revolution, equivalent in its production technologies and impacts to the Agricultural Revolution, or the Green Revolution. However, unlike the hunters and gatherers of 10,000 years ago, or the small, subsistence farmers who moved into hybrid crops 50 years ago, aquaculture involves a skills-knowledge base that is totally detached from that of the wild fishery. For fishers who have generations of acquired skills and knowledge that links land and sea, related to boats, tides, winds and fishing grounds, as well as being entrepreneurs and innovators, transferability of their distinctive knowledge and experience is not an option. The salmon aquaculture that has been established on the island of Grand Manan (population 2,700) requires scientists, manufacturers, truckers and investors. In describing the merits of aquaculture for the community, the government points to jobs. But these jobs are hourly paid, with only minimal skills or knowledge-based requirements, and with limited potential for advancement.

The rapid growth of aquaculture has been encouraged largely because of declining fish stocks around the world, related to over fishing and environmental degradation of marine ecosystems. The increasing scarcity of traditional wild fish stocks has been a source of political and social instability, as governments have sought to secure sovereignty rights over the high seas, and local communities experience loss of historically and culturally significant sources of food and income. The social and political consequences of environmental degradation and resource scarcity have been explored and theorized by several scholars, notably Homer-Dixon (1999), who has argued that environmental violence is an inevitable outcome of resource depletion. He argues that the most likely mitigating factor would be ingenuity, a concept similar to technology as posited 50 years ago by Zimmerman. Using this argument, Homer-Dixon's idea is that aquaculture offers the potential of an alternative food supply that would preclude conflict and violence in fishing communities.

Other scholars have challenged this simplistic position, pointing out that violence is not a direct outcome of scarcity, and that other factors significantly affect the dynamics of new relationships and the possibility of violent outcomes (Peluso and Watts, 2001; Hartmann, 2001). For some researchers, the importance of globalization is in the new and extended connections that contribute to transformation of agro-food systems. For example, Byres (1996) would argue that the importance of the introduction of aquaculture to Grand Manan has to be viewed in terms of its new linkages to a global economy. In his examination of the impacts of new forms of agricultural production, Byres concludes that the politics and the productive relations of new structures are unimportant compared to the characteristics of inter-sectoral linkages between agriculture and the industrial sector. His arguments downplay the significance of profoundly different institutional structures that re-shape movements of goods and knowledge in the context of local communities. A fundamental shift in decision making, financial investments, scientific knowledge, trade determinants, pricing controls, and even markets, all impact upon local communities and are directly linked to globalization. While the new connections in themselves alter the ways in which small communities interact with the world, for the individual fishermen and their families, the new globalized connections are experienced in very specific ways.

Aquaculture on Grand Manan

Brief history of aquaculture

Aquaculture can be defined as the farming or husbandry of aquatic plants and animals, and so implicit in the activity is some degree of human intervention (Boghen, 1995). However, this hardly touches upon the profound implications that modern aquacultural technology has for social, economic, environmental and political systems. The technological change involved in the development of aquaculture is fundamental, representing a quantum leap and unlike either the gradual evolution of agriculture or the modernization within the wild fishery, two important differences distinguish this revolution. One is the significant integration with world market relations; and the second is the important role of scientific research. The international aquaculture industry is characterized by a global integration process with increasing competition across borders and continents (Borch, 1999). In North America it has become a major factor in food production and as a contributor to the economy only in the past several decades. Worldwide, aquaculture production has increased steadily in the past 20 years, more than doubling in value in the 1984-1994 period alone. At the same time, wildfish stocks declined precipitously, a factor that is increasingly important in encouraging the growth of the aquaculture industry.

In Canada today, aquaculture is dominated by finfish culture, including trout, char and salmon, accounting for almost 90% of total national sales in 2002. The gross value added to the Canadian economy by the aquaculture industry in 2002 increased by 13.7% over the previous year. New Brunswick, where the first Bay of Fundy salmon site was established in 1978, contributes almost one-third of the value (gross sales) of all aquaculture production in Canada. In 2002, salmon aquaculture generated $282.3 million in sales of products and services, up 1.9% from 2001. References to aquaculture in this chapter refer exclusively to salmon aquaculture, although there is some experimental work on Grand Manan in the culture of flounder, scallops and cod. On Grand Manan, one of New Brunswick's first pioneers in aquaculture, John L'Aventure, established the first site in 1980. Apart from a short-lived attempt by another entrepreneur at a new site in North Head (1988-1990), it was not until 1990 and 1992, over the strenuous objections of local fishermen, that two new companies with ownership based on the mainland, sub-leased weir sites from fishermen who had acquired rights to weir privileges. It was during this period that the provincial government formulated its first policy on aquaculture development.

The overarching purpose of the original policy was to impose order and equality, which implied support for the participation of local fishing interests. However, initially most investments by local residents were through leasing their weir sites (privileges) at attractive fees over four-year contracts. By the mid-1990s only three local men were truly invested in the industry, none of whom was a traditional fisherman insofar as new government regulatory structures had imposed definitions of core fishermen. This distinction is broadly reflected in the ownership patterns and contractual arrangements for all sites on the island, and has significance for changing social relations on Grand Manan (Marshall, 2001).

The critical criterion for site allocation in 1991 was tied to the traditional herring weir. Very similar requirements of shore access, shallow waters, protected areas, and regular flushing, made the weir sites ideal for the location of aquaculture pens. When a potential salmon farmer applied for a lease of the seabed, the application entered a centralized - making process, through the Department of Fisheries and Aquaculture of the New Brunswick government (DFA). Subsequent to 1992, another 15 sites were located around the island, bringing the total number to 18 in 1999. The period from 1997 to 1999 was an especially crucial time for the industry, partly related to the rapid expansion in scale of operations and partly as a result of the sudden spread of ISA (infectious salmon anaemia), a previously unknown virus that attacks the liver of the fish. As the industry grew it became apparent to government regulators that a wide variety of issues associated with organizational structure, regulation, and scientific understanding of environmental impacts and harvesting practices, needed careful scrutiny. By the end of 1998, the province declared a moratorium on any new site development, which was rescinded with the announcement of the new policy in October 2000.

Over the intervening two years, the government of New Brunswick, through its Department of Fisheries and Aquaculture (reorganized in 2000 to include Agriculture, DAFA), developed a radically new regulatory regime and strategy for site allocation. The overarching management tools identified for the new policy addressed issues of sustainable practices, with the establishment of single year class sites and a bay management area framework. But it also incorporated provisions that strictly limit access to the industry to already established fish farmers. Because of the extremely large financial investments required and the high risks associated with the relatively new technology, as well as their historic ties to the wild fishery, local island entrepreneurs had been reticent about committing resources to the new industry. As a result, the 2000 policy effectively prevented any other entrants, while consolidation, mergers, and re-alignments among those already situated began in earnest.

In January 2001, applications for an additional 15 sites were announced by the Provincial government, of which six had been established by the end of 2002. The process of expansion proceeded both with respect to the numbers of sites, and in terms of their capacity (areal extent, numbers of cages and total capacity). Although each site has a slightly different ownership structure and contractual arrangements, the basic corporate structure in May 2003 can be characterized as shown in Table 4.1.

Table 4.1. Grand Manan aquaculture sites - ownership, October 2004

	Location of majority owner
10 sites	Stolt Co-ownership (Norway) (* 3 with some Grand Manan investment)
3 sites	Heritage (Weston Corporation, Toronto)
4 sites	Cooke (St. George, New Brunswick)
1 site	Limekiln (St. George, New Brunswick)
3 sites	AquaFish, and Ocean Salmon (overlapping ownership) (St. George, New Brunswick)
1 site	MG Fisheries (Grand Manan)
2 sites	Fundy Aquaculture (John L'Aventure, mainland\island residency)

Because of the substantial investments required for establishing a site (estimated at $2 million in 2003) most site operators are now subsidiaries of one of three major marketing organizations: Heritage Salmon owned by Toronto's Weston family, Stolt Sea Farm Inc., a Norwegian multinational, or the mainland local company, Cooke Aquaculture. New technologies themselves become a determinate force altering relative transaction costs, providing opportunities for the redefinition of property institutions (Millar, 1995). When economic entrepreneurs identify alternative property-rights systems that enhance the potential value of a resource, they tend to enter the political arena and demand changes that will capture increased values. Unlike the historic weir sites around Grand Manan that have captured wild herring over generations, the aquaculture sites are being leased and developed by distant, mainland and international companies rather than by island families. Whereas the weirs operated within informal understandings evolved through community relationships, that became codified more recently in government regulations, competition for aquaculture sites among off-island companies and state control of leases has created a market for the weir sites that is effectively privatizing the marine commons (Marshall, 2001). According to interviews I conducted in 2002, the value of a suitable weir site in 1999 was about $50,000, but by 2002, a weir site could not be leased for less than $150,000 per year.

The vulnerability of small investors was illuminated by the insolvency of one Grand Manan investor in January 2004 following two episodes of storm damage to his site, compounded by plunging market prices for salmon. The site was taken over by Stolt Sea Farm, the Norwegian company that controls or owns most of the sites around Grand Manan.

In preventing the entry of new investors, the government was not only stabilizing the industry, it was also establishing territorial rights over specific areas of marine space. Unlike the licensing framework applied in the traditional fishery, this policy effectively aimed to create a new form of resource allocation, in the fact that marine property has become a marketable resource.

Another aspect of the 2000 policy driving the market valuation of fish farm sites is the requirement for single-year class sites, devised as a safeguard to improve the health of the fish and to control the spread of disease. This policy means that a single site, having between 10 and 30 cages, and perhaps up to 1 million fish, would have all fish on the same grow-out cycle. Therefore, in planning for buying feed, hiring labour, and operating boats and scows, there is an intense economic dimension that dictates the need for two sites that allow for balanced production flows. The recognition that preventing the spread of disease and minimizing financial risk necessitated single class (year) grow-outs, created instant demand for second sites for all companies in the context of a two-year growing programme and the economics of labour and capital control. Following the announcement of the new policy in October 2000, there was a race to find additional suitable sites. That the traditional weir sites had suddenly become a valuable and increasingly scarce resource, subject to intense competition, meant that there was pressure from those already in the industry, notably those who controlled the Salmon Grower's Association, to restrict entry to the industry.

The impacts on local populations of such fundamental changes in productive relations and ownership patterns have been examined in other aquaculture/farming contexts. In their study of shrimp farming, Stonich and Vandergeest (2001) argued that government goals associated with the broadening of the economic bases of coastal areas, the generation of local employment, and the enhancement of food security, were in fact all subordinated to profits for producers and input suppliers. As with salmon aquaculture, the dominant roles of industry and government proponents are ignoring the potential for ecological change, environmental destruction and social dislocation. Stonich and Vandergeest also point to the very different property regimes in the wild fishery and farming industries. Whereas wild shrimp become private property when they enter the net; farmed shrimp are private property throughout their life cycle. Shrimp farming in some areas such as Thailand has degraded coastal resources even as it has made the spatial areas a more valuable resource. Growing tensions between the corporate dominated shrimp farms and local populations is linked not to scarcity of wild fish but to the nature of the farming industry itself.

As the contractual and ownership patterns in Grand Manan's aquaculture industry have evolved over the period after 1995, but especially since 2000, it is apparent that control of the industry is dominated by the large Norwegian company, Stolt. There are only five islanders who have significant (and often precarious) direct financial investments in the industry, excluding consideration of the spin-off companies associated with feed, pens and machine shops. Not only is island ownership very thin, but those who are involved have not been core fishermen in the past. In other words, those who have committed capital to the industry are people who had worked at other types of work, such as fish processing and packing, repairing wharves, selling insurance, or operating a retail food company. In a community where status has always derived from being a successful fisher, there is a rapidly changing structure of success related to investment in aquaculture. One of the lobster fishers was characteristically unequivocal: 'Those guys couldn't make a living fishing, so they're ruining it for us!' This new industry is not only creating new property regimes, it is also causing profoundly new social divisions and, gradually, new perceptions of social status in the community. While the government may extol the virtues of job creation and increasing revenues, with gross sales rising from $25 million in 1996 to an estimated $100 million in 2002, the net benefits are accruing mainly to off-island interests.

The marine commons is being alienated from the traditional fishery and control by the adjacent community. As well, while aquaculture has brought many jobs to the community, especially for young men, the jobs require only limited skills and opportunities for advancement.

Within the past five years, salmon aquaculture has grown from an industry producing equal revenues to the traditional fishery, to exceeding it by a factor of two to three times. That social divisions based upon a status acquired through generations of participation in the vagaries and challenges of a diverse wild fishery will change is already becoming apparent. New productive relations that emphasize the roles of unskilled, young male labour, and exclude the traditional roles of women as book keepers and fish plant processors, also contribute to a radically new dynamic for social relations. New patterns of social cleavages and definitions of status, changing relationships with the sea, the nature of ownership and access to marine resources, and productive relations of work are all changing in ways that fundamentally alter the meaning of rural fishing community. Fish farms are industrial capital aimed at profit, and their rationale, in the context of a global market economy, takes little account of community interests and local control.

Rationalization and integration

Concurrent with the evolving ownership patterns of aquaculture sites have been structural changes affecting levels of vertical integration. The length of time to maturation, from the time the smolts are moved into the sea cages when they are about 100 g, until they are harvested at about 5 kg, is approximately 18 months. The importance of the single-year class mentioned earlier is that in alternate years the site would not have fish to harvest. Therefore, in order to maintain an even cycle of labour-equipment use, thereby maximizing efficiency, it is important for owners to have at least two sites. Apart from investments into boats and scows, cages and net manufacture, the major operating cost components are feed and labour. For independent owner John L'Aventure, it was crucial that he maintain control of the factors of production by refusing to buy feed from the major corporations such as Weston and Stolt. While there is scientific debate about the merits of grain-based feed for older fish, for smolts a high percentage of protein (herring) is a crucial ingredient. In Norway, fish meal and fish oil used in the mixing of salmon feed are regarded as globally tradable products. The conversion ratio of feed to fish weight gain is estimated at 3:1 which supports the argument of environmentalists that the feed itself constitutes a growing problem for marine degradation. In Norway, over half of the total herring tonnage caught (2 million tonnes) is directed into feed manufacture for the salmon aquaculture industry. There are concerns that the aquaculture industry may face a fish meal trap as a major resource constraint (Pillay, 1999). An important area of research is now investigating alternative feed mixes, including use of animal byproducts. Already added to the feed are antibiotics and food colouring to ensure the fish retain the pink colour that people associate with salmon. As a major cost factor, feed constitutes a key target for vertical integration in the industry. The fish are fed three times a day during high growth summer seasons, and once a day during the winter months.

On Grand Manan, there is one feed plant that sells directly to all sites during the crucial three- to six-month period that smolts are first introduced into the marine environment. The history of this feed plant and the new linkages established through it, provide clear evidence of the changing patterns of economic life on the island. Originally built by a mainland entrepreneur in 1997, the plant was repossessed by the bank, and taken over in 1999 by the largest fish broker on Grand Manan, who also owns a herring seiner.

His captive market of 24 aquaculture sites for three to six months of the early grow-out period has provided assured sales. This is combined with his capacity to ensure a continuous supply of fresh herring to the feed plant because he owns a herring seiner. A guaranteed market for herring, and steady work for his crew of seven, offer important economic benefits that are broadening and strengthening his business interests on the island. In March 2003, he leased a temporarily closed site in Dark Harbour from the holder of a weir privilege, as a way to further integrate his business interests and ensure a continuous market for his feed. The business acumen of this entrepreneur is linked into the social milieu in a number of ways that suggest the beginnings of social embeddedness for island-owned aquaculture interests. His participation on Municipal Council, viewed as a potential leader in the community, his donations to various causes, and his status as a major employer, all contribute to integrative networks that link social and economic interests. This case illustrates a degree of simultaneity and interpenetration of global and local processes that argue for a non-bipolar approach to investigations of agro-food systems (Robertson, 1994). Again, however, a note of caution related to expectations must be acknowledged. For in the summer of 2004, two complaints related to poor quality control of the feed caused two operators to move their business to mainland feed companies. The rapid rise of the industry on the island has occurred concurrent with very steep learning curves for local entrepreneurs.

Two other companies have attempted to expand vertically, both with the intention of assuring supplies and just-in-time delivery. In particular, Fundy Aquaculture (John L'Aventure) has sought to expand vertically in both directions, even as early as 1992 when he started a small net and cage building company, *Sea Nets.* Since then he has been the most proactive entrepreneur, focusing not only on numbers of sites or growing capacity, but also broadening his reach to develop a sustainable hatchery that can provide a reliable supply of smolts, and to build a processing facility that will enable him to ship final product directly to Boston buyers within hours of the harvest. In describing the new processing operation that opened at the end of October 2002, Manager Chris Rayner was very positive: 'It has been better than we hoped for; we're very happy' (Interview, May 2003). Unfortunately, the highly competitive market, particularly from Chilean salmon, caused lower prices and market demand in the USA, with the result that the new facility was closed down more than it was operational during 2003-2004. When in operation, approximately 16 people (mainly women) have full time jobs. Problems also affected the hatchery which took longer to bring into production than expected, mainly hampered by an extraordinarily cold winter, 2002-03. The first 3 g fingerlings from the Fundy broodstock were transferred to growing tanks in early July 2003 and the first smolts were transferred to ocean cages (at 100 g) in early summer 2004. The completion of their first 'production year' was hailed as a significant event by hatchery manager Brenda O'Donnell when I interviewed her in August 2004. The urgency to develop a fully integrated system was apparent in Rayner's description of the higher percentage of morts (mortalities) (approximately 8-10% as compared to 1%) using smolts that have been shipped from the mainland, and the economic benefits of rearing their own compared to buying from suppliers whose quality controls are unpredictable. Given that each smolt purchased from an external facility costs about $3.50, and each sea cage will grow-out about 20,000 fish, the economics of producing their own smolts is compelling.

The second company that has attempted to integrate backward and forward linkages has been less successful. With significant backing from Stolt, one local entrepreneur enlisted a family member (non-fisherman) to build a hatchery in 2001, which provided the first smolts for the sea cages in July 2002. However, while the product was excellent according to one buyer of the smolts, the partnership ran into difficulties, and it is now run as an independent supplier. In the development of several sites in partnership with Stolt since 1995, this entrepreneur has relied upon political connections and subsidies to leverage his involvement in several salmon farm sites.

As the industry began to enter a new phase of consolidation in 2003, specific problems that affected every site operator began to be addressed. At the top of this list was reliance upon the ferry service, which has limited capacity during the winter and spring months when only one boat runs three times a day. Incredibly, a provincial government that has encouraged and supported the growth of the aquaculture industry has done nothing to relieve the problems associated with the transportation of both smolts and final product. With an estimated demand for 400-500 ferry trips (four trailer trucks per trip); the impacts upon the local community in 2003 have been significant. The smolt deliveries, as well as all fresh fish (including salmon), have priority on the ferry, with the result that individual islanders are frequently left waiting for a ferry for five to six hours, rather than the half hour that had been the case until 2000. In the spring of 2003, one local woman left at the ferry that she had arrived early for, was overheard to exclaim: 'Damn aquaculture!' The demand for ferry transport by salmon companies was exacerbated in 2003 by the fact that the largest company (Stolt) changed its transport strategy, from depending upon its own two boats to transport freshly killed fish to the mainland to dependence upon the ferry. The change was made due to problems with transmission of the ISA disease via boats moving between the island and the mainland. That island tensions about this industry are being barely contained is due to the fact that virtually every family has someone involved in the industry. There is a widespread acknowledgement that, without aquaculture we would be in trouble.

Local foods and embeddedness

As pointed out earlier, local foods in this chapter are those foods that have been directly linked to historical livelihoods and cultural meanings in the community over generations. They include the produce of individual gardens, the wild game that is a significant source of winter protein for many families, and the diverse wild fishery that has been the basis of livelihoods for over 200 years. In their discussion of the environmental crisis, Fitzsimmons and Goodman (1998) suggest that agro-food networks represent a crucial entanglement of the nature-society relationship, and that the political economy of agro-food networks needs to be carefully examined. They argue that the shared corporeality of agro-food practices challenges the abstraction of agriculture as an economic activity. Like Whatmore and Thorne (1997), they suggest that actor network theory offers a way to understand the complex ways in which agency can be conceptualized as a collective capacity involved in relational networks. There is an acknowledgement of the inseparability of the nature-society relationship bound up in agro-food systems.

While the traditional fishery is deeply implicated in the global market economy, with complex networks of interactions and dependencies that impact upon livelihoods in a very immediate way, the local economy nevertheless retains intricate networks rooted in social structures based in the church and family webs of cultural identities. Deconstructing or unpacking the processes of interaction and destabilized relationships is not merely a matter of extricating global and local parameters of action. The concept of embeddedness provides one way of exploring the complexity of these global-local networks. What is important to understand, however, is the essentially dynamic character of the interactions, and that scale and level and magnitude of various relationships do not evolve in any linear and formal process. Like Bourdieu's habitus (Bourdieu, 1984), the global-local nexus constitutes a distinct but non-permanent state of interactive networks. There is an interpenetration of processes that denies the validity of bipolar opposites.

The idea that food supply chains are embedded in social and environmental relationships has been examined recently by Marsden *et al.* (2000) and Hinrichs (2003). It is an attractive idea in its affirmation of the obvious links between the daily, taken-for-granted activities and behavioural patterns of consumers, and the economic, organizational and technological imperatives of food production. Polanyi's notion that the modern industrial state dominated by the power of the market system represented an inevitable de-linking of social and economic worlds, a disembedding of the economy from social life, reflects concerns about the alienation of individuals from the structures within which they interact. It has been argued that increasingly globalized food systems are being disembedded from social relations in ways that disenfranchise and disempower local populations. According to Hinrichs (2003), atomized market relations are seen as a defining but negative feature of the global, while the local will manifest high levels of social capital and relations of care - in short, a more moral associative economy. The complexity and irony contained within the concept of embeddedness was explored by Granovetter (1985), who argued that social-economic relations need to be seen as a gradation of interactive networks. Denying both the over-socialized view that privileges behavioural patterns that have been internalized to the extent of precluding a role for agency, and the atomized perspective, Granovetter posits an ongoing process of negotiated realms of social and economic relationships that define a sort of habitus, dynamic and transformative in nature.

Polanyi had argued that earlier phases of human development were more positive insofar as economic activity exercised a social function and thus helped to constitute society as a whole. Labour was then embedded in social relations (Baum, 1996). Some researchers have focused on the ways in which decision making can be vested with the community, recently with growing interest in promoting purchase of local foods as a way to support local farmers (Winter, 2003). However, while local control is important, and shopping local can support neighbours, it is not these aspects of embeddedness that I want to assert. In referring to Polanyi's concept of embeddedness, I wish to underline the fundamental relationship between the activity of producing foods within a community setting (location) and a community culture. In other words, embeddedness is more than conscious decisions to buy locally or a 'defensive politics of localism' (Winter, 2003). The importance of local foods in family diets is in their direct links to livelihoods and sustaining cultural meanings related to lifestyles. For Grand Mananers, local foods have significance in daily and seasonal activities and employment. Earlier I defined local foods as including the produce of home gardens, the animals trapped or hunted, and the wild fishery. Specific examples of these are briefly described here.

Important foods in the diets of Grand Mananers include a range of wild and domestic foods that are tied into economic networks of exchange to varying degrees. While most families rely on vegetable gardens for summer produce, many also depend upon significant quantities of preserved vegetables such as green beans, peas, potatoes, tomatoes and beets through the winter months. Several local residents have larger gardens from which they sell directly to the community. A local farmers' market through the summer months is supported by tourists. Hunting and trapping are important activities that affirm and sustain a different but equally significant aspect of local identities. Considered to be recreational, the killing of deer (on the island), moose (on the mainland), and rabbits (on the island) accounts for significant amounts of protein for some families. Frozen or preserved in mason jars, this meat harvest involves time and shared ritualistic challenges that underpin important cultural traditions and community understandings. Outside of the market system, the activity is nevertheless deeply embedded in the social relations of the island.

The wild fishery, the core of the livelihood and survival of this community for over 200 years, represents the centre of economic activity and the market economy for the island. People's lives are defined by and within it. But the nature of these relationships is diverse, with a variety of levels of capital investment, entry costs, regulatory regimes, economic benefits and global market connections. At one end of the scale is dulse, a seaweed that is harvested at low tide from May to October, which accounts for about $1 million in gross revenues. Dulse is not only a commercial harvest that is shipped all over the world. Its meaning as a marker of cultural heritage is apparent in the gifts carefully stored in the freezer to be brought to mainland families at Christmas. Eaten as a snack, brown bags of dulse sit beside the men in their trucks and next to women quilting in their dining rooms. At the other end of the scale in terms of capital investment are the herring seiners, valued at over $1 million for the basic boat and equipment. But in terms of the aesthetic value and emotional involvement of individual fishermen, it is the herring weirs that represent the purest nexus of social, economic and cultural meanings. The unpredictability of the catch, the vagaries of hurricane winds that regularly destroy segments of the $100,000 weirs, and the beauty and rhythm of the herring harvest itself, create a sense of belonging and identity that link sea and land, livelihood and selfhood.

These activities, gardening, hunting and fishing, and the seaweed harvest, all sustain community relations. They are all producing local foods that have long histories and rich traditions tied to survival and shared experience. These local foods are well and truly embedded in community and social relations. Having said this, it is important to emphasize that some of these activities are part of the global markets and lengthening networks (Whatmore and Thorne, 1997) that characterize the food system. As pointed out earlier, the globalization of fish markets was initiated during the 1970s with the new ocean regime and the establishment of the 200-mile exclusive economic zone. The effective nationalization of the fishery resource stimulated this trend that was magnified by the enhancement of technology and increased scale of fishing operations. A third force was the rise of multinational corporations and the growing power of industrial actors at the market end of the food system (Arbo and Hersoug, 1997). The closure of the resource system has coincided with an opening of the trading system (Arbo and Hersoug, 1997). The traditional fishery on Grand Manan is characterized by a variety of globalized network designs. The events of 9/11 had an immediate impact upon the price of lobster on Grand Manan. Without any guarantee of flights, lobster prices plummeted to $2.50/lb for several weeks. By Christmas of 2001 it had increased to only $5.00/lb, compared to a year later when it was almost $8.00/lb. The main market for sea urchins is Japan; and for scallops is Boston, Massachusetts. High value crustaceans and shell fish are directly tied into global markets, their prices dependent upon the global economy.

The restructuring of Weston Corporation based in Toronto in 1998-1999 impacted upon the Grand Manan herring seiners and weir fishermen when Connors (a wholly owned subsidiary) closed its sardine packing plants in the USA, consolidating processing at its two Canadian plants, including Grand Manan. The end result for the island was a second shift at night that required more labour, necessitating recruiting trips to Newfoundland by the plant manager. While the original migrants in 1999 and 2000 were employed mainly in the Connors plant, by 2001 the expansion of aquaculture sites created an added demand for additional labour. The Newfoundlander presence on the island, directly associated with the restructuring of Connors and the expansion of aquaculture, represents a significant change in the social milieu of Grand Manan (Marshall, 2002). The incorporation of the sardine industry into global markets does not preclude its significance as local food with respect to both production and cultural significance for personal and collective identities. For salmon, on the other hand, the local location of the productive capacity has not provided any basis for embedded social meanings. The assurance of income and of jobs for youth, discouraging higher education and mobility off the island, certainly have profound implications for social relationships. But to argue that they represent embedded meanings is to stretch the meaning of the concept as formulated by Polanyi. While the traditional fishery (and sardine production) are embedded because of their reach within the community, historically, culturally, socially and economically, salmon aquaculture remains largely detached.

Future embeddedness?

Baumann (2001) in his discussion of local orders and global chaos argues that globalization has had one revolutionary effect, insofar as order itself has been devalued. But the crucial element in this acceptance of disorder, he suggests, is the devaluation of place and a rejection of the long term. The idea that long-term memories, commitments and shared experiences have lost their value means that fleeting forms of association become more useful to communities. For the industrialized structures of aquaculture, the loss of commitment and the dependence upon hourly paid, seasonal labour means that production is grounded in jobs and an economic return that does not necessitate a social network of support.

The main theme of this paper is the significance and impacts of new globalized food structures and high technology production systems, for one small, remote, fishing community. The recently established aquaculture industry on Grand Manan Island provides a window on the complexity of community responses, related to the changes in social and productive relations, market and industrial structures, and community understandings of lifestyle choices. On the one hand, significantly reduced groundfish catches and declining markets for pelagics such as smoked herring, have threatened the historic livelihoods of hundreds of island families. On the other hand, salmon farming is not a fishery alternative.

The transformation of the island economy has not been merely a fishing community moving into a different type of fishing. The change has been from a lifestyle and level of independent activity in which entire families contribute to the success of the venture, to individually hired, manual labour, paid by a company based on the mainland. One fisherman said cryptically, 'Its slave labour'. Not only is salmon a product divorced from any experience of fishing, as a food product to be consumed by island residents, it cannot be found in local supermarkets, except occasionally during the summer months when tourists request it. When I compared the prices of cod and salmon at a Montreal fish market in 2001, ($9/kg salmon; $15/kg cod) and conveyed the information to an island fisherman, his immediate comment was: 'You know which I would buy!?' (Not the salmon). For islanders, salmon is in no sense a local food. Despite the market connections, cod, pollock, haddock and mackerel, are all perceived as local, sought after for both income and local consumption. But, islanders do not eat salmon. Some 'don't like what goes into it'; others simply do not like it.

Unfortunately, for Grand Mananers who have lived for 200 years on the resources of the sea, it is becoming increasingly difficult to do so. Seasons are shorter; quotas restrict catches; fewer groundfish licenses are being used because the high cost of fuel results in any fishing trip becoming more costly than returns from fish sold; and the price of wild fish, even at the wharf, is the same price offered by the broker. In 2003 only one fisherman continues to go handlining. His fish are much sought after by islanders whose discriminating tastes recognize the freshness and quality of handlined as compared to gillnet fish. For many families on the island, the high prices for their traditional fish, driven up by scarcity and by world market prices, mean that even a weekly meal with fish is impossible. In a recent and as yet unpublished study of dietary patterns in two New Brunswick communities, including Grand Manan, Melissa Legrand found that the percentage of protein from traditional sources consumed by islanders was between 25% and

31%, depending upon the season. As a percentage of dietary intake, local foods constituted only about 15%. While her analysis is not yet complete, her interviews suggest evidence of a decline in consumption of local foods associated with high prices and increasing scarcity of wild fish. Interestingly, anecdotal evidence suggests a different conclusion. Interviews offer the possibility that considered on an annual basis, the existence of freezers has allowed long-term storage of fish, which has actually increased total consumption over the year.

The desirability of industrial farming has been hotly debated for decades. Nobody denies the significance of declining real prices for food, with Canadians spending only 13% of disposable incomes on food compared to 50% in places such as India. There is a sense in which society values our food system's triumph over hunger, which means that the acceptance of aquaculture as a desirable form of food production is likely to grow.

While the reality is that on Grand Manan both the traditional fishery and salmon aquaculture will continue to dominate economic structures, the expectation must be that one will be regarded as a means to an end, whereas the other will be an end that is richly experienced through the means. Whereas fishing will require vigilance and effective measures for conservation and control of access, for fish farming the main current issue is the maintenance of market share (Coull, 1996). That market share will be competed for within the networks of global interactions, in which local entrepreneurs will have very little input. In his analysis of the future for salmon aquaculture, a spokesman for Heritage Salmon (owned by Weston) said that the new technologies offer the possibility of expanded production within the same spatial areas being currently exploited. But, as with agriculture, the expectation is that labour demands will decrease with new mechanical feeders and more rapid biological growth rates. As the economic dimensions of aquaculture on Grand Manan become entrenched in the community, social relations will undoubtedly be transformed, and a degree of embeddedness will be enhanced. As a local food, however, it is unlikely that salmon will ever acquire the significance for the community that has defined the relationships of the traditional fishery.

Acknowledgements

I wish to acknowledge the funding support of the Social Sciences and Humanities Research Council of Canada, without which my research on Grand Manan could not be done. As well, I am most appreciative of the ongoing support and trust given to me by residents of the island and industry representatives, whose interviews constitute the crucial body of data for this chapter.

References

Apostle, R., Barrett, G., Holm, P., Jentoft, S., Mazany, L., McCay, B. and Mikalsen, K. (1998) *Community, State, and Market on the North Atlantic Rim.* University of Toronto Press, Toronto.

Arbo, P. and Hersoug, B. (1997) The globalization of the fishing industry and the case of Finnmark, *Marine Policy* 21, 121-142.

Arce, A. and Marsden, T. (1993) The Social Construction of International Food: A New Research Agenda. *Economic Geography* 69(3), 293-311.

Atkins, P. and Bowler, I. (2001) *Food in Society: Economy, Culture, Geography.* New York & London, Arnold.

Baum, G. (1996) *Michael Polanyi.* McGill Queen's University Press, Montreal.

Baumann, Z. (2001) *The Individualized Society.* Polity Press, Cambridge.

Bessiere, J. (1998) Local Development and heritage: traditional food and cuisine as tourist attractions in rural areas. *Sociologia Ruralis* 38(1), 21-34.

Boghen, A. (ed.) (1995) *Cold-Water Aquaculture in Atlantic Canada*, Second edition, The Canadian Institute for Research on Regional Development, Moncton.

Borch, O. (1999) New organizational forms within the aquaculture industry: the fish farming enterprize as a virtual organization, *Aquaculture Economics & Management* 3(2), 105-120.

Bourdieu, P. (1984) *Distinction: A Social Critique of the Judgement of Tastes*. Harvard University Press, Cambridge, Massachusetts.

Byres, T. (1996) *Capitalism from Above, Capitalism from Below*. Macmillan, London.

Coull, J. (1996) Toward a sustainable economy for the Shetland Islands: Development and management issues in fishing and fish farming. *Geojournal* 39(2), 185-194.

Davis, A. and Bailey, C. (1996) Common in Custom, Uncommon in Advantage: Common property, Local Elites, and Alternative Approaches to Fisheries Management, *Society and Natural Resources* **9**, 250-265.

Fitzsimmons, M. and Goodman, D. (1998) Incorporating Nature: Environmental narratives and the reproduction of food. In: Braun, B. and Castree, N. (eds) *Remaking Reality: Nature at the Millennium*. Routledge, London, pp.194-219.

Granovetter, M. (1985) Economic Action and Social Structure: The Problem of Embeddedness. *American Journal of Sociology* 91(3) 481-510.

Hartmann, B. (2001) Will the circle be unbroken? A critique of the Project on Environment, Population and Security. In: Peluso, N. and Watts, M. (eds) *Violent Environments*. Cornell University Press, Ithaca, pp 39-64.

Hinrichs, C. (2003) The Practice and Politics of Food System Localization. *Journal of Rural Studies* 19(1), 33-45.

Homer-Dixon, T. (1999) *Environment, Scarcity and Violence*. Princeton University Press, Princeton.

Ilbery, B. and Kneafsey, M. (1999) Niche Markets and Regional Speciality Food Products in Europe: toward a research agenda. *Environment and Planning A* 31, 2207-2222.

LeHeron, R. (1993) *Globalized Agriculture: Political Choice*. Pergamon, Oxford.

Marsden, T., Banks, J. and Bristow, G. (2000) Food Supply Chain Approaches: Exploring their Role in Rural Development. *Sociologia Ruralis* 40(4), 424-438.

Marsden, T., Munton, R., Ward, N. and Whatmore, S. (1996) Agricultural Geography and the Political Economy Approach: A Review. *Economic Geography* 72(4), 361-375.

Marshall, J. (2001) Landlords, Leaseholders and Sweat Equity: changing property regimes in aquaculture. *Marine Policy* 25, 335-352

Marshall, J. (2002) Between Belonging: *Habitus* and the migration experience. *The Canadian Geographer* 46(1), 63-83.

Matthews, R. (1993) *Controlling Common Property: Regulating Canada's East Coast Fishery* University of Toronto Press, Toronto.

Millar, C. (1995) *Salmon Aquaculture in Shetland and New Brunswick: A Comparative Study of Resource Regimes within a Moral Perspective*. Unpublished Ph.D. dissertation, University of Western Ontario, London.

Peluso, N. and Watts, M. (eds) (2001) *Violent Environments*. Cornell University Press, Ithaca.

Pillay, T. (1999) Resources and Constraints for Sustainable Aquaculture. In: Svennevig, N., Reinertsen, H. and New, M. (eds) *Sustainable Aquaculture: Food for the Future?* A.A.Balkema, Rotterdam, pp.21-27.

Polanyi, M. (1944) *The Great Transformation*. Beacon Press, Boston.

Robertson, R. (1994) Globalization or Localization? *Journal of International Communication* 1(1), 33-52.

Stonich, S. and Vandergeest, P. (2001) Violence, Environment, and Industrial Shrimp Farming. In: Peluso, N. and Watts, M. *Violent Environments*. Cornell University Press, Ithaca, pp.261-286.

Tregear, A. (2003) From Stilton to Vimto: Using Food History to re-think Typical Products in Rural Development. *Sociologia Ruralis* 43(2), 91-107.

Ward, N. and Almas, R. (1997) Explaining Change in the international agrofood system. *Review of International Political Economy* 4(4), 611-629.

Watts, M. and Goodman, D. (1997) Agrarian questions. Global appetite, local metabolism: nature, culture and industry in *fin-de-siecle* agro-food systems. In: Goodman, D. and Watts, M. (eds) *Globalizing Food. Agrarian Questions and Global Restructuring*, Routledge, London, pp.1-32.

Whatmore, S. (1995) From farming to agribusiness: the global agri-food system. In: Johnston, R. Taylor, P. and Watts, M. (eds) *Geographies of Global Change: Remapping the World in the late twentieth century*, Blackwell, Oxford, pp.36-49.

Whatmore, S. (2002) From Farming to Agribusiness: Global Agri-Food Networks. In: Johnston, R., Taylor, P. and Watts, M. (eds) *Geographies of Global Change*. Second edition, Routledge, London, 57-67.

Whatmore, S. and Thorne, L. (1997) Nourishing Networks: alternative geographies of food. In: Goodman, D. and Watts, M., (eds) *Globalizing Food: Agrarian Questions and Global Restructuring*. Routledge, London, 287-304.

Winter, M. (2003) Embeddedness, the new food economy and defensive localism. *Journal of Rural Studies* 19(1), 23-32.

Chapter 6

Alternative or conventional? An examination of specialist livestock production systems in the Scottish-English borders

Brian Ilbery and Damian Maye

Alternative geographies of food

There has been a clear and recent resurgence of interest within many developed market economies in foods of local and regional provenance (Ilbery and Kneafsey, 1998; Murdoch *et al.*, 2000; Parrott *et al.*, 2002; Hinrichs, 2003; Morris and Buller, 2003; Weatherell *et al.*, 2003; Ilbery *et al.*, 2004a). This interest is in part a response to consumer reactions to a range of environmental, ethical and health concerns associated with 'conventional' food supply systems that have become increasingly industrialized and global in reach. In an examination of consumers' confidence in Austria, Italy and the UK, Sassatelli and Scott (2001), for instance, interpret the growth in novel food markets by key actors - the state, consumers' movements, retailers and marketing boards – as confidence-building strategies to address deficits in disembedded trust, widening chains of interdependence, food scares and the introduction of GM technologies. Weatherell *et al.* (2003) note that the new 'concerned' consumer is now showing considerable interest in 'alternative' foods produced under more organic, environmentally-friendly and local supply systems. Among such consumers, the focus has shifted away from price, packaging and appearance towards higher quality and more natural products that can be traced to specific places. Accompanying this move from 'conventional' to 'alternative' has been a growth in food purchases from local (shorter chains) rather than supermarket outlets, including farmers' markets (FMs), box schemes, community supported agriculture and home deliveries (Festing, 1997; Cone and Myhre, 2000; Ventura and Milone, 2000; Morris and Buller, 2003; Renting *et al.*, 2003; Ilbery and Maye, 2004).

Many authors have referred to the distinctions drawn between 'conventional' and 'alternative' agro-food systems (Ilbery and Kneafsey, 1998; Lang, 1999; Murdoch *et al.*, 2000; Hinrichs, 2003). Some of these contrasts are listed in Table 6.1, where a number of binary opposites are depicted. Thus words such as 'quality', 'embedded', 'sustainable', 'traditional', 'natural', 'artisanal' and 'diversity' characterize alternative food production systems. As Parrott *et al.* (2002) demonstrate, the alternative food sector is surrounded by conventions of 'domestic worth', 'inspiration' and 'environment', in contrast to the 'market', 'industrial' and 'economic efficiency' conventions of conventional agro-food systems. However, in reality these binary opposites are not as simple and clear-cut as portrayed in Table 6.1. In fact, the distinctions between conventional and alternative are blurred, leading Hinrichs (2003, p.36) to claim that 'specific social and environmental relations do not always map predictably and consistently onto the spatial relation'. Thus, while organic food may be regarded as 'alternative', most organic sales still occur through 'conventional' supermarkets. Hinrichs advocates a more nuanced approach when dealing with relations between global (conventional) and local (alternative), especially as both are linked together in an overall agro-food system. In a case study of 'food system localization' in the very conventional agricultural state of Iowa, Hinrichs suggests that initial 'defensive localization', in which there is homogeneity and real resistance to external forces and protection against 'outsiders', begins to give way to 'diversity-receptive localization', where heterogeneity and difference within and outside a particular spatial scale allow the local to become embedded in the larger regional/national community.

Table 6.1 Distinctions between 'conventional' and 'alternative' food supply systems.

Conventional	Alternative
Modern	Post-modern
Manufactured/processed	Natural/fresh
Mass (large-scale) production	Craft/artisanal (small-scale) production
Long food supply chains	Short food supply chains
Costs externalized	Costs internalized
Rationalized	Traditional
Standardized	Difference/diversity
Intensification	Extensification
Monoculture	Bio-diversity
Homogenization of foods	Regional palates
Hypermarkets	Local markets
Agrochemicals	Organic/sustainable farming
Non-renewable energy	Reusable energy
Fast food	Slow food
Quantity	Quality
Disembedded	Embedded

Sources: Ilbery and Kneafsey (1998); Lang (1999); Murdoch *et al.* (2000); Hinrichs (2003).

Increasing academic interest in the alternative food economy (AFE) has led some social scientists to proclaim the emergence of 'alternative geographies of food' (Whatmore and Thorne, 1997; Murdoch *et al.*, 2000; Parrott *et al.*, 2002). These geographies revolve around changing production and consumption relations that give rise to new regional and local food 'complexes' (Murdoch *et al.*, 2000). Such complexes may (partially) develop into what Ilbery and Kneafsey (1998) describe as 'quality production areas' that are spatially removed from the 'hot spot' regions of productivist farming (Whatmore, 1995). As Parrott *et al.*, (2002 p.243) advocate, alternative geographies of food may be:

> 'associated with agriculturally peripheral regions precisely because such regions have, for a variety of reasons, failed to fully engage with the productivist conventions that have predominated in the agro-food system in the second half of the twentieth century'.

These sentiments are also echoed by Murdoch *et al.* (2000, p.108):

> 'an alternative geography begins to emerge because the local ecologies of quality food production are frequently to be found in areas that have escaped the industrialisation processes that underpin globalisation'.

Parrott *et al.* (2002, p.243) suggest that the 'promotion of regionally distinctive produce may offer an alternative development trajectory for regions less able to compete in a globalized, commodity-driven, market place'. They use the European Union's designations of regional origin (PDO/PGI), where food and drink products are embedded in the quality characteristics of their place of origin, to highlight a geographical concentration of speciality products in less favoured farming regions. Few would deny the 'stylised' inceptions developed here, an impression that joins the words 'alternative', 'quality' and 'lagging' together in seamless applause. In reality, the socio-spatial contours of the AFE make for a more rugged conceptual terrain, with a switchback of currents around conventional pillars. It is in this respect that Whatmore (2002), more generally, emphasizes the importance of 'hybridised spaces'. Thus one might question whether PDO and PGI labels can be used as an indicator of local, short chain production (Renting *et al.*, 2003), as many (e.g. Newcastle Brown Ale) are shipped around the globe in highly commercial and conventional forms. Equally, one can talk about alternative geographies of food in a global sense, notably through international Fair Trade and organic movements that challenge exploitative relations in agri-food systems (Raynolds, 2000; 2002). Goodman (2003) draws useful comparison between US and European alternative food practices, framing US alternatives more as social and oppositional movements (Allen *et al.*, 2003; Vos 2000), in contrast to the European Union's more endogenous reclaims on historical and cultural traditions of product and place.

The blurring between 'conventional' and 'alternative' food supply systems raises a number of important questions. For example, does a truly alternative food supply system exist and can lagging rural regions reinvent themselves in the context of local, quality food production? This chapter attempts to answer these and other questions by presenting some empirical findings from an on-going international research project on alternative food supply systems, the SUPPLIERS research programme into supply chains linking food Small and Medium Enterprises (SMEs) in lagging rural regions in the European Union.

In particular, the chapter focuses on the development of specialist livestock product chains in the lagging rural region of the Scottish-English borders. This particular 'alternative positioning' is far removed from commentaries on, for example, community agricultural schemes (e.g. food co-ops) or public procurement initiatives (Ilbery and Maye, 2004). The former are market driven, with concerns about consumer access and social improvement, and have little to do with value-added Supply Chains; they may not even use locally grown food. The latter are more likely part of a commodity-based system (e.g. beef) and thus differ from many of the specialist products studied here. These parameters – at least to us – are not always apparent in the 'alternative food' literatures, with social scientists and policy makers alike talking about 'local', 'locality', 'regional', 'alternative' and 'quality foods' in often interchangeable and undistinguishable terms. The unproblematic application of these different reference points may further increase contradictions and ambiguities. Before exploring the 'production worlds' of specialist livestock producers, it is first important to raise a number of conceptual points about alternative food systems.

Conceptualizing 'alternative' food production

Implicit in the notion of an alternative food supply system or network is the importance of key concepts such as quality, embeddedness, trust and local (Ilbery and Maye, 2004). However, conceptualizing such ideas is far from easy because as Goodman (2003, p.3) articulates 'quality, embeddedness and the local are all constructs whose material and symbolic meanings are vigorously contested'. For Robinson (2002), the word 'alternative' refers to the capturing of more value added at the primary end of the food supply chain. Nevertheless, this collective noun stretches beyond the purely economic, signifying also a realignment of 'value' and 'trust' between food producers and consumers and the creation of new forms of food politics and market governance (Whatmore *et al.*, 2003). Borrowing from Michel Callon and colleagues, Whatmore *et al.* (2003) represent the turn to 'alternative' as an archetypal case of the 'economy of qualities'. This denotes a moment in (food) market relations when the state of production, consumption and regulation become one fluid form under the spotlight of intense social reflection, enquiry and qualification.

Four concepts are central to understanding the development of alternative geographies of food:

1. *Short food supply chains (SFSCs)* The key characteristic of SFSCs is that foods reach the final consumer having been transmitted through a supply chain 'embedded' with value-laden information concerning the mode of production, provenance and distinctive quality assets of the product. In many cases, the number of nodes between the primary producer and the final consumer will also be minimized (Renting *et al.*, 2003). While this 're-connection process' is best demonstrated through forms of direct marketing and thus face-to-face contact between producer and consumer, Marsden *et al.* (2000a) identified two further types of SFSC: spatial proximity and spatially extended. The former is where products are sold through local outlets in the region, locality or place of production, so that the consumer is immediately aware of the locally embedded nature of the product at the point of retail. In contrast, the latter occur when products are sold to consumers (e.g. via the Internet) who are located outside the region of production and/or have no experience or personal knowledge of the area. Here, the key is to use product labelling and imagery to transfer information about those producing the food and the place of production to the consumer (see Ilbery *et al.*'s 2004b analysis of product labelling schemes).

2. *Social embeddedness* This propagates the idea that economic behaviour is embedded in, and mediated by, a complex and extensive web of local social relations. In the case of local (alternative) foods, both economic (e.g. price, markets) and social (e.g. local ties, trust) relations are vital for success. However, in stressing the role of social relations in generating the trust necessary for economic transactions to take place, it is easy to make the false assumption that social embeddedness relates just to alternative food systems (Krippner, 2001). In reality, all economic relations are socially embedded in a range of contrasting ways and so there are different degrees of embeddedness in all food supply systems (Winter, 2003). Nevertheless, social interaction between producer and consumer can make the difference between success and failure for local food businesses. This can take the form of acknowledgement, attention, respect, friendship and sociability, all of which can be subsumed within the concept of 'regard' (Lee, 2000). For example, Sage (2003) found the local geography of regard to be particularly important in the development of alternative food networks in southwest Ireland.

3. *The 'turn' to quality* Closely linked to the ideas of SFSCs, embeddedness and regard is the notion that local (alternative) foods are of higher quality than products

produced under more conventional farming systems. In other words, quality and locality are inextricably linked. Yet, 'quality' is a social construction and thus means different things to different people (Ilbery and Kneafsey, 2000). Product labelling is an important dimension of marketing quality and, in the case of local foods, labels tend to focus on different combinations of the three 'Ps': product (e.g. local cheese), process (e.g. organic) and place (e.g. Stilton) (Ilbery *et al.*, 2004b). The turn to quality has led to a growth in quality assurance schemes (QAS). However, the recent proliferation of QAS has been dominated by supermarkets, thus further blurring the distinction between 'conventional' and 'alternative' food supply systems (Morris and Young, 2000).

4. *Defensive localism* While it is beyond the scope of this paper to enter into the many debates surrounding the definition of terms such as 'local', 'locality' and 'regional', it is clear that the literature is divided about whether 'quality' or 'local' is more important in local food production (Winter, 2003; Weatherell *et al.*, 2003). Indeed, Winter (2003) uses the concept of defensive localism to suggest that the turn to local is more important than a turn to quality based on, for example, organic or ecological principles. Thus the turn to local is not just about alternative food systems; instead, it can cover different forms of agriculture (including conventional) and a variety of consumer motivations. In his survey of consumers in different rural districts of England and Wales, Winter (2003) found a clear preference for the purchase of local food over organic food. This reflected support for local farmers and the local economy, irrespective of whether the food was produced using conventional or alternative methods. Yet again, this warns against the danger of using binary opposites such as conventional and alternative.

It should be clear from the first two sections of this chapter that the processes affecting shorter ('local') food chains are complex and that there is considerable blurring between conventional and alternative food production systems. The review also suggests that a simple 'alternative geography of food' does not exist. However, while the literature on local and alternative foods is increasing quite significantly, there remains a lack of empirical evidence that moves beyond the casuistic of specific examples, especially in more marginal farming regions where it is suggested that alternative food supply systems might flourish. It is to these empirical dimensions that the rest of the chapter now turns.

Study region and methodology

The Scottish-English border study region comprises the county of Northumberland and the Borders region of Scotland. As shown by Figure 6.1 it is characterized by a dispersed, low-density population. A large proportion of the region is designated as Objective 2 by the European Union. Under the Agenda 2000 guidelines, Objective 2 status is awarded to lagging regions that are facing structural difficulties in an attempt to support economic and social conversion. The region is predominantly rural in character, with much of it renowned for its variety of landscapes and unspoilt environment, wildlife and natural habitat; large parts have poor quality agricultural land. Not surprisingly, therefore, the region is typified by upland, low-intensity livestock farming, notably beef and lamb, with some dairy and cereal farming (Ilbery *et al.*, 2004a). Farms in the region are, on average, large by national standards. For example, the mean farm size for Northumberland is almost three times that of the English average of 64 ha. This type of farm structure is a reflection of both the area's poor quality land and a certain type of land tenure, which is of landed estates with tenanted farms (Ward and Lowe, 1999).

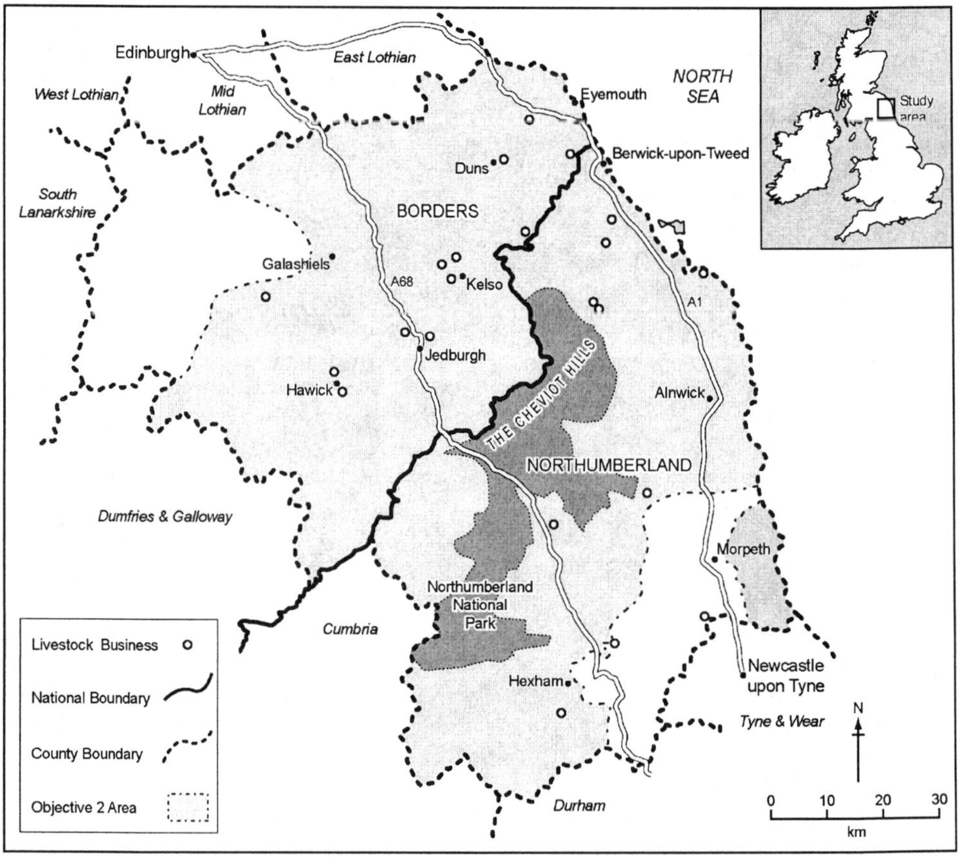

Fig. 6.1 Livestock surveys in the Scottish-English borders.

Based on Storper's (1997) four 'worlds of production, (standardized-specialized-generic-dedicated) the SUPPLIERS project focused on food products that are specialized (made with restricted technology and know how) and dedicated (niche markets that are locally embedded). Analysis of the two areas food and drink directories revealed SMEs producing 'speciality' food products. This information, together with regional reports, internet searches and advice from a Consultation Panel of key agency 'actors' specially convened for the project, led to the selection of three product types for detailed study: livestock products; fish and fish products; and bakery, confectionery and preserves. 'Livestock products' form the basis of this chapter and is an umbrella term encompassing dairy produce (e.g. milk, eggs, cheese, ice cream and yoghurt) and meat and meat-based products (e.g. beef, lamb, poultry, pork, game and pates).

Primary data were collected through in-depth interviews with SME food producers in the Scottish-English borders region. In total, 43 interviews were conducted, of which 20 related to livestock products. A further two on-farm butchers surveyed at a later stage in the project have also been included. Producers were selected through non-random 'purposive' sampling, a method which encourages a case to be selected because it illustrates a feature that is known, or assumed, to be relevant to the research (Robson, 1998); much is thus based on the researcher's judgement. Nevertheless, a range of criteria was considered when selecting producers; these varied from the type of product, size of business and ownership to production methods, the nature of upstream and downstream linkages, and the type and location of markets. In some cases, producers were asked to identify further producers that they considered of relevance to the study, thus adding a degree of 'snowball' sampling to selection procedures.

The interviews were conducted between November 2001 and February 2002 at a time and date convenient to the interviewee; in a majority of cases they took place at the SMEs location. Both structured and unstructured questions were used in a schedule comprising sections on business/entrepreneur characteristics; information and communication technologies; the role of institutions; food SMEs and rural development; and supply chain features. The latter section was particularly innovative in that it asked each producer to draw a supply chain diagram for their business, including both upstream suppliers and downstream links. This formed the platform for a series of questions about upstream and downstream features, as well as the chain as a whole. Most interviews lasted for at least one hour, reflecting the interviewees' interest in the project and especially the open-ended questions. With permission, all interviews were tape-recorded; they were later selectively transcribed and annotated. The transcriptions were then analysed and theme-based information from the transcriptions coded, ensuring comprehensive appraisal and comparisons to be drawn. In essence, this involved reading and re-reading the scripts to develop a 'feel' for the business, the products and their supply chains (SCs); the SC diagrams were particularly instructive here.

The following sections put 'empirical flesh' on the conceptual bones mapped out above. After briefly introducing the different livestock products, business and entrepreneurial biographies, the chapter explores the 'alternative worlds' of individual specialist livestock producers. The authors purposely avoid using 'generic' SC conflations (e.g. *a* red meat or *a* cheese SC) notable in some conventional agro-economic accounts (see, e.g. Stanford *et al.*'s. (1999) description of the Canadian meat supply chain). Instead, after a few general points, individual diagrams will be used to show heterogeneity and differentiation between individual specialist livestock producers.

Specialist livestock producers

Livestock products

A number of meat and dairy products are included under this umbrella term. Table 6.2 provides a 'flavour' of the products and associated characteristics identified by producers. In essence, most are part of the study region's 'quality food scene', appearing in local food guides, tourist pamphlets and so on. In Northumberland, six local surveyed livestock businesses are members of the regional food group (Northumbria Larder), recently formed in 2001 in response to the foot and mouth crisis.

Ilbery and Kneafsey (2000) argue that 'quality' products can be defined in terms of:

- association (the link with the place of origin)
- specification (raw materials, recipes or production methods)
- attraction (the taste, texture, flavour, appearance or presentation) and
- certification (a 'quality' mark or label).

Table 6.2 Livestock products.

Product range	Characteristics identified by producers
Cheese, yoghurt, cream, butter, ice cream, eggs, poultry, ostrich, game, beef, lamb, pork, haggis, meat paste	Free-range, rare/traditional breeds, organic, local products, traceable, good animal welfare standards, unique or novel products, natural ingredients, traditional production methods, handmade, vegetarian, healthy food

This definition can equally be applied here. Overall, establishing association, through traceability, is important for most, particularly meat livestock producers (LPs). As one red-meat producer explained: 'This farm prides itself on producing assured grass-fed lamb and beef. Traceability is very important...the animals are reared to very high welfare standards...this is reflected in the great taste of the beef and lamb that is produced' (LP08, *beef farmer-butcher*). Attraction – notably in terms of flavour, taste and strength – is also important, especially for 'specialist' dairy producers. One of the key ingredients here is milk (i.e. specification). In terms of certification, a few livestock producers have been awarded quality labels from recognized accreditation bodies (e.g. the Soil Association, Organic Farmers and Growers, Scottish Beef and Lamb Quality certificate). Some have also won coveted awards for their products (e.g. Scottish Eggs Quality Award, Ice Cream Alliance Competition, Nantwich International Cheese Show, National Festival of Meat). These awards, in some instances, are very important for product identity.

For many livestock enterprises, a key characteristic is to market and deliver the product directly to the public. In other words, the short *nature of the food chain* is an essential embodiment of the product's attraction. Producers often supply more than one product and more than one variation of a product, with different value added potentials. During interviews, some businesses suggested that their products were in fact not 'specialist'. Businesses also supply product(s) - some specialist and others not - for both alternative and conventional markets. Nevertheless, both types of product may still be sold to a very local market.

The business

The surveys included different business types as well as product differences. These can be summarized as: dairy and egg producers; on-farm slaughter/processor (poultry, ostrich); on-farm butchers (red meat); farmer-butcher; rare breed producers and butchers. A further distinction needs to be made between '*primary*' and '*secondary*' producers. The former comprise farmers that grow, produce and add value to their product either on site or at other premises. The latter are butchery enterprises that source primary raw material to produce foods such as a local haggis or meat paste. Crucially, 90% of surveyed businesses are 'primary' producers.

The majority (86%) are family run 'micro' enterprises. While a few employ as many as ten full-time personnel, others support only the entrepreneur and possibly one other. The outbreak of Foot and Mouth Disease (FMD) in 2001 had an immediate and damaging impact on most enterprises. In a few cases, LPs were forced to close down their business completely. Most LPs (81%) surveyed started their 'alternative enterprise' between 1991 and 2001. When questioned about the motivations influencing business formation, most indicated more than one factor as shown in Table 6.3. Many indicated that they established the enterprize as a means of diversifying away from the commodity-based market and subsidy treadmill that has become the hallmark of the conventional farming model. The following quotes typify this reaction:

> 'We are primarily an arable farm...The egg business was originally designed to utilise labour and act as a form of diversification. We wanted our business to be less reliant on subsidies and to be able to stand on its own two feet.' (LP20, *free-range egg producer)*

> 'it was probably set up for the wrong reasons...At the time, it was obvious that if we continued trading on world commodity prices it would not sustain a living on a small hill farm here in the uplands.' (LP01, *organic hill farmer/on-farm butcher*)

Table 6.3 Motivations for business formation.

Sector	Business motivations
Livestock products	• Diversification away from 'mainstream' agriculture (12) • Love of product (7) • 'Add value' to primary product(s) (6) • Control the food chain (5) • Produce 'quality' product (5) • Direct sales (5) • Promote local produce (4) • 'Lifestylers' (3) • Provide extra jobs/income (2) • Evolved over time (2) • Entrepreneurial spirit (2) • Part of career/personal development (2) • Part of family tradition • Concern about food safety

Note: The number in brackets indicates the number of mentions

Interestingly, many surveyed businesses started their 'alternative' enterprize usually on the back of European Union institutional grants, which are actually little different to previous subsidy and headage payment architectures (Lowe *et al.*, 2002). Allied to the above, LPs also talked about a desire to control and manage *their* SC, to find ways of 'adding value' to *their* products and to source and supply *their* own ('alternative') food outlets. As two meat producers now practising various forms of direct marketing explained: 'The initial thing was to add value to our own stock and to market it ourselves so that we had control of the end product. That was the philosophy behind starting the business' (LP02, *pig, beef and sheep farmer/processor/farm shop*). For some, particularly dairy producers, a 'passion' or 'interest' in the product is also essential:

> 'It was mainly interest to be perfectly honest, interest in a quality product... it was a labour of love to begin with...we worked on the product for 4 or 5 years before we started selling it. It's not for pure profit orientation, although it does help!' (LP17, *cheese producer*)

The entrepreneur

A high percentage (76%) of entrepreneurs originated from the Scottish-English borders region. Of the 24% from outside the region, four characteristics are worthy of note. First, most were new entrants into farming itself, with little background in 'conventional' farming practices. Second, the group adopted 'alternative' farm enterprises from the outset (e.g. organic, rare or alternative breeds). Third, the business was often 'a new venture' and for some a 'hobby enterprise'. Fourth, the entrepreneurial spirit of the group seems important; with many being actively involved in promoting quality food production and a local sustainable food system ethos (Sage, 2003). The modal age of business owners in the LP is 45-54. Most entrepreneurs are either in this age bracket or the 35-44 age bracket. Overall, a high proportion of entrepreneurs are male. However, many enterprises involve man and wife, or brother and sister, teams. As one female producer of ice cream explained: 'there are 3 brothers and a sister working on the farm here...it was an opportunity for us [husband and wife] to set up a separate business' (LP16, *ice cream producer*). A high percentage of entrepreneurs (81%) have completed some kind of business related training, e.g. business development/management, quality standards, marketing. Training is usually (part-)funded by national (e.g. Institute of Directors), sector specific (e.g. Milk Marque Product Development Centre) or local development agencies (e.g. Northumberland Business Link, Scottish enterprize Borders). Interestingly, a few producers refused to take part in training courses, citing them as inappropriate to their business. This attitude often reflects a wider resistance to develop the 'associational realm' between the business and available institutional support (Renting *et al.*, 2003). Within the sample, it is possible to identify different 'forms' of business strategy:

1. *Progressive*: the entrepreneur is highly motivated towards continuous growth and a movement upwards in terms of size, output and profit margins. Moreover, the owner is highly entrepreneurial, constantly looking to introduce new ideas, information and products.
2. *Planned*: the entrepreneur develops the business over a more gradual time scale, often with consolidation periods to assess development and progress.
3. *Passive*: the entrepreneur is happy to play a fairly submissive role in terms of business development and growth of the local food sector. In this instance, no clear strategy for the future is identifiable.
4. *Preservationist*: the entrepreneur does not want to grow the business further. Rather, the aim is to 'keep things as they are'. Most cottage industries fall under this bracket, often resisting growth because of additional development costs.

Thus, as one moves through the typology, the entrepreneur becomes less instrumental and market sensitive and more 'lifestyle' orientated. However, the latter strategies still require profitability and, for some, this may be the very factor constraining growth. Furthermore, these strategies would appear to influence SC confluences and socio-spatial relations, an issue that the chapter now addresses.

'Alternative worlds' of food production: exploring specialist livestock supply chains in the Scottish-English borders

The chapter adopts a 'whole chain' approach. What this means is that the following analysis will unfold in two stages: upstream and downstream. The analysis documents business inputs up to and including the retailer, but not the final consumer. To highlight key trends, the sector has been divided into two parts: the first introduces dairy, egg and poultry products (8 in total) and the second meat-based products (14)[1]. Meat businesses make up nearly two-thirds of the surveys and so dominate discussion.

Upstream

Figure 6.2 shows four SC diagrams, three for dairy and egg products and one for organic poultry. The business in Fig. 6.2a sources rennet and other ingredients from a supplier in Somerset. Milk is the largest input for both the cheese and ice cream producers (Fig. 6.2a and 6.2b). The owner in Fig. 2b sources honey and jams locally: 'The jams and meringues are homemade and it does make a difference to the ice cream, we wouldn't get the quality from more national suppliers I don't think.' (LP16, *ice cream producer*). For the egg producer, input supplies are fairly minimal, but again they are usually sourced from outside the region (Fig. 6.2c). For the organic poultry businesss (Fig. 6.2d), upstream links have developed as environmental circumstances have changed. Organic chickens are sourced from a local hatchery in the Borders. For reasons of quality, this local supply link – brought about because of foot and mouth - is not the preferred option. Meanwhile, organic feed is sourced from Yorkshire because of limited local supply.

Most dairy producers source milk direct from their own farms. However, other key ingredients are usually from non-local suppliers. This often involves using whichever node is necessary. For example, one specialist cheese producer sources rennet and other ingredients direct from The Netherlands. Starter culture is from a French-based company, but is bought from a UK intermediary. As the producer herself explained: 'there are very few people in Britain still doing it...For the cheeses we are making we have to go where the technology is. Some of the companies are so big that you cannot buy orders direct from them, you have to go through someone else' (LP17, *cheese producer*). For secondary inputs (e.g. cartons, boxes, packaging, clothing), dairy producers use a mix of local, regional (e.g. Newcastle, Edinburgh) and national (e.g. Middlesex, Northamptonshire) suppliers. Relations are largely verbal and price dominates as the key factor for selecting primary and secondary inputs. As one cheese producer alluded: 'The key criteria is price...I source my labels from Ireland because they are cheaper' (LP14, *cheese producer*).

The minimization of input supplies is also important for most primary meat businesses ranging from ostrich through to pork, beef and lamb, as shown in Fig. 6.3. Surveyed secondary producers source meat through local livestock agents. In all cases, relations are largely informal and verbally based. For most LPs, sourcing is determined by price: 'it's all to do with price, that's just the economic reality of the whole thing' (LP20, *butcher/Haggis producer*). Many red meat businesses have implemented some system of traceability in response to increasing European Union and National legislation imposed by the UK's Food Standard Agency concerning health, hygiene and food safety standards (Marsden *et al.*, 2000b). These regulations present a significant barrier for alternative livestock producers surveyed. Thus paradoxically, the watchdog designed to bare its teeth at the industrial food chain is here perceived more as a threat than opportunity towards improving local sustainable food production.

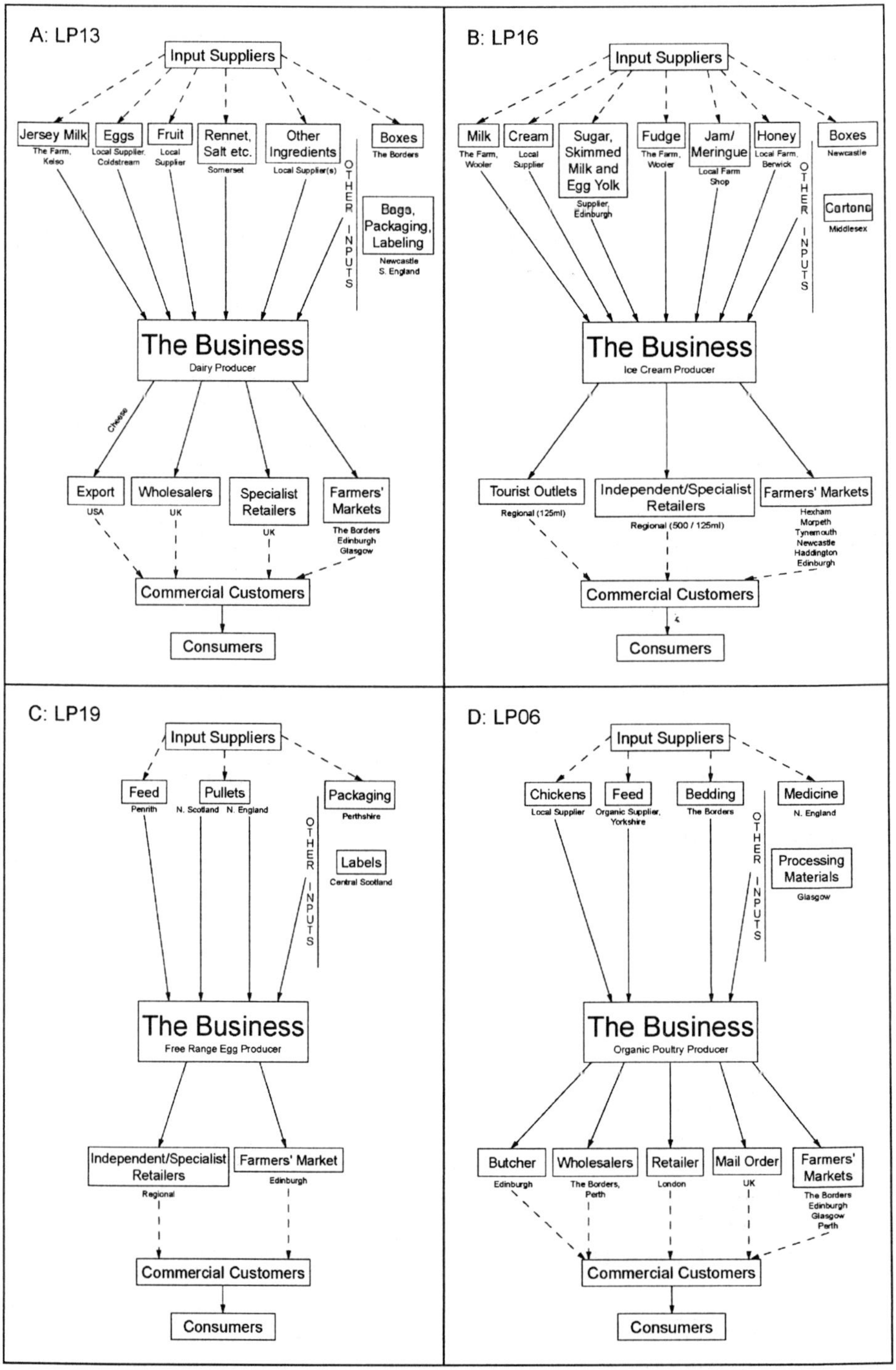

Fig. 6.2 SC diagrams for livestock products (dairy, eggs and poultry).

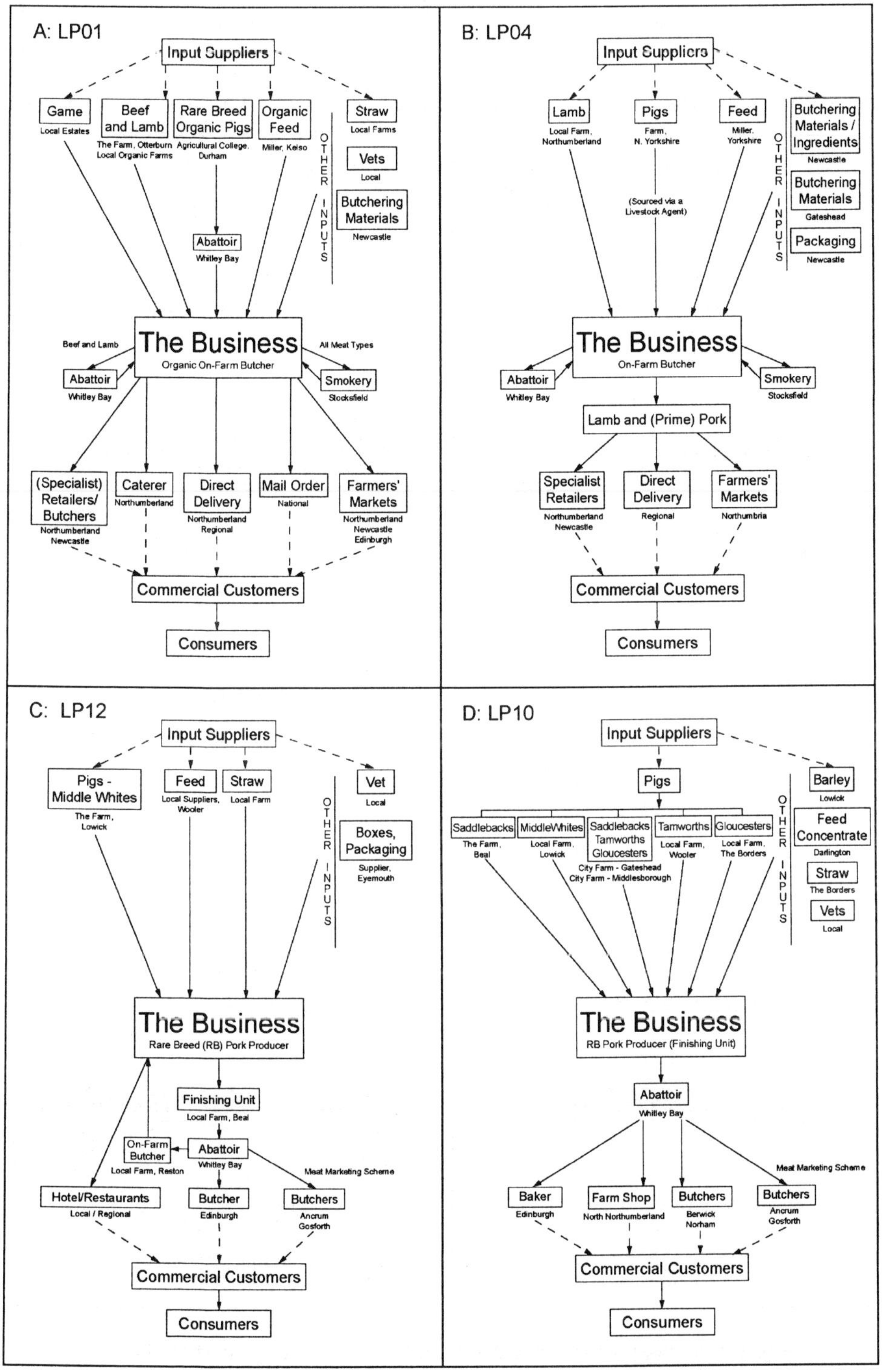

Fig. 6.3 SC diagrams for livestock products (meat).

Figure 6.3 shows two on-farm butchers and two rare breed SC examples. Where appropriate, reference will be made to other meat-based examples. In Fig. 6.3a, the principal raw material (organic beef and lamb) is produced on the farm. The breeds (Blackface sheep, Aberdeen Angus, Galloway-cross cattle) are native to the region and farmed in a traditionally extensive manner. Local branding and traceablity are used to promote and differentiate the meat products. The company's promotional material is also carefully crafted to target consumers looking for 'natural', 'wholesome', 'organic' meat: 'the product has always been around...the secret is how we market it, that is the key to our success' (LP01, *organic hill farmer/on-farm butcher*). As the business has developed, beef and lamb have been sourced from other local organic farms. Most recently, rare breed pork has been sourced from an organic farm in Durham (Fig. 6.3a).

In contrast, the other on-farm butchery in Fig. 6.3b has moved away from producing their own stock to sourcing all products from outside suppliers (pork and lamb). The reasons for the SC shift are twofold: first, the entrepreneur could source stock of the same quality; and second, the family wanted to devote time to butchering. Weaned pigs (the main input) come from a pig producer in North Yorkshire. The farmer sourced these through a livestock agency and agreements are contract based: 'the intensive nature of the industry demands this...most pig farms are large scale and linked up with big buyers' (LP04, *pig farmer/on-farm butcher*). The pig breeds are commercial and marketed on the grounds of high animal welfare standards (e.g. small herds, kept on straw, no use of antibiotics in feed). The business does not source organic or rare breed pigs (i.e. more niche market) for three reasons. First, from a farming viewpoint, the land is unsuitable. Second, from a marketing viewpoint, the local consumer would not support such a product. Third, from a supply chain viewpoint, some of the pigs are sent down conventional pork SCs and rare breeds are unsuitable because of their high fat content.

Like this example, three other on-farm butchers also sell 'commercial' products (e.g. Texel sheep) direct to the consumer. As one farmer explained: 'The breed itself isn't local...The process is the important thing' (LP03, *sheep farmer/on-farm butcher*). Thus quality here is maintained in terms of how the animal is produced (e.g. feed), accredited (e.g. FABBL, Countryside Stewardship Scheme), graded (e.g. Meat and Livestock Commission's EUROP system) and processed (e.g. pork is traditionally cured, beef and lamb are hung for an extensive period). Another on-farm butcher/farm shop explained that a local wholesaler sometimes supplies the business with meat. What distinguishes all these different on-farm butchery businesses is the 'shortened' nature of their SCs; they are further differentiated by their different upstream links, with many using conventional suppliers. The rare breed pork producers in Fig. 6.3c and 6.3d show two different upstream scenarios. In the first, inputs are minimal and predominantly local. This is similar to other rare breed pork producers in the survey.

In the second (d), the upstream part of the chain is more complex. This is because the business operates as a producer and finishing unit for the Rare Breeds Survival Trust (RBST) meat marketing scheme[2]. Consequently, the business sources pigs from other local/regional rare breed pork producers (including the business in Figure 6.3c). This 'producer network' was established to provide a more consistent end product; as the farmer himself explained:

> 'The idea was to bring the pigs into the one finishing unit. That guarantees a more consistent and reliable product for the butchers...The contacts I have with my pig suppliers are very close. It's very personal...they just want somewhere decent for the pigs to come' (LP10, *rare breed pork producer/finishing unit*)

From this analysis of upstream supply links, one can begin to draw out two or three key points. First, in terms of commonality, most upstream livestock links are largely informal and verbally based. Moreover, traceability is an important upstream component, particularly for meat producers. Second, a number of businesses use non-local, conventional supply links with price the governing SC relation. For these isolated micro businesses, with limited economies of scale, this usually involves using second or third level (intermediate) suppliers. The desired quality of inputs (e.g. rennet, organic feed) and the lack of local availability are key to explaining why businesses source from outside the region. In dairy, it is notable that the main product and ingredients for processing (e.g. milk, jams, eggs) continue to be sourced locally.

Most red meat producers also want to minimize supplies. However, contrasting pictures emerge. So while the organic (Fig. 6.3a) and rare breed producers (Fig. 6.3c and d) have developed local alternative links, other on-farm and off-farm butchers source commercial product from conventional supply lines (e.g. Fig. 6.3b). This latter group differentiate themselves by the shortened nature of the chain (downstream from the farm) and they sell commercial products to match associated demand as shown in Box 6.1. Organic and rare breed producers also work together (see Fig. 6.3d) to provide an alternative product for a market that requires some form of central finishing. This central unit instills a degree of standardization, consistency and guaranteed volume.

Box 6.1 Organic poultry farming at the cutting edge

In Fig. 6.2d, the organic poultry producer/processor (LP06) has established her own on-farm abattoir and cutting plant. Set up as a on-farm diversification strategy, the business uses a number of different outlets to guarantee sales: '...it's organic and so you restrict yourself'. Initially, the business established outlets by directly approaching specialist butchers/delis to see if they were interested in the product. Interestingly, a noticeable split emerged between customers in The Borders who were *not* interested in organic poultry and in urban centres (e.g. Edinburgh, Glasgow) who were. For the farmer, the competitive advantage of the product lies in the fact that it is produced and processed on the farm:

> 'it was completely traceable...they were dealing with me, the producer. They got to know me and they began to trust me...If I say no, we don't feed GM, or yes, they [the poultry] are kept outside they believe me.'

This establishment of *trust* is crucially important. Farmers' Markets (FMs), particularly those in the large urban catchments, are now the main outlet for the organic poultry: 'they are *foodie FMs* in the city...there are lots of people with lots of money looking for local activity'. The FMs provide a direct link with the consumer and are used as a barometer for new product development: 'if we make a new value added product we try it at the FM...it's a good way of testing the water'. In an attempt to supply the home market, serious consideration is now also being given towards producing a 'free-range chicken' in response to local consumer demands in The Borders.

Downstream

Dairy, egg and poultry products (Fig. 6.2a-d) are sold through a variety of outlets (supermarkets, wholesalers, specialist retailers, delis, health shops, tourist venues, FMs and mail order) and distributed at different scales (local, regional, national and international). Overall, local and regional outlets are the most predominant, with most producers delivering product directly. Much of the ordering is verbally based. Final customers are usually sourced by the producers themselves (e.g. 'cold sale', mail shots, telesales). Contacts are also established through word of mouth as the reputation of the business develops. Local independents are especially important. One cheese business, for example, supplies a village shop with cheese and promotional leaflets and visitors are encouraged to take a leaflet to their local cheesemonger. Businesses also attend food exhibitions. Generally, these are regarded as not being particularly helpful. For most dairy and egg producers, establishing a good personal relationship with customers is critically important. As one cheese producer explained:

> 'It's all about personal contact. With specialist products, it's not just about liking the product; it's about them liking you. If they don't like you they won't stock your product. Regular contact, good marketing skills, personal touch, and honesty are crucial.' (LP17, *cheese producer*)

The relationships between the dairy producer and retailer are, for some, a two way process: 'outlets need our products, they need local food' (LP14 *cheese producer*). The business in question is commercially oriented and supplies leading supermarkets in the region. The organic poultry business meanwhile uses a range of short chains to sell their produce (Fig. 6.2d and Box 6.1). Farmers' Markets (FMs) are an important outlet for most dairy producers, but for different reasons. In Fig. 6.2a, the specialist dairy producer has been extremely proactive regarding FMs: 'I still sell to wholesalers and shops, but I prefer to sell through the FMs because I get the retail price...At the moment we do 7 a month' (LP13, *dairy producer*). Meanwhile, the ice cream producer in Fig. 6.2b adopts a more cautious approach: 'we are a small business so we can't attend them all...if we did we would risk losing the high level of quality that we have set ourselves' (LP16, *ice cream producer*). Despite successes, FMs are regarded by some as expensive and time-consuming. Others regard them more as a marketing tool:

> 'FMs can be costly, it's more about PR. People come to taste and to ask where they can buy it locally. You don't always do it for the profit, you do it to open up new markets.' (LP17, *cheese producer*)

A variety of different downstream features can also be identified for meat products. Most producers began their discussion of this stage of the supply chain bemoaning the processing sector in the region (abattoir, cutting and packing, and meat manufacturing) as insufficient and under represented. Red meat producers in the South East Borders have access to two local abattoirs – Galashiels and Whitley Bay. An ostrich farm in The Borders has its own abattoir and processing facility. The real value here is the ostrich skins which are sent to South Africa for tanning before being sold on the global market. Consequently, the ostrich meat is of little concern or value and most is sold to wholesalers in the Borders, Edinburgh and Glasgow, as well as a few local FMs. For most other meat producers, the first stage in the downstream part of the SC *must* be the abattoir. Often this is the 'weakest link' (LP01) in the SC. Another farm business (not included in Figure 6.3) is the only small red meat processor in the Borders; this takes contract work (e.g. meat cuts, sausages, burgers) for local producers involved in direct sales. This has enabled the business to develop a strong hand in the Borders alternative livestock sector, but not without problems: 'We try to be gentlemanly. We don't use it to our advantage...[in the past] we have been accused of sending out the wrong stuff to them [other local producers]. We have been accused of stealing meat from carcasses before they get it back. We have been accused of sending our stuff to them' (LP02, *pig, beef and sheep farmers/processor/farm shop*).

Box 6.2: Traceability: selling the link?

YES
Organic beef and lamb
'The whole point about a differentiated product is that it is 100% squeaky clean, it has to be better than the supermarket. Our customers ask for it' (LP01, *organic hill farmer/on-farm butcher*).

Quality assured pork and lamb
'Traceability was our main selling point...People want to know where food comes from...Before FMD, people often came to see the place, they often came to see the animals and the prep room' (*LP05, former farmer/on-farm butcher*).

YES and NO
Cheese
'Some cheese wholesalers want complete traceability, but others are totally unaware of the problems that you could have. It's my job to tell them' (LP13, *cheese producer*).

NO?
Rare breed pork
'We sell direct and there are very few people that specifically ask for it [traceability]...They are buying on the basis of the product, they want the flavour and the taste quality...the fact that it is direct from the farmer is a big selling point...It's an add on, but it's not the primary. The primary is the quality and the direct buying. They enjoy the chat!' (*LP11, rare breed producer*).

Most red meat producers supply a very local/regional custom base. On the whole, retailer/customer demand for traceability – in response to increasing food scares and disease risks – is high, but not always 'formally' essential as shown in Box 6.2. Many producers reported a relatively stable supply and customer base, with some refusing to supply to certain outlets (e.g. hotel chains, wholesalers and large retailers). For the five on-farm butcheries, outlets for the product are mostly alternative/independent SCs (e.g. farm shops, specialist food retailers, local butchers). FMs provide a particularly important output and are used by some producers to shift surplus stock:

> 'We use them to help balance our stock. We, for instance, hardly ever take fillet steak or sirloin to FMs. We always take 2 packets of each so we can say that we have run out. We don't want to make them feel unimportant to us. We encourage them to try and put orders in. We take orders for them if they do want something like fillet steak or sirloin. We only take what we want to take and we find it totally different from the shop. At the shop they come and ask for what they want. At the FM they will usually take what is there.' (LP02, *pig, beef and sheep farmers/processor/farm shop*)

Indeed, some farmers are also looking to establish 'stable alternatives', notably through their own farm shops. Two on-farm businesses also still sell a high percentage of their produce through conventional marketing channels and rely on this these links to off-load the product that cannot be sold direct.

Turning to the downstream SC elements in Fig. 6.3, organic beef and lamb are sent from the farm in Northumberland to the abattoir (organically certifed – Fig. 6.3a). The meat is then collected and brought back to the on-farm processing unit. Here it is processed, alongside rare breed organic pork and game. As the farmer noted: 'we are trying to add value and control as much of the SC as we can' (LP01, *organic hill farmer/on-farm butcher*). Produce is also sent to a local smokery for processing. Like the organic poultry example (Fig. 6.2d and Box 6.1), the business uses a range of different short chains, which include face-to-face, spatially proximate and spatially extended chains (Marsden *et al.*, 2000a; Renting *et al.*, 2003). The latter takes the form of an impressive e-commerce website, set up using funding from One NorthEast (the Regional Development Agency). The success of the business is due largely to the drive and persistence of the entrepreneur and his ability to adapt, diversify and reinvent the product, tapping into available funding pots and promotional campaigns such as the Countryside Agency's 'Eat the View' scheme. Much of the produce is sold to local/regional retail outlets (Fig. 6.3a). The business secures many of the outlets itself and actively promotes the products through the local media, promotional material (e.g. leaflets, company video), food shows, the Internet and its on-going outlets. FMs have been especially important in this respect:

> 'I often get invited to FMs because I am an organic producer...they are a great marketing concept...we have leaflets and a video of the farm and the whole process...customers often buy bulk orders after they have tried samples at the markets.' (LP01, *organic hill farmer/on-farm butcher*)

For the business depicted in Fig. 6.3b, weaners are finished on the farm, sent to the local abattoir and returned to the farm and processed (e.g. cured) for sale. Again FMs act as an important 'testing ground' (LP04, *pig farmer/on-farm butcher*).

For the rare breed pork producers (Figs 6.3c and 6.3d), two different, but connected downstream features emerge. For the finishing unit (d), the principal outlet is wholesale via specialist butchers accredited to the Traditional Breeds Meat Marketing Scheme. At the beginning of each week, the farmer phones the different butchers, collects the orders and delivers the pigs to the abattoir along with the necessary paperwork (e.g. breed details, dates of birth, parentage). The abattoir then kills and processes the meat and delivers to the butchers. A copy of the paperwork is sent to the RBST and a certificate is sent to the butcher with rare breed accreditation. Cooperation with other rare breed producers is important for the business: '...we work for each other...our product is great because we have control of the SC...our real problem is transporting the animals and accessing the abattoirs' (LP10, *rare breed producer/finishing unit*). While some butchers have been assigned to the unit through the meat marketing scheme, others have been sourced directly by the farmer. In contrast to this scenario, other rare breed pork producers have opted for direct sale (e.g. door to door deliveries and FMs). The other rare breed producer in Fig. 6.3c works in partnership with the finishing unit. Consequently, the unit finishes and delivers some pigs to the abattoir, while the other producer markets the pork and invoices the final customer. The producer has also established a relationship with the local processing unit to produce sausage. These are sold direct to local country hotels and restaurants. Interestingly, this facet of the chain developed after the producer won an award for his sausages:

> 'When I won the award it helped to differentiate me, there was accreditation to prove the quality of my product...it gives you a story to work with...The award has given me access to publicity... I can play the local connection.' (LP12, *rare breed pork producer*)

Like another pork producer, agreements are verbally based and relations with suppliers are generally good. Some meat producers define relations in terms of business partnerships: 'It's a mutual thing. I need the outlets, but they need me to differentiate themselves' (LP05, *former pig farmer/on-farm butcher*). Thus for some producers '...the social and cultural embedding of relations between firms usually depends, not so much on trust *per se*, but on overlaps in their self-interest' (Sayer, 2001, pp.698-699).

Again, one can start to identify important SC issues relating to downstream movements for livestock businesses. First, most specialist livestock producers supply largely local/regional outlets. Second, most downstream supply links and contracts are verbally based and informal, with contact usually initiated by the producer; often the drive and personality of the entrepreneur dictate SC access. Third, many alternative producers must 'dip into' more conventional nodes. In the dairy sector, this may involve using transport couriers or finding wholesalers that deal with small-scale artisan producers (e.g. *Neals Yard Dairy*, London). In the meat sector, the most obvious link is the abattoir. For on-farm butchers, the key is to get the carcass back from the (usually small sized) abattoir to add value through producing meat cuts, cures, meat-based products and so on (i.e. 'economies of scope'). Product that is sent for smoking usually goes to small companies who themselves operate in the local Alternative Farming Enterprises (AFE). Lastly, most businesses supply to different types of outlets to disperse risk/vulnerability; accesssing short food chains is the order of the day. FMs are particularly important here, but for often quite different reasons; some enterprises use them as their main source of sale, while others regard them as a marketing venture or as a 'dumping ground' for surplus product. Short, alternative SCs are by no means explicit. Thus while a number of businesses sell some product through alternative chains (e.g. farm shops, FMs), they continue to rely heavily on conventional outlets to sell high proportions of stock. Equally, one dairy business is taking its products to supermarket shelves, outside newly established local food sourcing initiatives. As the ostrich farm example also demonstrates, money talks, with their alternative ostrich meat playing second fiddle to the more lucrative global ostrich skin trade.

Conclusions: 'alternative' or 'conventional'?

This chapter began by sketching the parameters of an alternative geography of food, and noted a significant blurring between alternative and conventional food systems. The rest of the chapter based on 22 semi-structured interviews, with individual SC diagrams as the formative methodological tool, has shed important light on the emergence and evolving nature of AFE in lagging regions. A key finding has been that many of the businesses operate very individual SC constellations, with produce regularly sold through Short Food Supply Chains (SFSCs) (Morris and Buller, 2003; Ilbery and Maye, 2004). The supply chain analysis also demonstrated the heterogeneity of producer initiatives developing under the headings of 'local', 'alternative' and/or 'quality food'. Moreover, the research has argued that 'soft' conceptualizations of AFE, notably notions of 'trust', 'regard' and 'social embeddedness' are appropriate, but need to be weighted against a 'hard' (agro-)economic context (Sayer, 2001).

Many businesses in this study have therefore been driven to their present alternative state as a strategy for survival and growth. Thus LPs have been forced to act 'instrumentally' to try to control their SC and add value wherever possible (Hinrichs, 2000). It is this context that has forced many entrepreneurs to shorten their SC and to change their marketing style. Furthermore, in the case of on-farm butchers, one observes a highly competitive edge between individual producers, an edge that often appears when the tape recorder is 'turned off'. Most entrepreneurs are local, although new entrants have been important in the study region's alternative food economy. Earlier in this chapter, important distinctions were drawn between the more commercially sensitive, 'progressive' and 'planned' alternative producers and the more 'passive' and 'preservationist'. The evidence suggests that the strategy and social capital of the entrepreneur can determine the nature of the SC (e.g. Fig. 6.3a, progressive).

What is also common is that many specialist producers are forced to 'dip in and out' of different conventional nodes, links that can sometimes take their business SC outside the region. Upstream, many businesses access whatever nodes they can, with price and quality the over-riding criteria. By looking upstream to consider with what the product is made it becomes clear that several 'alternative' producers are either obliged or choose to make use of conventional channels. Downstream, the most obvious intermediate example is the abattoir. Admittedly, these abattoirs are small in size and better set up to deal with small producers than larger operations that deal with large commodity volumes for multiple customers. However, even in the only two small abattoirs in the region, producers have on numerous occasions found it very difficult to process 'alternative' (e.g. organic cattle, Highland cattle and Tamworth pigs) products. Some livestock businesses are supplying alternative SCs, but also use conventional links for surplus product. Equally, some dairy producers are happy to supply their product to 'conventional' wholesale and retail links. Another dairy producer sells ice cream made from powered milk through local ice cream vans and independents. These findings certainly call into question the separate system dualisms presented in Table 6.1 and the rather stylized and utopian localization models coveted by environmental activists and academics such as Hines (2000) and Pretty (1998). Moreover, the supply chain diagrams highlight how the starting point of a SFSC is clearly not the point of production, but rather a series of upstream supply links. In this chapter, one has seen how 'alternative worlds' of livestock production are 'hybrid spaces', contingent upon 'conventional' links if they are to survive and do business (also see Ilbery and Maye, 2004).

In this lagging region at least, specialist livestock production of this kind is extremely small compared to the overall commodity-based system. The evidence of a turn to more 'alternative' geographies of quality food production in marginal regions, as articulated by some rural scholars (Parrott *et al.*, 2002), is piecemeal, in this region at least. The foot and mouth crisis provided an important litmus test in this respect. When questioned about the long-term effects the crisis might have on the local farming culture, many respondents noted that most farmers in the region had restocked as before, content to supply the mainstream commodity market. So, cases discussed in the chapter may stand as unique exemplars rather than as broad representations of agrarian change for the region. Consumer demand is also important, with some respondents noting fairly limited demand for local products, with particular reference made to the stagnant growth of some local FMs. This reportedly saturated local food market, a consequence of limited consumer responsiveness, deters other producers from changing production in a bid to add value and sell direct. Add to the equation the regulators' cookbook of do's and don'ts and initially attractive specialist short food supply chain proposals begin to taste bittersweet and for some even sour.

This chapter has focused on the producer end of the SC. Future work on the project will uncover downstream links with intermediaries (e.g. abattoirs, wholesalers) and commercial customers (e.g. retailers, caterers) who process and buy livestock products. Discussions have also concentrated on the food supply chain and case studies will explore the wider networks that make whole an individual producer business. This work will substantiate the producer accounts introduced here.

Acknowledgements

This article derives from the EU funded project: 'Supply chains linking food SMEs in Europe's lagging rural regions' (SUPPLIERS, QLK5-CT-2000-00841). Collaborating laboratories are: SAC, Aberdeen, UK (Co-ordinator); Coventry University, UK; University of Wales, Aberystwyth, UK; Teagasc, Dublin, Ireland; ENITA Clermont-Ferrand, France; University of Patras, Greece; SIRRT, University of Helsinki, Finland; and the Agricultural University of Krakow, Poland.

Notes

1. One poultry business surveyed at the producer stage acted predominantly as a wholesaler and has been excluded from the supply chain analysis.
2. The scheme was set up to establish an alternative chain of distribution for rare breed producers to sell direct to independent butchers outside the supermarket loop.

References

Allen, P., FitzSimmons, M., Goodman, M. and Warner, K. (2003) Shifting plates in the agrifood landscape: the tectonics of alternative agrifood initiatives in California. *Journal of Rural Studies* 19(1), 61-75.

Cone, C. and Myhre, A. (2000) Community Supported Agriculture: a sustainable alternative to industrial agriculture? *Human Organisation* 59, 187-197.

Festing, H. (1997) The potential for direct marketing by small farms in the UK. *Farm Management* 9, 409-21.

Goodman, D. (2003) The quality 'turn' and alternative food practices: reflections and agenda. *Journal of Rural Studies* 19(1), 1-7.

Hines, C. (2000) *Localization: a global manifesto*. Earthscan, London.

Hinrichs, C. (2000) Embeddedness and local food systems: notes on two types of direct agricultural market. *Journal of Rural Studies* 16(3), 295-303.

Hinrichs, C. (2003) The practice and politics of food system localization. *Journal of Rural Studies* 19(1), 33-45.

Ilbery, B. and Kneafsey, M. (1998) Product and place: promoting quality products and services in the lagging rural regions of the European Union. *European Urban and Regional Studies* 5, 329-341.

Ilbery, B. and Kneafsey, M. (2000) Producer constructions of quality in regional speciality food production: a case study from south west England. *Journal of Rural Studies* 16(2), 217-230.

Ilbery, B. and Maye, D. (2004) Food supply chains and sustainability: evidence from specialist food producers in the Scottish/English borders. *Land Use Policy* (in press).

Ilbery, B., Maye, D., Kneafsey, M., Jenkins, T. and Walkley, C. (2004a) Forecasting food supply chain developments in lagging rural regions: evidence from the UK. *Journal of Rural Studies* 20, 331–344.

Ilbery, B., Morris, C., Buller, H., Maye, D. and Kneafsey, M. (2004b) Product, process and place: an examination of food marketing and labelling schemes in Europe and North America. *European Urban and Regional Studies* (in press).

Krippner, G. (2001) The elusive market: embeddedness and the paradigm of economic sociology. *Theory and Society* 30, 775-810.

Lang, T. (1999) The complexities of globalisation: the UK as a case study of tensions within the food supply system and the challenge to food policy. *Agriculture and Human Values* 16, 169-185.

Lee, R. (2000) Shelter from the storm? Geographies of regard in worlds of horticultural consumption and production. *Geoforum* 31, 137-157.

Lowe, P., Buller, H. and Ward, N. (2002) Setting the next agenda? British and French approaches to the second pillar of the CAP. *Journal of Rural Studies* 18 (1), 1-17.

Marsden, T., Banks, J. and Bristow, G. (2000a) Food supply chain approaches: exploring their role in rural development. *Sociologia Ruralis* 40(4), 424-438.

Marsden, T., Flynn, A. and Harrison, M. (2000b) *Consuming interests: the social provision of foods.* UCL Press, London.

Morris, C. and Young, C. (2000) 'Seed to shelf', 'teat to table', 'barley to beer' and 'womb to tomb': discourses of food quality and quality assurance schemes in the UK. *Journal of Rural Studies* 16(1), 193-205.

Morris, C. and Buller, H. (2003) The local food sector: a preliminary assessment of its form and impact in Gloucestershire. *British Food Journal* 105, 559-566.

Murdoch, J., Marsden, T. and Banks, J. (2000) Quality, nature and embeddedness: some theoretical considerations in the context of the food sector. *Economic Geography* 76(2), 107-125.

Parrott, N., Wilson, N. and Murdoch, J. (2002) Spatializing quality: regional protection and the alternative geography of food. *European Urban and Regional Studies* 9, 241-261.

Pretty, J. (1998) *The living land: agriculture, food and community regeneration in rural Europe.* Earthscan, London.

Raynolds, L. (2000) Re-embedding global agriculture: the international organic and fair trade movements. *Journal of Agriculture and Human Values* 17, 297-309.

Raynolds, L. (2002) Consumer/producer links in Fair Trade coffee networks. *Sociologia Ruralis* 42(4), 404-424.

Robson, C. (1998) *Real world research.* Blackwell, London.

Renting, H., Marsden, T. and Banks, J. (2003) Understanding alternative food networks: exploring the role of short food supply chains in rural development. *Environment and Planning A* 35, 393-411.

Robinson, G. (2002) The alternative food economy: myths, realities and potential. Innovations in Rural Areas, Fourth Anglo-French Symposium on Rural Geography, University College Worcester.

Sage, C. (2003) Social embeddedness and relations of regard: alternative 'good food' networks in south-west Ireland. *Journal of Rural Studies* 19(1), 47-60.

Sassatelli, R. and Scott, A. (2001) Novel food, new markets and trust regimes: responses to the erosion of consumers' confidence in Austria, Italy and the UK. *European Societies* 3(2), 213-244.

Sayer, A. (2001) For a critical cultural political economy. *Antipode* 33(4), 687-708.

Stanford, K., Hobbs, J., Gilbert, M., Jones, S., Price, M., Klein, K. and Kerr, W. (1999) Lamb-buying preferences of Canadian abattoirs and producer marketing groups: implications for the Canadian supply chain. *Supply Chain Management* 4(2), 86-95.

Storper, M. (1997) *The regional world.* Guilford Press, London.

Ventura, F. and Milone, P. (2000) Theory and practice of multi-product farms: farm butcheries in Umbria. *Sociologia Ruralis* 44(4), 452-465.

Vos, T. (2000) Visions of the middle landscape: organic farming and the politics of nature. *Agriculture and Human Values* 17, 245-256.

Ward, N. and Lowe, P. (1999) *Agricultural change and rural development in Northumberland.* Centre for Rural Economy, University of Newcastle upon Tyne.

Weatherell, C., Tregear, A. and Allinson, J. (2003) In search of the concerned consumer: UK public perceptions of food, farming and buying local. *Journal of Rural Studies* 19(2), 233-244.

Whatmore, S. (1995) From farming to agribusiness: the global agro-food system. In: Johnston, R., Taylor, P. and Watts, M. (eds.) *Geographies of global change.* Blackwell, Oxford, pp.36-49.

Whatmore, S. (2002) *Hybrid geographies: natures cultures spaces.* Sage, London.

Whatmore, S. and Thorne, L. (1997) Nourishing networks: alternative geographies of food. In: Goodman, D. and Watts, M. (eds.) *Globalising food: agrarian questions and global restructuring.* Routledge, London, 287-304.

Whatmore, S., Stassart, P. and Renting, H. (2003) What's alternative about alternative food networks? *Environment and Planning A* 35, 389-391.

Winter, M. (2003) Embeddedness, the new food economy and defensive localism. *Journal of Rural Studies* 19 (1), 23-32.

Chapter 7

Agritourism and the selling of local food production, family and rural American traditions to maintain family farming heritage

Deborah Che, Gregory Veeck and Ann Veeck

Introduction

In advanced industrialized countries, small farmers have been challenged by increased global competition, falling commodity prices, and capital- and technology-intensive agricultural production. This chapter will first discuss agricultural restructuring, which pressures farmers to develop value-added alternatives such as agritourism in order to keep land in production and in the family. It will then discuss agrarian values and the appeal of rural landscapes which help drive the growth of agritourism. Then using focus groups of Michigan agritourism operators, the chapter will illustrate how agritourism successfully sells American tradition. First, agritourism sells the tradition of farming and local food production in its provision of much-needed education on where food comes from for those removed from a farming background. Secondly, agritourism sells traditional rural American landscapes and values. Agritourism uses its marketing of rural or small town virtues such as simplicity, service and a personal touch, to offset prices which for comparable items are higher than those at supermarkets. Successful agritourism operators sell these virtues as they assist customers and listen to individual family concerns and those about post-September 11 America. Finally, agritourism taps into family traditions, such as parents bringing children back to the same farms they visited as children and/or reliving their own farming background. According to its proponents, agritourism promotes family values and creates new family-oriented farm traditions by providing a safe, intergenerational, nature-based setting for family outings. Agritourism's success in selling tradition and constancy in an uncertain world generates both a financial cushion and a sympathetic urban constituency important in maintaining family farms in an era of agricultural restructuring.

Agritourism: a strategy for sustaining family farming?

> 'If it [agritourism] wasn't there, I would be gone. I would be in the city, Chicago, wherever, in an office job. I mean, the way the commercial ag-industry has gone over the last 10 or 15 years, there's no way I'd be able to stick around, so for me, I mean, it's [agritourism] been huge...It's

important to me to stay there, family-wise, and if you look around and you see what's happening to these family farms, I'm one of the very few people under 30 that are left in my area. You know, there's just a handful of people, and the other people are 50 to 60 years old. And what happens when they're gone? Who's going to take over? Or what's going to happen to the land?' (Sixth generation Michigan farmer)

Commodity producers in advanced industrialized countries have been impacted by agricultural restructuring. Price-cost squeezes driven by declining prices for agricultural commodities, a need to adopt capital-intensive technologies via intensive production on large-scale farms in order to be economically competitive, and global competition have disproportionately impacted smaller farms in Europe and the US (Kenney *et al.*, 1989). At the same time, public environmental concerns about industrial agricultural production and financial and political pressure to reduce agricultural subsidies have grown (Wilson, 1995).

Small Michigan farms which average 87 ha face these competitive problems, since capital and technology-intensive agricultural production favours larger farms further west. In addition, intensified global competition by foreign producers also has affected Michigan's commodity farmers. For example, Michigan apple production has been heavily impacted by foreign competition. US imports of Chinese apple juice concentrates, which increased by more than 1,200% between 1995 and 1998, have driven down prices for domestic processing apples. While the US International Trade Commission found that Chinese juice exports unfairly harmed the US apple industry and subsequently imposed a 52% tax on US juice makers using Chinese concentrate (Longcore, 2000), imports have compounded problems with severe weather, overproduction and market fluctuations. Michigan land in apple production has thus declined from 21,870 ha in 1997 to 19,035 ha in 2002 (Revere, 2002). China's joining the World Trade Organization (WTO) and the resulting Permanent Normal Trade Relations (PNTR) with China will likely intensify global competition. The Michigan Department of Agriculture (MDA) felt that PNTR with China would benefit Michigan agriculture by reducing Chinese tariffs on US agricultural products, loosening Chinese import regulations, creating anti-dumping provisions, and by lowering Chinese export subsidies (Michigan Department of Agriculture, 2002). However, China and the USA have similar climates and seasons, and like Michigan, China largely processes and packages for distant markets. Its extremely low labour and production costs may thus result in increased imports of competing commodities.

At the same time, farmland in the state's suburban and exurban fringe is valued more highly for residential and commercial uses than for fruit production. According to cooperative extension fruit specialist Phil Schwallier, Kent, Ottawa and Muskegon county apple orchards are worth $4,000 an acre (0.405 ha) for development, or double their market value for fruit growing (Longcore, 2000). Development pressures in exurban counties of the state's metro areas fuel this valuation difference. From 1990-2000, the population in the West Michigan fruit-belt counties of Ottawa, adjacent to Kent County and Grand Rapids, and Allegan, home of dormitory communities for Holland, Kalamazoo, South Haven and Grand Rapids, respectively grew 26.9% and nearly 17% (Walters, 2001). Population in formerly rural Livingston County, located between Detroit and Lansing, increased by 35.7%, the fastest in the state. Northern Lower Peninsula counties, known for cherry, peach, apple and grape production, but popular with retirees and younger families wanting to live far from the cities, also had high population growth rates. In contrast, Michigan cities such as Detroit, Flint, Lansing and Kalamazoo lost population, while others like Grand Rapids have grown slowly (Flesher, 2001).

Agritourism, another consumptive land use of farmland, may help farmers keep land in production by enabling them to profit from increasing demands for amenity countryside uses. Agritourism has been described as including 'farm-based accommodations (either fixed-roof or camping), farm-based meals, farm-based activities, agricultural festivals/events, attractions (e.g. museums, cooperative or corporate agri-business tours with retail opportunities) and farm-based retail opportunities where the travelling public interacts directly with the farm family/farm workers' (Hankins, 1997, p.6). While farmers have long engaged in activities now called agritourism, policies which try to coordinate agriculture with tourism as a way to conserve agricultural land are a more recent response to agricultural restructuring. The European Union has supported agritourism development to support farming regions as part of its broader rural tourism initiatives. Since 1991, European Union countries have spent $2 billion to subsidize agritourism development in rural farming areas which cannot compete in a global market with declining commodity prices that favour large-scale North American farms (Tagliabue, 1998). Since the 1980s, the number of farms participating in tourism has doubled in the UK, France and Italy, and in these countries, the number of farm accommodation units exceeds 600,000 (Hankins, 1997).

Similarly, in the USA where producers have long offered farm tours and had on-site retail, interest in agritourism has recently increased, driven by low prices, crop-damaging natural disasters, lifestyle migration to rural communities, and interest in replacing or supplementing traditional farm operations (Maetzold, 2000). The USDA has published an *Alternative Enterprises and Agritourism Resource Manual* and awarded grants to further specialty farm product development. In 2001, Michigan received $3 million in USDA funding to promote development of products such as chunky asparagus salsa, hard cider, fresh apple slices, wine barrels, cut flowers, brandy and organic fruit juices. Michigan's diverse and sizeable production of 124 agricultural commodities, more than any state except for California, supports specialty product development. Michigan ranked first in production of dry black beans, blueberries, tart cherries and processing cucumbers; second in celery; third in apples, asparagus, squash and fresh market carrots; and fourth in Concord grapes, fresh market cucumbers, plums and sugar beets (Hoogterp, 2002). This diverse crop production near large urban tourist-generating areas also provides the foundation for linking Michigan's second largest industry, tourism, with its third largest, agriculture, through agritourism.

Agritourism: tapping into American traditions and trends

In addition to diverse crop production, agritourism can tap into romantic, idyllic views of nature, rurality, and agriculture. Rural lands and rural life have a special place in American culture. The positive image of rural areas is associated with romanticism, in which nature invokes both an aesthetic and spiritual experience. As the USA underwent rapid industrialization in the nineteenth century, vacationing amid nature was believed to have a restorative quality against nervous disorders resulting from life in urban industrial centres. In this romantic view of nature, people sought solitude, privacy and a personal, semi-spiritual relationship with their environment as they visually consumed nature. Nature today encompasses both forested, mountainous wilderness as well as the agricultural countryside. The countryside is attractive as it appears to be closer to nature given its relative absence of people and seemingly non-mechanical and unplanned environment (Jacoby, 1997; Jasen, 1995; Urry, 1992). Agritourism draws on underlying romanticism in American culture as it offers experiential nature-based vacations with meaning, depth and spirituality (Feinsilver, 1998).

Agritourism's appeal also results from its connection to agrarian ideals incorporated into core American values. The privileging of the rural countryside derives in part from an agrarian ideal, in which it is believed that the true wealth of the nation is drawn from the land and that agriculture serves both political and social goals. Connectedness to the land, including the family farm, is associated with traditional American values such as self-reliance, simplicity and rural virtue (Bunce, 1998; Dorst, 1989). Dalecki and Coughenour (1992) found the following themes or tenets of agrarian life to be widely held by the American public: i) agrarian fundamentalism (i.e. agriculture is the most basic occupation in society, on which other professions depend), ii) agricultural naturalism (i.e., farming is a more satisfying occupation because it involves understanding and working with nature, societal ills result from urbanization and the separation of work from nature), iii) the family farm, and iv) economic independence and yeomanship. The independent yeoman with integrity and independence was considered the bulwark of American democracy. Agrarian beliefs are linked to social mores of God, family, human welfare, democratic progress of the nation, and self-reliant individualism. While agrarian values are most strongly held by those in agriculture and in lower educational and income groups, Dalecki and Coughenour found this system of beliefs which today is associated with freedom, independence and anti-urbanism to be widely held throughout American society. Consequently, agrarian themes and images can be used to generate support for farm support programmes (Dixon and Hapke, 2003) and for agritourism, which involves visiting the family farm and viewing a naturalistic lifestyle.

Agritourism is an increasingly important segment of the $3 trillion worldwide tourism industry. Certain psychographic and demographic trends favour future growth of the agritourism market. First, the American Recreation Coalition's 'Outdoor Recreation in America in 1998' survey shows continuing upward trends in outdoor recreation, including farm-based activities (Maetzold, 2000). Two important market segments, baby boomers and seniors have in part fuelled this upward trend in nature-based tourism. A Values and Lifestyles (VALS) survey conducted by Stanford Research International found that 30 million of the total 75 million American 'baby boomers' born between 1945 and 1963 have psychographic characteristics emphasizing environmental concern, social awareness, a global view and personal growth. These 'green' consumers, who are well-travelled, well-educated, professional, and have high income levels, are considered the core of the US ecotourism and agritourism market (Center for Tourism Policy Studies, 1994). Seniors are also a key market for agritourism, given their disposable income and leisure time. Agritourism appeals to the growing number of family-oriented tourists (i.e. baby boomers and seniors travelling with children and grandchildren respectively) by providing hands-on, educational activities that involve both children and adults (Mason, 2000 and Ragsdale and Real, 2000). It also attracts the increasingly urban and suburban boomer and senior populations who are a few generations removed from the farm and who hold nostalgic and romanticized views of rural, agricultural areas that contrast with negative views of urban areas. Such rural places are viewed as repositories for certain traditional American ideals, peoples and landscapes (Bunce, 1998; Hinrichs, 1996; Willits, 1993; Willits *et al.*, 1990).

As agritourism taps into long-standing romantic and agrarian traditions as well as recent trends towards environmental, family-oriented travel, it has been identified as a way to market rural landscapes and livelihoods in order to support agricultural production. For Michigan farmers, impacted by cost-price squeezes, foreign competition, and demands for alternative land uses, but offering diverse products and picturesque settings near urban areas, agritourism can be the means to maintain family farming heritage.

Methods

Focus groups of agritourism producers were convened during February 2002 as part of a project funded by the US Department of Agriculture and administered by the MDA to determine the impact of agriculture-based destinations on Michigan's tourism economy. Representatives from the MDA, Michigan State University Extension, and Western Michigan University invited a range of individual agritourism producers to the focus groups so that the diversity of Michigan's agricultural products (i.e. apples/cider, wine, peaches, cherries, asparagus, pumpkins, squash) would be represented. Primarily farmers selling fruits and vegetables and resulting value-added products were selected for the agritourism focus groups, as these small to medium-sized producers were over-represented in Michigan agritourism relative to larger commodity-oriented corn (maize), soybean and dairy farmers. In addition to representing agricultural product diversity, the focus groups also were set up to ensure the geographical diversity of producers, customer bases, marketing techniques and residential and commercial pressures facing farmers. Three focus groups which focused on different regions of the state as shown in Figure 7.1 were convened: i) southwestern Michigan, which attracts South Bend, Indiana and Chicago tourists, ii) the Traverse City area, which attracts longer-distance Chicago and Detroit metro area tourists and second-home owners; and iii) Central and Southeast Michigan, which mainly attract nearby metro area suburban customers, but which in the case of the latter faces intense development pressures from expanding greater Detroit.

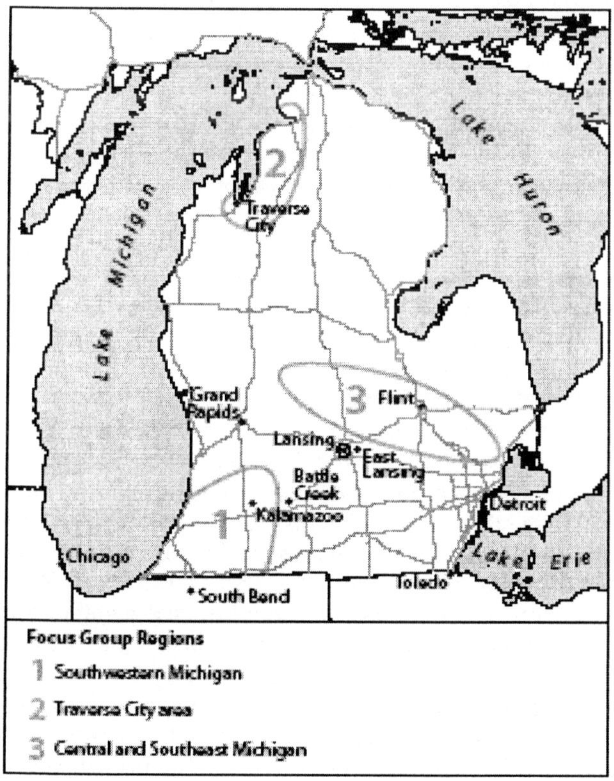

Fig. 7.1 Michigan Agritourism Focus Group regions.

The focus groups were used to identify key ways in which agritourism draws on traditions and trends in American society. Participating agritourism producers were asked about their experience with agritourism, general perceptions and knowledge of other agritourism sites, potential of agritourism (i.e. benefits and downfalls), what constitutes successful and unsuccessful agritourism operations, promotion activities, and the impact of agritourism in Michigan. Key concepts regarding agritourism's selling experiences associated with American traditions and values and creating new family traditions were identified from video and audio tapes and transcripts of the focus groups. The chapter now turns to three agritourism activities discussed within the focus groups, namely: i) selling rural traditions of food production associated with agricultural naturalism, ii) selling traditional American and Michigan rural landscapes, values, and service; and iii) selling and creating family-oriented farm traditions.

Selling traditions of rural food production

Agritourism, which involves experiencing life on the farm and the desire to be in touch with nature, incorporates agricultural naturalism as it connects urban and suburban visitors with traditions of rural food production. Agritourism generates needed revenues by allowing visitors to take in farm sights, smells, and activities. For the average American, who has little or no contact with farms, food production, or animal husbandry, the small, independent farm is becoming a popular leisure destination (Nicksin, 1999). People desire to stay on or visit a farm, learn where food comes from, come in contact with the soil, and see live animals. Agritourism satisfies the increasing interest for real-life, educational experiences (Nolan, 2000) as John[1], a farm marketer, noted:

> 'my daughter had sheep, and so we brought them up and penned them within a few yards of our selling stand. And you've got all those tiny lambs running around. I mean, it was just fantastic, the reception we had for, you know...from the public...There's a lot of...we don't really realize, being farmers ourselves, how many people have no contact with animals. You know, well...well where does...what is a sheep good for? You know, what do you do with a sheep? You know, and I mean then can you explain it'

According to Ed, another northern Michigan farmer, school group visits to farms can also be used to educate about food production and to generate word-of-mouth fuelling future adult expenditures:

> 'We had some of the preschool groups come out, like the Holy Childhood, they call it the Bright Tykes, and we have them come out three or four times a year and we'd give them cider and cookies or something, and they'd come out and feed the animals and they came and, you know, picked beans one day or, you know, just something that involved, you know, school kids. And the teachers are actually, you know, saying this is...this is where food comes from. It's being grown. And then they go and they cook in the school and, you know, turn it into something. I thought it was different, and it also gets the kids going home and talking about the farm and then the parents come out.'

The opportunity to learn about farm animals and crops attracts customers to the farm. Customers can get a better understanding of agriculture from the farm animals, the land, and from the experience of picking an apple off the tree or a pumpkin off the vine. Agritourism destinations benefit since interacting with farm animals and experiencing an actual farm draw visitors as much as farm-fresh produce does.

Fig. 7.2 The farm as tourist destination
Photo by Gregory Veeck

Agritourism also allows producers to address misconceptions Americans have about food production and food quality, and thus to stress the importance of the independent farmer. Visitors unaware of how food is produced may be less comfortable with food just produced on the farm, than with the mass produced, packaged supermarket varieties, as one farmer, Ed, recalled:

> 'I just collected the eggs and I come in with a bucket of eggs, and a family was in the store, and the father brought the little girl over and said, 'Oh look, these eggs just came from the chickens and he's washing them,' and they bought a dozen eggs and they said, 'Well, we're going to eat these in the next couple days. They'll still be alright to eat, right?'...but it was this idea of, these eggs had just come from the chicken so they were different from the eggs you get in the store. ...Yeah, I mean, it's just that they weren't the same eggs to us, I mean, and you're right. They aren't. But, I mean, I said, 'Well, how old do you think the eggs are in the store when you get them? And they kind of looked, and...and thought about it, and said...Oh well.''

While Americans positively regard the family farm and their naturalistic working with nature, because most people are separated in time and space from an agricultural heritage, they may not understand how food is produced. Agritourism provides a venue for farmers to promote their quality food production to visitors who seek both leisure and learning. Agritourism operators can thus make them long-term customers. By interacting with agritourists, farmers can also inform visitors about farming lifestyles and concerns. As farmers make up only 2% of Michigan's population, urban and suburban guests can be crucial advocates for farm-friendly policies and legislation.

Providing personal service of a past, simpler era

Agritourism markets personalized, relationship-oriented service, which while it may be reminiscent of an earlier time or small-town America, is essential to survival in the current era of agricultural restructuring. Researchers have found that even though national retail chains and Internet vendors can offer lower prices, rural retailers can successfully draw customers by emphasizing personalized service, cultivating community goodwill and adapting service strategies to meet local consumer needs (Besser, 1999; Miller and Kean, 1997; Taylor and Archer, 1994). Personal service and relationships are critical in agritourism, which involves direct contact between the producer and the consumer (Busby and Rendle, 2000). Successful agritourism operators help customers find and use products, make them feel welcome, and personalize their experience. While supermarkets can beat the farm markets on price, their staff often cannot match the knowledgeable service about different varieties of produce. Service, which extends beyond the business transaction, is critical to agritourism, as one farmer, Gene the Pumpkin Man, explained:

> 'The pitfalls are the customers themselves, because you got to learn how to treat the customer, but in nine-tenths of the...99% of the deal they'll...if you treat them how you would like to be treated, and you pat them on the back, and you laugh with them, and you get to know *them*, and you get to understand *them*, you don't have a whole lot of pitfalls. It's when you sit out there and all you're worried about is that's $18.50 and...out you go. No, no, no, no. You can't do that...The pitfalls is where you can't have...when it's strictly a business and you can't have that personal relationship with your customers. That's where the pitfall is. You can't be a commercialized used car salesman on a farm. You've got to be a farmer first and a salesman second.'

For Gene, personalized service has generated repeat customers who bring their friends and relatives, positive word-of-mouth, and free publicity in regional travel magazines.

Although agritourism operators interact with customers on a commercial basis, they have a long-term personal relationship with visitors. As retired farm marketer, Sharon, explained:

> 'They [customers] like that personal contact, cause they can go to the grocery store now, who's doing a lot better job of merchandising than we used to do in the farm markets, but they don't get the personal contact. And you've got to be visible, and you've got to talk to them, and not be afraid of them, and visit with them. They'd like...they'd like you to listen to their problems too...You'd be surprised. They become like family.'

Long-term customers and agritourism operators truly become like family. As word got around that the aforementioned Gene was seriously ill, several hundred people relayed concern about his well-being and whether he would continue his business which they patronized regularly.

While the personal relationship between farms and their customers seemingly belongs to the pre-big box store past, modern technology can be employed to maintain it. One agritourism operator, Ed, has effectively used e-mail to maintain personal relationships with customers from a distance, which can then be reinforced during farm visits:

> 'I can tell you one of our biggest things that has worked for us is through the internet, having an e-mail list where people will come into the farm and sign up for an e-mail list, and then we send out recipes, farm updates, you know, when a new calf is born on the farm, anything, you know, and most of them are funny stories, but makes a lot of people feel connected with the farm. You know, they may be living in the city. We have people in every continent, you know, I think all fifty states (are) now on the e-mail list. I think it's around four thousand people. And so, sending out these farm updates makes people feel connected, well to Harbor Springs (Michigan) and also to a farm... It's a way to remind them that you are still there. You know, they know the names of the animals when they come to visit, which is a very positive thing. You know, where, if they stop once and then, you know, they may forget that you are there, or whatever, but to be able to, for free, send out e-mail to four thousand people, that is basically, I mean, it's advertising, but people like it, and they choose to be on the list.'

Agritourism's strength is in its marketing rural or small town virtues such as simplicity, service and a personal touch. Successful agritourism operators sell these virtues as they take time to assist customers and listen to their concerns. Such personalized service and relationship building help explain the success of rural agritourism businesses given large retailers' cheaper, commoditized products.

Selling Americana, core values and certainty in an uncertain, post-September 11 world

Agritourism's selling of core American images and themes resonate in post-September 11 America. After September 11, 2001, leisure travel patterns emphasized short-distance car trips and family ties relative to those a year earlier (Pfenning, 2002). Agritourism, which offers contact with the family farm, the farmer, and food production outside large urban areas, fits these travel patterns and taps into the current wave of patriotism and long-held agrarian values. Visiting and supporting the family farm and the 'solid' yeoman through on-site expenditures uphold these unyielding symbols of American democracy. One farmer and cider mill operator, Katy, described how agritourism and farm visits were connected to agrarian values re-emphasized after September 11:

> 'I found after September 11, I wasn't sure what to expect that next weekend. There were no sports events and I didn't know if people were going to just huddle up at home and feel safe and, I was surprised that the

weather cooperated, that's important for all of us, but our numbers were up, so I think it is a return of some of the basic values that we've had for generations. I know at our farm, just seeing the farm animals and being able to touch them...although some people don't know a calf from a dog and even a camel, but they're going to learn that there. They have that experience, back to nature, back to generations ago. I guess it's a family experience, but the animals...and it's not just the...the children, I mean, I even have seniors during the week come in and they want to go in and see the farm animals, probably going back to when they were younger as well with their experience, but I definitely feel that some basic values and some family values that we've maybe drifted from that, especially now I think we've turned back to some of those since September 11.'

The family farm is the perfect setting to re-engage the agricultural naturalism of working with the land, as well as agrarian-connected social mores of God, family, and democratic progress of the nation. During this time of crisis and uncertainty, agritourism embodies the 'family' values Katy suggests the nation has drifted from, possibly towards competing liberal values of radical individualism (personal freedom and achievement), racial equality, and leisure associated with an affluent life style during the economically booming 1990s.

Agritourism also links with another element of American agrarian tradition - the family farmer whose integrity and independence is seen as the bulwark of American democracy. For farm visitors after September 11, visitors connected the strong, independent family farmer with democracy and freedom, as one farmer, Gene (shown in Figure 7.3) recalled:

> 'And, of course, September 11, tremendous amount of people wanted to talk about September 11, so I saw everybody from the reservist Marine that says we're going after them to the retired Air Force Colonel, and I think there's only one above Colonel, I think that's General, and we talked...I'm just a little peon pumpkin planter, and big deal. And most of them could buy and sell me and never flinch, and yet, they're talking to down-earth and they want to know what makes me tick, and I tell them.'

Agrarian movements have historically been associated with socioeconomic groups such as farmers, farm-related groups, and lower income and educational groups (Dalecki and Coughenour, 1992). As Gene owns a Michigan centennial farm which has been in his family for over 117 years, he not surprisingly holds agrarian values as he linked God, the farmer and America, saying, 'God bless the American farmer... Greatest nation on earth, and keep it that way. And the farmer is one of the ones that are going to do it. That's the truth. The grass roots'. Yet in an uncertain post-September 11 America, which witnessed key symbols of American economic and military power being attacked, the symbolic strength of its family farmers resonates widely throughout American society.

Lastly, agritourism offers cherished rural landscapes that represent Americana and safe idylls. The agritourism operators in our focus groups believed they offered visitors a wholesome, safe environment. In post-September 11 America, safety and homeland security concerns are increasingly emphasized. Homeland security involves protecting the US food supply from terrorist threats. Under the scenario of terrorists tampering with the food supply, large-scale corporate farms and food processors would be more likely targets than small-scale agritourism operators. Food from the latter may thus be perceived as safer. Likewise, their rural locations are outside the large cities thought to be likely targets of terrorist attacks or where crime is believed to be rampant. Farms are idyllic pockets of Americana, providing some certainty in an uncertain world, as one farmer, Peter, explained:

> 'One thing nobody mentioned was the fact that, not only is there a comfortable atmosphere, but it's also, sort of safety in a herd numbers. If they're coming out with their kids, they're seeing the very same thing in the minivan next to them. They're going from a place where they know what the crowd of people is going to be like that they're going to meet, and be in contact with this family, kids from their school or from those things, they're...a place that they won't go to in the city or in...some of the county parks, but the farm has become the...a haven of more of a specific type of a family. You know exactly what they'll meet. No knives and accoutrements, or anything,...just the standard atmosphere of almost like a church setting. There's everybody they're going to know.'

Agritourism, the American farm, and farmer are associated with core American values. They provide reassuring solidity in a world where terrorism or violence can break out at any time.

Fig. 7.3 Americana and the family farmer

Selling and creating family-oriented farm traditions

Agritourism attracts urban and suburban baby boomers and seniors who are a few generations removed from the farm; who hold nostalgic and romanticized views of rural, agricultural areas; and who have their own family-oriented farm traditions such as farm visits. Farms which have had on-site retail and activities for decades prior to the coining of the term agritourism appeal to parents who want their children to experience the same farms they visited as children. These farms sell tradition to generations of repeat customers, as Jan, an agritourism operator, observed:

> 'And those are who you want to hype…to impress because they're going to want to bring their kids back. I don't know how many times I hear, oh, I came down as a kid and now I'm bringing my child back and, so, it's really the tradition and, like you said, if you market for the youngsters, but have everything available that mom and dad and grandma and everybody else want while they're there, you've got the perfect…perfect mix. Where we are I think, giving people an experience, it's a family experience, and it's…just like everyone else said, it goes on for generations. My grandfather brought me here, my parents brought me here, and hopefully, they'll bring their children, and on and on.'

Agritourism also appeals to individuals who grew up on farms or visited relatives' farms as youth. Agritourism destinations can allow them to relive poignant memories of their own family farm heritage as Gene, the pumpkin farmer, recounted:

> 'This old gentleman came along. He was 85 if he was a day, and he looked all around and I watched him, and he was just having a ball, looked at the cattle and the whole works. Pretty quick, he came over and put arm around my shoulder and says, thanks son. I says, well what… what for? I didn't…you know, you didn't buy anything. And he stood there and wept, and he says I left the farm when I was 18. This is the first time I was ever back to visit a farm, and I want to thank you, and the man stood there and wept.'

To augment the heritage appeal for those who grew up on farms, visited farms as youth, or are part of the larger American populace that holds agrarian values centred on the family farm and its role in food production, some Michigan agritourism destinations display old farm equipment and antiques and hire Amish staff.

In addition to tapping family-oriented farm heritage, agritourism can create new family-oriented farm traditions. Agritourism appeals to the growing number of family-oriented tourists (i.e. baby boomers and seniors travelling with children and grandchildren respectively) who are looking for safe, intergenerational, learning, bonding experiences. As overscheduled suburban parents and children have limited family time together, farm vacations can provide a relaxed setting for them to bond and create memories. Agritourism appeals to boomer parents who want vacations to be growth, educational and family bonding experiences. Portico Research found that one of the most important motivators of family travel is the creation of shared memories (Sakakeeny, 2002). Agritourism can provide the setting for all types of families to create such shared memories as Ed explained:

> 'Another thing to add on that is, you notice, I mean, people are really just looking for something to do together, or something to do as a family. We started a scavenger hunt around our pasture…We had hundreds of people that came and, you know…I mean, people coming…you'd think it would be just with their kids, but, you know, older couples. I mean, just people looking for something to do that's, you know, that's social. Not just watching TV or going to a movie and sitting there and, you know. But, you know, this is something that…I mean, people come and go three or

four times...the same people...I mean, it's just...you know, it's amazing.'

Such farm visits are driven by activities for children and the desire for family bonding activities, which according to one farmer, Peter, is a shift from 20-30 years ago when customers were mainly older people who knew how to cook and use raw produce. But now families with children aged 10 years and under drive agritourism. These families are attracted to farms' entertainment or tourist value and have money to spend on pleasing young kids. Agritourism provides opportunities for people to get out into nature, experience or re-experience farms that are part of their and/or America's heritage, and create or re-live traditions as they enjoy themselves as a family as shown in Fig. 7.4.

Fig. 7.4 Creating family memories on a family farm.
Photo by Gregory Veeck

Conclusions

Agritourism sells traditions of local food production, family, and rural Americana to maintain Michigan's and America's family farming heritage. Given the declining price for commodities and increasing value of land to 'grow houses on,' farmers must add value to their agricultural produce. Agritourism allows them to do this by tapping into agrarian and romantic ideals in American culture as well as trends towards hands-on, educational, intergenerational vacations. Agritourism also links with Michigan traditions. Agritourists who spend money on autumn pumpkins, cider, doughnuts and apple picking help preserve Michigan land and agricultural heritage embodied in the family farm, apple orchard and cider mill visited as part of another state tradition, autumn leaf colour touring.

Agritourism helps preserve the institution of the family farm as it provides urban and suburban visitors with an increasingly rare, hands-on experience with nature and food production. Perhaps Eileen, an agritourism business owner, best expressed the mutual benefits of agritourism for farmers and their visitors:

> 'Benefits, personally, I'm still on the same four corners that I've been all my life. It has allowed me to have two of my children living next door, even though they're not involved in the farm, it's all part of that rural peace that I love. My father put me on a tractor when I was seven, you know, which was not unheard of in that day, you know, so I love a newly ploughed field and all of that. So, it has allowed me to do that...it has allowed me to share that with people. They come...you can go to a park in a city and it can be quiet and peaceful, but it's artificial. They can come to the farm and it's real.'

Acknowledgements

The author thanks the US Department of Agriculture and the Michigan Department of Agriculture for financially supporting this research, Margaret Pearce for cartographic assistance and Gregory Veeck for the use of photographs.

Note

1. Names of focus group participants have been changed to maintain confidentiality.

References

Besser, T. (1999) Community involvement and the perceptions of success among small business operators in small towns. *Journal of Small Business Management* 37, 16-29.

Bunce, M. (1998) Thirty years of farmland preservation in North America: Discourses and ideologies of a movement. *Journal of Rural Studies* 14, 233-247.

Busby, G. and Rendle, S. (2000) The transition from tourism on farms to farm tourism. *Tourism Management* 21, 635-642.

Center for Tourism Policy Studies (1994) *Ecotourism Opportunities for Hawaii's Visitor Industry*, Department of Business, Economic Development and Tourism, Honolulu.

Dalecki, M. and Coughenour, C. (1992) Agrarianism in American society. *Rural Sociology* 57, 48-64.

Dixon, D. and Hapke, H. (2003) Cultivating discourse: The social construction of agricultural legislation. *Annals of the Association of American Geographers* 93, 142-164.

Dorst, J. (1989) *The Written Suburb: An American Site, an Ethnographic Dilemma*, University of Pennsylvania Press, Philadelphia.

Feinsilver, E. (1998) Back to the land: Agri-tourism on rise; The flip side of the near-disappearance of small family farms: A taste of farm life becomes a marketable commodity. *Greensboro News & Record* 2 October 1998, B1.

Flesher, J. (2001) Suburban migration up statewide. *Kalamazoo Gazette* 29 March 2001, A1-A2.

Hankins, C. (1997) Agritourism: Profit and preservation for Tennessee's rural landscape. *Tennessee's Business* 8, 4-11.

Hinrichs, C. (1996) Consuming images: Making and marketing Vermont as distinctive rural place. In: DuPuis, E. and Vandergeest, P. (eds) Creating *the Countryside: The Politics of Rural and Environmental Discourse*, Temple University Press, Philadelphia, pp.259-278.

Hoogterp, E. (2002) Michigan bounty-long on diversity, short on federal aid; Most farm support goes to states producing wheat, corn, soybeans, cotton and rice. *The Grand Rapids Press* 15 December 2002, F1.

Jacoby, K. (1997) Class and environmental history: Lessons from 'The war in the Adirondacks'. *Environmental History* 2, 324-342.

Jasen, P. (1995) *Wild Things: Nature, Culture, and Tourism in Ontario 1790-1914*, University of Toronto Press, Toronto.

Kenney, M., Lobao, L., Curry, J. and Goe, W. (1989) Midwestern agriculture in USA Fordism: from the New Deal to economic restructuring. *Sociologia Ruralis* 29, 131-148.

Longcore, K. (2000) The core of the problem: Three hard years push West Michigan apple growers out of the business. *The Grand Rapids Press* 23 September 2000, A1.

Maetzold, J. (2000) *Alternative Enterprises and Agritourism Resource Manual*, USDA Natural Resources Conservation Service, Washington, DC.

Mason, P. (2000) Trends in family travel. Paper presented at Trends 2000: Shaping the Future, The 5th Outdoor Recreation and Tourism Trends Symposium, Sept. 17-20, 2000, Lansing, Michigan.

Michigan Department of Agriculture (2002) *New Markets.* Web page, [accessed on 14 February 2003]. Available at www.michigan.gov0,1687,7-125--14963--,00.html.

Miller, N. and Kean, R. (1997) Reciprocal exchange in rural communities: Consumers' inducement to in shop. *Psychology & Marketing* 14, 637-661.

Nicksin, C. (1999) Dude farms. *Country Journal* 26, 16-19.

Nolan, E. (2000) Farmers' Marketing - Some families who grow corn, peaches, pumpkins and such to feed the rest of us now depend on another cash crop: agritourism. *The News Journal* 9 October 2000, record no. 008410E752E4D64C90DA0.

Pfenning, A. (2002) Travel post-9/11: Where are we now? *Travel Weekly* 61, 12.

Ragsdale, O. and Real, T. (2000) Attracting baby boomers to your fair or festival. Paper presented at Michigan Festivals and Events Association Eighth Annual Convention, Nov. 2-5, 2000, Kalamazoo, Michigan.

Revere, P. (2002) Bumper price; Apple growers get more for less in non-lean year. *The Grand Rapids Press* 5 October 2002, A1.

Sakakeeny, K. (2002) Nice family, nice niche: Surveys say most travel is family travel. *Travel Weekly* 61(1), 79.

Tagliabue, J. (1998) Preserving a heritage via bed and barns; European governments subsidize agritourism. *New York Times* 13 August 1998, D1.

Taylor, D. and Archer, J. (1994) *Up Against the WalMarts: How Your Business Can Prosper in the Shadow of the Retail Giants*, American Management Association, New York.

Urry, J. (1992) The tourist gaze and the 'environment'. *Theory, Culture, and Society* 9, 1-26.

Walters, B. (2001) Urban exodus, rural boom; Kalamazoo shrinks as area township populations swell. *Kalamazoo Gazette* 29 March 2001, A1, A5.

Willits, F. (1993) The rural mystique and tourism development: Data from Pennsylvania. *Journal of the Community Development Society* 24, 159-174.

Willits, F. Bealer, R. and Timbers, V. (1990) Popular images of 'rurality': Data from a Pennsylvania survey. *Rural Sociology* 55, 559-578.

Wilson, O. (1995) Rural restructuring and agriculture - Rural economy linkages: A New Zealand study. *Journal of Rural Studies* 11, 417-431.

Chapter 8

Re-imaging agriculture: making the case for farming at the agricultural show[1]

Lewis Holloway

Introduction

Many in the farming sector, in the UK at least, are experiencing a sense of 'crisis', combining economic pressure with catastrophic events such as the 2001 outbreak of foot and mouth disease (FMD) and a feeling of being beleaguered by an ignorant public and unsympathetic politicians. There is a perception that the non-farming public holds farming and farmers in low esteem, is increasingly susceptible to negative media representations of agriculture and mistrusts the ways food is produced. This chapter explores one mode of response by the farming sector to perceptions of agricultural crisis in the UK, focusing on agricultural shows as sites and events where attempts to 're-image' agriculture are occurring. While it is recognized that the image of some parts of farming, in particular the organic sector, is positive, there is concern over negative public perceptions of 'mainstream' agriculture. Some actors in mainstream farming are thus attempting to mobilize farmers into taking an interest in reconstituting the image of farming by first, associating it with, for example, food quality and environmental conservation, and second, by responding to a perceived need to educate a non-farming public represented as misinformed about farming. Agricultural shows are important to this re-imaging process. First, they consist of spatial and temporal nexuses of agricultural and non-agricultural functions, entities and people. Second, they give the agricultural societies which run shows potential access, in situations where they have the ability to stage-manage the presentation of agriculture, to large numbers of non-farming visitors. Finally, they can become bases for more extended strategies of connection with non-farming publics and, importantly, they can encourage actors in the farming sector to reconsider their own relationships with non-farmers. These factors contrast with other ways in which farming has entered into a struggle to defend a position of power in the countryside, ranging from the engagement of representative agencies such as the National Farmers Union in national policy making (Clark and Jones, 1998); to forms of direct action and protest (Woods, 2003).

Based on interviews with show managers and others involved in agricultural shows, the chapter develops the following themes. First, shows have become part of a struggle by some sections of the farming industry to legitimise its status and practices against widespread criticism. The way in which agricultural societies fulfil their objective of 'promoting' agriculture has thus partly changed, from an investment in the development of agriculture, to engaging in public relations for farming. Accordingly, shows present a particular vision of farming, emphasizing the successes of expertise in food production and conservation. Second, despite an emphasis on re-imaging agriculture to a non-farming public, in some ways this re-orientation of show events can also influence how farmers imagine themselves. There is thus a degree of reflexivity involved in the process of re-imaging agriculture, as show managers consider their own, and farmers', position in relation to non-farming publics. Third, although agricultural societies stress the importance of establishing reconnection between farmers and non-farmers, a farming/non-farming distinction is reproduced in showing farming to the public. In particular, non-farmers are represented as lacking knowledge of farming and food, and as inhabiting 'lifeworlds' distinct from farmers. While strategies of 'educating' non-farmers seek to inform, they are implicated in presenting a particular and partial image of farming which maintains the distancing of non-farmers from many of the 'realities' of agriculture. Finally, the chapter suggests that, given attempts to re-image agriculture, further research into the effects of these attempts on public perceptions of farming would be of value in considerations of public-sphere debates about the future of the countryside and farming. In addition, research into the effects of re-imaging farming on farming identities and practices should also prove valuable in understandings of rural change.

The changing role of agricultural shows

Agricultural shows emerged in the late eighteenth and early nineteenth centuries in association with an increasing application of scientific theory to farming. Run by newly formed agricultural societies, their main concerns were with the promotion of 'progress' and 'improvement' in fields such as livestock breeding, equipment manufacture and tillage practices. Shows were used to display the latest 'advances', and to facilitate their dissemination from the elite landowners and farmers who sponsored their development, to 'ordinary' farmers (Fox, 1979; Goddard, 1981, 1989; Hudson, 1972; Overton, 1996; Wilmot 1990). While these functions have tended to continue to be regarded by agricultural societies as central to their activities, more recently, they have focused on incorporating 'education' of the public into their mix of agricultural business, competition, spectacle and consumption (e.g. of food and clothing) which has tended to attract large numbers of non-farming visitors. They are increasingly interested in encouraging non-farming publics to learn and think about farming in particular ways, by presenting a particular image of farming, and so influencing public debates over farming's future. This moves UK shows towards the focus of many North American agricultural events, whose primary reason for existence has become to 'educate' and entertain non-farming, largely urban, audiences (Henning 1998).

The changing position of farming in the British countryside has been the subject of much debate. Some rural geographers have been interested in attempting to outline and explain what they have seen as a fundamental shift from a 'productivist' agriculture, heavily supported by EU policy and focusing on maximizing agricultural outputs, to a 'post-productivist' context where a continuing emphasis on production is tempered by pressures on farm incomes and a need to account for the environmental effects of production-driven agricultural techniques (Ilbery and Bowler, 1998). While notions of a straightforward 'post-productivist transition' have been challenged by authors calling for greater attention to geographical, historical and discursive specificities (Evans *et al.*, 2002; Wilson, 2001), it is evident that, from the perspective of farmers at least, their 'natural' position of authority in rural areas has increasingly come under threat. The countryside has instead become seen as multifunctional, not simply agricultural, space (Lowe *et al.*, 2002).

As Halfacree (1999, p.69) suggests, 'British farmers now both feel and indeed are in a much less secure and certain position as regards their role in both agriculture and rural life generally'. This sense in which farmers might be understood as being displaced from an earlier position of power in rural areas has prompted Seymour *et al.* (1997) to argue that farmers might be understood as an 'othered' rural group. They suggest that a new moral consensus centred on environmental protection has emerged, in which farming is associated with environmental damage rather than rural stewardship and benign authority in rural matters. Seymour *et al.* (1997) claim that, while it might be 'provocative' to list farmers alongside those more often referred to as rural others, the emerging dominance of bureaucratic and service class discourses relating to the management of rural areas is in effect decentring farmers from their previously high status. Contesting this process of 'othering' is perhaps one way of understanding the attempts to re-image agriculture discussed in this chapter.

The chapter now turns to consider agricultural societies and shows as contributors to a struggle to re-image agriculture. Shows are used to stage encounters and exchanges between farming and the non-farming public which are increasingly rare in societies where many experience a distancing between themselves and the way their food is produced. Research with show organizers and others, conducted between 2001 and 2003 as part of a project examining the changing role of agricultural shows in the UK[2], has indicated some of the ways in which this staging has recently become associated with an explicit strategy of attempting to influence public perceptions of agriculture. The following sections therefore consider the strategies employed to re-image agriculture at shows. They draw mainly on interviews with senior managers of nine of the larger agricultural societies based in the English Midlands, south west England and eastern England. Interviewees were promised confidentiality; respondents are therefore not identified, but have been allocated a number given in brackets after quotations. In addition, interviews were conducted with representatives of a small number of cattle breed societies and pedigree breeders, and observations were made at several of the shows. Finally, reportage of shows in the UK trade magazine *Farmers Weekly* (FW) is used. The following section begins by outlining some of the general reorientations being made by agricultural societies, before looking at three specific dimensions of the re-imaging strategies taking place in shows.

'Interpreting' farming to non-farming publics

Most show managers claimed that the value of shows was in the ways they were able to present a positive image of farming to the non-farming public. This was important in countering the negative perceptions of agriculture that they felt pervaded non-farming media and groups. Thus, one manager said, 'I think they are an event that tries to reflect what's going on in the countryside today, but I'm afraid that the countryside is becoming ... associated with unfortunate incidents, you know, pollution, environment, BSE, scrapie ... which are affecting public perceptions of the countryside, badly as far as agriculture goes' (1). According to another, 'the farming community and the general consumer seem to have become somewhat remote and distanced from each other and there's suspicion and concern about quality and standards, so I think we've got to bridge that gap ...' (2). Shows are thus thought to have a role in a process of re-establishing a lost connection between farmers and non-farmers. A crucial way of effecting such a reconnection has been to present farming as a series of spectacles. As one manager suggested, 'in shows and exhibitions you want to provide something you can't see anywhere else, and certainly the biggest parade, when we ... parade all the winners ... nothing like it in the world, so it is actually a spectacle' (2).

The presentation of farming as spectacular does, however, have a subtext. Modern farming is presented as the culmination of a progressive, modernizing process. Emphasis is placed on the increasing levels of mechanical power, 'improved' livestock, enhanced yields and high level of what is taken to be 'efficiency', in modern farming. At one show, the manager described 'a display in the Grand Ring, called Farming's Big Guns, where we will bring in a lot of big farm machinery' (3). In many cases, modern machines were placed alongside 'vintage' equipment, drawing attention to the sense of 'progress' being suggested. Simultaneously, such displays were part of a strategy of entertainment. For many managers, there was an explicit feeling that a combination of education and entertainment - 'edutainment' - was a valuable strategy for engaging with, and informing, the public, in common with the approach often used in North America (Henning, 1998).

Many shows have long had a remit of bringing together the rural and the urban, or the agricultural and the non-agricultural as, for example, shown by Hall (1987) in relation to the Yorkshire Agricultural Society. However, the evidence from the interviewees downplayed the connection of urban and rural communities and instead envisaged shows as an opportunity to effect a re-imaging of agriculture. The need to make a case for farming, and to enrol the non-farming public into particular ways of thinking about farming, was felt by most managers. For example, one claimed that shows had to 'celebrate and promote farming to the wider general public. And unless you make that jump, I don't think there's much of a future for you' (5). Another society intended to subtitle its event a 'Festival of Food and Farming', rather than simply holding an agricultural show. This change was discussed in terms of a repackaging; so that 'the contents are still basically the same, but there's an outer package that has to be changed a little' (4). There was also evidence of some divisions between agricultural societies, according to how much of a 'traditional' show they felt themselves to be, or were perceived as by others. The level of 'tradition', in this sense, was taken to refer to the degree of orientation of the show to agricultural audiences, and many show managers made the point that although their show was being re-oriented in order to present a particular image of farming, it retained a 'traditional' focus on the farming 'community'. These managers referred sometimes critically to shows that were felt to have lost this 'traditional' focus.

Countering this process of moving away from the 'traditional' agricultural orientation, some managers argued that the presence of actors from the farming sector (including farmers and agricultural trade stands) was important in defending farming, and establishing a positive agricultural image. In this view, farming and farmers need to remain central to shows. Accordingly, in his history of the Three Counties Agricultural Society, Lewis (1996, p.114) argues that 'This opportunity of presenting a true picture of the agricultural industry, to the lay person, at a time when there is so much misinformation about, must be an important function of the Society. This will not happen unless true farmer interest is maintained.' There are two important points here. First, there is an understanding that there is a 'truth' about farming, which can, and should be, communicated to non-farming publics in response to the perceived 'untruths' currently affecting public perceptions. Second, it is taken as a moral imperative that people should know about the sources of their food. Here, many interviewees emphasized their sense of the public's 'ignorance' in relation to food; and there was felt to be a gulf of understanding between the farming and non-farming communities. As one manager said, 'we think we do have an educational role, and that it is necessary and desirable that people should understand that milk comes from somewhere before it comes out of bottles' (6). It was also clear that the perceived knowledge differential is between farmers and non-farmers, rather than simply between rural and urban dwellers. The following comment suggests that even many rural dwellers are, in the term used by the interviewee, 'ignorant', and that non-farmers in general need to be informed about the origins of their food; 'we want them to go away with the knowledge that food does not come out of polystyrene trays, and cling film wrapped ... we actually want them to realise, and it is unbelievable living in the country, how many people think that milk comes out of a cardboard box' (3).

The specific strategies for education, often referred to as 'interpretation', adopted by agricultural societies, are important in examining how they are engaged in a re-imaging processes. Three are briefly reviewed here: re-establishing farmer-consumer connectedness by focusing on food, the use of educational material aimed at children, and the role of livestock.

Connectedness

A preview of the 2003 Royal Bath and West Show (FW, 23.5.03) cited the society's chief executive as saying, 'Our audience is largely urban people leading frenetic urban lives. If we can persuade them to visit the show and leave with just one thought about how the countryside works then we will be on the way to achieving that government buzzword - reconnection'. A similar story is told about the 2003 South of England Show, whose chair of the agricultural, education and development committee (cited in FW, 30.6.03) says that, 'We thought it would be good to explain some of the connections between what visitors see and how agriculture works.' In another example, a preview of the 2002 Royal Show (FW, 24.5.02) suggested that the show was to be restructured into integrated spaces such as a 'Food Production Area', replacing segregated spaces for machinery, livestock, etc. Here, 'the farmer's role will be presented in the context of the entire food chain'. The re-imaging of farming involves persuading the public that farming is about 'food production in a sustainable countryside' (FW, 21.6.02). Connecting consumers to farming through the medium of food, and associating farming with the notion of sustainability are important in discursively associating farming with a range of positive ideas.

Persuading the public of farming's importance is, here, central to its re-imaging. The founding principle of many agricultural societies, to promote agriculture, in this context takes on a new emphasis. Rather than simply promoting agricultural 'progress', the promotion here is of farming as a legitimate and valuable set of practices. As one manager said:

> 'The mission statement is to promote agriculture. The vehicle for that promotion is predominantly the [named] Show. It is there to show the public what agriculture is about, that agriculture is more important than they would ever imagine in the sense that all the food they get from Tescos, Asda, and whoever, actually has to start from somewhere, you know, the 'from field to plate' concept is part of the show.' (7)

At one level, particular show exhibits could be used to 'inform' non-farming publics about the sources of their food, often through the notion of a 'food chain'. As demonstrated by the following quotation, images of a straightforward link between farming practices and food items are constructed:

> 'Through exhibits like 'Chicken and Egg' and things like that, it can tell people where a chicken comes from. Bread making, can show wheat to loaf. One exhibit that was about three years ago was absolutely brilliant where it was, so many cardboard milk cartons, at the back end of a cow, with a pile of silage at the beginning of the cow, and this much silage will make this much milk out of that cow ... And also showing people how farmers have diversified - ostriches, wild boar, all type of things like that. And the local produce and

> the farm food fairs ... they are bringing fresh produce, local produce to the public, they're showing it, portraying it, and it's all part of the education.' (8)

However, in other cases, it was recognized that there was a more subtle and complex strategy for 'informing' the public. Here, it was recognized by show managers that 'education' in itself would be unlikely to attract the interest of the desired audiences, and that such 'education' would need to be incorporated into the entertainment dimensions of the show.

> 'our role is to promote agriculture, and in the past we've said, well, we'll promote agriculture by telling everybody how marvellous farming really is. Well, people will not come to a show to be educated, if you say 'come to the [named] Show and be educated', they won't come, you have to say 'come to the [named] Show and have a good time', well, when you're having a good time there will be a lot of subliminal messages going on ... you might not realise you're being educated but that is the underlying rationale behind the show.' (5)

Specific case studies of food chains were thus deployed at various shows, involving the staging of encounters between the public and elements of what was presented as a food chain leading straightforwardly from field to consumer. One, for example, used potatoes. In interview, the manager stressed the need for 'different' ways of educating the public;

> 'We've got a feature this year which is based on potatoes. So we're going to look at potatoes for the whole food chain, we picked potatoes because they look similar in supermarkets to they do when they come out of the ground ... at one end of the feature we're going to have them growing, and the other end it's going to be chips and snacks and tasting. We're going to try and drive the consumer that way down the chain [i.e. from product to field], down towards the understanding production issues ... so that you get the issues of quality assurance and traceability all the way through, and if you can demonstrate to the consumer that you can actually tie back that bag of crisps to that field, and how it was looked after, then hopefully that will give a good story.' (2)

This illustrates some of the important aspects of image construction being undertaken. First, the familiar is related to the unfamiliar as consumers are shown how everyday foods are processed and grown. Second, consumers' sensual engagement with food, through sight and taste, is used to make the experience pleasurable. Third, there is a focus on reassurance, showing consumers how food is translated between the different parts of the food chain in ways which maintain an assured sense of quality and traceability. Finally, there is the idea that the food chain can be told as a narrative of food movement from field to plate in a way which tries to connect consumers to food production - they become part of the same process. The consumer, as well as the bag of crisps, is thus tied back to the field.

Some show managers, in recognition of public sphere debates and perceptions of public distrust of some aspects of agriculture, were prepared to go beyond representing farming as unproblematic. One thus described how he wanted to use the show to disseminate information on controversial topics such as agricultural biotechnologies or agro-chemicals. Yet, while suggesting that he was interested in presenting a balanced view to consumers, like most interviewees he wanted to emphasize what he saw as the benefits of 'conventional' farming:

> 'I'm really keen to give them a realistic vision of what farming is about and what farmers do ... so I think actually, the most simple stuff, you want to say 'this is what we do, this is why we plough the field, this is what we're growing, and this is what we do with it, and these are the, cos we do need chemicals and this is the reason why, and if we don't this is what it looks like ... OK, so you don't like fungicides but would you rather eat, you know, mouldy wheat or would you rather eat healthy, clean wheat?' (5)

Like several others again, this manager argued that farmers were partly to blame for the negative public image of farming, and for numerous misconceptions about farming practices, agro-chemicals, etc. Given this, another manager suggested that the solution was to restructure the idea of the agricultural show, to extend its reach in time and space, so that the educational and informational functions of the show would take place away from the show as an event and showground as a site. 'We completely changed the concept of the show being two days of the year. We should have activities that were going on 365 days of the year' (9). For this manager, the promotion of agriculture needed to become a constant activity for agricultural societies, rather than just at show events.

Education

A number of strategies structured specifically around engaging with schoolchildren were for many societies a key dimension of extending the spatial and temporal reach of the show. First, in many cases guided tours and schools days were an important part of the show event. Children could be brought into direct contact with farmers, livestock, and so on, in ways which could combine information with the presentation of particular messages about farming. At one show, for example, around 6,000 children were taken on guided tours over the four days of the show. Second, many show societies had prepared educational materials for use in local schools, ensuring these were related to the National Curriculum to encourage their use. As far as the show societies were concerned, it was necessary for them to become directly involved in educating children about farming, since most teachers and parents lacked the appropriate knowledge themselves. Indeed, targeting children draws in associated adults, allowing them too to be targeted with a particular narrative about farming, and enrolling them into the process of representing farming to children in particular ways. Third, many shows included children's competitions, exhibited and judged during the show. These tended to draw on classroom activities in the period preceding the show, and the show event itself was thus the culmination for the children of an extended period of work related to farming. Here, such competition could again be used in order to construct a particular image of agriculture. For example, the schools competition at one show had required children to submit poetry on the theme of 'Conservation in British agriculture', or to make models illustrating 'My happy pond'. The partial image of farming that such themes present contributes to a wider dissimulation of some of the 'realities' of modern farming at agricultural shows.

Focusing on children has clearly become an important mechanism for agricultural societies to influence public perceptions of agriculture. Particular images of farming, associating it with environmental protection and healthy food, emerge from the educational material and events, while other aspects of farming (in particular, intensive livestock husbandry) are not included. The potential of such educational activities to contribute to a re-imaging of agriculture is certainly recognized by agricultural societies. The Royal Agricultural Society of England, for example, manages the Farming and Countryside Education (FACE) project, which is, according to a promotional leaflet, 'dedicated to helping children understand food, farming and the countryside'. FACE aims to work with sponsors from the agricultural sector to influence children's perceptions. Potential sponsors are encouraged to send funds by presenting FACE as an opportunity for them to influence perceptions of agriculture; 'If you care about what children think and feel about food and farming, perhaps you would consider making a financial contribution to FACE ... your contribution could make a world of difference to the way children view their food, the countryside and the people in it.'

Livestock

Livestock were central to most show managers' vision of how shows could be used to present a particular image of farming. Livestock are valued because they are a significant attraction to the farming audience (and are thus an important part of retaining farmer interest in shows). Livestock are also a key element of the presentation of farming as a spectacle, and they have a function in the educational strategy of connecting consumers to the processes of food production. Opinions differed as to the actual significance of livestock for non-farming audiences. For some managers, their 'difference' to most people's experience meant that they were an important spectacle: 'it's important to the consumer, because particularly those that don't live in the countryside it's harder and harder to come face-to-face with a cow these days' (2). Alternatively, another manager suggested that 'I actually firmly believe that for the visiting public, the cattle add a flavour and style, they're not actually a major reason for going, but without it you do lose the character' (5). Whichever of these views on the significance of livestock is correct, show managers are increasingly involving livestock in their attempts to inform non-farming publics. This has included setting up information boards illustrating the agricultural importance of different livestock breeds, and encouraging judges of livestock classes to give commentary in a form that was accessible to a non-farming audience, alongside more technical comments aimed at farmer audiences. In some cases, livestock exhibitors were being asked to become directly involved in a process of establishing communication with non-farming visitors. For example, one manager said 'We're going to ask individual competitors to actually bring their animal forward, to a breed description board, to actually remain there for half an hour, and actually discuss why his bull is a lovely Charolais bull, or whatever' (1).

For these managers, then, the breeders themselves, and their associated breed societies, are important resources in re-imaging agriculture. The spectacle of valuable animals in 'show condition', parading in the ring, or accessible (along with their handlers) to the public in the 'lines', can be used by agricultural societies as a powerful counterbalance to more negative images of livestock production. In this way, show livestock perhaps dissimulate much of modern, intensive livestock farming practice, presenting an alternative positive image of the human-animal relationships in farming. Thus, a representative from a cattle breed society said 'I think from a farming point of view the reassurance of welfare, cattle management, good husbandry, nutrition, health, etc. it's a great opportunity to show the general public how caring farmers actually are, for their stock.' Exhibitors themselves made similar points, suggesting that contact with an interested public, particularly if visitors were allowed to wander around the livestock lines and meet farmers, was an important means of improving the image of farming. A breeder of Shorthorn cattle thus said that showing was an opportunity to demonstrate that they care about the animals and the broader farm environment, 'and that we enjoy the cattle and the job. We're not perpetual moaners, hopefully, cos that comes across sometimes if you're not careful, that farmers are perpetual moaners.' According to such exhibitors, therefore, negative perceptions of agriculture can be confronted by bringing visitors, livestock and farmers into contact.

Here, however, there was a tension between different strategies for re-imaging agriculture. Contact with, and display of, livestock was presented as a useful mode of 'reconnection' between the public and farming. Similarly, promoting a sense of connectedness between the non-farming public and farming as a source of food was understood as a valuable way of effecting a positive re-imaging. Yet, bringing those two strategies together, and directly associating livestock with food, was seen as problematic by several show managers. According to one, for example, 'it's difficult to do because you can't show them a cow, and say 'this is where you get your beef from' ... it's seeing the live animal and the dead animal, and I think people are very squeamish about that. They shouldn't be, but this is a message you've got to be very careful about' (3). The image thus created of farming tends hide particular aspects of livestock production, and the transformation of appealing or spectacular living animals into food.

'Interpreting' farming to farmers

The preceding section has illustrated some ways in which agricultural shows have attempted to become involved in a process of re-imaging agriculture, suggesting that particular narratives of agriculture have been invested in by show societies as something to present to a non-farming public. However, other processes noted by show managers have also become important in influencing how re-imaging might be taking place, and illustrate a reflexivity in show managers' changing understandings and constructions of the position of farming. Two in particular are discussed here.

First, in parallel with the emphasis being placed on 'interpreting' farming to non-farming publics, a need for farmers to be (re)educated in some ways about farming is stressed. Farmers are being encouraged by agricultural societies to rethink their position in relation to consumers, and to play a more active role themselves in the presentation of particular images of farming. There is a recognition here among show managers that it is not simply that the non-farming public has become distanced from food production, but that food producers have experienced a distancing from the consumers of the food they produce. As one manager put it, the show is about bringing the non-farming public into contact with farming, but also about bringing farmers into contact with the consumers of their products. Shows are about:

> 'bringing [farmers and the public] together in an environment where the two normally never meet ... the farming community is completely so isolated nowadays, you know, a lot of farmers spend their time on a tractor, late hours, you know, they don't have the people on the farm, very isolated ... [and] ... the consumer on the other hand, the urban population, don't have that relationship with the people on the farm'. (9)

Another manager argued that part of his job was to educate producers and exhibitors into seeing themselves as part of a food chain. He described this as the 'biggest change in 150 years, basically, you know, there are some very switched-on, progressive farmers, and there are some that don't see themselves in food production, and it's that we need to change' (2). For many, farmers, along with shows, needed to become involved in public relations on behalf of agriculture, and part of the agricultural societies' task was to encourage farmers in this direction. Farming was thus, according to these respondents, having to learn how to manipulate the media in a context where many stories presented the sector in negative ways. One manager commented, 'It's about good PR ... if you're being quite cynical you've got to make sure that the consumer's on your side. And the only way you can do that is by good PR, by continual, you know, positive reinforcement that agriculture ... needs to survive' (7). This need to legitimize farming to the public was a reason for farmers to support shows. For example, it was argued that 'the main reason [for farmers to attend the show] is because they need to be part of our promotion to the great British public about farming and they are the best people to do it' (5). There was a feeling that farmers were not good at communicating, and often contributed to negative perceptions of farming. In response, according to these managers, farmers need to be persuaded that they are not simply food producers, but part of a supply chain. Part of their job should be to appeal to consumers, and providing training in this was being adopted by agricultural societies. One manager, for example, described how she had been involved in a conference designed to persuade farmers to think about the way they represented farming: 'we actually ran a dialogue, in January [2002], to try and explain to farmers how they can promote themselves better, rather than coming across as whinging farmers, they've got to be a lot more positive in the message that they're getting over' (2). The idea of training some farmers to act as farming representatives in that particular county had emerged from this event.

Second, however, and paradoxically, is that despite the suggested significance of using shows to (re)connect the non-farming public to farming and develop positive images of agriculture, many shows' mainstream agricultural content has experienced decline. Two reasons for this were given. First, there were fewer companies operating in the agricultural supply business, due to company amalgamation. The field of potential exhibitors had thus severely declined. Second, the high costs of companies' attendance at shows were causing many companies to question the necessity to attend. Show managers had in many cases tried to counter this by reducing fees, yet the major part of the costs lay in staffing trade stands, a cost necessarily borne by the company. Ironically, then, given an emphasis on showing the non-farming public the 'truth' or 'reality' of farming, there has apparently been an at least partial evacuation of mainstream agricultural content from many shows. This evacuation was related by many managers to a trend towards specialist, technical events devoted solely to potential farming customers of agricultural companies. These concentrated events, known simply as Cereals, Grassland etc. usually run on farms rather than showgrounds, have been established by agricultural societies in accordance with their charitable aims, yet have simultaneously had the effect of reducing the agricultural content of their general agricultural shows. The agricultural content of shows has, according to some managers, become refocused on broader themes, such as business management, diversification, conservation and environmental protection, and farm-based recreation and leisure business options, rather than on specialist techniques and equipment. The simultaneous placing of agriculture within what one manager called the 'softer' (7) vision of countryside pursuits has also affected the image presented of farming at shows, developing a soft narrative of farming as associated with supposedly rural lifestyles and landscapes.

Given this situation, some managers suggested that the separation of 'real' farming content from the general agricultural show might jeopardize the objective of establishing a (re)connection between farming and non-farming communities. They were thus concerned with maintaining the agricultural content of the show in ways which both benefit farmers and contribute to the re-imaging of farming. This again could be seen as part of educating farmers and the agricultural industry that part of their job must be to engage in PR. One manager thus said:

> 'We can actually have a series of specialist events within the show, and at the same time we can make sure that we actually target the consumer ... So what we're seeing strategically is, if we can convince the farming community, and

> the farming community realises it's got to build its relationship back with the consumer, now it's lost that trust ... then it's quite easy if you get them on board, and they see what you're doing, it's then quite easy to keep your agricultural content.' (9)

Another argued that for farming to present an 'upbeat' image of itself shows needed a strong agricultural trade stand presence. Trade stands needed to simultaneously offer farmers technical advice and be able to communicate with the public in such a way as to contribute to the re-imaging of farming. With reference to trade stand image and content, an interviewee thus said 'it's got to be encouraging, it's got to tell a story' (8). According to this manager, the loss of agricultural content at general shows, partly driven by agricultural societies themselves as they have focused on specialist farmer-oriented events, has been detrimental to other strategic aims of reconnection and re-imaging. In this context it is interesting that, according to Henning (1998), many North American shows have experienced a more dramatic degree of evacuation, yet this is seen in positive terms as an opportunity to focus on their primary aim of 'educating' the non-farming public.

Conclusions

It has become increasingly clear that some actors in the farming sector are seeing it as important to establish processes of re-imaging agriculture to the public, in ways which they feel will increase public understanding of UK farmers and farming. There is a persistent feeling that such a re-imaging will be associated with a heightened sympathy for farmers, which could be translated into greater consumption of UK-grown food. Processes of 'informing' non-farming publics have therefore been engaged in by agricultural societies, in explicit attempts to establish a 'connection' with non-farmers and to represent farming in specific ways. The image of farming thus being constructed is one which stresses environmental awareness and stewardship, the positioning of farming as important to the conservation of an attractive countryside, and the production of welfare-friendly, high quality, safe food. At times it also presents the continuation of a viable farming sector as important to a sense of national identity and security.

The sites and events of agricultural shows present important opportunities for agricultural societies to engage in re-imaging. Although some have stressed the necessity of extending the temporal and spatial 'reach' of shows beyond their limited time-space occurrence, shows are key nexuses for disparate farming and non-farming publics, activities and discourses, at one moment in the year. Such meetings permit show societies to stage encounters with farming entities (e.g. machinery, animals, as well as farmers themselves), practices (e.g. livestock showing, conservation activities) and discourses (e.g. of agricultural progress, efficiency, and centrality to the countryside). In these ways, non-farming publics are both shown and told about farming in particular selective ways, contributing to the constitution of a particular image of farming as part of a struggle to defend, legitimize, or even recoup, a powerful position for agriculture in a contemporary countryside seemingly redefined as 'multifunctional'. Yet for many agricultural societies, this is also part of a reflexive recognition of the power of non-farming consumers (of food and the countryside) to be able to influence what happens in farming and farm policy, through their consumption practices and involvement in public sphere discourse and political change. Agricultural societies are thus staging events in which particular forms of showing, telling, reconnection and re-imaging can occur in order to influence public perceptions of farming. As one show manager commented:

> 'I want people to go away with the impression of having had a good day out, but also having learned something about the countryside, that the food being produced is wholesome, is actually safe, and if they've learned something like, you know, milk doesn't actually come from milk machines, it actually comes from cows ... then that's a positive outcome.' (1)

Non-farming publics therefore become subject to a re-imaging of agriculture, one which also constructs an interconnectedness between farming and maintenance of attractive rural landscapes, and tries to develop a sense of connection between non-farmers and the farming which produces their food and countryside.

However, despite such efforts to establish connections between non-farmers and farming, in several ways a practical and discursive distancing is reproduced between the two groups at show events. First, there is an inevitable difference between what is shown to non-farmers at show events, and the mundane, everyday experience of farming on farms. In particular, the mode of the spectacular presents farming as a series of discrete entities (impressive livestock or large machines), rather than seeing those entities as parts of practices which extend over longer time periods and require the enrolment of complex networks of other entities, knowledges and practices. Second, attempts by shows to counteract this issue by focusing on ideas of 'food chains' and connecting farming to food products indicate a recognition that a move away from the spectacular is an important part of 'informing' and 'educating' non-farming publics. Yet here too, partial representations of farming are constructed, focusing on particular 'acceptable' farming practices and leaving out those which might be regarded as unacceptable or controversial by non-farmers. At the same time, these processes of educating non-farmers tend to rely on assumptions of their 'ignorance', and a perception during the discourses that farmers and non-farmers inhabit very different 'lifeworlds', separated by a wide gulf of understanding. Finally, the partial evacuation of agricultural content from many shows, despite show managers' concern with maintaining this 'traditional' element of shows, again means that non-farming publics' encounters with farming miss out some of the 'realities' of agriculture. In these ways, then, agriculture is still represented to the non-farming public as something, somewhere, and somebody else despite, or perhaps because of, processes of attempted re-imaging.

Following this research into agricultural societies' efforts to re-image farming, further research will be important to determine how non-farming show visitors do make sense of what they have been shown and told of farming at agricultural shows, and through the various other strategies which agricultural actors are engaging in to re-image and legitimize farming. Further research into the extent to which these strategies 'act-back' on farmers' understandings of their own positions and their own sense of identity would be valuable in considering this important aspect of debate over a changing countryside.

Notes

1. Parts of this chapter are reprinted from: Holloway, L. (2004) Showing and telling farming: agricultural shows and re-imaging British agriculture. *Journal of Rural Studies*, 20, 319-330, with permission from Elsevier. I would like to thank Elsevier and the editor of *Journal of Rural Studies* for permission to reproduce some sections of that paper here.
2. Research for this chapter was conducted as part of a project on agricultural shows funded by the Nuffield Foundation Social Sciences Small Grants Scheme.

References

Clark, J. and Jones, A. (1998) Agricultural elites, agrarian beliefs, and their impact on the evolution of agri-environment policies: an examination of the British experience, 1981-92. *Environment and Planning (A)* 30, 2227-2243.

Evans, N., Morris, C. and Winter, M. (2002) Conceptualising agriculture: a critique of post-productivism as the new orthodoxy. *Progress in Human Geography* 21, 313-332.

Fox, H. (1979) Local farmers associations and the circulation of agricultural information in nineteenth century England. In: Fox, H. and Butlin, R. (eds) *Changes in the Countryside: Essays on Rural England, 1500-1900.* Institute of British Geographers Special Publication No. 10, pp.43-63.

Goddard, N. (1981) Agricultural societies in Mingay, G. (ed) *The Victorian Countryside, Volume I.* Routledge and Kegan Paul, London, pp.245-259.

Goddard, N. (1989) Agricultural literature and societies in Mingay, G. (ed) *The Agrarian History of England and Wales, Vol. VII 1750-1850* Cambridge University Press, Cambridge, pp.361-383.

Halfacree, K. (1999) A new space or spatial effacement? Alternative futures for the post-productivist countryside. In: Walford, N., Everitt, J. and Napton, D. (eds) *Reshaping the Countryside: Perceptions and Processes of Rural Change.* CAB International, Wallingford, pp.67-76.

Hall, V. (1987) *A History of the Yorkshire Agricultural Society 1837-1987.* B. T. Batsford Ltd, London.

Henning, J. (1998) *The Future Role and Direction of Agricultural Shows.* Report for the Nuffield Farming Scholarship Trust, Uckfield, East Sussex.

Hudson, K. (1972) *Patriotism with Profit: British Agricultural Societies in the Eighteenth and Nineteenth Centuries.* Hugh Evelyn, London.

Ilbery, B. and Bowler, I. (1998) From agricultural productivism to post-productivism. In: Ilbery, B. (ed) *The Geography of Rural Change* Longman, London, pp.57-84.

Lewis, J. (1996) *Three into One: The Three Counties Agricultural Society 1797-1997.* Baron Birch, Whittlebury.

Lowe, P., Buller, H. and Ward, N. (2002) Setting the next agenda? British and French approaches to the second pillar of the Common Agricultural Policy. *Journal of Rural Studies* 18, 1-17.

Overton, M. (1996) *Agricultural Revolution in England: The Transformation of the Agrarian Economy 1500-1850.* Cambridge University Press, Cambridge.

Seymour, S., Lowe, P., Ward, N. and Clark, J. (1997) Environmental 'others' and 'elites': rural pollution and changing power relations in the countryside. In: Milbourne, P. (ed) *Revealing Rural Others: Representation, Power and Identity in the British Countryside.* Pinter, London, pp.57-74.

Wilmot, S. (1990) *'The Business of Improvement': Agriculture and Scientific Culture in Britain, c.1700-c.1870.* Historical Geography Research Series no. 24, Historical Geography Research Group, Institute of British Geographers, London.

Wilson, G. (2001) From productivism to post productivism ... and back again? Exploring the (un)changed natural and mental landscapes of European agriculture. *Transactions, Institute of British Geographers* 26, 77-102.

Woods, M. (2003) Deconstructing rural protest: the emergence of a new social movement. *Journal of Rural Studies* 19, 309-325

Chapter 9

Stewardship, 'proper' farming and environmental gain: contrasting experiences of agri-environmental schemes in Canada and the EU

Guy M. Robinson

Introduction

A common feature of agriculture throughout the developed world in the last two decades has been a growing concern with the adverse impacts of farming upon the environment. Although it is possible to recognize schemes aimed at ameliorating these impacts dating to the 1970s, most countries have only established distinctive environmentally-friendly farming measures from the mid-1980s; some even more recently. These measures have varied considerably, though a common element has been the relatively small amounts of financial assistance associated with them when compared with that going towards support of production (e.g. only around 3% of the budget of the Common Agricultural Policy (CAP) of the European Union (EU) in the late-1990s went towards agri-environmental schemes). Indeed, so-called environmentally friendly farming has often been seen as a very limited adjunct to the main purpose of farming, namely to maximize output of food and fibre, primarily through reliance on use of industrial-style methods (Robinson, 2004).

In the EU, the growth of agri-environmental schemes dates to the late 1980s when the principle of paying farmers for undertaking desirable conservation-related activities was established. Pioneering this approach in the UK was the Environmentally Sensitive Areas scheme, launched after the 1986 Agriculture Act, and eventually leading to the designation of 44 Environmentally Sensitive Areas (ESAs) covering 17% of the land surface. The same basic principle, that of payments to farmers who volunteer to implement specified programmes, has been followed in England from 1991 by the Countryside Stewardship Scheme (CSS). In this scheme, farmers anywhere in the country are eligible to participate providing they had land with characteristics enabling them to fulfil the obligations of the specified programme.

In North America, agri-environmental measures have either tended to be less prominent than in the EU or have adopted different interpretations of the idea of attaining environmental gains. In the USA, there has been the large-scale Conservation Reserve Program, part of the 1985 Farm Security Act, introduced to set aside over 18 million ha over a 10-year period. Under this and similar schemes, 31.6 million ha were converted to 'conserving uses' in the 1980s (Bedenbaugh, 1988) with over 375,000 farmers enrolling on the programme, though many farmers have perceived this and other set-aside as the antithesis of 'proper' farming (Potter, 1998). Indeed, farmers have often responded by setting aside their poorest land while intensifying production on the rest of their holding (Ervin, 1988). Hence 'slippage' has occurred whereby the output of cereals has not fallen in proportion to the amount of land retired, an experience subsequently repeated in the EU's own set-aside scheme and in some agri-environmental programmes where it has been possible to enter only parts of a farm into a scheme (Robinson and Lind, 1999).

Meanwhile, in Canada in the mid-1990s, Hilts (1997) referred to a small number of mainly fairly localized agri-environment schemes (Table 9.1). Seven years on from the production of this list, it appears that one of them, namely the Ontario Environmental Farm Plan (EFP), is forming the blueprint for programmes in other provinces, e.g. in British Columbia (from February 2004) and Prince Edward Island. Hence it is most opportune to consider the nature of the EFP and to draw some significant contrasts between its rationale and operation and counterparts in the EU.

Table 9.1 Agri-environmental schemes in Canada in the mid-1990s

(a)	The Island Nature Trust in Prince Edward Island
(b)	Corporate conservation agreements in Nova Scotia
(c)	Local programmes operated by community groups in Quebec
(d)	Landowner agreements negotiated by the Muskoka Heritage Foundation in Ontario
(e)	The Natural Heritage Stewardship Programme in Ontario; this has involved voluntary agreement from 2000 landowners to protect approximately 24,000 ha of woodland and wetlands
(f)	Landowner agreements to conserve wildlife in the three Prairie provinces
(g)	In British Columbia there are stewardship agreements with farmers in the Fraser delta and with landowners around the Cowichan estuary
(h)	The Environmental Farm Plan (EFP) programme in Ontario

Source: Hilts (1997)

This chapter concentrates primarily on the notions of 'stewardship' and 'environmentally friendly farming' contained in the EFP, making some contrasts with ideas underpinning European agri-environmental schemes. The information about the operation of the EFP is based largely upon informal interviews with Ontarian farmers, primarily conducted by the author at Canada's Outdoor Farm Show at Woodstock in September, 2002 and at Orillia Farmers' Market in the same month (see Fig. 9.1). Other informal interviews with farmers and observations of farms were conducted in eight different counties across southern Ontario and with key informants from both provincial government and farmers' organizations within the province. These interviews covered both adopters and non-adopters of the EFP, but do not constitute a representative sample of farmers - instead they provide a guide to attitudes towards the scheme prevalent in farming communities and government in the province.

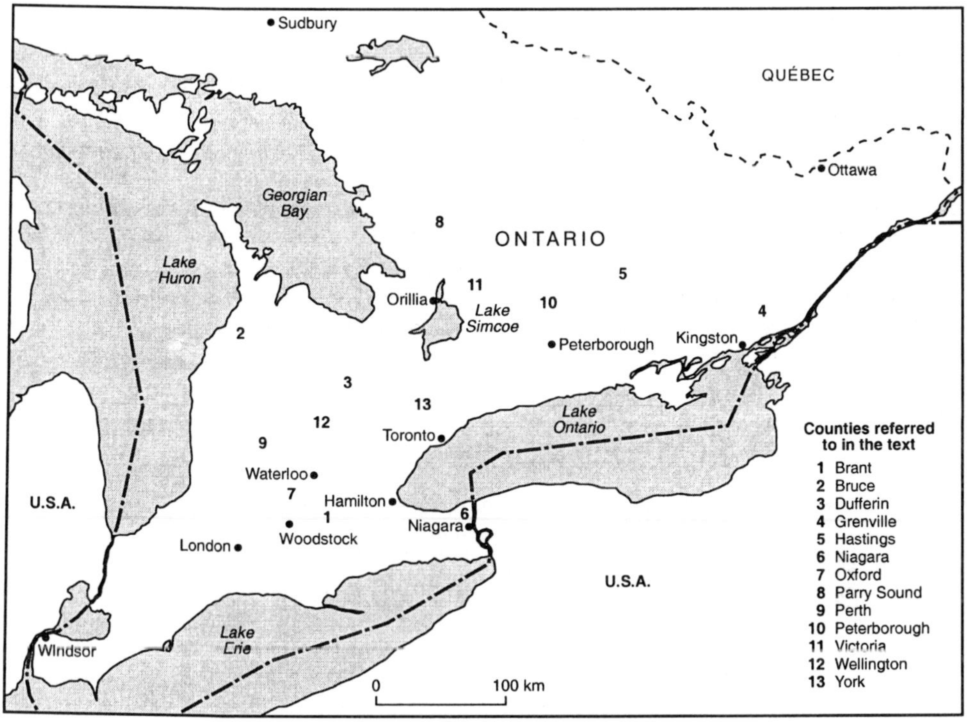

Fig. 9.1 Ontario - location map.

The Ontario Environmental Farm Plan (EFP)

Outline and operation

The EFP programme was formulated in 1991 by the Ontario Farm Environmental Coalition (OFEC), a grouping of Ontario farm organizations, to set an agenda for the farming community to adopt with respect to environmental concerns associated with agricultural production practices. It identified key issues as involving water quality (with special reference to agricultural nutrients), soil quality, air quality, agricultural inputs and natural areas such as wetlands and woodlots. It recommended that every Ontario farmer should develop and implement an environmental farm plan tailored to the needs of their own farm. These concerns were translated into a series of worksheets for participating farmers to complete, under five broad headings: farm productivity and profitability, aquifer degradation, surface water degradation, farm health and safety, and air pollution. The principal issues under these headings are detailed in Table 9.2.

Table 9.2 Worksheets in the Ontario Environmental Farm Plan.

Soil and site evaluation	Water wells
Soil management	Pesticide storage
Nutrient management in growing crops	Fertilizer storage
Manure use and management	Petroleum products storage
Field crop management	Disposal of farm wastes
Pest management	Treatment of household wastewater
Stream, ditch and floodplain management	Storage of agricultural wastes
Wetlands and wildlife ponds	Livestock yards
Woodlands and wildlife	Silage storage
Energy efficiency	Milking centre wash water
Water efficiency	Noise and odour
Horticultural production	

Following a pilot programme in 1993 in seven selected counties, Ontario's EFP programme was launched, funded through Agriculture and Agri-Food Canada's Green Plan Program. This scheme provides up to Cn$1500 per farm business to help farmers implement new management practices that effectively address environmental problems. As an additional incentive each year, the 12 best (practical, simple and economical) individual plans receive a small cash prize (Cn$1000). The individual plans are reviewed and approved by a local peer review committee. This peer review process was introduced to meet Agriculture Canada's funding requirement that a system be put in place to monitor the quality of the individual EFPs. To participate in the scheme, farmers have to attend a one-day EFP workshop and then develop their own environmental farm plan.

Participation in the EFP is voluntary and, if followed in full, involves a six-stage sequence from the introductory workshop to the implementation of the Plan, as indicated in Table 9.3. The workshops encourage those farmers who subsequently complete worksheets to pursue actions to minimize environmental risks, usually on the basis of self-identified actions (Bidgood, 1994). Through their individual plans, farmers highlight environmental strengths on their farm, identify areas of environmental concern and set goals to improve environmental conditions according to their own timetable.

Table 9.3 The six-stage sequence of the EFP.

	Stage	Actions
1	Introductory workshop	Site evaluation; assess potential concerns
2	Complete farm review	Review farm operations; complete relevant worksheets
3	Second workshop	Consider possible actions; learn how to develop a realistic plan
4	Complete action plan	Identify actions for all 'fair' or 'poor' rated situations; develop timetable for action
5	Peer review	Add suggestions/ask for changes; return plan to farmer; send information anonymously to OFC
6	Implementation of plan	Put plan into action; re-evaluate each year

In terms of farmers making an environmental evaluation of any given situation, a four-point scale is used, from 4 (best - conditions that protect the environment or have least potential for environmental damage) through 3 (good) and 2 (fair) to 1 (poor - conditions have the highest potential to affect the environment). In effect, farmers self-evaluate the environmental performance of their own farm in terms of environmental risk. In assessing the level of risk, existing environmental conditions (e.g. slope, soil type, drainage conditions) and farming practices are considered. Each situation is described in worksheets for 23 selected topics (and 260 individual questions), with indications of what to expect for each of the four categories. The worksheets cover the key concerns referred to above and to other topics such as pest control, use of water, disposal of wastes and cropping practices. An example, from the soil management worksheet, is shown in Table 9.4. In the worksheets, where relevant, conditions that violate provincial legislation are highlighted, e.g. septic pumpage disposal not performed by a licensed contractor. Participating farmers provide a self-rating for the selected worksheets for up to five sites, five fields or field groups. The worksheets themselves were designed by over 100 individuals working co-operatively through 23 committees consisting of farmers and representatives of farming organizations (both governmental and non-governmental). Through the province-wide system of workshops, over half of Ontario's farmers have now participated voluntarily in the programme, though with considerable sectoral and spatial variation.

Table 9.4 Farmers' self-assessment of soil resources: How do you rate on soil resources, pollution, wastes, pesticides?

Variable	Best (4)	Good (3)	Fair (2)	Poor (1)
1 Potential for soil compaction	very low	low	moderate	high
2 Effects of harvesting	Wide choice of harvest dates; good natural or tile drainage	Good choice of harvest dates; good natural drainage	Limited choice of harvest dates; fair natural drainage	No choice of harvest date; poor natural drainage
3 Crop rotations and cover crops	Regular rotation with soil-building crops and cover crops	Occasional rotation with soil-building crops and/or cover crops	Occasional rotation but little plant material retuned to soil or left on surface	No rotation or cover crops and little plant material returned to soil

Source: EFP worksheets.

Outcomes

By September, 2002, there had been Cn$48 million in on-farm improvements identified in the EFP. For every dollar received for support of measures in the EFP, it is estimated that farmers have spent three or four more dollars from their own pocket (Klupfel, 1998, p.16). The aim is to extend the EFPs towards a target of 40,000 individual farm plans implemented (approximately two-thirds of the province's farms), with additional measures identified for farmers to concentrate their plans on particular issues, notably on water quality.

By July 2002, 14,200 farmers (23.8% of the province's registered farmers) representing 1.1 million ha of farmland out of a possible total of 5.5 million ha had proceeded through the peer review stage of EFP (Fig. 9.2). There had been over 23,000 participants (38.4% of all farmers) in workshops. However, around 38% of participants in workshops have taken no further part in the scheme, a high drop-out rate that is prompting further thought within OFEC regarding the possible motivations for farmers to prepare and then implement environmental farm plans. The availability of assistance and encouragement from other farmers may be a significant motivating factor (Furman, 1997).

Fig. 9.3 shows that uptake has tended to be greatest in eastern Ontario, with lower levels of uptake in the province's agricultural heartland in the south-west. The highest uptakes per county as a proportion of possible adopters are for Granville county (81.4%) in the east, Peterborough county (75.8%) and Parry Sound county (73.6%) in central Ontario, and Dufferin county in the south-west (85.3%). In the south-west of the province, eight counties had an uptake of under 30%, with York (19.1%), Niagara (13.8%), Brant (14.0%) and Oxford (14.6%) under 20%. Victoria county (17.4%), east of Lake Simcoe, also recorded an uptake rate below 20%.

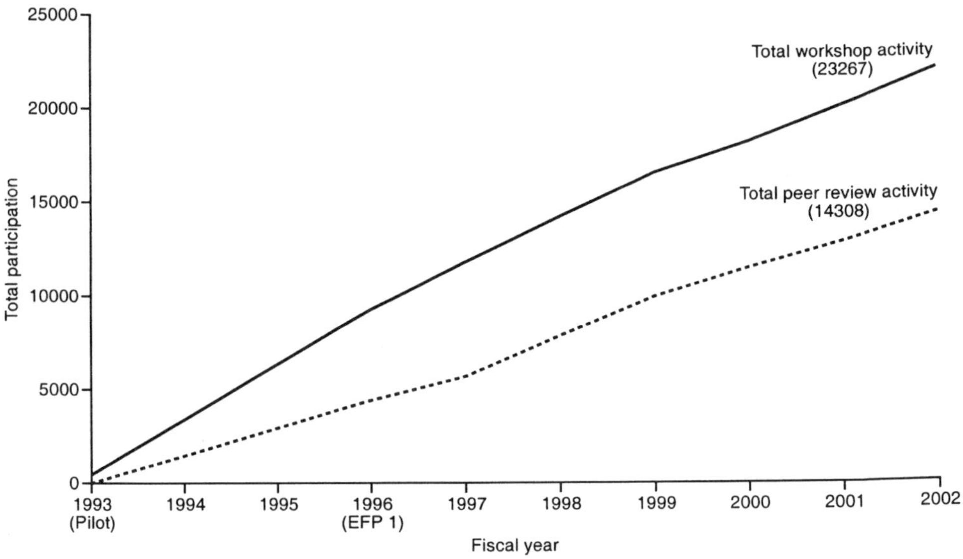

Fig. 9.2 Environmental Farm Plan uptake, 1993-2002.

In part, this spatial variation in uptake can be explained through the influence of the distribution of different farm types. For example, very low uptakes are associated with the chief horticultural areas in the province, the lowest rate of all occurring in the principal fruit and wine-producing county, Niagara. As shown in Table 9.5, it is the cattle enterprizes that are most well represented, accounting for 48% of completed worksheets, whereas they account for only 38% of farms in the province. In contrast, the fruit and vegetable producers are under represented in the completed worksheets, as are the 'other' category, which mainly represents mixed and non-traditional specialist enterprises, e.g. wine producers, tobacco growers. There is a discernible tendency for greater participation by the province's livestock producers and lower rates of participation among the province's crop and fruit and vegetable producers. The relationship between participation and farm size is not statistically significant, though both the smallest and the very largest farms seem to be less likely to participate.

Table 9.6 summarizes data from the Ontario Soil and Crop Improvement Association (OSCIA) for 9,991 worksheets completed by farmers between April 1993 and August 2002. It highlights the importance of soil management, water wells and storage of agricultural waste as farmers' major concerns. However, it also demonstrates the wide range of actions covered in EFPs. Further evidence of this range comes in OSCIA statistics concerning EFP-related actions taken by the farmers: 600 no-till planter drill purchases, 220 double-wall farm fuel tanks, 600 watercourse and woodland fencing projects to restrict livestock movement; 250 transplanting projects; 200 household septic system replacements; 1,600 drinking water well upgrade or plugging projects; and 300 foam markers purchased for sprayers.

Table 9.5 Participation in the EFP by sector (% of total).

Sector (n = 9,991)	EFP [all enumerated farms]	Ontario Agricultural Census (1996)
Field crop	31	30
Pork	4	4.5
Dairy	15	14
Beef	33	24
Fruit/vegetables	4	6
Poultry	2	3
Other	11	18.5

EFP percentages based on 9,991 farmers' identification of major farm enterprize in worksheets. (Source: OSCIA).

Table 9.6 Range of activities undertaken within the EFP, 1.4.93 – 31.7.02.

Activity	Percentage of all activities
Soil management	25
Water wells	17
Storage of agricultural waste	10.5
Stream, ditch and floodplain management	7
Livestock yards	6.5
Storage of petroleum products	5.5
Treatment of household waste water	4
Energy efficiency	4
Pest management	3.5
Pesticide storage and handling	3
Field crop management	3
Nutrient management in growing crops	2
Woodlands and wildlife	1.5
Disposal of farm wastes	1
Wetlands and wildlife ponds	1
Water efficiency	1
Fertilizer storage and handling	1
Noise and odour	1
Milking centre wash water	1
Horticultural management	0.5
Silage storage	0.5
Manure use and management	0.5

Source: OSCIA

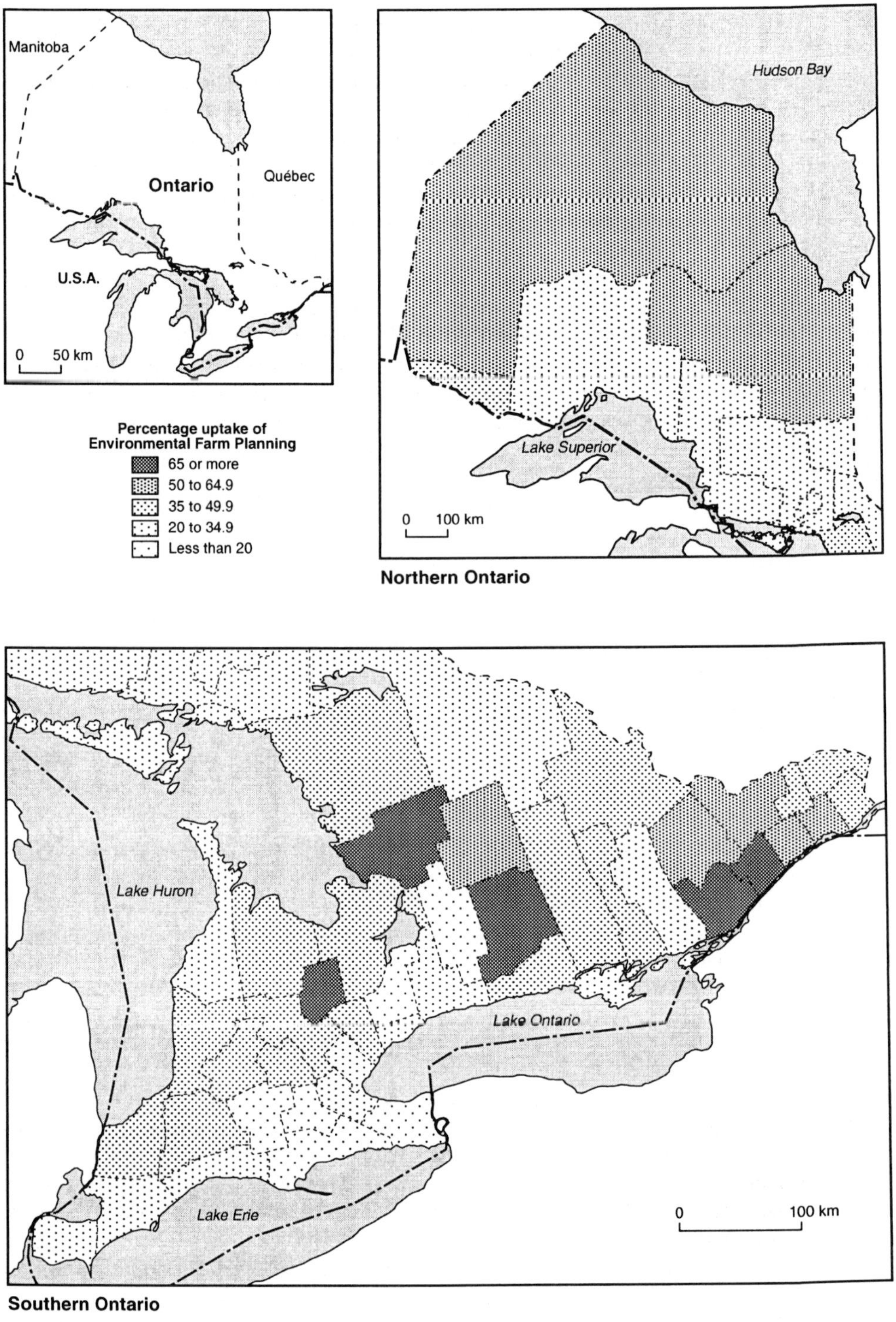

Fig. 9.3 Uptake of the Environmental Farm Plan (by county).

In terms of the primary actions taken by participants in the EFP during the first 9 years of its operation, OSCIA's classification of actions (detailed in Table 9.6) revealed that soil management accounted for one-quarter of all activities, in part related to the fact that, in Canada as a whole, soil erosion reduces farm income by over Cn$1 billion per annum (Klupfel, 1998, p.20). Particular soil management concerns among participating farmers interviewed by the author related to the application of conservation tillage to reduce soil erosion, greater use of crop rotations, planting of cover crops and shelter belts, introduction of tile drainage, careful use of farm machinery to avoid soil compaction, and application of cropping practices that retain a high amount of organic matter.

From the outset, water quality has also been a key concern in the EFP, and just under one-quarter of the actions taken by participating farmers were associated with water management and water quality, especially linked to use of water wells (17%). The latter reflects the growing concern for the quality of groundwater both for human consumption and for irrigation and livestock consumption. There has been concerted lobbying for farmers to upgrade existing wells, and install new and deeper wells. The significance of issues relating to soil management and water quality is born out in the sample survey by Fitzgibbon *et al.* (2000a, 2000b). This research showed that, in terms of the types of conservation actions pursued, respondents acted upon more than 60% of the worksheet concerns identified under the headings of soil erosion and quality, water consumption, water well contamination and the physical degradation of surface water.

Table 9.6 also shows that management of waste on the farm has been a significant concern, including storage of toxic materials and waste products as well as disposal of wastes. Together these concerns accounted for around one-quarter of the actions taken, some of which can be closely linked to those associated with maintaining and improving water quality. In informal interviews of farmers conducted by the author, the management of farm waste was a recurrent theme. Here a clear link was established with the incident at Walkerton (100 km north-west of Waterloo) in May-June, 2000 when seven deaths and illness in over 2,300 people, including children who may endure lasting effects, were caused by *E. coli* O 157:H7 and *Campylobacter jejuni* contamination of water supplies to the town by manure-based run-off from farmland (O'Connor, 2002). The fact that it was unsafe practices at the town's pumping station that were found to be responsible at a public inquiry had not prevented much prior speculation about mishandling of farm waste as the culprit. Nor had it stopped farmers from considering the potential for issues relating to environmental liability to lead municipalities to introduce greater regulatory controls through by-law provisions relating to the establishment and management of livestock facilities. The potential for nuisance suits against farmers was also raised by this incident (Caldwell, 2001). Key issues, now more apparent in the public arena following Walkerton and other incidents elsewhere in North America, include the land base available for disposal of manure; proper use of manure and nutrient components; type of manure storage; distance from manure storage to non-farm uses; methods of manure disposal; and the size and type of livestock operations (Henderson, 1998).

In a sample survey of 179 participants throughout the province, Fitzgibbon *et al.* (2000a) found that the chief positive responses of farmers to the EFP were increased awareness of farm conservation issues, its educational value and the identification of potential environmental risks. Just over half the sample had revised or updated their action plans and nearly two-thirds had undertaken environmental actions in addition to those stated in their own EFPs. Two-fifths claimed that they had taken actions to stop 'an environmental peril'. For this sample of farmers, the cost of additional environmental actions undertaken at their own expense was Cn$1,537,443 (or Cn$8,589 per farm). These actions primarily referred to large capital expenditure involving new machinery, tile drainage and instigating no-till systems. In general, farms under 40 ha and horticultural farms were less likely to undertake additional actions. The main actions pursued involved improvements to soil and water quality, and general improvements to family health and safety. Relatively few in the sample (15.4%) were members of conservation-related organizations. This finding may support the view that the scheme is being taken up largely by farmers who see it primarily as supporting better management of their farms rather than as something that explicitly yields a conservation or environmental benefit. While 'better management' and 'environmentally beneficial' are not mutually exclusive outcomes, the author's informal interviews with farmers suggested that it is the former that they most readily associate with the scheme.

In assessments of the EFP's operation a key feature, according to the farmers interviewed by the author, was the voluntary nature of the process and its confidentiality. The available financial incentives were not seen as a significant factor in encouraging participation, but greater financial rewards were seen as a possible improvement to the scheme. 'Keeping government away' and no enforcement of conservation measures via legislation were seen as key aspects of the scheme's future development, though more guidance for conservation actions was also deemed important (see also Klupfel, 1998, p.61, p.98). Possible links were noted to future adherence to ISO14 000 standards as a means of enhancing marketing opportunities.

Despite the very positive views of farmers towards the EFP, it must be remembered that, as of September 2002, three-fifths of Ontario's farmers had not yet attended EFP workshops and that three-quarters had not submitted an EFP for peer review. This level of non-involvement suggests that barriers to participation are prevalent (see Smithers and Furman, 2003). Informal interviews with farmers suggested that prime barriers were perceived costs of environmental actions and the farmers' personal priorities, in which environmental actions were often deemed as relatively unimportant (see also Fitzgibbon *et al.*, 2000a; Holmes, 1998). It appears that a crucial factor in helping to determine the extent of participation in the scheme is this notion of what farmers regard as a central element in their job. Many Ontarian farmers seem to regard the measures that are integral to the EFP as either 'extras' that do not really apply to them or as simply unnecessary or even as things that they pursue already without needing to enrol in a special plan. In contrast, the participants in the EFP may have a somewhat different concept of what is 'normal' or 'proper' farming practice, and hence they have been more willing to sign up and to be seen by their peers as being 'good stewards' of the land. Therefore, the pattern of participation may reflect the operation of peer pressure among the farming community, with that pressure varying spatially and sectorally.

The EFP and notions of 'stewardship'

There are clear differences between the nature of the EFP and the majority of agri-environment schemes operating in the EU. These differences relate primarily to the underlying aims of the various schemes. In the EU, the dominant intention has been to encourage extensification in the form of a move towards lower input - lower output production that places less stress upon the environment. Hence, the schemes have encouraged farmers to undertake environmental actions that are part of a move away from industrial-style farming methods. Typical actions have included creation of and/or better maintenance of farm woodland, the creation of buffer strips between cultivated fields and field boundaries to support wildlife, reductions in the amount of ploughing activity (especially to encourage birds at key times of year), reduction in the use of stubble burning and stubble removal once grain crops have been harvested, adherence to traditional cultivation or pasture management practices, setting-aside portions of arable land from cultivation, and restricting 'industrial' practices such as draining wetlands, removing trees, and applying artificial fertilizers and pesticides (Tables 9.7 and 9.8).

Table 9.7 Objectives of the first Scottish Environmentally Sensitive Areas (ESAs)

Objective	Breadalbane	Loch Lomond	Machair of the Western Isles	Stewartry	Whitlaw/Eildon
Protect rough grazing	X	X		X	X
Protect unimproved, enclosed land in valleys	X				
Repair farm dykes/hedges	X	X		X	X
Regenerate farm woodland					
Harmonize new developments with landscape amenity	X	X		X	
Limit modern agricultural practices			X		
Encourage traditional crofting techniques			X		
Protect machair from over-stocking			X		
Encourage positive conservation of machair			X		
Protect wetland areas/loch shores from damage				X	X
Restrict pesticide usage on edges of arable fields					X
Protect archaeological earthworks					X

Source: Robinson, 1994.

In contrast, Ontario's EFP essentially involves 'sensible' management of the existing farming system. It includes some of the same measures as found in EU agri-environment schemes, though its primary intention is not to promote extensification, but to encourage 'good stewardship' of the land in terms of minimizing negative environmental outcomes arising from productivist practices. Indeed, it was aptly described by a livestock farmer interviewed at the Woodstock show as 'simply representing the promotion of better care for the land'. Two quotes from different farmers at the show illustrate a common view of the EFP as being sympathetic to the productivist model of farming: 'I have not reduced inputs to my land, but I take more care about how I handle pesticides and how I handle the manure from my (dairy) herd'; and 'I have introduced a no-till regime, but my yields have not been (adversely) affected. I am still using the same tractors and combines.'

Table 9. 8 Items of work in England's Countryside Stewardship Scheme.

Managing grassland	**Managing fen and reedbeds**
Lowland hay meadows	Managing fen
Upland hay meadows	Managing reedbed
Lowland and Culm pastures and old pastures	**Managing lowland heath**
Upland in-bye pasture	Maintaining existing lowland heath
Upland rough grazing enclosed pastures	Enhancing existing lowland heath
Upland rough grazing	Re-creating heath
Chalk and limestone grassland	**Access Agreements**
Upland limestone grassland	Open
Recreating grassland on cultivated land	Linear - footpaths
- Supplement for raised water levels	Bridleways
- Supplement for use of native seed	Access for the disabled
	Access for educational purposes

Source: Simpson and Robinson, 2001

The more widespread nature of these views about 'good stewardship' among Ontarian farmers is supported in research by the Environics Research Group (2000) in a survey of the attitudes of over 1200 landowners to land stewardship. This survey concluded that environmental issues were both of significant interest and concern to a large number of landowners, including farmers. In particular, there was concerted appreciation of 'stewardship activities' for the future health and productivity of the land. Farmers in the survey placed emphasis on a largely utilitarian view of stewardship while also indicating the potential for undertaking environmental actions if these were accompanied by financial rewards. A link between caring for the environment and 'good management' was evident, but the attitudes expressed were not in favour of extensification or low-input, low-output farming, rather they related to a mixture of 'keeping the land economically productive' and 'voluntarily conserving the natural environment' (p.24). Stewardship as voluntary conservation rather than for direct financial gain was most likely to be pursued by the better-educated farmers (p.26).

Similarly, the farmers interviewed by the author in different parts of the province did not view the EFP as a challenge to their 'normal' farm practices, but in general as sensible precautionary measures designed to represent good stewardship of the land. Several extended this view by adding such comments as, (cereal producer) 'it's much better to do these things voluntarily rather than being made to do so by the government', and (beef producer) 'I'd rather volunteer to put in a new septic tank than be told (by government) that I had to have a new one because I was breaching some law or other'. This concern with the prospect of 'punitive' environmental regulations being introduced appeared to be widespread and was a strong motivating force underlying participation in the EFP. However, it was invariably allied to positive reasons for participation.

The 'positives' were generally tied to expressions of what the farmers regarded as 'good stewardship'. For example, 'I have upgraded the well on my farm because it will give better quality water to the stock' and 'I've put in new tile drains; they'll improve crop growth'. The EFP's emphasis on 'sensible' handling of waste, fertilizers, pesticides and petroleum products was generally regarded as the fostering of good practice and was linked by the farmers both to their own positive view of how they farmed and to their standing within the farming community. Clearly the EFP workshops and the subsequent actions taken by farmers have become 'the norm' in many counties - so not participating and not being concerned to apply the latest ideas about 'good management' are increasingly being seen as reasons to be held in lower esteem within the local farming community. Nevertheless, the high rates of non-conversion of workshop participation into plan formulation and action, coupled with relatively low attendance at workshops in some districts, suggests that the EFP has not yet appealed to a significant cross-section of the province's farming community.

One point explored in the interviews was the potential for a link between participation in the EFP and financial rewards from other programmes. Such a link was generally regarded favourably. This connection was also commented upon in Van Osch's (1995) small sample survey of farmers in Oxford County, which revealed that two-thirds of farmers supported the view that farmers should have participated in the EFP to be eligible for receipt of any government financial assistance. The support for regulatory cross-compliance may suggest that this particular sample group had a high proportion of the early participants in the EFP. Indeed, only half of the sample felt that it was imperative for the EFP to remain reliant on voluntary participation. A concern for and interest in the environment were seen as prime motivations for participation in the EFP.

The possibility of a link between participating in the EFP and receiving financial assistance in other schemes has been raised in another province-wide programme, Ontario Stewardship, initiated in 1995 by Ontario's Ministry of Natural Resources (OMNR), to encourage private landowners to care responsibly for their land, again emphasizing encouragement rather than legislation or statutory controls. Hence, the emphasis has been upon education, collaborative arrangements and peer networking via 39 Community Stewardship Councils, which receive financial support from OMNR. Each Council has between eight and fifteen members from volunteer landowners, land interest groups and representatives from key agencies. The Councils are based on the upper tier municipalities and regional bodies. They have local autonomy, operate on a non-profit basis and are non-incorporated organizations. OMNR provides a staff person to each Council and some seed money. The Community Stewardship Councils have been strong promoters of the EFP, commonly through sponsoring workshops (MSC, 2002).

Informal interviews with farmers who are strong supporters of the EFP suggest that such farmers are also quite likely to be involved in the Ontario Stewardship programme or support its ideals, including its notions of environmental stewardship. Tangible benefits for farmers through Ontario Stewardship were identified as the transfer of knowledge (especially of practical management possibilities), the broadening and strengthening of their chain of contacts, and greater access to key agencies and associations. Examples from Middlesex County, in southwest Ontario, indicate the type of actions involving farmers participating in the programme: tree or grass buffer zones along streams, establishing tall-grass prairie habitat, reforestation, reducing pollution of watercourses, improving woodlot management, participating in the Clean Water Project (a rural water quality initiative that provides technical and financial assistance to improve and protect water quality on farms), and participating in a Healthy Futures programme, in which from 50% to 70% of the costs of a range of environmental improvements are met (including improving septic systems, erosion control structures, fragile land retirement, restricting livestock access to water courses). Farmers are only eligible to participate in the Healthy Futures and Clean Water programmes if they have completed an EFP. Similarly, the Christian Farmers' Federation of Ontario has established an Ontario Environmental Initiatives Loan Fund Programme in which farmers' eligibility to receive loans is dependent upon their preparation of an EFP Action Plan.

From the preceding discussion, it is apparent that Ontario's EFP does not actively promote moves away from productivist (industrial) models of farming. Instead, it encourages some modification of existing agricultural systems rather than a set of specific prescriptions for extensification and sustainability. It is also the antithesis of a state regulation approach to sustainability favoured in some circles, although such regulation can be seen to have increased gradually in Ontario in recent years.

The EU's Agri-environmental Regulation 2078/92, promoting wider use of agri-environmental measures, advocated a standard model in which a voluntary contractual agreement between farmers and the state would be concluded, in which farmers agreed to produce environmental goods in return for specified payments. In contrast, the EFP has incorporated voluntarism, but without formal agreements between farmers and the state. Indeed, such regulation has been strongly resisted in favour of a form of self-regulation operated via inputs from farmers' organizations.

The EFP and most agri-environmental schemes in the EU provide opportunities for environmental gains to be incorporated into agreed farm management plans. However, such plans do not necessarily deflect farmers from pursuing a productivist model of farming. Even in schemes in the EU that are promoting extensification, farmers may reduce inputs on the land entered into environmental agreements while intensifying on the rest of their land (the halo effect), and hence the championing of 'whole farm' approaches (Morris and Cobb, 1993). However, such approaches are geared largely to the establishment of a new model of farming, which is a different aim to that of the EFP, itself essentially pursuing a whole farm approach.

Furthermore, in the EU, under Agenda 2000 a more substantive system of 'cross-compliance' was introduced whereby full receipt of arable area, headage and other payments became conditional upon respect of environmental standards (Beard and Swinbank, 2001). This and subsequent reforms will allow member states to make direct aid payments conditional on farmers fulfilling nationally-defined environmental management standards. Penalties or even complete withholding of payments can be made for non-compliance. A 'modulation' proposal allows member states to redirect up to 20% of compensation payments per farm, as a function of the level of employment on the individual farm or overall prosperity of the holding through funding shared with the European Commission, for 'accompanying measures' such as agri-environment programmes, afforestation and early retirement schemes (Falconer and Ward, 2000; Lowe and Ward, 1998). This proposal has been termed the 'double dividend' in that resources previously tied up in price support can now be available to be reinvested in environmental schemes (Potter and Goodwin, 1998).

Decoupling of the link between farmers' subsidies and the quantities of crops they grow or livestock they keep has been further advanced in the most recent CAP reforms agreed by the EU agriculture ministers in June 2003. Countries or regions will retain some level of production subsidies, notably on cereals (c. 25% of the current level), cattle and sheep, based on the average of the farmers' last two annual subsidies. The reduced subsidy should encourage reductions to sheep and cattle numbers, promoting environmental benefits and improvements in the quality of produce. The rate at which decoupling will be advanced will vary between member states, the UK Agriculture Secretary of State, Margaret Beckett, promising to decouple all English farmers' subsidies at an early date. The proposal is to pay a fixed 'single farm payment' dependent on land being maintained to an acceptable level. Food producers will have to meet 18 cross-compliance requirements on the environment, animal welfare and food quality standards. 'For the first time, the main subsidies are explicitly linked to compliance with EU standards covering the environmental, public and animal health and animal welfare. Farmers also have to maintain land in good agricultural and environmental condition as defined by the Member States. The Member States will have to inspect a sample of the farms each year on a systematic basis to ensure that standards are met.' (DEFRA, 2004).

Beckett has claimed that these reforms would 'reconnect farmers to their markets' and enable them to 'produce for the market, not the subsidy' (*Daily Telegraph*, 27.6.03, p.14). However, countries can delay decoupling until 2007. Under cross-compliance, member states will be allowed to redirect up to 10% of their CAP payments into a 'national envelope' to encourage farming in less favoured areas. This feature is intended to reduce the threat of farmers deserting uneconomic land. Farmers will also be subject to progressive cuts (modulation) in their payments: from 3% in 2005 to 5% in 2007, to pay for rural development schemes. However, UK farmers will be modulated at a higher rate in accordance with the farming and food strategy designed to revitalize the countryside after the foot-and-mouth epidemic. In addition, member states must set up a farm advisory service that will be available to farmers to help them meet their cross-compliance obligations by 2007.

Meanwhile, the advance of cross-compliance, as now occurring in the EU, is something about which many Ontarian farmers express strong reservations. Even those who have adopted the EFP, who may have gained some experience of measures based on cross-compliance and who have a positive view of the environmental benefits associated with the EFP, fear enforcement of environmental standards by government and reject extensification as it represents the antithesis of 'proper' farming. Indeed, the discussion above suggests that the contexts under which environmental benefits are sought from agriculture in Ontario and the EU may continue to diverge, with the result that Ontario's 'soft green' EFP is likely to contrast with a 'harder green' in large sections of EU agriculture induced by decoupling, cross-compliance and modulation.

Acknowledgements

The author gratefully acknowledges support from the Department of Foreign Affairs and International Trade, Canada, who funded the research project upon which this paper is based, and the Centre for Land and Water Stewardship, University of Guelph, who provided logistical and other assistance.

References

Beard, N. and Swinbank, A. (2001) Decoupled payments to facilitate CAP reform. *Food Policy*, 26, 121-46.

Bedenbaugh, E.J. (1988) History of cropland set aside programs in the Great Plains. In: Mitchell, J.E. (ed.) *Impacts of the Conservation Reserve Program in the Great Plains.* USDA Forest Service, Rocky Mountain Forest and Range Experiment Station, Fort Collins, Colorado, no. GTR RM-158, pp.14-17.

Bidgood, M. (1994) *A Study of Actions by Simcoe County Environmental Farm Plan Participants.* Ontario Soil and Crop Improvement Association, Guelph, Ontario.

Caldwell, W. (2001) A municipal perspective on risk management and agriculture. *Great Lakes Geographer*, 8, 31-40.

DEFRA (2004) Common Agricultural Policy Reform http://www.defra.gov.uk/farm/capreform/agreement-summary.htm (accessed on 25 November, 2004).

Environics Research Group (ERG) (2000) *Surveys of farmers, ranchers and rural landowners: attitudes and behaviours regarding land stewardship.* ERG, Toronto.

Ervin, D.E. (1988) Set-aside programmes: using USA experience to evaluate UK proposals. *Journal of Rural Studies*, 4, 181-91.

Falconer, K. and Ward, N. (2000) Using modulation to green the CAP: the UK case. *Land Use Policy*, 17, 269-78.

Fitzgibbon, J., Plummer, R. and Summers, R. (2000a) *Environmental Farm Plan indicator survey.* OSCIA for the OFEC, Guelph.

Fitzgibbon, J., Plummer, R. and Summers, R. (2000b) *Environmental Farm Plan indicator survey: systems analysis.* OSCIA for the OFEC, Guelph.

Furman, M. (1997) Environmental farm planning: understanding prospects for sustained change. Unpublished MA thesis, Department of Geography, University of Guelph.

Henderson, H. (1998) Noxious neighbours, *Planning*, 64 (11), 4-9.

Hilts, S.G. (1997) Achieving sustainability in rural land management through landowner involvement in stewardship programmes. In: Ilbery, B.W., Chiotti, Q. and Rickard, T. (eds.) *Agricultural Restructuring and Sustainability: A Geographical Perspective.* CAB International, Wallingford, pp.267-77

Holmes, E.A. (1998) Barriers to participation in the Ontario EFP program. Unpublished MA thesis, University of Guelph, Guelph.

Klupfel, E.J. (1998) Developing an understanding of and recommendations for promotion of the Ontario Environmental Farm Plan. Unpublished MSc thesis, Department of Rural Extension Studies, University of Guelph, Guelph.

Lowe, P. and Ward, N. (1998) Regional policy, CAP reform and rural development in Britain: the challenge for New Labour. *Journal of Regional Studies*, 32, 469-74.

Middlesex Stewardship Committee (MSC) (2002) *The Middlesex Steward.* Middlesex County, Ont.

Morris, C. and Cobb, P. (1993) Agriculture and conservation – the whole farm approach. *Ecos*, 14 (3/4).

O'Connor, Hon. D.R. (2002) *Report of the Walkerton inquiry: the events of May 2000 and related issues.* Ontario Ministry of the Attorney General, Toronto.

Potter, C. (1998) *Against the Grain: Agri-environment Reform in the USA and the European Union.* CAB International, Wallingford.

Potter, C. and Goodwin, P. (1998) Agricultural liberalization in the European Union: an analysis of the implications for nature conservation. *Journal of Rural Studies*, 14, 287-98.

Robinson, G.M. (1994) The greening of agricultural policy: Scotland's Environmentally Sensitive Areas (ESAs). *Journal of Environmental Planning and Management*, 37, 215-25.

Robinson, G.M. (2004) *Geographies of Agriculture.* Pearson, Harlow.

Robinson, G.M. and Lind, M. (1999) Set-aside and Environment: A Case Study in Southern England. *Tijdschrift voor Economische en Sociale Geografie*, 90, 296-311.

Simpson, S. and Robinson, G.M. (2001) Agri-environment policies, post-productivism and sustainability: England's Countryside Stewardship Scheme. In: Molinero, F., Baraja, E. and Alario, M. (eds), *Proceedings of the Second Anglo-Spanish Symposium on Rural Geography, University of Valladolid*, July 2000, University of Valladolid, Spain, Section 3-3, pp.1-16.

Smithers, J. and Furman, M. (2003) Environmental farm planning in Ontario: exploring participation and the endurance of change. *Land Use Policy*, 20, 343-56.

Van Osch, K. (1995) Planning for agriculture in the countryside through farm plans. Unpublished MES thesis, Department of Geography, University of Waterloo, Waterloo.

Chapter 10

Stemming the urban tide: policy and attitudinal changes for saving the Canadian countryside

Hugh J. Gayler

Introduction

Saving the countryside by fighting traditional suburbanization and urban sprawl, promoting the greater intensification of existing urban areas, redeveloping brownfield sites, providing better public transportation, and protecting the rural and agricultural resource base beyond the city have become important goals in the twenty-first century in various jurisdictions throughout North America. It has increasingly been recognized that present urban expansion policies and practices are economically, socially and environmentally unsustainable; and there is a need to educate government and the public about the long-term benefits of reform, to change attitudes towards a new urbanism and to provide for a more inclusive urban environment. These reforms, and the momentum behind them, have been termed 'Smart Growth', and different policies have been devised to put these initiatives into practice in various parts of Canada and the USA.

Unfortunately, reform will not come easily. The unsustainability of traditional urban expansion has been known for a long time, but only lip-service has been paid to it. Meanwhile, land-hungry practices continue because a development industry and a middle-class public are firmly wedded to them. The traditional ingredients, including spacious sites, one and two storey building, car dependency, maximizing personal and business freedoms and profit, and scant regard for the rural resource, continue unabated. Moreover, these features are a fundamental part of the culture of the ruling group, embedded in lifestyles for the last half century. Turning all of this around will not just happen through programmes such as 'Smart Growth', especially when the public scarcely understands what it means, but through more long-term education leading to attitudinal change.

Chinks in the armour of traditional methods of growth, however, can be seen and more than lip-service is being paid. This chapter will first examine the nature of the problem, and then using a case study drawn from Canada's Niagara Wine Region, it will examine urban pressures and the failure to protect valuable agricultural land. A critical element has been an agricultural industry, in particular the grape and wine and horticultural industries, that is showing renewed strength and a desire to be better protected from urban development. The wine industry itself and the various associated agritourism developments have raised the whole profile of how important it has become to save the countryside. Meanwhile, government policies, which on paper at least seemed restrictive towards urban expansion, have shown over time to be less than effective; and now, through measures such as 'Smart Growth', there is greater confidence that more aggressive policies to control sprawl will be put in place. However, the Golden Horseshoe of southern Ontario, of which Niagara is a part, is a fast-growing area, various economic and social pressures for traditional urban expansion are ever present, and there are renewed calls for more draconian measures to restrict urban development.

Nature of the problem

Urban development in North America since the Second World War has been characterized by a particularly land-hungry suburbanization that has been demanded by the public and sanctioned by government (Bourne, 1996; Fishman, 1987; Garreau, 1991; Jackson, 1985; Muller, 1981). It has been fuelled by various technological and business changes that have led to the decentralization of most urban activities from central cities. Meanwhile, a growing, middle-class, car-owning and home-owning population has sought the convenience, spaciousness and illusions to countryside that developers have attempted to provide in an ever-expanding suburbia. These developments have occurred in two principal ways. The first, and numerically the most important, has been serviced developments that are contiguous to the existing city. The second has very often been an unserviced, scattered development throughout the countryside close to the city (Bryant, Russwurm and McLellan 1982; Coppack, 1988). Both types of development have been equally invasive of the countryside and its agricultural economy, natural resources and rural population; the very term urban development or suburbanization is increasingly seen negatively as urban sprawl, while scattered development is commonly referred to as 'death by a thousand cuts.'

This 50-year-old suburban and ex-urban development model has come under increasing fire in the last few years from planners and other professionals, politicians, various activists intent on saving the countryside, as well as a development industry that has sought more traditional and compact urban designs, often on brownfield sites (Jamieson *et al.*, 2000). This model is generally deemed to be economically, socially and environmentally unsustainable. The costs of building and maintaining new services in low-density areas on the edge of the city result in ever-increasing tax burdens. The suburb, let alone the area beyond, is socially exclusive; for example, poor or non-existent public transportation results in extreme difficulties for, or the exclusion of, anyone who does not own or drive a car. Environmentally, low-density development encourages greater car and energy use, more pollution, and the continual loss of agricultural land, open space, tree lots and other sensitive natural areas.

The driving force behind restricting urban development in many areas has frequently been pressures on critical rural resources, including high quality agricultural land and significant (and often internationally recognized) physical features. In the fast-growing area of southern Ontario, these pressures are not simply occurring on Grade 1 soils and Canada's best agricultural land, but unique lands where a combination of aspect, soil and microclimatic conditions combine to give a spatially limited opportunity to grow specialty crops. The Niagara Wine Region, the northern part of the Niagara Peninsula between Lake Ontario and the Niagara Escarpment, is one of these unique areas (Figs 10.1 and 10.2), and one of only two small areas in Canada, the other being the Okanagan Valley in British Columbia, where it is possible to have sizeable, commercial, tender-fruit and grape and wine industries (Chapman, 1994; Shaw, 1994, 2002). Given this small resource base and the recent emergence of a quality wine industry, there has been a strong lobby for government to preserve these unique lands.

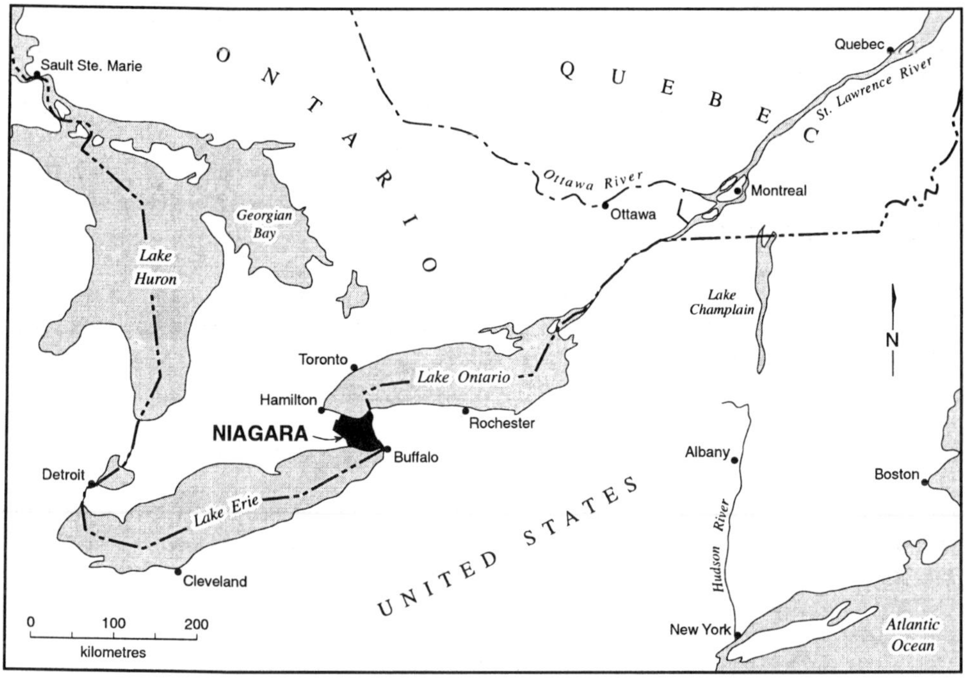

Fig. 10.1 Location of the Niagara Region.

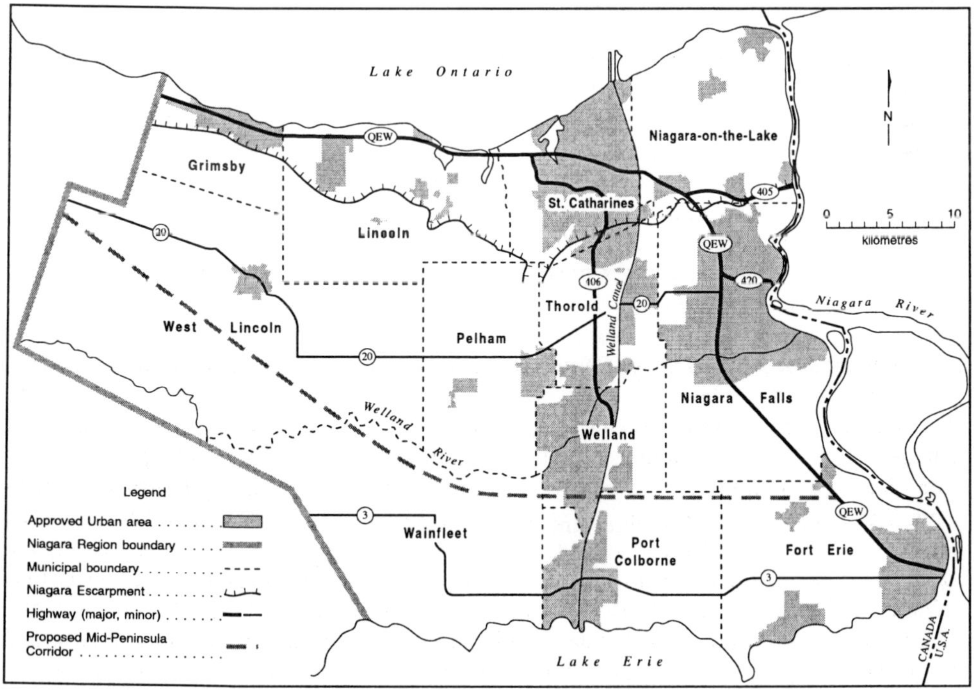

Fig. 10.2 Niagara Urban Areas (Source: Regional Municipality of Niagara, 2002).

An important aspect of saving the countryside from various urban pressures is the need to change attitudes relating to rural development. In Ontario, as elsewhere in North America, there are strong provincial policies on paper to preserve high quality agricultural land (Government of Ontario, 1978, 1996), and regional and local plans that are approved by the Ontario government are expected to identify and respect such areas. However, in spite of restrictive policies at the regional and local levels, the so-called 'death by a thousand cuts' continues unabated, and the Ontario government rarely intervenes. This inaction results from a long-held desire, some would consider it a right, for landowners to be able to do what they want with their land and for communities to be able to develop in any way they see fit. Regional and local politicians may in general embrace policies to preserve high quality agricultural land, but fail to see the irony in allowing so many exceptions to the rule based on carefully crafted and convincing arguments from landowners and their lawyers and planning consultants. In many instances, the arguments can be backed up by the economic and physical difficulties of farming; and in most instances, it is safe to say, that the economic returns from some type of urban-related development outweigh those of agriculture.

A further difficulty concerning urban-related development in the countryside has been the desire to diversify the employment base away from agriculture and to adopt economic development policies that will result in new industries and services (Panschow, 2002). While much of the new employment could well have remained in urban areas, it is seen as a means of reducing rural unemployment and the need to either commute or permanently migrate to urban areas. The resulting higher rural incomes afford an opportunity for other services to be attracted or remain viable. All too often, however, the desire to reverse the economic fortunes of rural areas encourages local politicians to favour an 'any growth is good growth' principle, which can result in conflict with rural residents, many of whom may be migrants from the city, who do not want the noise, pollution, increased traffic, visual intrusion and loss of countryside that non-residential development will bring. Increasingly, as will be seen in the Niagara Wine Region case below, agricultural change and its economic improvements can be viewed with the same degree of hostility. Many rural residents fantasize about what a rural environment should be for them to enjoy, and the noise, smells, traffic and visual intrusion of the agricultural operation frequently interfere with that fantasy. Since the process of land division and the resulting rural-residential lot invariably begins with the farmer, it could be argued that the farmer is responsible for any resulting conflict. On the other hand, it also points to an ever-increasing need for government to restrict scattered development in the countryside and any further conflict.

Urban pressures and the failure to protect agricultural land in Niagara

The unique agricultural lands of the Niagara Wine Region have faced some of the worst urban sprawl in all of Canada. It was a problem that was identified very early in the post-war development of Niagara as scattered residential subdivisions extended beyond the narrow confines of towns and cities into the various rural townships (Irving, 1957; Krueger, 1959; Reeds, 1969). Responsibility for this sprawl can certainly be extended to small tender-fruit and grape farmers looking for a quick financial fix and a car-owning public seeking a residential opportunity in the 'countryside'. However, these were accompanied by a lax or non-existent planning system and little in the way of professional management or judgement at the local level. Indeed, local politicians in these townships favoured most kinds of urban development as means of improving the rural tax-base (Gayler, 1979, 1994, 2004).

Towards better planning for Niagara

It was urban sprawl and the needless loss of valuable agricultural land that hastened changes in the local government structure. The intent was to bring rural (and increasingly suburban) administrations under better urban governance where the 1946 Ontario Planning Act was being put into effect. At first, this reform was done by the age-old annexation of all or part of rural townships by adjacent towns and cities. This land and tax grab did not engender good feelings between municipalities and was destined to mask the wider public good that would come from better local government and land-use planning. As a result, in the 1960s, there were local moves for the total reorganization of local government in Niagara. An Ontario Government Commission was appointed and, after considerable local input, it recommended a reduction in the number of municipalities from two counties and 26 local municipalities to a fully two-tier system of one region and 12 local municipalities (Krushelnicki, 1994; Niagara Region Local Government Review Commission, 1966). The Ontario government essentially endorsed the Commission's recommendations (although there was growing local opposition to regional government), and the Regional Municipality of Niagara and its 12 cities, towns and townships came into being on 1 January 1970.

Since better land-use planning was such an important reason for local government reform, it is perhaps not surprising that the establishment of a Regional Plan, and the eventual conformity of local plans to it, became a leading objective of the new Regional Municipality of Niagara. The Plan was scheduled to be completed in December 1973, although final completion was not achieved until almost eight years later. The final controversial chapters of the Plan had to be imposed on the Region by the Ontario Municipal Board (OMB), the Ontario government's quasi-judicial appeals court on planning matters (Gayler, 1982a, 1982b; Jackson, 1982; Regional Municipality of Niagara, 2001). The delay and controversy resulted from a clear inability by regional and local governments (representing the interests of their landowners) to set realistic urban-area boundaries, those areas within which serviced urban development would take place for the next 20-30 years. For landowners, being included in these boundaries held out the prospect of many more millions of dollars in sales for urban development compared to the continuation of agriculture. The Regional Plan became an amalgam of local plans and initiatives; and when all the land available for future development was totalled by the Region, the population that could be accommodated was 640,000, compared to a projected figure (for 1991) of 510,000. As new census projections were made available, the figure continued to fall; and the 1977 projection of 415,000 for 1996 is close to today's actual figure.

Between 1974 and 1981, one third of the land proposed for development had been removed, but there were still excessive amounts of land available. The OMB in its 1981 decision defended these lands on the basis that it gave the local municipalities of Grimsby, Lincoln, St. Catharines, Niagara-on-the-Lake and Pelham the opportunity to adjust to permanent urban-area boundaries and for the Region to redirect development to the poorer lands south of the Niagara Escarpment (Fig. 10.2). The legal status of permanent boundaries was always questioned, and it was removed from the Regional Policy Plan in 1997. Meanwhile, slow growth in the Niagara Region has resulted in land still being available for development in the northern municipalities and redirection being largely unnecessary.

The OMB also approved policies for the areas outside the urban-area boundaries, and development had to be agriculturally-related for the highest quality, good tender-fruit and grape lands (Fig. 10.3). For the first time, rural areas were to be formally planned, and the freedom for landowners to do what they wanted with their land was now curtailed. This change was something that has never been well received; and in the last 20 years, while the worst excesses of unserviced urban sprawl have certainly disappeared, there have been various attempts to circumvent the Regional Policy Plan with a few large development proposals. Also, a host of smaller proposals (the so-called 'death by a thousand cuts') have been approved where legitimacy was often open to question.

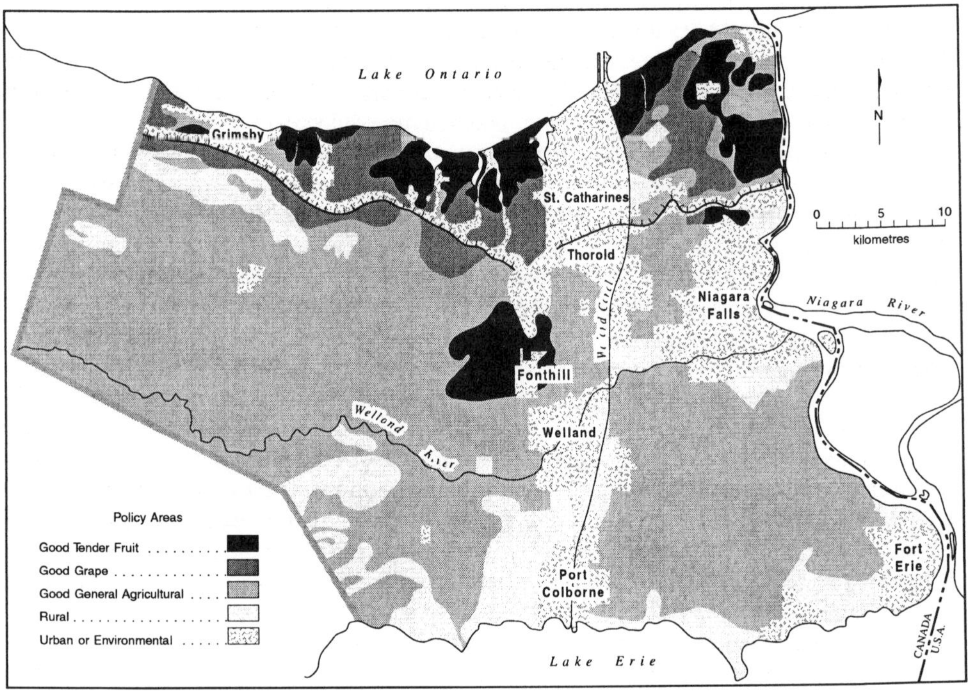

Fig. 10.3 Niagara Agricultural Areas (Source: Regional Municipality of Niagara, 2001).

The continuing saga of 'death by a thousand cuts'

From the point of view of preserving valuable agricultural land, Niagara has been blessed by slow growth. Since 1966, the percentage growth figure has consistently been lower than that for the rest of southern Ontario, as well as that for the province and country as a whole. While approximately a third of the Niagara Wine Region has been lost to urban development in the last 50 years, the remaining two-thirds has been saved. However, a number of development scenarios continue to arise which lead the professional and agricultural communities, as well as the public at large, to be concerned about the long-term future of these valuable agricultural lands.

First, Niagara continues to invite urban sprawl by allowing full-time farmers to have retirement severances. In 1981, the OMB deleted this policy from the Regional Plan, but Regional government appealed the decision to the Ontario Cabinet and retirement severances were restored. Originally, the right was given to farmers who had farmed full-time for 20 years prior to December, 1973; and thus over time, there would be no farmers left who would qualify. But in the recession of the early 1990s, with farmers facing undue hardships, approval was given in Amendments 60 and 111 for anyone who had been farming full time for 20 years to have such a severance (Regional Municipality of Niagara, 1991, 1997). Not only are these building lots scattered through the countryside, but the large number of small tender-fruit and grape farms results in even more farmers being eligible compared to other parts of southern Ontario. Also, as there is no requirement that the retiring farmer actually builds a house, many lots turn over as a real-estate deal providing the farmer with quick cash fix. Even when a retired farmer does build, it should be recognised that one day this property will be sold, and likely to a non-farmer.

Second, while the Regional Policy Plan directs that these good tender-fruit and grape lands be only used for agriculture and agriculturally related activities, it has never been defined what constitutes agriculturally related. As part of Amendment 111, farmers were given the opportunity to diversify their operations and increase their family income by developing secondary, and possibly non-agricultural, businesses. While there are important social and economic benefits from this policy, it does, however, raise the prospect of the secondary business becoming the primary one and/or various land use and development pressures occurring. It is increasingly the case that businesses that have been traditionally accepted as agricultural are being questioned as to their continued status. Niagara's expanding greenhouse industry conflicts with other agricultural uses as well as with urbanites living in the countryside; and since the soils on which the greenhouse sits are not used, it is argued that the industry could just as well be located on poorer soils to the south of the Niagara Escarpment or in an industrial park in an urban area. It is also argued that grape growing is undeniably agricultural, but the increasing number and size of wine-making operations, warehousing, retailing and hospitality facilities on the farm are raising the alarm bells of urban uses in the countryside, and the associated problems of access, traffic, noise, parking and amenity loss. The question asked by nearby residents is could these uses be located in commercial areas in the city? Politicians and planners have consistently supported the agricultural community and argued that these activities are either agricultural or agriculturally related.

Third, the difficulties of being successful in farming (especially in times of recession), compared to selling out to some urban-related function, lead to a much more invidious situation: the demand for economic severances. Examples abound where farmers will argue that they cannot continue to farm economically; but rather than sell to someone who can (especially when it can be seen that neighbouring farmers are viable), blame is levelled against physical properties or other factors beyond the farmers' control. The intent is to ask for a land use change to accommodate an urban, and higher paying, use; and the sympathy vote and exception to the rule are sought. In the early 1990s, Amendment 60 was initially a policy that would have allowed all 900 or so tender-fruit farmers to sever anywhere from one to seven one-acre lots from their farms as a means of increasing income. Later, the privilege of doing so was extended to all Niagara farmers (Regional Municipality of Niagara, 1991, 1995a). This farcical situation was a clear invitation to an oversupply of building lots, falling land prices and the potential for another round of urban sprawl far into the future. A neo-liberal government in Ontario, which could be expected to be sympathetic, had little hesitation in turning down the Amendment. However, it has not stopped a number of individual applications for what can be seen as economic severances.

Fourth, the ease by which an economic severance can be approved can be related to the land use that is proposed for the site. Agriculturally related activities would not be a problem; but where relatedness cannot be defended, the argument is used that the land use is compatible with agriculture, that nearby agricultural operations will not be unduly affected. A wide variety of land uses that fit this category are attracted to rural areas for space and cost reasons, in spite of the fact that urban residents are by far the larger users of the activity. Included here are churches, social clubs, sports facilities, especially golf courses and driving ranges, cemeteries and garden centres. To counter the argument that valuable agricultural land is still being lost, it is often stated that the land use does not preclude its return to agriculture at a later date. However, preservationists are quick to point out that land values have risen with an urban use designation and frequently top soil has been removed or altered through pesticide and fertilizer use, both of which can make a return to agriculture economically impossible.

Fifth, rising standards of living and changing lifestyles continue to fuel a growing demand for rural-estate residences. These uses epitomize the worst of low-density living and waste of rural land; but they represent the satisfying of certain personal needs and aspirations, and society is loathe to inhibit them. In Niagara, as elsewhere, rural-estate development has been channelled into areas that are unsuitable for agriculture, but the cry can then be heard that the location chosen has various scenic and aesthetic qualities or natural resources that will be put at risk. Moreover, no amount of good planning can stop a person's right to buy a whole farm and then not farm. Niagara's small farms are seen as ideal for gentrifying and hobby-farm conversions, although it is to many owners' credit that land that is not needed has been rented to neighbouring farmers.

It is the large number of these urban-related developments in the countryside, the so-called 'death by a thousand cuts,' that causes the greatest concern; yet they attract little publicity, are too often seen as compatible with neighbouring uses, and for the most part both the buyer and seller of the land elicit a good deal of sympathy from the local community. Caldwell and Weir (2002), in a study of severance activity in Ontario in the period 1990-2000, found that Niagara had the second highest number of rural non-farm lots created on agricultural land of the 33 counties and regions of Ontario. Also, of the 939 lots created in Niagara, 395 were on good tender-fruit and grape lands, and the majority of these were not agriculturally related in any way. It was argued that this long and painstaking research, and the resulting publicity it has received, is necessary because the Ontario Ministry of Agriculture and Food no longer reviews severance applications and monitors the loss of agricultural land.

The small number of very large development proposals in Niagara, on the other hand, does attract publicity and political and professional attention. There have only been two significantly large proposals outside the urban-area boundaries involving valuable agricultural land in the last 20 years, and both were eventually refused. The first was a 35-acre autoplex and farmers' market at a freeway intersection in rural west St. Catharines, and involved a relocation of car dealerships with on-site recreational, restaurant and retail activities and a future winery (Miller O'Dell Planning Associates, 1997; Regional Municipality of Niagara, 1992). The implications in terms of loss of valuable agricultural land, servicing arrangements and the westerly extension of St. Catharines' urban-area boundary resulted in the development being pared down to a small (25,000 sq.ft) seasonal farmers' market and winery, both of which were permissible under the Regional Policy Plan. The second development proposal was a culinary centre and accommodation in the form of over 50 chalet bungalows on the bench of the Niagara Escarpment in Lincoln, adjacent to the owner's cottage winery (Regional Municipality of Niagara, 1998). This proposal was certainly in keeping with the agritourism initiatives on the part of the growing number of wineries. However, it failed to take into account that the Policy Plan only permits bed and breakfast accommodation outside urban-area boundaries, there were various servicing and aesthetic problems, and the tourist accommodation could easily be converted into permanent rural residences (Gayler, 2003). In spite of professional judgements against the proposal, both local and regional governments approved, as did the Niagara Escarpment Commission, in whose planning area the development lay. Unlike OMB decisions, the Commission's decisions can be appealed to the Ontario Cabinet, and this condition not only gave local activists the opportunity to do so, but to see their appeal supported by the Cabinet. For a pro-development Ontario government to support these objections was regarded as a total surprise. It was unclear whether the government was simply making a one-off political statement concerning unsuitable rural development, or whether it was genuinely part of an attitudinal change towards better support of its own provincial policies concerning the preservation of agricultural land.

Large development proposals that cannot be contained in urban areas will continue to threaten valuable agricultural land. Niagara is presently being promoted by both the public and private sector as a superior golf destination in North America, and while the new courses have been built on lower quality land to the south of the Niagara Escarpment, there are concerns that landowners to the north may want to take advantage of this growth. A second potential development presenting some concern is the commercial and tourism expansions associated with the Niagara District Airport. This facility, located close to the Queen Elizabeth Way in the southern part of Niagara-on-the-Lake, is small, very under-utilized and has never sustained a regular passenger service. However, Niagara's growth as a tourist destination and potential for other economic growth have attracted a proposal for office and hotel developments and associated services (Regional Municipality of Niagara, 2003a). The town of Niagara-on-the-Lake rejected the proposal for agricultural, residential and servicing reasons, and the Niagara Region supported the town. A subsequent appeal by the developers to the OMB was turned down (Ontario Municipal Board, 2004).

Changing urban-area boundaries in Niagara

It is the recent changes to urban-area boundaries in Niagara, as well as possible future changes, that have triggered the re-evaluation of where Niagara is going as a region, what particular assets it wishes to highlight and preserve, and what makes the area distinctive from other parts of southern Ontario (Regional Municipality of Niagara, 2000a, 2000b). Niagara may be consistently a slow-growth area, but some towns and cities in the region have come perilously close to developable land running out on their urban fringes. Grimsby, Lincoln, St. Catharines and Niagara-on-the-Lake are in this category, with any expansion to the boundaries seriously compromising valuable agricultural land, while Pelham has already had a large and controversial expansion approved.

The potential for the supply of developable land to be eventually depleted afforded Niagara the opportunity to enact a policy that favoured the redirection of future urban growth to areas south of the Niagara Escarpment. A general area between Niagara Falls, Welland, Thorold and Fonthill was chosen in the early 1990s, and a specific location in the southern part of Thorold, close to Welland and Highway 406 on poor agricultural land, was designated in 1999 (Regional Municipality of Niagara, 1999a; see Fig. 10.2). This area has since attracted virtually no private-sector interest, which is hardly surprising when at the same time the town of Pelham, supported by the Region, approved a 550-acre expansion to the urban-area boundary on the east side of Fonthill (Ontario Municipal Board, 2000; Regional Municipality of Niagara, 1999b).

This area in Fonthill includes some of Canada's most valuable agricultural land, and its approval for urban development made a mockery of the Regional Policy Plan. The decision was appealed to the OMB; but, as is so often the case, government is seen to be backing the development industry, leaving the wider public good, let alone the Regional Policy Plan, to be defended by a precariously financed public-interest group, in this case PALS (the Preservation of Agricultural Lands Society). It was a true 'David and Goliath' situation with the hired guns of government and the developers able to run roughshod over the limited evidence put forward by PALS. The OMB dismissed the appeal, citing the conflicting professional evidence of how valuable the agricultural land actually was and the need (as set out in the Ontario Planning Act) for communities to be able to accommodate future growth.

The OMB decision in Fonthill jeopardized any 'teeth' being given to a Regional redirection policy, and today there are numerous examples of actual or potential urban-area boundary changes, many of which are destined to consume valuable agricultural land. The pressures for these boundary changes are locally based, representing public and private boosterism, municipal pride and an opportunity to increase the local tax base. There is already continuous strip development along the service roads of the Queen Elizabeth Way for the 110 km distance from Toronto to Grimsby; and the prospect of similar development occurring along the highway between Grimsby and Niagara Falls remains very real.

St. Catharines presents a particularly troublesome picture in terms of urban-area boundaries. It can only expand westwards once the present boundaries are fully developed in the next few years; and this expansion would be principally into an area of orchards and vineyards. City Council is on record as being opposed to the westerly expansion; meanwhile, the Region has viewed Niagara-on-the-Lake and Thorold as de facto suburbs of St. Catharines with the capability of taking population overflow. However, the eroding local tax base of St. Catharines and the failure so far to pool taxes across the Region, either by agreement or local government reform, have resulted in St. Catharines reviewing a boundary expansion as a possible growth option.

Changes in Niagara's agricultural industries

There is no doubt that the current demand to restrict urban development and sprawl in the countryside is strengthened as a result of some attitudinal shifts in the agricultural community. Ten years ago, a vocal minority of farmers, especially tender-fruit farmers, were clamouring for every last severance on their farms and looking for any kind of urban development as a way out of economic difficulty. While there was no public support for this option, there was recognition that farmland can only be saved if society helps to save the farmer, and the public as a whole needs to share in the responsibility of doing so. One option that was decided on were agricultural easements, a one-time subsidy to a farmer in return for excluding any form of urban development in perpetuity (Regional Municipality of Niagara, 1995b). However, the lack of sufficient provincial funding resulted in the programme being cut back and only the farmers under greatest threat from urban activities became eligible; and before any payments could be made, the Ontario government changed in 1995 and the incoming government, citing a budget deficit, abolished the programme.

Since 1995, economic difficulty on the part of some farmers and their continual dependence on subsidies from all government levels have been overtaken by economic improvements in the agricultural industry. There is a growing demand from farmers and other agricultural interests, in particular the wine and horticultural industries, to see that Niagara's finite agricultural land resource is better protected against urban developments.

The impetus behind this change has been the complete overhaul of Ontario's wine industry (Bramble and Darling, 1992). Until the 1970s, there had been a 100-year legacy of bad wine in Ontario. The local *Vitis lambrusca* variety made wines that were judged by the cognescenti to be undrinkable, were constantly panned in the media and were the butt of jokes. The turnaround came in the 1960s when the industry imported French hybrid and *Vitis vinifera* stocks from Europe and was able to grow them successfully in Ontario. The improvement in wine quality led to heightened demand from a growing middle-class market, and it set in motion a restructuring of the wine industry (Chapman, 1994). In 1975, a new cottage winery, Inniskillin, was opened in Niagara-on-the-Lake, the first to be established since the 1920s.

The industry has never looked back. The old wine industries either closed or were amalgamated into two large conglomerates that control over 80% of production in Ontario. Meanwhile, the number of cottage or boutique wineries has expanded to over 60 in Niagara (and still growing), 50 of which have been established since 1990. As a result of the restructuring of the wine industry, an award-winning quality product has emerged that Canadians want to purchase, often at higher prices compared to similar imports. The setting up of the Wine Council of Ontario has contributed considerably to promoting Ontario wines and a wine culture. Meanwhile, the establishment of the Vintners' Quality Alliance (VQA) not only put in motion a quality control mechanism, but assured consumers that they were purchasing a product made from 100% local grape varieties.

A key feature behind raising the profile of Niagara wines has been the ancillary developments in agritourism (Gayler, 2003; Ontario Wine Council, 2002; Telfer, 2001a, 2001b). Wine and the various spin-offs from it are a tourist destination. Cottage wineries are approved on the basis of vineyard acreage, manufacturing plant, warehousing and distribution facilities; but they are also allowed an ever-growing list of tourist attractions, including winery tours and tasting, wine and gift shops, restaurants, cookery schools, banquet and conference facilities, bed and breakfast accommodation, and arts performance spaces. The 60 or so wineries are all located to the north of the Niagara Escarpment, in the area traditionally known as the Niagara Fruit Belt that is now being promoted as the Niagara Wine Region and Wine Route (Fig. 10.4). The potential for tourism growth is enormous. Niagara Falls with its recent growth as a gambling destination attracts upwards of 15 million visitors per year; Niagara-on-the-Lake, as a historical town, adds a further 3 million visitors. Also, some 8 million people live within a 100-mile radius, and an easy day's drive, of Niagara.

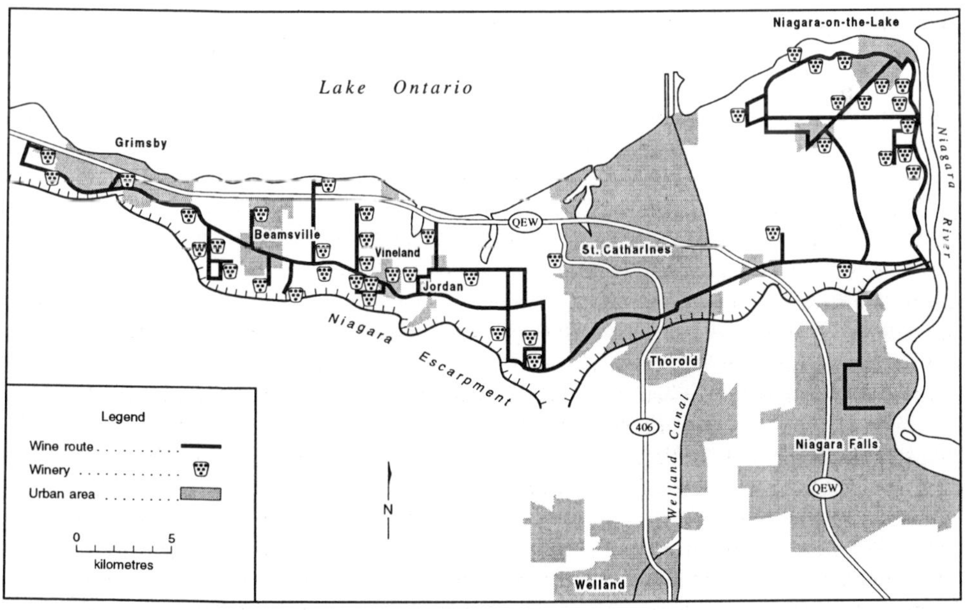

Fig. 10.4 Niagara Wine Region (Source: Ontario Wine Council, 2002).

An important element in this experience is the promotion of the rural setting in which the wineries find themselves (Gayler, 2003). It is a local planning requirement, in fact, that the wineries are set among a minimum size of vineyard; and the surrounding countryside, in particular the Niagara Escarpment and the different river valleys, has also become part of the image that is promoted. While many of the winery buildings can best be described as functional and hardly memorable, there have been concerted attempts to develop very distinctive buildings by which to promote particular wineries including French chateau, neo-Georgian white clapboard and airport terminal look-alike. A new winery in Lincoln, Le Clos Jordan, has retained the world famous Canadian architect, Frank Gehry; and should his design concepts become reality, Niagara's wine experience will likely secure a global position. A further aspect of the wine experience is to tie wine tasting and the winery restaurants into various off-site cultural events. One of the largest of these is the Shaw Festival in Niagara-on-the-Lake. This theatre company attracts over a third of a million patrons each year; and its promotional literature and various events regularly feature local wineries.

The cottage wineries have such a large, multi-million dollar investment through the Niagara Wine Region that they have become increasingly committed to the area, in spite of the fact that cool-climate viticulture research can demonstrate that various grape varieties can be grown in other Lake Erie and Lake Ontario locations (Shaw, 2003). Thus, urban expansion, worse still urban sprawl, are the last things that are desired; not only do they ruin the image that wineries are trying to project but they introduce a different constituent with a different agenda. Already, local residents complain about nearby winery activities, in terms of noise, smell, effluent risks, traffic, and the dominance of the factory-like buildings. It is not surprising, therefore, that winery owners are now at the forefront of lobbying different levels of government to protect their industry and its land base from urban developments.

The grape and wine industry is not alone in seeing growth. A more worrisome development has been the recent changes in the horticultural industry, where the area under glass has doubled in Niagara in the last 15 years (Gayler, 2001). Although at least a third of this capacity is to be found south of the Niagara Escarpment, there has been no success in attempting to persuade the industry or politicians to accept future expansion only in this area. Existing greenhouse operators in the Niagara Wine Region exercise their right to expand, while new operators cite various access factors (the Queen Elizabeth Way), infrastructure needs and microclimatic and cost advantages for developing there. However, traffic and pollution can raise the ire of neighbours, while acre upon acre under glass, especially if lit up at night, is not the image of the countryside that ex-urban migrants have come to expect.

Growth and change in the agricultural industry in Niagara have strengthened the resolve of many that there needs to be better protection from urban developments in the countryside. In the process, however, it has resulted in the solidarity of the agricultural community being weakened, and the realisation that there are landowners who still want the freedom to do what they want with their land. While attitudes towards urban expansion and sprawl are changing, there remains perhaps a long educational process in order to convince most people.

Introduction of Smart Growth principles

An important part of this educational process has come through the introduction in Niagara of 'Smart Growth' principles. The concept of 'Smart Growth' was originally formulated in the USA among the professional planning community and has been developed and put into practice in a large number of jurisdictions (Smart Growth Network, 2003; Porter *et al.*, 2002). Essentially, it states that the low-density, car-dependent, suburban and ex-urban model of urban development that has prevailed for the last 50 or more years is not sustainable, and that urban development should adopt design principles that discourage greenfield development and the physical expansion of the city, encourage intensification and brownfield redevelopment, lessen the physical and visual impact of the car, promote more mixing of land uses and thereby lowering journey times, and encourage walking and better neighbourhood interaction.

Niagara's introduction to 'Smart Growth' in 2000 was the result of a 10-year review of the 1981 Regional Policy Plan. This review focused on what made Niagara a special place in which to live, what distinguished it from other areas in southern Ontario, and what policies needed to be enacted to maintain and enhance that special place (Regional Municipality of Niagara, 2000a, 2000b). A series of public meetings across the Region identified an interest in retaining the distinctiveness of Niagara's different urban and rural communities and taking urban pressures off its valuable agricultural lands in order to better protect the wine industry. The interest in 'Smart Growth' as a means of achieving these objectives recognises that the 1981 OMB-imposed Regional Policy Plan, and its restrictive policies in terms of protecting valuable agricultural land, have not been embraced consistently by local and regional politicians (the Fonthill expansion debacle noted above was on many people's minds!). 'Smart Growth' was a new approach which aimed to re-order the way in which cities developed so as to take away so much of the urban pressure that was brought to bear on the rural areas.

In Niagara in 2001, the Regional Council formed the Smarter Niagara Steering Committee, nominally a Regional Chair's committee but reporting to the Planning Services Committee, a standing committee of Council. It consisted of Regional councillors, professional staff and ten members of the public, including this author, drawn from a wide variety of interest areas and appointed by Council for a 3-year term. The Committee's mandate has been to formulate 'Smart Growth' principles that recognize the local situation and laws and customs governing past practice[1] and to effect policy and social and economic changes, at both the local and regional levels, that will help put these principles into practice.

An important aspect of 'Smart Growth' is the potential for the redevelopment of brownfield sites (Regional Municipality of Niagara, 2003b). Niagara has a long legacy of heavy industry, but, since the 1970s, deindustrialization has seen many of these industries closing or downsizing and much vacant or underutilized land as a result. However, the potential for the redevelopment of this land for other uses could not be realized until the Ontario government passed legislation which would facilitate the clean-up of often heavily polluted properties and reduce the risks of previous owners being liable for damages from any subsequent claims. Local municipalities in turn are able to promote the clean-up of properties by the private sector; the cost can be subsidized through the difference between the old and new tax assessment (monies that would never have been received had the property not be cleaned up) being returned the owner for a limited period of time. Opportunities for this approach are currently being put in place on a former paper mill site on the edge of the downtown in Thorold.

A further aspect to the work of the Smarter Niagara Steering Committee has been the recognition that urban expansion and the loss of agricultural land can be slowed by the revitalization and intensification of existing urban areas. Two decaying areas in central and east St. Catharines have been identified, and given the two-tier planning system, the resulting projects were embraced by both the City of St. Catharines and the Region (City of St. Catharines and Regional Municipality of Niagara, 2003). The intent of the projects was to engage residents, business owners, politicians and interested stakeholders in an ideas charrette-type setting, to see how more residential and mixed use developments could be introduced and what infrastructure improvements would be necessary in order to improve the quality of the neighbourhoods.

An agricultural preserve for Niagara

A more volatile aspect of the 'Smart Growth' debate in Niagara has been the suggested permanent urban-area boundaries in the Niagara Wine Region and the development of a Mid-Peninsula Transportation Corridor to take urban pressures away from this northern area.

Permanent boundaries, giving an agricultural preserve beyond, are the recognition by the wine industry, politicians and various professional groups that the restrictions on development in areas of good tender-fruit and grape lands imposed by the 1981 Regional Policy Plan have not been well endorsed locally, as Caldwell and Weir (2002) ably demonstrate. Blame rests with the ease by which Official Plan amendments are approved locally and regionally under the Ontario Planning Act with little or no political or professional intervention at the Ontario government level. These permanent boundaries would be better cast in stone at the provincial level, as for example an Act of Parliament, and thereby would not be subject to the vagaries of local politicians, developers and landowners. While change to the boundaries is always possible at the provincial level, the process would be more cumbersome, costly, time-consuming, transparent and almost certainly less likely. The Niagara suggestion is no doubt based on experiences in British Columbia, Quebec and California; in the last example, the draconian measures enacted in the Napa Valley Wine Region 30 years ago have been credited with saving the industry from possible extinction.

A panel discussion between advocates and opponents of an agricultural preserve at the 2002 Smarter Niagara Summit, a public information and educational forum, degenerated into an ugly confrontation between farmers and the two advocates (a provincial politician representing St. Catharines, who had earlier introduced such a measure in the Ontario legislature, and a Niagara-on-the-Lake wine maker). Grape farmers among the audience clearly did not wish to be locked into an agricultural preserve when their livelihoods and the economic future of their particular agricultural sector were in doubt. They were hurting from contracts with wineries that were less than assured in the long-term, a reflection of Free Trade and the generous level of cheap but good imported wine that is allowed to be blended with Canadian wine and still be called Canadian (albeit without the VQA label).

In order for the Summit to not be completely derailed, the Regional Chair promised to set up a Task Force of farmers, farm organization representatives and politicians to work towards successfully addressing the various issues. A subsequent breakfast forum of some 40 interested stakeholders, held to identify the main issues and form a small Task Force, showed that attempting to preserve Niagara's best farmland would be a long and difficult process (Zimmerman, 2003). As always, the key issues did not focus on preserving Niagara's best farmland for all time so that future generations would benefit from that resource. There was little discussion on what an agricultural preserve would mean, where it would be designated, and what would be prevented from being there. Rather than a land-use issue, and thus part of Smart Growth's mandate, the issues that came to the forefront focused on the economic viability of the agricultural sector and the opportunity for the farmer to make enough money to stay in business. Included in the concerns here were continuing and new government subsidies to support their business; commitments from wholesalers, consumers and society as a whole to support farm produce; the entry of young farmers into an highly capitalized enterprize with increasing land costs; and the deleterious effects of speculation, urbanization and rural non-farm residents on the ability to farm successfully. The underlying rallying cry was that the only way to save the land is to save the farmer first; and since the latter was far from guaranteed, very few farmers were willing to give up their right to sell land to the highest bidder. By highest bidder, they did not mean from solely agricultural interests.

The Task Force deliberated for over a year, and its final report can be described as a surprising win-win compromise between the two groups for or against non-agricultural developments beyond urban-area boundaries (Regional Municipality of Niagara, 2004). The support of farmers for preserving agricultural land was secured through the adoption of, and commitment to promote, various economic measures that would improve the long-term viability of agriculture and deflect the never-ending desire to seek urban development.

Conclusions

There is presently considerable interest, as well as various actions, to contain urban expansion and sprawl in both Canada and the USA. However, a case study of the Niagara Region in Canada demonstrates that, even where some of the country's most valuable agricultural land is at risk and most restrictive containment policies are nominally in place, it has still not prevented a steady erosion of land for urban-related activities. Niagara's emergence as an important wine region has brought into focus the need for a better protection of this land, which is being implemented through the introduction of 'Smart Growth' principles, in particular the intensification and redevelopment of existing urban areas. However, the way forward to a more permanent solution of the problem has come from a change of government in Ontario in October, 2003 and the appointment of a Greenbelt Task Force to recommend a strategy for preserving the best agricultural and resource areas in southern Ontario. In Niagara, there is great optimism that the comprehensive land-use and economic approach to secure the land-base and the livelihood of farmers for all time will be endorsed by the Ontario government (Greenbelt Task Force, 2004a, 2004b).

Notes

1. The Smart Growth principles adopted by Niagara Regional Council are as follows:
(1) Mix Land Uses; (2) Take advantage of compact building design; (3) Create a range of housing opportunities and choices; (4) Create walkable neighbourhoods; (5) Foster distinctive, attractive communities with a strong sense of place; (6) Preserve open space, farmland, natural beauty, and critical environmental areas; (7) Strengthen and direct development towards existing communities; (8) Provide a variety of transportation choices; (9) Make development decisions predictable, fair and cost effective; and (10) Encourage community and stakeholder collaboration in development decisions.

Unfortunately, the success of the Smart Growth initiative has resulted in any number of developments adopting the Smart Growth moniker in order to be seen as responsible, progressive, sustainable and futuristic. The Ontario government has established a Smart Growth Secretariat in the Ministry of Municipal Affairs and Housing, as well as Panels in different parts of the province. The report of the Central Ontario Smart Growth Panel, of which Niagara is a part, is not really a Smart Growth design but a long-term growth strategy to handle the approximately 3 million more people expected in the Greater Toronto Area in the next 30 years (Central Ontario Smart Growth Panel, 2003).

Acknowledgements

My thanks to so many Niagara people who have contributed information, ideas and opinions in the course of this research. They include farmers and greenhouse operators, winemakers, politicians, planners, and other members of the Smarter Niagara Steering Committee.

References

Bourne, L.S. (1996) Reinventing the suburbs: old myths and new realities *Progress in Planning,* 46, 163-84.

Bramble, L. and Darling, S. (1992) *Discovering Ontario's wine country.* Stoddart, Toronto.

Bryant, C. R., Russwurm, L. H. and McLellan, A. G. (1982) *The city's countryside.* Longman, Toronto.

Caldwell, W. J. and Weir, C. (2002) *A review of severance activity in Ontario's agricultural land during the 1990s.* University of Guelph School of Rural Planning and Development, Guelph.

Central Ontario Smart Growth Panel (2003) *Shape the future.* COSGP, Toronto.

Chapman, P. (1994) Agriculture in Niagara: an overview. In Gayler, H. J. (ed.) *Niagara's changing landscapes.* Carleton University Press, Ottawa, pp.241-77.

City of St. Catharines and Regional Municipality of Niagara (2003) *Queenston Neighbourhood and Hartzel Road Smart Growth Studies.* St. Catharines, Ontario.

Coppack, P. M. (1988) The evolution and modelling of the urban field. In Coppack, P. M., Russwurm, L. H. and Bryant, C. R. (eds.) *Essays in Canadian urban process and form III: the urban field.* Department of Geography, University of Waterloo, Waterloo, pp.5-27

Fishman, R. (1987) *Bourgeois utopias: the rise and fall of suburbia.* Basic Books, New York.

Garreau, J. (1991) *Edge city: life on the new frontier.* Doubleday, New York.

Gayler, H. J. (1979) Political attitudes and urban expansion in the Niagara Region. *Contact Journal of Urban and Environmental Affairs,* 11, 43-60.

Gayler, H. J. (1982a) The problems of adjusting to slow growth in the Niagara Region of Ontario. *The Canadian Geographer,* 26, 165-72.

Gayler, H. J. (1982b) Conservation versus development in urban growth: conflict on the rural-urban fringe in Ontario. *Town Planning Review,* 53, 321-41.

Gayler, H. J. (1994) Urban development and planning in Niagara. In: Gayler, H. J. (ed.) *Niagara's Changing Landscapes.* Carleton University Press, Ottawa, pp.241-77.

Gayler, H. J. (2001) *Fighting success: rural-urban conflict and the development of Canada's greenhouse industry.* Paper presented to the Association of American Geographers Annual Conference, New York City.

Gayler, H. J. (2003) Agritourism developments in the rural-urban fringe: the challenges to land-use and policy planning in the Niagara Region, Ontario. In: Beesley, K. B., Millward, H., Ilbery, B. and Harrington, L. (eds.) *The new countryside: geographic perspectives on rural change.* Brandon and St. Mary's Universities, Brandon and Halifax, pp.179-96.

Gayler, H. J. (2004) The Niagara Fruit Belt: planning conflicts in the preservation of a national resource. In: Lapping, M. and Furuseth, O. (eds.) *Big Places Big Plans.* Ashgate, Aldershot, pp.55-82.

Government of Ontario (1978) *Food land guidelines: a policy statement of the Government of Ontario on the planning for agriculture.* Government of Ontario, Toronto.

Government of Ontario (1996) *Provincial Policy Statement.* Government of Ontario, Toronto.

Greenbelt Task Force (2004a) *Toward a Golden Horseshoe Greenbelt: Discussion Paper.* Ontario Ministry of Municipal Affairs & Housing, Toronto.
Greenbelt Task Force (2004b) *Toward a Golden Horseshoe Greenbelt: Advice and Recommendations.* Ontario Ministry of Municipal Affairs & Housing, Toronto.
Irving, R. M. (ed.) (1957) *Factors affecting land use in a selected area in southern Ontario.* Ontario Department of Agriculture, Toronto.
Jackson, J. N. (1982) The Niagara Fruit Belt: the Ontario Municipal Board decision of 1981. *The Canadian Geographer,* 26, 172-76 .
Jackson, K. T. (1985) *Crabgrass frontier: the suburbanization of the USA.* Oxford University Press, New York.
Jamieson, W., Cosijn, A. and Friesen, S. (2000) Contemporary planning: issues and innovations. In: Bunting, T. and Filion, P. (eds.) *Canadian cities in transition.* Oxford University Press, Toronto, pp.462-78.
Krueger, R. R. (1959) Changing Land Use Patterns in the Niagara Fruit Belt. *Transactions of the Royal Canadian Institute,* 32 (2), 39-140.
Krushelnicki, B. (1994) The progress of local democracy in Niagara: the evolution of regional government. In: Gayler, H. J. (ed.) *Niagara's changing landscapes.* Carleton University Press, Ottawa, 325-49.
Miller O'Dell Planning Associates (1997) *Background report to an application for a zoning by-law amendment regarding the Seventh Street market.* St. Catharines, Ontario.
Muller, P. O. (1981) *Contemporary suburban America.* Prentice Hall, Englewood Cliffs NJ.
Niagara Region Local Government Review Commission (1966) *Report of the Commission.* Ontario Department of Municipal Affairs, Toronto.
Ontario Municipal Board (2000) *Decision relating to official plan amendments to allow urban expansion in Fonthill.* File PL980963.
Ontario Municipal Board (2004) *Decision relating to proposed amendment to zoning by-law 500A-74, Niagara-on-the-Lake.* File Z020117.
Ontario Wine Council (2002) *Official guide to the wineries of Ontario.* St. Catharines, Ontario.
Panschow, C. (2002) Economic development and strategic environmental assessment: a tool for evaluating sustainable development in the Niagara Region. Unpublished Master of Applied Environmental Studies research paper, University of Waterloo.
Porter, D.R., Dunphy, R.T. and Salvesen, D. (2002) *Making Smart Growth Work.* Urban Land Institute, Washington DC.
Reeds, L. (1969) *Niagara Region Agricultural Research Report.* Ontario Department of Treasury and Economics, Toronto.
Regional Municipality of Niagara (1991) *Amendment 60 (revised agricultural policies).*
Regional Municipality of Niagara (1992) *Autotrend transportation centre.* File RE am 52.
Regional Municipality of Niagara (1995a) *Modified land use regulations to support the Niagara agricultural industry.* PD 114-95.
Regional Municipality of Niagara (1995b) *Niagara tender fruit lands program: proposed administrative details.* Report by the Agricultural Easement Committee to the Ontario Minister of Agriculture, Food and Rural Affairs.
Regional Municipality of Niagara (1997) *Regional Policy Plan amendment 111: revised agricultural policies.* DPD 110-97.
Regional Municipality of Niagara (1998) *Regional Policy Plan amendment 120: New World culinary centre Town of Lincoln.* DPD 47-98.
Regional Municipality of Niagara (1999a) *Regional Policy Plan amendment 138: Port Robinson West urban area, City of Thorold.* DPD 144-99.
Regional Municipality of Niagara (1999b) *Proposed expansion of the Fonthill urban area, Town of Pelham.* DPD 136-99.
Regional Municipality of Niagara (2000a) *Niagara – A special place today and...tomorrow?* Publication 95-1.
Regional Municipality of Niagara (2000b) *Smart Growth for Niagara: a new approach to development.* DPD 118-2000.
Regional Municipality of Niagara (2001) *Regional Niagara Policy Plan.* Office Consolidation Publication 91.

Regional Municipality of Niagara (2002) *A history of urban area boundaries in Niagara.* DPD 69-2002.
Regional Municipality of Niagara (2003a) *Proposed hotel development, Niagara District Airport lands.* DPD 21-2003.
Regional Municipality of Niagara (2003b) *Smarter Niagara – Brownfields incentives program.* DPD 49-2003.
Regional Municipality of Niagara (2004) *Final Report – Agricultural Task Force - Securing a Legacy for Niagara's Agricultural Land: a Vision from One Voice.* CAO 13-2004.
Shaw, A. B. (1994) Climate of the Niagara Region. In: Gayler, H. J. (ed.) *Niagara's Changing Landscapes.* Carleton University Press, Ottawa, pp.111-38.
Shaw, A. B. (2002) A climatic assessment of the Niagara Peninsula viticulture area of Ontario for the application of wind machines. *Journal of Wine Research,* 13, 143-64.
Shaw, A. B. (2003) (personal communication).
Smart Growth Network (2003) www.smartgrowth.org. Accessed 28 May, 2003.
Telfer, D. J. (2001a) Strategic alliances along the Niagara Wine Route. *Tourism Management,* 22, 21-30.
Telfer, D. J. (2001b) From a wine tourism village to a regional wine route: an investigation of the competitive advantage of embedded clusters in Niagara, Canada. *Tourism Recreation Research,* 26(2), 23-33.
Zimmerman, D. M. (2003) (personal communication).

Chapter 11

Vulnerability and sustainability concerns for the US High Plains

Lisa M. Butler Harrington

Introduction

Conditions of sustainability and vulnerability to disruption have become important topics as attention about human impacts on the environment has grown. Global environmental change, shifting conditions of hazards and response, and globalization concerns have all given rise to global, regional and local considerations of the maintenance of human and natural systems. The ability of systems to cope with change, whether on the order of local natural disasters or globally shifting climate conditions, are a growing area of research focus. In rural environments, as in others, broad consideration of sustainability and vulnerability can develop into reflection about the interconnected socio-economic and environmental systems.

Sustainability, vulnerability, resiliency and adaptability are intertwined concepts related to the ability of systems to adjust and maintain themselves in the face of change. Ideas of sustainability have been a focus in resource and land management issues since the 'Brundtland Report' was released (World Commission on Environment and Development, 1987). Sustainability issues are oriented towards long-term management of resources in manners that are consistent with development and stability. Kates *et al.* (2001) conceived a sustainability science, which focuses on understanding 'the fundamental character of interactions between nature and society...[which] must encompass the interactions of global processes with the ecological and social characteristics of particular places and sectors'. Cutter (2003) likewise has promoted vulnerability science as contributing to understanding of 'those circumstances that put people and places at risk and those that reduce the ability of people and places to respond to environmental threats'. She sees vulnerability science as parallel to sustainability science and emphasizes the need for an integrative approach to 'social, natural, and engineered systems'.

Vulnerability and sustainability both can be considered in a variety of settings, and with a variety of emphases. It is important to be aware that sustainability of one system or part of a system can be incompatible with sustainability of others, and that mitigation of one hazard may produce greater vulnerability to another (and perhaps lessened sustainability). Although vulnerability studies generally focus on risks that are felt most acutely in the short term, and sustainability research tends to focus on conditions promoting longer-term stability or instability, vulnerability can also be approached from a longer-term perspective and can be associated with vulnerability to environmental or societal stressors that are not usually described as hazards or risks. From this perspective, a vulnerable or sensitive system generally would not exist in conditions of long-term sustainability: perturbations to which the system is sensitive will lead to generally negative effects. Normatively, societal desires are for decreased vulnerability accompanied by increased sustainability of natural (biophysical) and socio-economic systems. Although change should be expected, a certain level of stability tends to be desirable.

While much of the research related to hazards vulnerability and the adaptability or resiliency of human populations focuses on concerns in more densely populated areas, predominantly rural regions should not be neglected. In particular, there are extensive agricultural areas, with scattered communities of various sizes, to consider (e.g. Antle *et al.*, 2004). These areas contribute much of the nation's food production and balance of trade, as well as directly supporting their inhabitants.

This chapter reviews the concerns associated with sustainability and vulnerability of a US Great Plains sub-region, the High Plains of southwestern Kansas, to both long-term stresses and potential shorter term hazards. Natural stresses on agriculturally-based socio-economic systems of the High Plains include drought, tornadoes, hail and severe winter weather. Socio-economic stresses and technological hazards also exist. The major concern regarding sustainability of the region is the capacity of the socio-economic system to adapt to declining groundwater resources.

Purpose

Geographers, historians and others have been considering environmental change and resource management issues for the Great Plains, including the High Plains/southwestern Kansas, for some time (AAG, 2003; Bloomquist *et al.*, 2002; Borchert, 1971; Hewes, 1958; Hudson, 1996; Kincer, 1923; Kromm and White, 1992, 2001; Lawson, 1976; Opie, 2000; Popper and Popper, 1987; Riebsame, 1991, 1994; Rosenberg, 1980, 1986; Sherow, 1990; Sindelar *et al.*, 1995; Warren *et al.*, 1982; White, 1994; White and Kromm, 1995). Concerns often have related to natural resource and social sustainability, including such issues as groundwater resource decline (the Ogallala/High Plains aquifer system), depopulation, climate change and variability and soil conservation.

Recently attention also has been focused on the vulnerability of linked environmental and social systems to global change (e.g. Allison and Hobbs, 2004). Change can include stresses induced by weather and climate change and extremes; shifts in organization of the agricultural sector of the economy (vertical integration and scale changes); social change through increasing immigrant populations in cities, but decreasing rural and small town populations; and depletion and degradation of natural resources like groundwater and natural gas.

Cutter (2003) noted a need for integration of multiple hazards in vulnerability studies. Qualitative consideration of the range of stressors applicable to a particular location is a necessary preliminary step in the integration of vulnerability and/or sustainability concerns to be approached through either 'vulnerability science' or 'sustainability science'. This chapter presents a review of historical and developing concerns related to sustainability and vulnerability to disruption in the US High Plains in general, and southwestern Kansas in particular.

Study area

This assessment of place-based sustainability and vulnerability concerns is related to the multi-site Human-Environmental Regional Observatories (HERO) project, which is focused on interactions between land use change, the vulnerabilities of people and places, and climate variation and change. The 19-county Kansas HERO study area, also known as the High Plains-Ogallala (HPO) HERO, is located in the semi-arid High Plains and in the heart of the former Dust Bowl. It encompasses about 41,120 km^2 (Fig. 11.1).

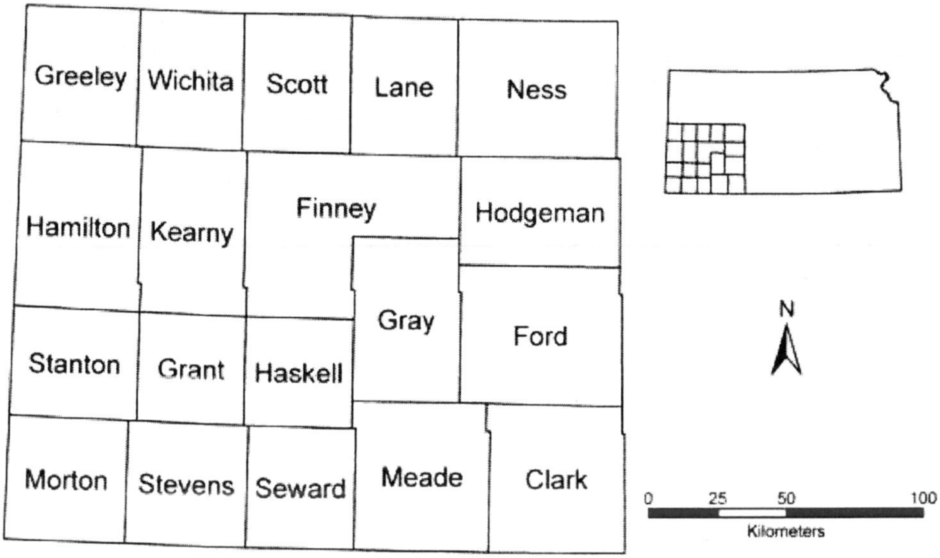

Fig. 11.1 Southwestern Kansas study counties in the central USA.

The HPO HERO has relatively little local surface water, with most reliance for water supplies based on alluvial groundwater and, more importantly, the High Plains aquifer system (including the Ogallala formation). The two most important rivers in the region are the Cimarron and the Arkansas. Elevation ranges from 760 to 1070 m above sea level. Soils are generally good for agricultural production, although some areas of sandier soils are relatively unsuitable for crops. Wind erosion potential is of greater concern than water erosion in this region (Leathers and Harrington, 2000). Oil and gas fields underlie the area, with important but declining production.

The region's economy is largely based on agriculture and agricultural support services. The top five counties in agricultural sales for Kansas are in the HPO HERO study area; these five accounted for almost US$2 billion in 2002 (USDA ERS, 2004). Widespread irrigation as a basis for agricultural stability commenced with the tapping of the Ogallala Aquifer in the 1950s and 1960s (Kromm and White, 1992), with irrigated cropland in the study area expanding from about 50,060 ha in 1950 to 420,000 ha in 1969. With a sub-humid climate, a large part of the crop production is based on irrigation: 96% of freshwater use in the study area is for irrigation. Locations not overlying the High Plains aquifer system also are dependent on groundwater drinking supplies, although there may be insufficient groundwater to support farming. Even though irrigated agriculture often is the focus of attention for the region, most of the land is devoted to dryland crops and range. Very large feedlots account for a small portion of the total area, but are economically very significant. A number of the feedlots for beef cattle have capacities ranging from 25,000 to over 100,000 head. The cattle, in turn, flow to large packing plants in the study area. The product, boxed beef, is transported to destinations across the globe.

Crop farmers settled much of the High Plains in the late nineteenth century, but variable precipitation and repeated droughts have resulted in periodic land abandonment. The region's population grew and the agribusiness economy expanded rapidly in the 1970s and 1980s (White, 1994). The HPO HERO area currently is inhabited by about 156,000 people and over 2.25 million cattle. Human population densities in the 19 counties vary from 0.9/km^2 to 11.3/km^2. Recent population changes (1990-2000) in study area counties have ranged from -14.4% to 22.5%, with seven counties showing declines. The three principal settlements are Garden City (2000 population: 28,451), Dodge City (25,176), and Liberal (19,666). The most significant growth has occurred in the three counties containing these regional support centres. Hispanic-identified persons now make up one-third of the study area population; those identified as Asian comprise about 2% of the total.

Socio-economic and environmental changes have occurred in response to the introduction of irrigation, the growth of feedlots, and the expansion of the meat packing industry in the High Plains. A number of supporting industries also have arisen, including agricultural services and trucking. The region has seen important shifts in population, including growth, redistribution and changing ethnicity. Rural counties have tended to lose people to the more urban areas of Garden City, Liberal and Dodge City.

Sustainability and vulnerability

Sustainability and vulnerability to environmental and socio-economic perturbations remain open questions for the southwestern Kansas High Plains, as well as for the US Plains in general. As Cross (2001) has pointed out, there are ways in which rural areas and small cities remain more vulnerable to hazards than 'mega-cities'. The greater numbers of people potentially at risk in very large urban areas logically lead to high levels of attention in terms of hazard research, policy development, and mitigation activities, as well as development of 'sustainable cities' programmes. However, it also is important to recognize the status of more rural regions regarding long-term sustainability and vulnerability of socio-economic and biophysical systems.

The best-recognized natural disaster to have affected the region thus far is the drought and Dust Bowl of the 1930s. Besides drought, extreme weather events constitute other hazards to the southwestern Kansas population and economy. Hazards affecting the High Plains, while certainly disruptive, seldom can be considered to create 'disaster' conditions, as disaster is sometimes interpreted. The losses in the High Plains seldom include great loss of life or damage to physical structures, simply because structures and population are dispersed. However, environmental and economic losses can be devastating, either on a spatially and temporally limited basis (tornado strikes) or spread over space and time (drought). Dryland crop producers are particularly vulnerable to climate variability. 'Furnace winds', tornadoes, hail and severe winter weather can all cause extensive damage either to agricultural resources or to human life and structural property.

Decline of the groundwater resource constitutes a socially-constructed hazard that has developed through an attempt to mitigate dry conditions. Depletion of regional natural gas reserves represents unsustainable resource exports for the area as well. Loss of such resources is likely to lead to decreased economic diversity, and thus greater vulnerability for the region. Since the HPO HERO exports almost all of the wheat and beef, and most of the natural gas and petroleum it produces, the region is vulnerable to pricing variability and global market shifts. Although not a hazard in the usual sense, or at least not a natural hazard, economic conditions also constitute a point of vulnerability for agricultural zones. Domestic and international market conditions affect the income and success of farms and feedlots in southwestern Kansas, as they do for other agricultural regions and products. Crop and livestock disease are historic concerns in agricultural regions; agroterrorism is a recent addition to biological hazard concerns. These conditions, their impacts and social responses are described below.

Hazards, stresses, impacts and response

Drought

As has been noted (Cross 2001; Cutter *et al.,* 2003a), agriculture is especially vulnerable to climatic hazards, and this region is highly dependent on agriculture. Perhaps the clearest hazard to rural agricultural regions is drought, but droughts 'are very difficult to measure in extent, property damage, and death toll' (Burton, *et al.*, 1993, p.12). Negative health effects related to drought, especially related blowing dust, are likely, but records tracking such effects are rare and incomplete.

The best-recognized natural disaster[1] to have affected the region thus far is the drought and Dust Bowl of the 1930s. Other important disruptions were created by droughts in the late 1880s/early 1890s and the 1950s. Important Great Plains drought 'midpoints' occurred in 1892, 1912, 1934, and 1953 (Borchert, 1971). The 1950s drought was, in some ways, more extreme than that of the 1930s, although it received less attention. Much lesser droughts have occurred, including one in the 1970s. The early 2000s also have seen a period of extreme dryness. One reason for this situation may be that differing climatic conditions led to much less wind and, therefore, an absence of the extensive soil erosion and environmental damage that occurred during the Dust Bowl years of the 1930s (Borchert, 1971); soil conservation practices also had been greatly improved since the Dust Bowl.

Loss of soil associated with wind erosion during the Dust Bowl, and both wind and water erosion in subsequent periods, also causes concern with agricultural sustainability over the long term. Modern soil conservation practices and federal cropland reserve programmes can serve to mitigate the effects of drought on both agricultural resources (soils) and farmer income (Nellis *et al.*, 1997), although the effectiveness of the Conservation Reserve Program (CRP) has come into some question (Leathers and Harrington, 2000). Through this programme, farmers can temporarily retire erodible or marginal lands in exchange for federal payments. Likely related to both CRP withdrawals and groundwater depletion, irrigated acreage in the study area fell between 1978 (at 659,550 ha) and 1997, when Agricultural Census estimates showed about 626,970 ha irrigated in the 19 county areas.

While the best way to mitigate a lack of precipitation for agriculture has been to pump water from below, there also has been an attempt to alleviate dry conditions through weather modification. In 1975, the Western Kansas Groundwater Management District (GMD #1) implemented a programme of cloud seeding to increase area rainfall, decrease crop-damaging hail, and to serve as a demonstration project (WKWMP, 2001). Although data do not clearly indicate a rainfall benefit associated with cloud seeding, the programme has continued. There is some belief that the sparsely distributed rain gauge system in western Kansas cannot accurately reflect rainfall benefits (WKWMP, 2001). The justification for continuing the programme has been a significant reduction in hail damage: according to the Kansas Water Office (2002), 'this programme...has reduced crop hail damage by 49% in the target area'.

Groundwater decline

Years with well below median annual rainfall have negatively affected the economic well-being of the region in the past, but farmers in southwestern Kansas and other parts of the Great Plains have been able to adapt and overcome this climate limitation through irrigation (AAG, 2003; Easterling *et al.*, 1993; Kromm and White, 1992; Rosenberg, 1986; Sherow, 1990). River-based irrigation began in the 1880s. Widespread tapping of the Ogallala aquifer began in the 1950s, with development of better understanding of the aquifer system and improved technology. Over the longer term, reliance on 'fossil' water with extremely low rates of recharge is likely to make the region more vulnerable to drought conditions. Earlier this century, some communities received their water through rail-based shipments: Kismet received its water via railroad until the mid-1930s, when a city-owned well was installed (Byrne and McLaughlin, 1948).

Use of groundwater over the core of the study area has resulted in the creation of a thriving agricultural economy centred on production of feed, livestock raised in feedlots, and meat processing. However, there is no question of the status of Ogallala Aquifer use: it is not sustainable. Withdrawals over decades, sometimes at rates averaging 38 cm/year or more, but with recharge of less than 5 cm/yr (and through most of the study area less than 2.5 cm/year) (KGS, 2000), means that the use of the region's 'fossil water' has been a form of mining (Fig. 11.2). Once used, this groundwater resource is essentially gone. Although not sustainable in the long term, there have certainly been moves towards prolonging the life of the resource through increased water conservation aided by changing technologies (e.g. subsurface drip irrigation) and other management shifts (metering and crop changes). More recent rates of decline have averaged less than 15 cm/year (KGS, 2000), largely due to improvements in irrigation technology and water use efficiency.

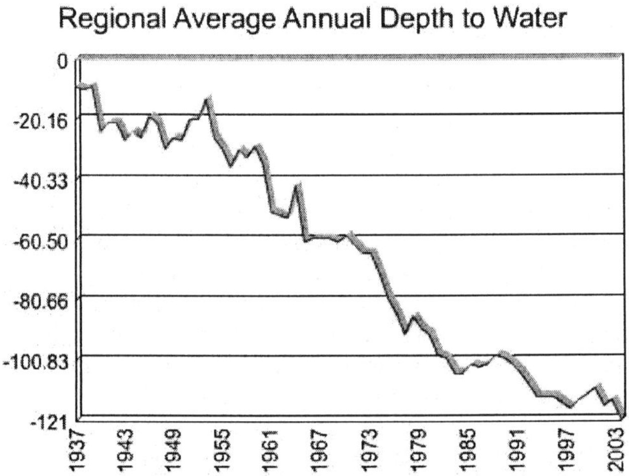

Fig. 11.2 Finney County winter regional hydrograph, indicating groundwater depletion based on multi well measurements. Depth in feet. (Source: KGS, 2003 (http://hercules.kgs.ku.edu/geohydro/wizard))

Heat/humidity/furnace winds

Extremes in heat or the combination of heat, low humidity, and/or wind, which create desiccating conditions for plants, as well as human discomfort with potential health problems for vulnerable individuals, are summertime meteorological hazards in the High Plains. Leathers (2001) and Leathers and Harrington (2001) described the incidence of these conditions across the Great Plains states and western US. In 44 years of records, Dodge City has recorded 7,020 hours of high heat, defined as temperatures reaching 35°C (95°F) or more, for an average of 160 hours/year (Leathers, 2001, p.162). 'Furnace winds', identified as events with wind velocity >5 metres/second, temperatures >35°C, and relative humidity <30% (Leathers, 2001, p.5), represent extremes of desiccation potential. The Dodge City area receives one of the highest frequencies of furnace wind events (Leathers, 2001, p.91, 165; Leathers and Harrington, 2001), averaging almost 106 hours/year.

Although high humidity combined with high temperatures can be a hazard for livestock and humans because of a reduced ability to self-cool, this condition generally has not been a problem in the Kansas High Plains. Indeed, low humidity has been one of the reasons that the area has been attractive for raising livestock. Although high humidity/high heat combinations (resulting in high heat indices) have thus far not constituted a problem for the area, climatic change could result in an increased incidence of high humidity in the summertime. Harrington (1998) has identified increasing spring season dewpoint temperatures (i.e. higher humidity) with irrigation increases in the Dodge City area.

Tornadoes/hail/severe storms

Tornadoes are associated with Kansas in the popular conscience, due both to Kansas' real location in 'Tornado Alley' and to Dorothy's travels to the Land of Oz. Other storm-related hazards include non-tornadic severe storms with hail or strong straight-line winds. These are real hazards, but the worst effects are often very localized. Although individual communities can experience terrible effects from a tornado, and a swath across a county can be impacted by hail damage, severe thunderstorms are less likely to cause great regional stress than drought or winter storms.

Cutter (2003), Mileti (1999), and Thomas (2001) note the incomplete nature of most hazards data. This paucity is certainly the case for meteorological hazards in Kansas. Not only are historic records of events incomplete, but realistic economic damage estimates also can be difficult to obtain (Cutter, 2003). However, National Climatic Data Center records (NCDC 2003, using NOAA *Storm Data*) at least give an indication of relative frequency of tornado, hail and wind events. For the period 1950-2002, 484 tornadoes are on record for the 19 county study area. The greatest number of these events were considered 'weak' tornadoes (F0-F1), as should be expected, and resulted in relatively little economic damage; only four were recorded as 'violent' (F4). Several tornadoes were estimated to have caused US$2.5-5 million in property damage. Four deaths were caused by the 484 tornadoes over the 52-year period. Instances of damaging straight-line winds also have been recorded in the area. In terms of damage potential, a major tornadic 'direct hit' on one of the larger communities can cause much greater loss of life and property than in a small community. However, a natural disaster for a small community could cause near total destruction of the town or village, while a larger community would be more likely to recover.

The greatest incidence of severe storm events is with hail: 371 hail events (hailstones >2 cm diameter) occurred in Ford County alone from 1955 to 2003. With several events having hail of 11 cm diameter in the study area, hailstones up to at least 14 cm have been recorded. NCDC data clearly are incomplete for damages from hail, but a single hailstorm can cause hundreds of thousands of dollars in losses. Insurance helps to mitigate losses, but insurance data for crops and livestock are difficult to obtain.

Floods

Floods generally have not been a major hazard in this region, although flooding and flash flooding can occur in some locations. A flash flood in 1997 caused US$2 million in property damage in Dodge City, for example (NCDC, 2003). Smaller communities also can experience flood losses: a flash flood in Lakin caused approximately US$100,000 in property damage and US$50,000 in crop damage in 1997 (NCDC, 2003). Small stream and urban flash flooding has been the greater hazard in the region in recent decades, but a change in water management might influence other flood risks. Colorado now releases additional water into the Arkansas River, as compared to periods before the mid-1990s, as a result of litigation by Kansas. The 1998 water year showed some of the highest flow rates on record (US Water News, 1999).

Winter storms/cold

Winter storms, severe cold and blizzard conditions can cause economic losses through damage to agriculture, both in terms of crops and livestock. They also, of course, can cause human fatalities when people are caught unprepared, but in the HPO HERO human deaths due to these hazards have been relatively rare (see NCDC records). Again, records are incomplete; NCDC data do not indicate an October 1997 blizzard that killed thousands of cattle. Up to 20,000 cattle were estimated to have been killed by blizzard conditions, with further economic losses expected due to stress-related weight reductions. One feedlot operator estimated close to US$2 million in losses (AP, 1997).

Extreme cold can cause crop losses, particularly if it occurs at important stages in a crop's development. According to one NCDC event record, extreme cold in early April, 1995 caused US$2.5 million in crop damage because it occurred during a critical point in wheat growth.

Climate change

Global climate change resulting from human-related sources of greenhouse gas emissions is widely accepted in the scientific community (IPCC, 2001; Houghton *et al.*, 2001; Fletcher, 1997). Scenarios based on the CGCM1 and HADCM2 models indicate increased maximum and minimum temperatures over the Great Plains, with the relatively greater increase in minimum temperatures associated with decreased nighttime cooling and warmer winters (Ojima *et al.*, 2002). Both models indicate decreased precipitation over the High Plains, which is cause for concern in a region of already marginal rainfall.

There have been indications that many inhabitants of this region mistrust climate change predictions, and are unlikely to take action to mitigate changes before they become evident (Harrington, 2001; Cutter *et al.*, 2003b). Inhabitants often express attitudes of adaptability and resilience, however, based on regional recovery from the Dust Bowl and 1950s drought periods. An ability to adopt new technologies and shift agricultural methods may stand the region in good stead, although outsiders might identify a lack of serious consideration of potential climate change as potentially contributing to regional vulnerability.

Disease and agroterrorism

In recent years, there has been increasing security surrounding the development of more and more intensive 'factory farming' and feedlot operations, with particular sensitivity to disease susceptibility in the raising of hogs (pigs) and poultry. In addition, access to meat packing plants is tightly controlled. The vulnerability of modern monocultural agriculture to the spread of disease has been noted for some time. The recent spread of additional diseases and pests within the USA (e.g. karnal bunt wheat fungus and the soybean aphid) and of genetically modified (GM) crops add to national concerns about the ease of spread of potentially catastrophic biomaterials. Potential agroterrorism has become a concern since the terrorist attacks of 11 September 2001 (Harrington, 2003).

Related to both livestock and crops, bioterrorism is of greater concern for production agriculture than chemically based attacks, due to the potential for disease-causing organisms to quickly propagate themselves through large populations prior to detection. Among livestock, anthrax, bovine spongiform encephelopathy (BSE–'mad cow' disease), foot and mouth or hoof and mouth, and brucellosis are some of the most widely recognized serious diseases (with anthrax and BSE also acting as human pathogens). A number of other pathogens also have been noted as potential bioweapons (Ban, 2000; CNS, 2002; Kohnen, 2000; Rogers *et al.*, 1999; Whitby, 2002). Crop diseases can include wheat rust and other pathogens that have been researched in biological warfare programmes. In the case of crops, fungal diseases are perhaps the largest concern (with more likely spread via wind), but attacks by viral and bacterial pathogens, as well as insects, also could cause widespread disruption.

Social and economic conditions

Socio-economic conditions affect the level of vulnerability people have in any particular place. One factor is the level of economic diversification, which can buffer the local economy during times of stress. The structure of agriculture in the region also can create conditions of vulnerability to stresses, both short and long term. For example, with a few very large packing plants, much of the employment is tied up with a few employers and locations of employment. The region felt significant stress regarding loss of employment and questions of stability following a fire that destroyed one of the packing plants in December 2001.

Agricultural regions can be hard-hit during times of low market prices for agricultural products, and much of southwestern Kansas also is sensitive to energy prices because of the need to pump groundwater for irrigation. Low market prices have coincided with high levels of production in the past, including the period just preceding the Dust Bowl period. Plentiful grain and low prices actually contributed to Dust Bowl conditions, as farmers ploughed more and more marginal land in order to produce enough wheat to maintain sufficient income. Under most forms of modern intensive farming, energy in the form of fossil fuels (and energy used for agrochemical production) can represent a significant cost to farmers. In the case of southwestern Kansas, the cost of irrigating is influenced by equipment costs and the cost to pump out groundwater, which in turn is affected by depth to water table and energy prices.[2] Agricultural enterprises in the HPO HERO thus may be vulnerable to losses both from low market prices and high fuel costs.

Mapping by Cutter *et al.* (2003a) has indicated medium to high vulnerability in much of southwestern Kansas with relation to several factors, including economic status, age, employment and ethnicity. Factors that contribute to regional vulnerability to both natural hazards and global change trends include a relatively large minority population. For the HPO HERO, poverty rates for 1999 ranged from 5.1% to 16.9%. Eleven counties have seen increased poverty rates, while the state has had a decrease; 14 counties are above state poverty rate. Shifting socio-economic circumstances (ethnicity) sometimes is cited as a concern (e.g. in reference to education needs). Other demographic and infrastructural changes also are some cause for local concerns, such as loss of population, ageing, and shrinking access to health care and education in smaller communities (see Table 11.1). Such changes make small communities more vulnerable to certain economic and sustainability concerns, although the region as a whole may be less vulnerable as the larger communities grow in both populations and ability to fund community needs such as health care, emergency response and water.

Locally expressed concerns

Interviews with key informants, particularly professionals with NRCS and state extension offices, have yielded a variety of concerns about southwestern Kansas and its sustainability. Interviewees were selected as 'key informants' knowledgeable about the region. Interviews most focused on sustainability concerns were conducted in late 2000, although related interviews have been conducted in this region for a number of years by several researchers (Kromm and White, 2001; White and Kromm, 1995; Harrington, 2001; Harrington and Lu, 2002; Harrington *et al.*, 2003; Cutter *et al.*, 2003b). Key informant interviews in 2000 were 'unstructured,' but focal topics were addressed through questions about: (1) major issues impacting the future of southwestern Kansas (with respect to economic development, regional sustainability, and changes in land use/land cover); (2) major concerns confronting the region that might make it sensitive or vulnerable in the future; (3) people/agencies/organizations that will have the greatest impact on changing the landscape of southwestern Kansas; (4) existing policies, laws, or regulations that have had an impact on change in the region; (5) changes to the landscape that have occurred over the past 20 years and changes likely over the next 20 years; and (6) what types of weather or climate events might cause change in land use.

Drummond and Marsden (1999) described what they believed to be the most useful approach to sustainability studies, the conception of 'realism' or a 'realist approach' recognizing multiple causalities. Accordingly, they assert that informal information-gathering techniques, especially flexible, unstructured interviews are especially useful. The researcher also should 'identify and investigate, not a representative sample of the population, but rather those agents with significant causal powers' (p.39). Although Drummond and Marsden (1999) emphasized a realist approach in contrast to a positivist approach, it could be argued that better understanding of both directions of environmental change and causes of change might be achieved by combining approaches, as has been done with respect to global change in local places research (AAG, 2003).

General observations regarding the status of southwestern Kansas and concerns regarding the sustainability of the socio-economic system arising from these interviews are summarized in Table 11.1. As indicated, sources of vulnerability/sustainability concerns include some issues that are well-recognized both inside and outside the region, such as the level of human reliance on limited water resources and depletion of non-renewable fossil fuels, social conflict, and demand/prices for agricultural products. Some resource specialists cite continued enlargement of agricultural enterprises and vertical integration as long-term concerns. Such restructuring in agriculture, in a region so heavily dependent on agriculture, can result in both changes in the direction of flow of farm income, and in further depopulation of some rural and small town locations. Other issues have been well recognized in other regions, such as 'urban sprawl' and agricultural land conversion, but are rarely recognized as a High Plains concern by outside observers.

Table 11.1 Southwestern Kansas: status and sustainability issues from key contact interviews.

Status

- recognition of agricultural basis for regional economy
- perception of lack of viable alternatives
- likely trend towards huge [corporate] operations, move away from smaller family farms
- importance of 1985 Farm Bill, 'Freedom to Farm' in affecting land use
- importance of Cropland Reserve Program (CRP); change in emphasis of CRP to a wildlife focus

Concerns

- water depletion and water quality
- effects of rising fuel prices (and flat or decreasing commodity prices)
- 'banks control all the farm decisions'
- belief that with diminished groundwater and/or crops to support cattle feeding, feedlots and packing facilities will leave
- population losses in smaller communities
- ageing farm populations and lack of younger people to take on farming
- lack of sufficient access to medical care in smaller communities/counties
- draw of larger communities (if still living in small community) for shopping and employment opportunities (commuting or relocating) (One interviewee also commented that 'Walmart is killing these small towns.')
- some farmers are unwilling to use less water and risk losing some of their rights, although they might generally support conservation[3]
- conversion to more water-conserving practices (sub-surface irrigation) is desirable, but expensive
- urban sprawl/housing development/agricultural land conversions (nearer larger communities like Dodge City, but including developing dormitory communities)

Adjustments

- there is/will be some conversion to dryland farming (or grazing); alternative crops are being considered
- there have been some changes in crop rotations

Adjustments (Table 11.1) are oriented towards declining Ogallala/High Plains aquifer levels. In addition to physically declining groundwater availability, rising energy costs are reducing the practical availability of groundwater: more energy is required to pump from deeper levels, and energy costs have been increasing at the same time. Actions taken to adapt to pressures being felt in the region include adoption of increasingly efficient technologies and some agricultural diversification, including more use of dryland crops and shifts in crop types. In some areas, groundwater already has become unavailable in practical terms. Shifts to more dryland crops (e.g. wheat, sunflowers, and 'dryland corn') take place as access to groundwater for irrigation diminishes.

Areas without significant groundwater resources continued to rely on dryland agriculture and cattle grazing through the twentieth century. Locations lacking groundwater resources also are the places that have lost population to 'booming' communities with good access to water (see White, 1994). In a sense, they have been vulnerable to competition from cities that have grown with irrigation-reliant industries.

Conclusions

For consideration of long- and short-term health of any region, numerous local characteristics and potential stressors can be identified. This statement is true even for regions that might be considered 'simple' in structure and economic dependence at first glance. The generally rural, agricultural Great Plains sub-region is no exception. For southwestern Kansas, there are a number of longer-term stresses and historic natural hazards. Further research should address specific levels of risk for the study area: the risk of a particular location experiencing a tornado in any given year is quite low; the risk of experiencing drought is higher, but all risks need to be analysed quantitatively as well as qualitatively.

In addition to a variety of potential hazards, there also are a number of characteristics that can be thought of as contributing to vulnerability, and those that may mitigate vulnerability or contribute to sustainability. Some may do both, as is the case with groundwater use: it mitigates vulnerability in the short term, but may undermine sustainability and contribute to vulnerability in the long term.

Natural resource dependence, particularly dependence on groundwater resources, represents the best-recognized factor that contributes to vulnerability of the region to environmental and economic stress. Irrigation use of the Ogallala/High Plains aquifer system is not sustainable. Although not sustainable in the long term, there have certainly been moves towards prolonging the life of the resource through increased water conservation aided by changing technologies (e.g. subsurface drip irrigation) and other management shifts (metering, crop varieties).

One mitigating factor has been an attitude of adaptability and resiliency. Although the periodic recurrence of drought has been a factor that has negatively affected the environmental and economic well-being of the region in the past, farmers in southwestern Kansas and other parts of the Great Plains have been able to adapt and overcome this obstacle, creating thriving agricultural economies through irrigation and diversification that includes dryland crops and livestock feeding. Irrigation itself has been a mitigating factor related to the naturally dry and variable climate of the region, helping farmers to adapt to conditions. However, the very factor that has reduced vulnerability in the short term (a few decades) may actually increase vulnerability in the long term (see Dregne, 1980), as the resource upon which the region has become reliant is depleted and agriculture is forced to change again in response to this stress.

Although the low population of the region can lead to the view that the area has relatively low vulnerability, simply because any stress would affect fewer people, this feature reflects an unfortunate tendency to consider damages as a total dollar amount or total number of casualties. From this perspective, any rural region is likely to appear much less vulnerable either to natural hazards or global change factors than an urban region. However, rural regional health may be just as vulnerable, if not more so, given relative wealth, single sector dependence and response capacities. As researchers begin approaching the suite of factors that can create vulnerabilities and sustainability concerns, it will be important to consider not only integration of multiple stressors and human-environment interactions, but also varying measures of impacts in different locales (see Cutter, 2003). *Per capita* vulnerability may actually be higher in rural regions, depending on hazards and other stressors.

In sum, sources of vulnerability and sustainability concerns include both long-recognized local environmental conditions and other developing conditions in the region and globally. Some issues are well recognized both inside and outside the region, such as the level of human reliance on limited water resources and depletion of non-renewable fossil fuels, and increasing reliance on a few very large agro-industrial enterprises for employment. Locals see a lack of economic alternatives to agriculture, some see lack of control of decision making, and are keenly aware of depletion of the groundwater resource. Other issues recognized in other regions, such as 'urban sprawl' and agricultural land conversion, are rarely recognized as a High Plains concern by outside observers. Social issues also come up related to sustainability, related to out-migration from the smaller communities and changing social make-up.

Some actions are being taken to prolong use of the groundwater resource, and to adapt to lower water availability. However, other means of adjustment to changing environmental and global economic conditions need to be developed. Southwestern Kansans have a history of adapting to environmental extremes and economic pressures. The declining availability of groundwater, a potential for increasingly erratic weather associated with global climate change, and economic globalization may cause further stresses, however. How, and how well, southwestern Kansans will adapt to further change remains unclear: resiliency and adaptation are likely in the short run, but long-term sustainability is questionable.

Acknowledgments

This work was supported by a grant from the National Science Foundation (NSF) and the National Oceanic and Atmospheric Administration (NOAA) [BCS-9978052], and by prior research conducted under grants from the National Aeronautics and Space Administration (NASA) through the Association of American Geographers, and from the National Institutes for Global Environmental Change. Input from collaborators on these projects is gratefully acknowledged, particularly that from John A. Harrington, Jr.

Notes

1. It should be recognized that this event was not, strictly speaking, just a 'natural' disaster, but a disaster created by the interplay of drought and human activities. Large areas had been cleared of vegetation and were left more vulnerable to erosion during a major drought period.
2. Water is heavy, and the cost to raise it to the surface can increase dramatically with depth.
3. It is important to note that perceptions among residents of southwestern Kansas can vary somewhat from the reality. For example, the concern with 'use it or lose it' and irrigation water rights are not always consistent with current regulations (see Ernst, 2002).

References

Allison, H.E. and Hobbs, R.J. (2004) Resilience, adaptive capacity, and the 'lock-in trap' of the Western Australian agricultural region. *Ecology and Society*, 9(1), 3. Http://www.ecologyandsociety.org/vol9/iss1/art3/index.html. Accessed 27 January 2005.

Antle, J.M., Capalbo, S.M., Elliott, E.T. and Paustian, K.H. (2004) Adaptation, spatial heterogeneity, and the vulnerability of agricultural systems to climate change and CO_2 fertilization: an integrated assessment approach. *Climatic Change*, 64, 289-315.

Association of American Geographers Global Change and Local Places Research Team (AAG) (2003) *Global Change and Local Places: Estimating, Understanding, and Reducing Greenhouse Gases.* Cambridge University Press, Cambridge.

Associated Press (AP) (1997) Nearly 20,000 cattle claimed by blizzard. *Topeka Capital Journal: C-J Online.* 30 October. Http://cjonline.com/stories/103097 Accessed July 2004

Ban, J. (2000) Agricultural biological warfare: An overview. *The Arena,* 9, 1-8. Http://www.cbaci.org Accessed July 2003.

Bloomquist, L., Williams, D. and Bridger, J.C. (2002) Sublette, Kansas: Persistence and change in Haskell County. *Persistence and Change in Rural Communities.* CAB International, Wallingford, pp.23-43.

Borchert, J.R. (1971) The Dust Bowl in the 1970s. *Annals of the Association of American Geographers,* 61(1): 1-22.

Burton, I., Kates, R.W. and White, G. (1993) *The Environment as Hazard.* Second edition. Guilford, New York.

Byrne, F.E., and McLaughlin, T.G. (1948) *Geology and Ground-water Resources of Seward County, Kansas.* Kansas Geological Survey Bulletin 69. Kansas Geological Survey, Lawrence, Kansas.

Center for Nonproliferation Studies (CNS), Monterey Institute of International Studies (2002) *Agro-Terrorism: Agricultural Biowarfare: State Programs to Develop Offensive Capabilities.* Http://cns.miis.edu/research/cbw/agprogs.htm Accessed July 2003.

Cross, J.A. (2001) Megacities and small towns: different perspectives on hazard vulnerability. *Environmental Hazards,* 3, 63-80.

Cutter, S.L. (2003) The vulnerability of science and the science of vulnerability. *Annals of the Association of American Geographers,* 93(1), 1-12.

Cutter, S.L., Boruff, B.J. and Shirley, W.L. (2003a) Indicators of social vulnerability to hazards. *Social Sciences Quarterly,* 84(2), 242-261.

Cutter, S.L., Mitchell, J.T., Hill, A., Harrington, L., Kaktins, S. Muraco, B., DeHart, J., Reynolds, A. and Shudak, R. (2003b) Attitudes toward reducing greenhouse gas emissions from local places. In: AAG *Global Change and Local Places: Estimating, Understanding, and Reducing Greenhouse Gases.* Cambridge University Press, Cambridge, pp.171-191.

Dregne, H.E. (1980) Task group on technology. In Rosenberg, N.J. (ed.) *Drought in the Great Plains: Research on Impacts and Strategies.* Water Resources Publications, Littleton, Colorado, pp.19-42.

Drummond, I., and Marsden, T. (1999) *The Condition of Sustainability.* Routledge, London.

Easterling, W.E., Crosson, P.R., Rosenberg, N.I., McKenney, M.S., Katz, L.A. and Lemon, K.M. (1993) Agricultural impacts of and responses to climate change in the Missouri-Iowa-Nebraska-Kansas (MINK) region. *Climatic Change,* 24, 23-61.

Ernst, H. (2002) Use it or lose it: the cat with nine lives. *HydroGRAM,* Winter, 29-30.

Fletcher, S. (1997) Global climate change treaty: Summary of the Kyoto Protocol. *Congressional Research Service Report for Congress 98-2 ENR* 22 Dec. http://www.cnie.org/nle/clim-3.html (Accessed July 2002).

Harrington, J. (1998) A climatology of changing dewpoint temperatures: Dodge City, Kansas. *Preprints, Ninth Symposium on Global Change Studies.* American Meteorological Association, Boston, pp.114-116.

Harrington, J. Jr., Kromm, D., Harrington, L., Goodin, D. and White, S. (2003) Global change and southwest Kansas: local emissions, non-local determinants. In: AAG *Global Change and Local Places: Estimating, Understanding, and Reducing Greenhouse Gases.* Cambridge University Press, Cambridge, pp.57-78.

Harrington, L.M.B. (2001) Attitudes toward climate change: Major emitters in southwestern Kansas. *Climate Research,* 16(2), 113-122.

Harrington, L.M.B. (2003) Bioweaponry and agroterrorism. In: Cutter, S.L., Richardson, D.B. and Wilbanks, T.J. (eds.) *Geographic Dimensions of Terrorism.* Routledge, New York, pp.187-197.

Harrington, L.M.B. and Lu, M. (2002) Beef feedlots in southwestern Kansas: Local change, perceptions, and the global change context. *Global Environmental Change,* 12 (4), 273-282.

Hewes, L. (1958) Wheat failure in western Nebraska, 1931-1954. *Annals of the Association of American Geographers,* 48(4), 375-397.

Houghton, J.T., Ding, Y., Griggs, D.J., Noguer, M., van der Linden, P.J., Dai, X., Maskell, K. and Johnson, C.A. (eds.) (2001) *Climate Change 2001: The Scientific Basis.* Cambridge University Press, Cambridge.

Hudson, J.C. (1996) *The Geographer's Great Plains.* Occasional Publications in Geography Department of Geography, Kansas State University, Manhattan Kansas.

Intergovernmental Panel on Climate Change Working Group I (2001) *Climate Change 2001: The Scientific Basis.* Cambridge University Press, Cambridge.

Kansas Geological Survey (KGS) (2000) *An Atlas of the High Plains Aquifer.* Http://www.kgs.ukans.edu/HighPlains/atlas/ Accessed March 2004.

Kansas Geological Survey (KGS) (2003) WIZARD database. Http://hercules.kgs.ku.edu/geohydro/wizard Accessed April 2004.

Kansas Water Office (KWO) (2002) *Weather Modification.* Http://www.kwo.org/Programs/weather_modification.htm Accessed August 2004.

Kates, R.W., Clark, W.C., Correll, R., Hall, J.M., Jaeger, C.C., Lowe, I., McCarthy, J.J., Schnellnhuber, H.J., Bolin, B., Dickson, N.M., Faucheux, S., Gallopin, G.C., Grübler, A., Huntley, B., Jäger, J., Jodha, N.S., Kasperson, R.E., Mabogunje, A., Matson, P., Mooney, H., Moore, B. III, O'Riordan, T. and Svedin, U. (2001) Sustainability science. *Science,* 292, 641-642.

Kincer, J.B. (1923) The climate of the Great Plains as a factor in their utilization. *Annals of the Association of American Geographers,* 13(2), 67-80.

Kohnen, A. (2000) *Responding to the Threat of Agroterrorism: Specific Recommendations for the USA Department of Agriculture.* BCSIA Discussion Paper 2000-29 ESDP Discussion Paper ESD-2000-04. John F. Kennedy School of Government, Harvard University.

Kromm, D.E. and White, S.E. (eds.) (1992) *Groundwater Exploitation in the High Plains.* University Press of Kansas, Lawrence Kansas.

Kromm, D.E. and White, S.E. (2001) Regional change in the American Ogallala High Plains. In Kim, K., Bowler, I. and Bryant, C. (eds.) *Developing Sustainable Rural Systems* PNU Press, Pyongyang, pp.67-74.

Lawson, M.P. (1976) *The Climate of the Great American Desert. Reconstruction of the Climate of Western Interior USA, 1800-1850.* University of Nebraska Press, Lincoln.

Leathers, N.J. (2001) *A Furnace Wind Climatology for the Central USA.* Dissertation. Kansas State University, Manhattan Kansas.

Leathers, N. and Harrington, J. Jr. (2001) Furnace winds of the West. *Journal of the West,* 40(3), 48-57.

Leathers, N. and Harrington, L.M.B. (2000) Effectiveness of conservation reserve programs and land 'slippage' in southwestern Kansas. *The Professional Geographer,* 52(1), 83-93.

Mileti, D.S. (1999) *Disasters by Design: A Reassessment of Natural Hazards in the USA.* Joseph Henry Press, Washington.

National Climatic Data Center (NCDC) (2003) Storm Events for Kansas. Http://www4.ncdc.noaa.gov/cgi-win/wwcgi.dll?wwevent~storms. Accessed 12 June 2003.

Nellis, M.D., Harrington, L.M.B. and Sheeley, J. (1997) Policy, sustainability, and scale: the US Conservation Reserve Program. In: Ilbery, B., Rickard, T. and Chiotti, Q. (eds.) *Agricultural Restructuring and Sustainability: A Geographical Perspective.* CAB International, Wallingford, UK, pp.219-231.

Ojima, D.S., Lackett, J.M. and the Central Great Plains Steering Committee and Assessment Team (2002) *Preparing for a Changing Climate: The Potential Consequences of Climate*

Variability and Change–Central Great Plains. Report for the US Global Change Research Program. Colorado State University, Fort Collins.

Opie, J. (2000) *Ogallala: Water for a Dry Land.* Second edition. University of Nebraska Press, Lincoln.

Popper, D.E. and Popper, F.J. (1987) The Great Plains: From dust to dust. *Planning,* 53(6), 12-18.

Riebsame, W. (1994) The historical bias for growth and intensification in Great Plains agricultural development. *Progress in Rural Policy and Planning,* 4 ,45-53.

Riebsame,W.E. (1991) Sustainability of the Great Plains in an uncertain climate. *Great Plains Research,* 1(1), 133-151.

Rogers, P., Whitby, S. and Dando, M. (1999) Biological warfare against crops. *Scientific American,* 280(6), 70-75.

Rosenberg, N.J. (ed.) (1980) *Drought in the Great Plains: Research on Impacts and Strategies.* Water Resources Publications, Littleton, Colorado.

Rosenberg, N.J. (1986) Adaptations to adversity: Agriculture, climate, and the Great Plains of North America. *Great Plains Quarterly,* 6, 202-217.

Sherow, J.E. (1990) *Watering the Valley: Development along the High Plains Arkansas River, 1870-1950.* University Press of Kansas, Lawrence, Kansas.

Sindelar, B.W., Montagne, C. and Kroos, R.R.H. (1995) Holistic resource management: An approach to sustainable agriculture on Montana's Great Plains. *Journal of Soil and Water Conservation,* 50(1), pp.45-49.

Thomas, D.S.K. (2001) Data, data everywhere, but can we really use them? In Cutter, S.L. (ed.) *American Hazardscapes: The Regionalization of Hazards and Disasters.* Joseph Henry Press, Washington, pp.61-76.

US Department of Agriculture Economic Research Service (USDA ERS) (2004) State Fact Sheets: Kansas. Http://www.ers.usda.gov/statefacts/KS.HTM Accessed 27 January 2005.

US Water News (1999) Arkansas River near Kansas-Colorado state line flows at record levels. US Water News Online, July. Http://uswaternews.com/archives/arcsupply/9arkriv7.html Accessed August 2004.

Warren, J., Mapp, H., Ray, D., Klatke, D. and Wang, C. (1982) Economics of declining water supplies in the Ogallala Aquifer. *Ground Water,* 20(1), pp.73-79.

Western Kansas Weather Modification Program (WKWMP) (2001) *2001 Target Area/2001 Final Report.* Http://users.pld.com/hailman/WKWMPpage.htm Accessed August 2004.

Whitby, S.M. (2002) *Biological Warfare Against Crops.* Palgrave, Basingstoke.

White, S.E. (1994) Ogallala oases: Water use, population redistribution, and policy implications in the High Plains of western Kansas, 1980-1990. *Annals of the Association of American Geographers,* 84 (1), 29-45.

White, S.E. and Kromm, D.E. (1995) Local groundwater management effectiveness in the Colorado and Kansas Ogallala region. *Natural Resources Journal,* 35, 275-307

World Commission on Environment and Development (1987) *Our Common Future.* Oxford University Press, Oxford.

Chapter 12

Environmental ghost towns

Chris Mayda

Introduction

Ghost towns are the busted remains of what were once thriving rural communities. Towns such as Bodie and Calico in California, and Bradian in British Columbia used to have active economies based on single extractive industries. When the local resources were exhausted or the market for their product declined, the residents were forced to abandon the town. Though the town may have been polluted due to industry, these issues were unknown and were not the prime reasons for abandonment. Although some old ghost towns have reinvented themselves to attract tourists with nostalgia or an attractive location, these are the ghost towns of yesteryear.

Today, a new type of ghost town is being created. These towns, like their predecessors, are dependent on the primary sector. However, their demise is not due to a local economic failure, such as the mine playing out, but, instead, is based on a larger picture of failure. These towns are being abandoned because of environmental deterioration caused by human interaction with the landscape, and they are some of the first to experience what may be a wave in the future. These small towns are becoming ghost towns because nature is too expensive, or too large an issue, to control. They are dying a natural death (Davies, 1998), but are not going quietly or cheaply. Unlike earlier ghost towns, these towns are seeking remuneration to pay for their relocation.

A rural small town has neither the ability to make significant repairs to the environmental damage that is causing their ruin, nor the financial resources to move the entire town to a new location. They are often at the mercy of being poor and used, and are now seeking environmental justice. America has resettled towns in the past due to progress (Cowery, 2004), 'acts of God', (Tighe and Mannies, 1993) and neighbourhoods due to eminent domain (Willmann, 1977), but not specifically due to environmental degradation. Part of the issue with environmental ghost towns is the proof that they are victims of environmental degradation. The variables involved with the interaction of law and science, the politics of culture and environmental justice shade each of the case studies. At what point do we know it is 'global warming', that air pollution is detrimental to health, or that the toxic pollution will affect our health?

This chapter describes the abandonment, or future abandonment, of three towns: Shishmaref, Alaska on the Pacific Rim; Cheshire, Ohio in Midwestern America; and Great Harbour Deep on the Atlantic seaboard of Newfoundland. Global warming, over-fishing, and air pollution are the environmental reasons why these towns will be abandoned, but they could be the tip of an extinction of rural towns through environmental causes in the future, especially as precedence is established. Shishmaref is being inundated with water due to global warming, Cheshire relocated due to its proximity to an energy plant, and Great Harbour Deep suffered the collapse of the Grand Banks fisheries.

As a humanistic geographer, my approach in this chapter considers people and their interaction with the land. I believe that culture, emotional attachments, and feelings rank equally with scientific measurements as paths to understanding our relationship with the environment. The case studies rely on both new accounts and fieldwork data collected in Cheshire, Ohio and Great Harbour Deep in 2002.

Shishmaref

Shishmaref is an indigenous Inupiat settlement located on the Arctic coastline, 8 km (five miles) off the northwest shore of the Seward Peninsula. The community is located on Sarichef, a 5 km (three-mile) wide barrier island in the Chuckchi Sea north of the Bering Strait (Fig. 12.1). Permafrost holds Sarichef together and the island is suffering severe stress and erosion as global warming weakens its physical integrity. Studies have shown that the permafrost regions bordering Arctic coastlines are among the areas most threatened by global warming (Nelson *et al.*, 2002). The three to five degree warming in the region over the past century may seem small, but the results are beginning to add up and change how life is lived in villages such as Shishmaref.

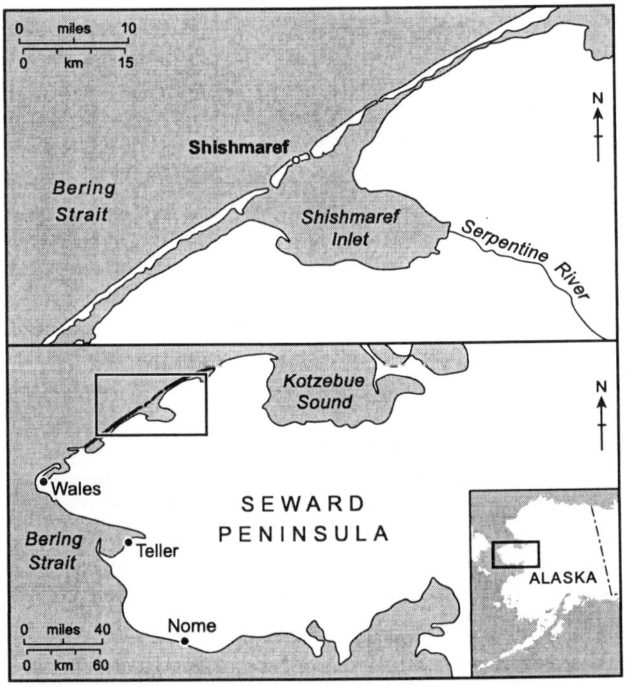

Fig. 12.1 Location of Shishmaref, Alaska.

The Arctic region, including Alaska, has been called the bell weather for climatic change, and is experiencing a variety of changes in weather. In 2001, Alaska had its warmest winter on record (NOAA, 2005). Current debates question the extent of this warming and whether it is anthropogenic or part of a natural cycle. Scientific studies and modelling of variables in the Arctic region show that Alaska is experiencing climate change, has warmed in the past century, and has experienced an accelerated rate of warming in the past quarter century (Batten and Welch, 2004; Barber and Hanesiak, 2004; Oelke *et at.*, 2004; Rasmussen and Conway, 2004; Lloyd *et al.*, 2003; Rupp *et al.*, 2000). Proving global warming has certainly been part of the problem of proof for Shishmaref, as no one really knows if we are in a period of climatic change or the reasons for these changes.

Shishmaref's problems over the past 30 years are also the result of a shift in settlement patterns. For over 400 years the Inupiat followed a seasonal cycle of settlement that brought them to the barrier island during the fishing season (Berardi, 1999). Shishmaref was originally a seasonal Inupiat settlement called Kigiktaq (renamed Shishmaref by Russian explorers in 1816). By 2000, Shishmaref was a permanent settlement occupied by 562 people who continue to live in a predominately subsistence, indigenous culture dependent on fishing and reindeer products.

Permanent settlement has brought benefits as well as problems for Inuit tribes. Modern infrastructure, communication and educational benefits, in particular, brought villagers to settle permanently on the island. But permanent settlement has affected the wildlife in the local area negatively, bringing extra stresses to the subsistence population. Further, environmental degradation, including shifts in vegetation, soil, water quality and infrastructure due to the climatic shift, have also caused problems for permanent settlements.

Powerful autumn storms roar in from the Bering Sea causing erosion and threatening the continued existence of Shishmaref and other small settlements in the region. The island's low sand dunes have eroded continually, consuming the ocean bluffs. In the past, sea ice would freeze creating a barrier protecting the coastline from the Arctic waves, but recently the sea ice is thinning due to warming temperatures and the barrier does not form until later in the year (Dyurgerov and Meier, 2000). Some homes have already been moved, while other homes built 20 m (50 feet) from the shore 30 years ago are now shoreline homes and threatened. What was once a 1.6 km (1 mile) wide island is now only 1 km (half to three-quarters of a mile) wide. Not only were homes lost, but also many lost their food supply, which they stored in caches buried below the sand and above the permafrost (Stanton, 1997). This circumstance is especially devastating to a hunter and gatherer community dependent on wildlife and their catch from the sea.

In October, 1997 a severe storm compounded current problems in Shishmaref when more than 30 feet of bluff on the north shore was destroyed which forced the relocation of 14 homes. In 2001 and 2002, powerful storms in October eroded the sea wall and undercut bluffs, leaving homes perched to fall in the ocean (O'Harra, 2001; Associated Press, 2002). After the 2002 storm five homes had to be moved away from the eroded shoreline. Further storm events in 2003 and 2004 eroded more shoreline.

In 1953, the Army Corps of Engineers first issued warnings to the villagers that they may need to relocate. But it took severe storms in 1973 for the villagers to vote for relocation. They have been continually thwarted by political complications. A study in 1994 by the University of Wisconsin and the University of Alaska suggested three options for the village: building a multimillion-dollar sea wall, dredging sand from a lagoon to rebuild the beach, or relocation. Relocation was agreed on as the best solution. Since then several million dollars have been spent on stop-gap sea walls, sand bags and other measures to protect the village, but relocation funds have not been available.

The way of life for Arctic natives has undergone severe cultural and economic changes in the past century, which has resulted in government efforts to allow them economic freedom. In Alaska, the Alaska Native Claims Settlement Act (ANCSA) of 1971 was established to manage community land and resources. But still, many Arctic groups have lost much of their heritage and now the encroaching global warming threatens to destroy their lifestyle. The villagers do not want to lose their village identity. They share a common Inupiat cultural background and are seeking to resettle as a community. They have located a site on the mainland called Tin Creek about 19 km (12 miles) from their current location. Moving there will allow them to continue their hunting and gathering way of life, but at a cost.

In December, 2004, The Army Corps of Engineers Report on Shishmaref gave four relocation estimates (Army Corps of Engineers, 2004). One was to remain in place and continually invest in fighting the sea, a second was to move to nearby cities (Nome or Kotzebue), and a third was to move to Tin Creek. Estimates ranged from CA$93 million (move to Nome) to over CA$179 million (move to Tin Creek).

The relocation of the entire town would be outwardly the most expensive relocation cost. But when cultural, social and environmental costs are added into the traditional balance sheet, the costs of relocation read differently. The residents value their way of life and holding the community together. Moving them into another town where they will lose how they live could have much larger costs than money alone, and so they have chosen to stay together and move as a community, despite the high cost. Senator Ted Stevens (2002) has repeatedly tried to get funding for the relocation and been unsuccessful.

The changing climate and the resulting changes in permafrost have created flooding and erosion problems in many native Alaskan villages. Studies of increased warming indicate that both the projected costs and the number of claimants will continue to grow. In 1974, the estimates for relocation of Shishmaref were CA$50 million. The lack of funds in 1974 pales in comparison to budget issues in the deficit spending era of 2005. There is not any money to move houses on the eroded bluffs, let alone an entire town. Fundraisers at the University of Alaska at Fairbanks brought in CA$10,000, not nearly enough to begin the relocation process. The General Accounting Office (GAO, 2003) report encapsulated much of the problem within its title: 'Alaska Native Villages: Most are affected by flooding and erosion, but few qualify for federal assistance.' Within the opening paragraph of the GAP Report are estimates that 86% of all Alaskan Native villages have flood and erosion issues, and that the cause is 'due in part to rising temperatures'.

Alaskan native villages, such as Shishmaref, are in a unique situation in relation to towns in the lower 48 states for obtaining relocation funds. Their climatic circumstances are more extreme. They are usually very small and discrete, indigenous cultural entities where social disruption would add to previous deleterious effects of modernization. Cost/benefit analysis by the Army Corps of Engineers and the GAO show that the benefits would be less than the costs (Army Corp of Engineers, 2003; GAO, 2004). Clearly the cultural, social and environmental implications of relocation, if added to the cost/benefit analysis (which does not account for these 'external' costs) would allow these villages more latitude in maintaining their lifestyle. However, the current system of cost/benefit does not usually consider items that are not 'accountable'.

The GAO report offers four alternative structures but indicates room for additional action (Direct the Corps and NRCS to include social and environmental factors in their cost/benefit analyses, p.42), as the current system does not usually address the social, cultural and environmental impacts of many actions.

The GAO and Army Corps reports recognize the environmental issues and their severity. However, the report conclusions and the necessary money to help even one of these villages, are very far apart. The 2005 US budget is cutting domestic programmes not adding them. This leaves the villages literally 'hanging' there waiting for economic assistance, when every year more of their infrastructure and land is washed away. The subject needs to be acted on soon or a catastrophic 'answer' may result, but money is not available for even the least expensive move.

Shishmaref is only one of the many villages that will need assistance if global warming continues. The expense for relocating these villages would be beyond the scope of the US budget, so further study need address what can be done about native cultures that continue to exist in a globalized world. Are the resources available to save the cultures and if not should the people be assimilated and if so how? Beyond social and cultural questions are the first hand experience with the results of scientific investigations into global warming and what the future may hold for its effect on human culture.

Climate change in the form of global warming has been well documented over the past 50 years, but more recent studies tie this climatic shift from global warming to another ice age, perhaps within our lives (Broeker, 1997; Dickson *et al.*, 2002; Gagosian, 2003). Perhaps towns like Shishmaref have more to teach us than how to find the necessary funds to resettle, but instead to investigate climate change and how to protect our cities and towns from further destruction.

Cheshire

In the early nineteenth century, farmers and coal miners settled in Cheshire, Ohio on the banks of the Ohio River across from West Virginia and 137 km (85 miles) southeast of Columbus. In 2000, the town had 221 residents, with about two-thirds living in the same house as in 1995, and 42% living in the same house since 1980. Their homes were generally older with 68% built before 1940, and a median value of US$62,300 almost half the US median value (US Census, 2000). The town was a lower middle-class, stable town with long-term residents, many of whom had lived much of their life in Cheshire. Despite this stability, in August 2003, the town was almost totally abandoned after the residents accepted resettlement funds of US$20 million from the nearby coal burning utility plant, American Electric Power (AEP).

Cheshire utility plants and coking ovens have been a part of a major mid-western coal network during the last part of the twentieth century. In 1971, the Cheshire area was the third largest receiver of coal in the Midwest after Chicago and Detroit (King *et al.*, 1971). The Gen. James M. Gavin Power Plant, one of the largest coal-fired plants in the world, came online in 1974, making the coal use in the area even larger. Burning 25,000 tons of bituminous coal a day the Gavin plant generates enough electricity for 2 million homes. Since the Clean Air Act of 1971, dirtier burning bituminous coal has often been replaced with less polluting sub-bituminous coal and has helped to clean up polluted areas. In 1992, AEP bowed to political pressure and kept burning bituminous coal to save 800 Ohio mining jobs, rather than switch to cleaner burning Powder River Basin Coal (Columbus Dispatch, 1998). Instead of burning the less sulphuric sub-bituminous coal, the plant proceeded with pollution restriction testing (US Department of Energy, 2005).

The plant emits pollution-creating nitrogen oxide, sulphur trioxide, and mercury emissions, allegedly contributing to global warming and acid rain. New environmental laws, which the company has worked within, have helped lessen pollution and acid rain for the northeastern region, but at the expense of the local town, Cheshire (AEP, 2000, 2002). In 2001, AEP installed Selective Catalytic Reduction (SLR) systems in order to reduce nitrogen oxide emissions. The result was an unexpected increase in sulphur trioxide (a by-product of burning coal), which created a noxious, ground hugging blue plume that periodically enveloped Cheshire (Citizens against pollution v. AEP, 2004). Cheshire's proximity to the plant did not allow any dissipation of pollutants. The plume, coloured soot, coal ash and sulphuric odours had affected Cheshire and had possible consequences for the residents' health (Fig. 12.2).

Fig. 12.2 The Gen. James M. Gavin Power Plant at Cheshire.

For many years, residents had complained about the pollution spewing into their homes and Cheshire became a 'poster child' for the environmental movement and its concerns for clean air. In Ohio, however, state and legislative forces have been adverse to the environmental movement and have not given the Environmental Protection Agency (EPA) the necessary muscle to clean up the polluting activities. At one point, Ohio governor James A. Rhodes labelled the Ohio EPA as a 'funds-impounding, red-tape-ensnarled, nit-picking harasser of local governments and private industry' (Hawthorne, 2000).

Complaints and environmental activism began in the 1970s, and AEP took action to lessen pollution levels in the 1990s. Scrubber installation reduced about 90% of the pollution at the Gavin plant. In 1990, it released 374,920 tons of sulphuric oxide. In 2001, sulphuric oxide was reduced to 46,300 tons. It remains, however, in the top ten of sulphuric acid polluters as well as high in nitrogen oxide and mercury emissions (Hawthorne, 2000).

During this time, neighbours to the plant complained about the acid, ash and respiratory problems. They were not taken seriously, often told that the problem was not the plant, but 'pollen, pet dander and road dust' (Hawthorne, 2000). Their rural location and relatively poor town were classic cases of class based environmental justice (Girdner and Smith, 2002).

In 1999, the states of Connecticut and New York filed suit against ten AEP power plants, including the Cheshire plant. The Attorney General of Connecticut stated, 'Our main goal is not money, but cleaner air and improved public health. These companies have flagrantly defied the law by overhauling the plants without installing pollution control equipment' (Connecticut, 1999). AEP was not a good neighbour for the village of Cheshire and so in 2001 the stand off ended as residents contacted attorneys to review their options. Clearly each side felt they were being asked to do more than was reasonable.

In order to quieten the residents and to rid AEP of liability, the utility agreed in 2002 to purchase the town's homes, rather than clean up the plant's pollution problems. The price for their homes was two to three times the 1999 appraised value (US$45,000 to US$50,000 each). AEP says they bought out the residents for plant expansion, but each resident who accepted the buyout was required to sign a contract not to sue the company for any illnesses arising from the plant's emissions (AEP, 2002; Lane, 2003). Nearly all (90%) of the residents took the offer, and resettled, not as a whole town, but individually to different locations. The abandoned homes were bulldozed. A few elderly residents voted to stay allowing the town to remain valid. Some seniors have sold their home to the plant and will live in Cheshire rent-free until their deaths. But AEP did not purchase any homes outside the incorporated town boundary.

When AEP settled the original lawsuit in 2001, only the legal town was purchased. In 2004, the unincorporated residents sued American Electric claiming that the buy out did not solve the pollution problem. They continue to suffer from the sulphuric acid releases (Citizens against pollution vs. AEP, 2004; Hunt, 2004). AEP has not substantially reduced the sulphuric acid levels. As the largest plant in Ohio, the amount of coal burned and the resultant pollution are not acceptable, especially when low-sulphur coal is available. The question remains whether AEP will be pressured to support Ohio jobs over the pollution. Politics could easily enter the equation again if AEP changes to sub-bituminous coal. Since 2001, Ohio lost 246,500 jobs (third highest in the nation) (Yudken and Baugh, 2004). Will it be willing to lose 800 more?

Cheshire and Shishmaref both share environmental issues that are beyond the scope of their small towns. Both towns are experiencing a sense of loss at having to move from their long held homes. However, when Cheshire residents moved, they did not move as a community as Shishmaref desires. There was no talk in Cheshire of the entire community moving together, though most wanted to remain in the county or even the school district. Though most of Cheshire shared a common European ancestry (98%), and long time residence in the community, they did not share the bonds of the Native American Inupiat or their subsistence economy.

A certain politics of culture, where culture and politics meet, is prevalent in these two case studies. No tidy packages of region or identity can be interpreted to come to a resolution in their cases (Lewis, 1991). The political, social, environmental and cultural regions each overlap and integrate in different patterns that are not easily assimilated into an answer. Each town shares common ground. Among the issues are environmental concerns, dissolution of another iconic small town, loss of identity and political reasons that these concerns are not met.

Another cultural issue that has yet to be addressed is the difference in cultures within these two towns. Shishmaref is a Native American Inupiat community, whereas Cheshire is 98% white. The Inupiat continue to seek retribution from the federal government, and use their unity of culture to continue their lives as a cohesive community. But for other Native American communities assimilation into modern society may be preferable to a zoo-like outsiders distinction. The outrageous cost of resettlement is due to both the harsh climate as well as the preference to be resettled as a community. Unravelling the threads of this tortured cultural politic is still too close for any objectivity.

In Cheshire, appealing to the government brought more derision than relief, and so they chose to go directly to the source (which, of course, the Inupiat cannot do) and threaten litigation. Here the politics of culture had as much to do with public relations as it did with bottom line profitability, both now and in the future if an outbreak of some cancer proved linked to the plant. Though the town was of modest means, they were able to bring the corporation to a 'win-win' settlement. The residents received ample money to move, and the corporation was able to expand their facilities. Whether either is entirely truthful is another threaded knot that will be answered only with time.

The next case study, Great Harbour Deep, Newfoundland has also resettled at the request of the residents, though in Newfoundland there is much precedence for resettling. But Great Harbour Deep is a cross-section of the other two case studies. It too has resettled due to environmental conditions, but like Shishmaref, it has been an isolated culture that has developed its own steadfast traditions, while at the same time it is a white, but maligned culture that appealed to the Canadian government for resettlement funds.

Great Harbour Deep

Newfoundland is a forbidding rock that was settled because of the abundance of fish in the Grand Banks. The cold Labrador Current and warm Gulf Stream blend over the shallow water of the continental shelf, creating the Grand Banks, an ideal location for fish propagation, especially cod. Agriculture did not suit the barren and often treeless shores of Newfoundland, but the sea provided the means of existence for the small towns, called 'outports' (out from St John's), that dot the fjorded shores. Cod were caught, dried and then sold to merchants. The reliance on cod as a means of earning a living was so important in Newfoundland that the fish was called 'Newfoundland currency'.

In the early twentieth century, over 1400 outports hugged the coastline and faced the sea. Most outports had fewer than 200 residents, most of whom worked within the inshore fishery industry. By 1992, the fishery had collapsed, a moratorium was called and the remaining outports lost their major source of employment.

When Newfoundland became a province in 1949, the first Premier, Joey Smallwood, favoured a social and economic agenda to attain the Canadian standard of living. This initiative involved providing a higher level of education, health care, transportation and public service to the people. As the cost of providing these services for all the outports would have been prohibitive, Smallwood encouraged a controversial programme called 'resettlement' in order to agglomerate populations into a tidier, economically viable settlement.

The provincial government earmarked many of the smaller isolated outports for abandonment. Forty-six outports were abandoned voluntarily from 1945 to 1953, but no standard for resettlement existed until 1954 (Copes, 1972). At that time, a Centralization programme provided removal grants under CA$1,000 per family to communities willing to abandon. From 1954 to 1965, the provincial programme relocated about 7500 people from 115 communities (Matthews, 1976). A Federal-Provincial Program instituted in 1965 was more successful, moving 16,000 residents from 119 communities in the following 5 years. The Smallwood era ended in 1972 and so essentially did resettlement. Only 30 communities resettled between 1970 and 1975. The lack of success in the later years was largely due to even higher unemployment levels and public criticism of the programme. Many felt the programme was purely an economic vehicle and therefore a failure because it did not take the lives and cultures of the resettled into account. Many felt the largest price paid (and this is still a popular sentiment) was the loss of the traditional culture.

Newfoundland traditional culture in the outports remains largely a subsistence type economy, not as ancient as the Inupiat, but still uniquely adapted to their circumstances over several hundred years. Those who lived in outports were bound in a tight-knit community where fishing, or for women the fish plants, became a social network that bound people together (Sinclair, 2003). Though considered poor by many outsiders, their sharing, bartering and the fact that they built and owned their homes without mortgages made them wealthier than might first be observed, and created a community culture that evolved into tradition. Still, the lack of jobs in the outports forced many Newfoundlanders to leave the province and their home, if only for a time. Many returned after a few years on the mainland, often with a different attitude, a more worldly air and enough money to incur envy among those who had not gone 'out there' (Gmelch and Richling, 1988). Traditional culture has changed in Newfoundland due to this migration and repatriation, resettlement, and since modern transportation and communication. Though resettlement was harsh, and often involuntary for residents, life was about to change regardless.

Although the resettlement programme had ended, Newfoundland's economic woes did not. Fishing was the major source of economic production, but the population was almost singularly dependent on it. Sustainable catches were not possible with the number of fishermen and the allowable catches (Copes, 1972). The fishery supported a majority of the outports until 1950, but then a change took place as international trawlers began to ply the Grand Banks and fishing catches tripled by 1968 and then declined to a pre-trawler tonnage in 1977, when the 320 km (200 mile) fishery zone was imposed in Canada, protecting much of the Grand Banks fishery for Canadian fishers. A slight increase by the Canadian fishery followed and then a decline until the collapse of the fishery in 1992 (Sinclair, 2003). During the time of the fishery decline, less and less fishers could support their families and a dependence developed on other fishery activities, reliance on women's labour, retrenchment and reliance on social welfare (Pettersen, 1996).

In 2002, the cod stock remained depleted at 2% of pre-collapse stocks (Lilly, 2002), which has not recovered in 2005. With the moratorium an immediate 25,000 fishers were out of work (Sinclair, 2003). Despite government programmes to help support those out of work, over 46,000 Newfoundlanders have left the province to find work since the moratorium (Statistics Canada, 2001). Those remaining in outports, especially in the more isolated outports, have little opportunity outside of social welfare to diversify beyond fishing. A majority of those dependent on the fishing industry continue to suffer serious economic loss as a result of the cod moratorium (Sinclair, 1996, 2003).

New resettlement

The moratorium on the cod fishery increased speculation that some isolated outports would be resettled. Great Harbour Deep, located on the eastern side of the Northern peninsula, was one of the most likely to resettle (Fig. 12.3). The town had no road connecting it to the rest of the province and was dependent on government-subsidized air transportation in winter, and a ferry service in the summer months. It was the last remaining outport from Jackson's Arm to the south or Englee, 120 km (80 miles) to the north. During the resettlement period of the 1960s, people moved to Great Harbour Deep, though it was never a designated growth centre (Department of Community and Social Development, 1967; Government of Canada, 1967).

The population grew from 140 in 1951 to 334 in 1966 (Population characteristics, 1971). However, since 1991, Great Harbour Deep lost 40% of its residents, its fish plant and its movie theatre. In 2002, it was struggling with only 120 people, a lodge and small general store. The fishery had deteriorated with only a few regulated weeks, and the loss of the fish plant had left the women with no form of employment for all but the few who participated in the fishery. The school had only 25 students and went up to grade 8. High school students had to leave the outport during the school year and live in the nearest town with a high school over 200 km (120 miles) away (Mayda, 2004).

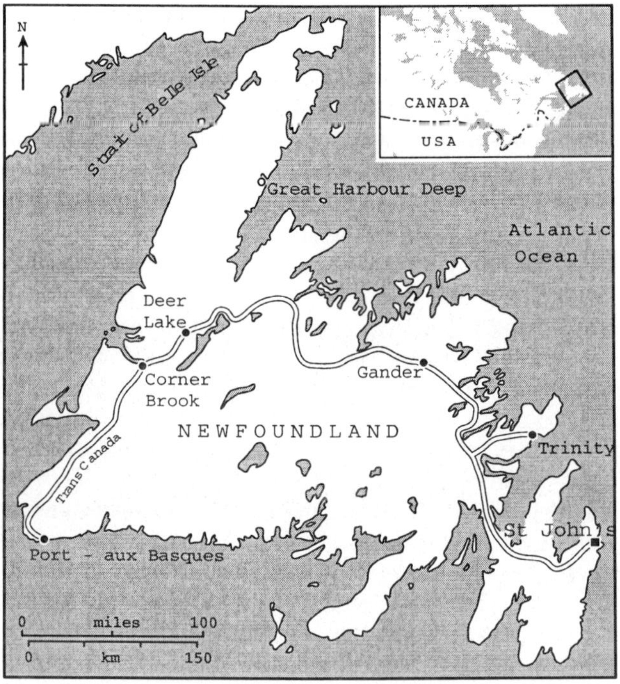

Fig. 12.3 Location of Great Harbour Deep.

In 1999, the Town Council of Great Harbour Deep was well aware of the shrinking population of their town, as well as the shrinking possibilities that a road would connect them to the rest of the island. Using the Newfoundland history of resettlement to their advantage and not able to relocate without compensation, the town council petitioned the provincial government to be ‘resettled’. The town asked for CA$150,000 per family. They were offered CA$100,000 per family. In April, 2002, the resettlement request was granted.

The government agreed to the resettlement and a CA$5 million payout because it felt the scheme was cost effective. Great Harbour Deep was expensive to maintain, especially as the number of residents dwindled. The medical and transportation services (planes and ferries), the education facilities, and food supplies were government supported. The town was also afforded airfreight subsidies to reduce costs of retail food in isolated areas. With these expenditures no longer necessary, the government believed it would recoup the resettlement costs in less than 5 years.

The government granted the town its resettlement funds in April 2002, and the town abandoned its site during the summer and early autumn months (Fig. 12.4). On 31 October

2002, all outport services were cut. The lodge chose to remain open on a limited basis, supplying its own utilities, phone service, and transport to the abandoned town. In 2004, the lodge remained open during some of the reduced fishing seasons and winter sports months. Some residents use their homes during short fishing seasons, but the town is now abandoned for most of the year and cut off from the power grid.

If all of Newfoundland is taken into account, the Smallwood resettlement period was successful technically and economically. People are no longer in desperate situations as the safety net of Canada has provided an entry to the middle class. However, the individuals who were among those who moved often were not able to improve their lives after the move (Iverson and Matthews, 1968). Often it was their children who benefited, as they had access to a high school education and the possible chance of other employment than in the fishery. It is too early to tell if this will be the case in Great Harbour Deep, but many of the residents chose to live in towns where their children could be near a high school, rather than choosing another outport location. Though some of the fishers have returned to Great Harbour Deep for the remaining heavily regulated fishing seasons, others are seeking change in larger towns and hope for their future in Newfoundland. None of the residents I am aware of moved out of the province.

Fig. 12.4 The resettlement of the residents of Great Harbour Deep.

Conclusion

Resettlement is uncommon in North American towns. Each of the case studies discusses a resettlement based on environmental deterioration. Residents of Shishmaref want to resettle as a unit and maintain it cultural integrity. So far this has not developed and reports (GAO, 2003; Army Corps, 2004) have been largely technical rather than social in their approach, much as other resettlements, such as in China's Three Gorges (Croll, 1999). Cheshire resettled as a result of possible litigation over air pollution and the corporation's decision to halt bad publicity. Other litigation questions if the corporation is making economic, political, or social decisions in relation to the town. Unlike the ghost towns of the past, the Newfoundland resettlement was organized and part of a political decision to better use governmental funds as well as better the lives of the inhabitants.

Each of these towns has been in the news due to their dire straits. Each has been fighting their particular environmental problems since at least the 1970s. It is only after years of formal complaints and litigation that attention has finally been enough to do something within these towns. Each of the towns is a small town in rural US or Canada with an under-represented group of people, a group who are often made to deal with environmental justice issues. Though two of these towns have won their battles, the cost has been high.

Each of the towns discussed is a part of a declining rural USA and Canada. With less than 20% of the population, sparse density, and a tumbling rural economy, rural towns have struggled to be heard within their states, provincial or federal political structure. They are often less affluent and unable to have their voice heard in the more dominant urban landscape. Rural social and economic development is in a state of steady decline, and yet, at the same time, suffers some of the most egregious environmental deterioration despite an urban belief in its 'authentic' environmental conditions (Opie, 1998).

Broader political, economic, and social contexts influenced what each of these towns could do. Thus Shishmaref is caught as an isolated indigenous town in a state and country that are stretched to their maximums. Though Shishmaref began seeking resettlement during the relatively wealthier 1970s, the Alaskan economy of 2005 has been steadily declining, largely because of decreased oil production. The economy of the USA has pulled back on domestic spending in favour of war and defence spending. Cheshire received little support from Ohio since their plight began in the 1970s and chose another route in 2001, pursuing litigation with the corporation that was causing their problem. The corporation responded by settling with the town once this action was taken and, by 2002, the town was purchased and vacated. Great Harbour Deep used their history of resettlement as an emotional lever to bring their case to the provincial government, and because the economics of the deal made sense for the government, they were able to resettle in a fairly short period of time (1999-2002) when compared to the long-term problems of Shishmaref.

Each of these towns has been dealt with technically and economically, with little care to the social and environmental issues at hand, or the psyche of the residents and their cultures. The combination of little habitable land remaining in the world, a growing consumer population, the environmentally inefficient global economy and the waste it produces, makes it seem inevitable that more environmental ghost towns will or already have appeared. With the environmental problems within rural America growing, it is easy to envision many future ghost towns seeking remuneration due to global forces destroying their local environment.

References

American Electric Power (AEP) (2000) *AEP announces plans to use urea-based system to reduce nitrogen oxide emissions at Gavin plant.* [Accessed 12 January, 2005] Available from: www.aep.com/newsroom/newsreleases/default.asp?dbcommand=displayrelease&ID

American Electric Power (2002) *AEP will invest $7 million to address community concerns over plume appearance at Gavin Plant.* [Accessed 12 January 2005] Available from <http://www.aep.com/environmental/performance/activities/default.asp?dbcommand=DisplayRelease&ID=887&Section=Environmental&colorControl=on>

Army Corps of Engineers. (2004) *Shishmaref Partnership: Shishmaref Relocation and Collocation Study. Preliminary costs of alternatives.* Tetra Tech, Seattle.

Associated Press (2002) State Assesses storm damage to northwest coast. *The Associated Press and Local wire*, 18 October.

Barber, D. G. and J. M. Hanesiak, (2004) Meteorological forcing of sea ice concentrations in the southern Beaufort Sea over the period 1979 to 2000. *Journal of Geophysical Research C: Oceans 109.*

Batten, S.D., D. W. Welch, (2004) Changes in oceanic zooplankton populations in the north-east Pacific associated with the possible climatic regime shift of 1998/1999. *Deep-Sea Research Part II: Topical Studies in Oceanography* 51, 863-873

Berardi, G. (1999) Schools, settlement and sanitation in Alaska native villages. *Ethnohistory,* 46, 329-359.

Broecker, W. S. (1997) Thermohaline Circulation, the Achilles' Heel of Our Climate System: Will Man-Made CO2 Upset the Current Balance? *Science* 278, 1582-1588

Citizens Against Pollution (plaintiff) v. American Electric Power, Co., Inc. (defendant) (2004) Civil complaint filed in the USA district court for the Southern District of Ohio, eastern division at Columbus. May 12.

Columbus dispatch (1998) Merger Politics: PUCO is wrong to micromanage utilities. July 20.

Connecticut Attorney General's Office (1999) Lawsuit filed against out-of-state power plants. Press Release. [Accessed 14 February, 2005] Available from http://www.cslib.org/attygenl/press/1999/util/power.htm

Copes, P. (1972) *The resettlement of fishing communities in Newfoundland.* Canadian Council on Rural Development, Ottawa.

Croll, E.J. (1999) Report from the field: involuntary resettlement in rural China: the local view. *The China Quarterly* 158. 468-483.

Davies, R.O. (1998) *Main Street Blues: The decline of small-town America.* Columbus, Ohio State Press.

Department of Community and Social Development (1967) *Memorandum on Reception Centres.* Department of Community and Social Development, Newfoundland.

Dickson B. I. Yashayaev, J. Meincke, B. Turrell, S. Dye,and J. Hoffort. (2002) Rapid Freshening of the Deep North Atlantic Ocean Over the Past Four Decades, *Nature* 416, 832-837.

Gagosian, R. B. (2003) Abrupt climate change: should we be worried? Woods Oceanographic Institute, Presented at World Economic Forum, Davos, Switzerland, January 27.

General Accounting Office (GAO) (2003) Alaska Native Villages: Most are affected by flooding and erosion, but few qualify for federal assistance. Report to Congressional Committees, December. GAO 04-142.

Girdner, E.J., J. Smith (2002) *Killing Me Softly: Toxic Waste, Corporate Profit, and the Struggle for Environmental Justice.* Monthly Review Press, New York

Gmelch, G. and B. Richling. (1988) 'We're better off here': Return migration to Newfoundland outports. *Anthropology Today* 4, 12-14.

Government of Canada (1967) *Draft of Agreement between the Governments.* Government of Canada, Ottawa.

Dyurgerov, M.B. and Meier, M.F. (2000) Twentieth century climate change: evidence from small glaciers. *Proceedings of the National Academy of Sciences of the USA of America* 97, 1406-1411.

Hawthorne, M. (2000) Cheshire: death of a village. *Columbus Dispatch* 10 November, 2000.

Hunt, S. (2004) AEP Pollution doesn't stop at Cheshire. *The Columbus Dispatch*, 13 May.

Iverson, N. and D. R. Matthews. (1968) *Communities in Decline: An examination of household resettlement in Newfoundland.* Newfoundland Social and Economic Studies, 6. St. John's: Newfoundland: Institute of Social and Economic Research, Memorial University of Newfoundland.

King, L., Casetti, E., Odland, J., and Semple, K. (1971) Optimal transportation patterns of coal in the Great Lakes region. *Economic Geography,* 47, 401-413.

Lane, M. B. (2003) Cheshire will live on. *Columbus Dispatch.* 5 February, 2003.

Lewis, M. (1991) Elusive societies: a regional-cartographical approach to the study of human relatedness. *Annals of the Association of American Geographers* 81, 605-626.

Lilly, G. (2002) Conversation and meeting, Department of Fisheries, St. Johns Newfoundland.

Lloyd, A.H., T.S. Rupp, C. L. Fastie, A.M. Starfield. (2003) Patterns and dynamics of tree line advance on the Seward Peninsula, Alaska. *Journal of Geophysical Research D: Atmospheres* 108,

Matthews, D. R. (1976) Communities in Transition: An examination of Government initiated community Migration in rural Newfoundland. Unpublished PhD thesis, Minneapolis, University of Minnesota.

Mayda, C. (2004) Resettlement in Newfoundland: Again. *Focus* 48, 27-32

Nelson, F.E., O. A. Anisimov, Ni. I. Shiklomanov. (2002) Climate change and Hazard Zonation in the Circum-Arctic Permafrost Regions. *Natural Hazards* 26

NOAA (National Oceanic and Atmospheric Administration) (2005) Winter 2001: Warmest winter on record in Alaska [Accessed on 14 January, 2005] Available from: http://www.noaanews.noaa.gov/stories/s599.htm.

Oelke C., T. Zhang, M.C. Serreze (2004) Modelling evidence for recent warming of the Arctic soil thermal regime. *Geophysical Research Letters* 31

O'Harra, D. (2001) Storm erosion of bluff puts 10 homes in danger on Shishmaref. *Anchorage Daily News*, 10 October.

Opie, J. (1998) Moral geography in High Plains history. *Geographical Review* 88, 241-258.

Pettersen, L. T. (1996) Crisis management and household strategies in Lofoten: a question of sustainable development. *Sociologia Ruralis* 36, 236-248.

Rasmussen, L.A., H. Conway. (2004) Climate and Glacier Variability in Western North America. *Journal of Climate* 17, 1804-1815.

Rupp, T. S., F. S. Chapin III, A. M. Starfield. (2000) Response of subarctic vegetation to transient climatic change on the Seward Peninsula in north-west Alaska. *Global Change Biology* 6, 541-555.

Sinclair, P. R. (1996) Sustainable development in fisheries-dependent regions? Reflections on Newfoundland cod fisheries. *Sociologia Ruralis* 36, 224-235.

Sinclair, P. R. (2003) 'A very delicate world': fishers and plant workers remake their lives on Newfoundland's Bonavista Peninsula after the cod moratorium. *Maritime Studies* 2, 89-109.

Stanton, D. (1997) Storm batters village; families forced out in Shishmaref. *Anchorage Daily News*, 5 October.

Statistics Canada (2001) Net Interprovincial Migration by Age group, Newfoundland and Labrador' Economics and Statistics Branch. Newfoundland Statistics Agency. Newfoundland.

Stevens, T. (2002) Senator Steven's Report to Alaskans. [Accessed on 14 January, 2002] Available from http://stevens.senate.gov/report011502.htm.

Tighe, T., J. Mannies (1993) Homeowners angry over flood rules. *St. Louis Post-Dispatch,* August 27.

US Department of Energy (2005) *Phase II field-testing of advanced mercury control technology.* Office of Fossil Energy.

Willmann, J. B. (1977) Harris warns realtors of relocation ills. *The Washington Post,* November 16.

Yudken, J. and B. Baugh. (2004) Ohio Job exports. AFL-CIO Industrial Union Council, Washington DC, September 22.

Chapter 13

Interpreting family farm change and the agricultural importance of rural communities: evidence from Ontario, Canada

John Smithers

Introduction

Family farming, despite some longstanding concerns for its future, remains a highly visible part of Canada's food production system and rural mosaic. In historical terms, the family farm sector has been integral to the development and prosperity of rural areas across the regions of Canada. This importance is particularly true in southern Ontario where family-based farming underpinned the initial settlement of the region and formed the backbone of local economies communities throughout the province's development (Fuller, 1985; Sim, 1988). In large part, farming provided the economic and social lifeblood of rural communities and those communities thrived on business done with farmers and the contribution of farm people to social institutions.

However, in the early 2000s, conditions are changing rapidly, both in terms of the nature of family farming and in the 'place' of agriculture in the wider frame of rural economic and social life. With respect to the former, much attention has focused upon the phenomena of intensification and expansion as a structural response to change, and on the economic failure of many farms leading to a decline in the number of farms and farmers. In the latter case, questions are now arising concerning the ways in which farming remains connected to, and dependent upon, rural communities as places of exchange, service and support. Such questions arise from a growing recognition that contemporary family farms are increasingly inclined to look beyond the local community for the satisfaction of business and household needs.

Today, casual examination of farm–community linkages might suggest that farming and community are as often in conflict as they are in accord. In Ontario, several well-publicized instances of community opposition to certain types of agricultural development, the occurrence of various environmental impacts from agriculture, and the progressive economic and social transformation of many rural communities away from agricultural service are some of the more visible indications of how farm-community linkages may be evolving. However, a more considered view is that among these specific instances of decoupling, many of the more mutually supportive types of interaction undoubtedly remain. For those wishing to promote stronger and more positive linkages between farm and town, there is value in understanding better the ways in which contemporary family farms are looking beyond local communities, and more importantly, the ways in which they remain dependent upon, and contribute to these places. Such insights can aid in the formulation of strategies to promote local collaboration and the sustainability of both farms and rural communities.

This chapter explores evolving interactions and (inter)dependencies between family farms and rural communities in light of long evolving and rapid recent changes in both sectors. Particular attention is given to the phenomenon of diversification in family farming, not only with respect to commodity emphasis and mix, but also to the increasingly apparent distinctions between farm business and household circumstances. It will be argued that the latter holds importance for understanding how specific farm business development trajectories, and some farm households, are increasingly linked with, and dependent upon, rural communities, while others are less so.

The remainder of the chapter is organized in four sections. The first two sections provide an overview of the main components of the family farm – rural community dynamic. These sections review literature pertaining to the causes of change in the farm sector, particularly as evident in the North American experience, as well as farm – community interaction and the community-level implications of agricultural change. The third and fourth sections adopt an empirical focus and report on two parts of a recent analysis that documented family farm business development trajectories in a selected farming system in southern Ontario and assessed the nature and importance of interactions between farms and the rural (settlement) community. In a broader sense, the research sought to unravel how linkages are 'playing out' at the level of a local farm and rural community system in order to better understand those aspects of community that remain important for farmers, and identify those that are changing.

Evolution and restructuring in the farm sector

Canada's agricultural systems have undergone a significant transformation during the past several decades, the nature and the causes of these changes are similar to those observed in other developed economies. In the main, labour and management intensive systems of farming have been replaced by systems of production characterized by a high degree of mechanization and intensification. Capital and technology have been widely substituted for labour, with greatly increased use of fossil fuels, chemical fertilizers, pesticides, animal antibiotics and other purchased inputs (Marsden, 1998; NRC, 1989; Troughton, 1997). The result has been an unprecedented increase in the productivity and efficiency of food production systems over the past several decades (Pierce, 1990; Surgeoner and Dalrymple, 1993).

Change has also occurred in a range of structural characteristics. These are now well documented in many jurisdictions and include a tendency towards larger (and fewer) farms, with a corresponding reduction in the number of farmers and farm employment opportunities, intensification of production, increasing specialization in commodity choice and spatial concentration of production (Pierce, 1994; Ilbery and Bowler, 1998).

The developments cited above reflect the evolving nature of farming as a productive activity. Examination of agriculture in its social, economic and institutional context reveals a second set of structural changes. As suggested at the outset, the history of farming in Ontario reflects a once strong connection between agriculture and rural communities. Yet in Ontario today, there is mounting evidence that farming and rural communities are effectively becoming decoupled from one another, with the critical relationships more often being *vertical* than *horizontal* (Marsden, 1998; Smithers and Joseph, 1999). The highly productive farms that now dominate the agricultural scene are increasingly linked with, and dependent upon, agri-business, government (through agricultural policy and programmes) and financial institutions for their markets and critical inputs. Hence, they are less dependent upon local rural communities as places of exchange and service provision. In the context of farm-rural community linkages, the effect of this decoupling ranges from ambivalence and benign neglect of one another to specific instances of conflict relating to land use, land management or the changing social composition of rural areas.

Environmental regulation must also be seen as a force in reshaping relations between farm and non-farm interests. In particular, heightened public scrutiny and sensitivity towards environmental quality, and the arrival and growing political influence of increasing numbers of non-farm people and interests in the countryside has led to a rise in complaints and conflict, often culminating in political pressure to alter farming practices (Buttel, 1997; Caldwell, 1998; Robinson, 1997). In many jurisdictions, such pressures have resulted in so-called 'Right to Farm' legislation designed to protect farmers from nuisance litigation and safeguard their ability to employ so-called 'normal farming practices' (DeLind, 1995; Penfold *et al.*, 1989). However, continuing intensification in some sectors, especially intensive livestock farming, and its potential detrimental environmental impacts has contributed to greater regulation of agriculture not only at national, but also regional and local scales (Caldwell, 1998; Furuseth, 1997; Ward *et al.*, 1995).

Documenting diversity

The family farm sector in Canada and in many other developed economies exhibits fundamental diversity, in enterprize types, in its structural characteristics and in the role that farming plays in the social and economic lives of families. In rural geography, this diversity has been captured via proposed 'pathways' of farm development (Bowler, 1992; Pierce, 1994; Whatmore *et al.*, 1987). In one direction, there is evidence of a trend towards continuing, even intensifying, industrialization. Such tendencies represent a continuation of the pattern of agricultural development over the past several decades in most industrialized agri-food systems. Among the best-known and most contentious examples in North America are developments in the pork sector. However, less extreme examples exist throughout the family farm sector on farms where expansion and intensification have been chosen as the means of sustaining the enterprise.

The trend to expansion, while widespread in many areas represents only one trajectory of development. Those commonly suggested in the various pathways conceptualizations include: the production of new non-traditional agricultural commodities, the development of non-agricultural economic activities on the farm, the mixing of off-farm employment with farming, persistence in traditional forms of production and disengagement either through reversion to part-time farming or abandonment. Among these considerations, much attention has been given to diversification as an adjustment strategy (Bowler *et al.*, 1996) and to off-farm work (Ilbery *et al.*, 1996; 1997; Shucksmith and Smith, 1991). In the case of the latter, it is increasingly recognized that off-farm work is vital in many cases for the survival of the farm business itself (Evans and Ilbery, 1993; Fuller, 1990). Further, the incidence of off-farm employment may not only reflect financial need among farm families, but also relates to proximity to labour markets (Bryant, 1992; Coombes and Campbell, 1996; Edmond and Crabtree, 1994), the desire for career-mixing among some farmers, and the increasing participation of women in the workforce (Shucksmith and Smith, 1991).

Changing farm and rural community connections

Research in geography, sociology and other disciplines has also explored the causes of change in farm–rural community interaction with specific reference to the process of agricultural change. Acknowledgement of the impact of change on the community side is more limited.

The so called industrialization of agriculture has been widely implicated as a core contributor to the economic and social decline of rural communities. Trends towards a reduction in the number of farms, and the loss of critical mass through the exit of many farmers have been tied to dwindling support for local businesses and traditional community institutions in many regions (Berry, 1977; Gregor, 1982; Gertler, 1999). Goldschmidt's (1978) study of two California communities demonstrated how the structure of agriculture had implications for community well-being, and argued that farm systems based on large-scale corporate agriculture led to decreased community vitality in comparison with those based on a number of smaller-scale family farms. Many subsequent studies have tested this basic hypothesis with generally similar results (Bird *et al.*, 1995; Loboa *et al.*, 1993; Durrenberger and Thu, 1996). However, it is noteworthy that, for the aforementioned studies, the farming method under scrutiny, 'industrial agriculture', was set apart cleanly from its alternative – the family farm. Analysis of both long-term trends and recent developments in Canada's farm sector suggests that such a distinction is poorly founded as the characteristics of industrialized farming now pervade in family-based farming – some of which occurs at a very large scale indeed (Pierce, 1994; Troughton, 1992).

As noted earlier, the phenomenon of agricultural intensification has also become associated with an increased frequency of community conflict, particularly in the case of large-scale animal confinement agriculture. While farm expansion and the intensification of production are seen by some as a necessary adaptation to current economic conditions in agriculture, an alternate view holds that such farms pose unacceptable levels of environmental risk with poorly demonstrated economic benefits to regions. Further, for some, they contravene a normative view of what family farming should be. Consequently, it is not only newcomers and non-farm rural interests who are opposed to large scale intensified farming, but also many family farmers (Thu and Durrenberger, 1998; Grey, 2000).

Apart from such value-based issues, researchers have explored the importance of several more tangible linkages between farming and rural communities. The importance of off-farm community-based employment has been shown as a critical factor in the development and persistence of many family farms (Edmond and Crabtree, 1994; Fuller, 1990; Van den Bor *et al.*, 1997). Community characteristics pertaining to the quality and quantity of employment opportunities may then hold implications for the economic viability of farms and the perceived quality of life among farm families (McCoy and Filson, 1996). A second example of linkage relates to the traditional presence of farmers and farm families in the social, recreational, and community organizational life of rural settlements. Here the traditional importance of farm-based contributions to community life, and the importance of communities to farmers are well recognized (Green, 1985; Everitt and Annis, 1992; Ramsey and Smit, 2001). Economic exchange represents a third form of interaction and occurs through purchasing for both the farm business and the household. Contemporary structural trends in farming, ongoing economic pressures and the continuing decimation of the importance of distance in economic life have altered the nature and significance of these exchanges for both farmers and communities alike. There is a growing belief, and some empirical evidence to suggest, that an increasing number of farmers now 'shop around' outside of their local rural community to find better prices, a behaviour that has ramifications for the economy of the community (Wilson, 1994, 1996). Thus it may be useful to assess in the context of specific localities, farmers' economic attachments and loyalties. Finally, it is important to remember that the family farm is both an economic and social unit (Moran *et al.*, 1993) and that decisions on how to adjust production or (re)deploy the resources of the household are influenced by this duality (Wilson, 1994; 1996). Recognizing and unravelling this relationship represents an important pre-condition for understanding both the nature of farm development trajectories and the reasons for, and possible implications of, underlying decisions.

Farm–community linkages in North Huron County, Ontario

The remainder of the chapter reports on an analysis of family farm interactions with rural communities in North Huron County, Ontario. The broad purpose of the research was to document and interpret diversity in family farming and in farm-community linkages – as seen from the farm side. In the first portion of the analysis, farms were allocated to one of six farm business development trajectories to distinguish developmental tendencies and farm circumstances in the region. The second component of the analysis explored the ways in which farmers identify 'their community' and look to local rural communities for services, goods and social interaction.

Case study region and research approach

The study focused on the townships of Ashfield and Colborne in North Huron County (Fig. 13.1). The area is situated on the eastern shore of Lake Huron and lies well outside the urbanized core of the province. These townships, like the rest of Huron County, have a long history of family farming and have retained a strong focus, both in economic and social terms, on agriculture.

The largest settlement in the study area, Goderich (population 7853 in 2001), is the county Seat of Government and serves as the 'central place' in the region. The town is a significant regional tourism destination based on both historical significance and the amenity of the lakeshore. Indeed, much recent economic development and planning has focused on promotion of the town as a vacation destination. A second locally significant settlement is the town of Lucknow, (population 1215 in 1996) located on the northern boundary of the study area approximately 20 km inland from Lake Huron. Lucknow exhibits substantially less economic diversity than Goderich, however, it is significant from an agricultural perspective. While limited in the range of economic functions, the town features a number of basic retailing activities, many of which are farm-oriented. Indeed, it is the presence of a significant number of agriculturally oriented businesses and services that most clearly distinguish Lucknow from other communities in the study area and the region.

Two strategies were employed in the collection and analysis of data. First, a profile of the study region was developed using published census data and information obtained from interviews with local agricultural advisers (Ontario Ministry of Agriculture and Food; Huron County Federation of Agriculture) and published reports in the local print media. This information provided an overview on agricultural trends and issues in the vicinity. Second, data for the farm-level elements of the project were obtained via telephone interviews with family farmers in the Townships of Ashfield and Colbourne. A random sample of 120 family farms was obtained from the Municipal Assessment Roll and, in the end, interviews were conducted with 70 consenting study participants. Use of this telephone-administered questionnaire satisfied the need for a relatively broad-based sample of local farmers to capture the range of farm business trajectories in evidence in the study region and to provide somewhat generalizable data on farm, farm family and farm community characteristics.

Fig. 13.1 Location of Ashfield and Colborne in Huron County, Ontario.

Findings I: revealed trajectories of family farm development

The research adopted the concept of pathways or 'trajectories' of family farm business development as a means of understanding the nature and causes of farm-level change, and as way to explore farm–community linkages in a more differentiated manner. In the first instance, there was a need to capture the specific nature and circumstances of family farming in North Huron, and to adopt a dynamic view of farm business development that integrates both farm and the farm household as important influences on change. Rather than fitting an a priori typology to the study, the approach was somewhat more inductive with an attempt to discover and characterize development trajectories in the data. However, the classification exercise was founded upon three basic elements from the outset: the 'direction' of farm business development (growth, decline or static), the concentration of effort on farming in comparison to the diversification of activities (off-farm work), and the context or circumstances determining family farm change. The first two criteria were more categorical in nature, while the third was more contextual and served to reveal circumstances and motivations that guided decisions – especially those relating to external forces and to the life course of the farm and farm household.

The research confirmed the existence of three broad directional tendencies in the study sample: expanding, stable and contracting. However, subsequent analysis of other data, particularly those pertaining to the second and third criteria described above, suggested a necessary bifurcation of each group, yielding six more specific 'trajectories' of farm business change in the townships of Ashfield and Colborne. Notwithstanding some within-group variation that reflects the specific circumstances of family farms, the resulting characterizations captured some clearly identifiable general patterns in the data and were found to accommodate each farm in the sample (Table 13.1).

Table 13.1 Family farm development trajectories in Ashfield and Colborne townships

Expanding *n* = 34 (48.6%)		Stable *n* = 15 (21.4%)		Contracting *n* = 21 (30.0%)	
Farm focused	Assisted growth	Persisting	Hobby	Forced down	Winding down
23	11	9	6	8	13
32.9%	15.7%	12.9%	8.6%	11.4%	18.6%

The largest broad group within the sample was found to be actively pursuing expansion of the farm enterprize (48.6%). However, a fundamental distinction was found to exist between those who were fully focused on and relatively self-sufficient in, farm development, and individuals who were relying on other income streams to underwrite the costs of growth. Determinations were made according to not only the presence of other paid activities, but also the stated role that they play in supporting the farm. Those farmers who were found to be '*Farm Focused*' (32.9%) indicated that their family's main source of income was derived from farming and that the farm business was wholly capable of supporting both it and the household without any secondary activities. For these respondents, involvement in OGA by family members was a largely discretionary activity. In terms of the context or explanation offered (criterion three) for pursuing expansion rather than focusing greater attention on OGA, respondents unanimously expressed a desire to first and foremost be farmers without the need of another career. In these cases, the concept of farm self-sufficiency and farm enlargement and intensification were synonymous.

A second group, the '*Assisted Growth*' trajectory (15.7%), was also attempting to expand the size and scale of their farm businesses, and shared many of the same farming aspirations of the previous group, but were reliant on other sources of income to subsidize this growth. All farmers in this group expressed a desire to have the farm contribute a larger proportion of their household income. They placed a priority on farming and all members of this group expressed a hope that the farm would eventually be the only source of income for their family. Several members of this group indicated that they required other income sources at the current time because of the economic conditions in agriculture, while others espoused the logic of expansion as the only route to viability given what were, in their eyes, inevitable future developments in the food production sector.

The third and fourth trajectories were derived from a broader group of farms that was characterized as static in terms of its development efforts. Among those farmers who were found to be in this situation a clear distinction emerged between those who were working hard merely to persist and those who were effectively farming for lifestyle reasons. Farmers that were classified as operating in the '*Persisting*' trajectory (12.9%) indicated that they were dedicated to farming and maintaining productive farm businesses. However, these farmers were found to be more critically dependent on additional sources of income to keep the farm afloat. The majority of respondents believed that, given their scale of operation, their farm should be able to support a family, and attributed its failure to do so to external conditions in agriculture – most often commodity prices. However, a number of these respondents also subsequently cited other reasons why their farm business was unable to completely support their family, reflecting both business and household circumstances.

In contrast, circumstances and motivations for respondents in the '*Hobby*' farm trajectory (8.6%) were much different. For these individuals, farming was not the principal economic activity of the household, but rather was pursued for personal enjoyment and lifestyle reasons. All respondents in the '*Hobby*' trajectory acknowledged that they were farming in a residual sense only with several specifically stating that they did not consider themselves to be 'real farmers' any longer. This outlook is in stark contrast to respondents in the '*Persisting*' trajectory, where most expressed resentment that they had to supplement their farm income with other work in order to maintain their principal occupation.

The fifth and sixth farm trajectories were derived from farms that were found to be contracting in size and intensity of production. These farms were quite homogeneous in terms of their concentration of effort on farming, as most respondents had very low average percentages of household income derived from agriculture. Here the important distinction was rooted in knowledge of the past circumstances and decisions that placed farmers in these two groups.

The group characterized as '*Forced Down*' (11.4%) consisted of respondents who had attempted to survive solely on revenue generated by the farm business and had been unable to do so. When asked why they had scaled back farmers identified a wide array of economic factors, both external and within their farm business, as having conspired against them. Particular attention focused on debt, low commodity prices, and the high cost of maintaining a productive farm business as factors that spelled the end of any growth aspirations. Several individuals expressed significant frustration over their lengthy, and ultimately unsuccessful, attempt to survive in the current economic climate and shared that the decision to downsize was not necessarily one that they had chosen but one that they had been forced to take:

> 'The bank asked me to sell, and then told me to sell. It was either lose part of the farm, or lose it all.'

Conversely, respondents in the '*Winding Down*' trajectory (18.6%) present a much different story. Those farms allocated to this trajectory were operated by farmers in or nearing retirement. These individuals were relying, partially or wholly, on pensions and other income streams. However, a small number were still farming actively and in one case, deriving nearly 80% of household incomes from farm receipts. Nonetheless, all members of this group were clear that they were into or entering a deliberate 'wind down' for purposes of impending retirement.

This initial portion of the research confirmed the existence of some quite distinctive farm business 'models' in North Huron County, a region that is widely regarded as among the most agriculturally focused in Ontario and Canada. While many farmers remain committed to, and active in, agricultural expansion and intensification, a significant proportion has elected (or been obliged) to make other choices. It is noteworthy that a large fraction of the sample was either presently engaged, or had previously been committed to expansion beyond the size needed to first establish the farm business. However, the majority expressed concern over the external conditions for agriculture and the future of family farming. These sentiments revolved around an increasing awareness that it was becoming more difficult to farm without a substantial source of off-farm income. A majority of the farmers believed that the only way of surviving without off farm money was to achieve better economies of scale through (financed) expansion: a trend that, for some, led away from what they envisioned the family farm to be. Interestingly though, off-farm work was frequently characterized as a necessary evil with the exception of some of the '*Hobby*' farmers. Off-farm work was accepted by many as a means to an end, but not seen as 'the way things ought to be'. Thus, across the sample there was something close to a consistent vision for family farming with the farm supplying adequate returns and labour opportunities to sustain the family. Respondents were differentiated by their ability or inability to act on this core vision of family farming rather than their farming goals per se.

Findings II: Assessing community linkages and allegiances

A second component of the analysis considered the ways in which the farmers look to local rural communities for services, goods and social interaction. The analysis explored whether and how these patterns of interaction are changing, and explored the importance of farm business trajectories and farm household considerations in accounting for revealed patterns. Issues were explored around three broad avenues for involvement reflecting themes noted in the literature and suggested by reconnaissance work in the study area: (i) participation in community organizations and institutions, (ii) purchasing, and (iii) perceptions of community support for farming. Each of these themes was explored with reference to specific forms of involvement and in each case the linkage (or lack of it) was assessed with regard to its current status as well as how this level of involvement may have changed or be changing.

Community involvement

Farm family involvement in the community was explored in relation to four types of potential activity: community organizations, churches, schools and organized sport. Farm families were found to be involved in a wide variety of activities, with nearly half of the respondents having at least one family member involved in community organizations, church, and sport. Fewer respondents had family members involved with local schools (Table 13.2).

Within this generalized picture of community involvement, there were apparent distinctions between farm business trajectories that hint at potential connections to farm business development trends and household circumstances. Involvement with community organizations, specifically those agriculturally related, was concentrated among respondents in the '*Farm Focused*', '*Assisted growth*' and '*Persisting*' trajectories. Previously it was noted that these farms tended to be larger and displayed a stronger reliance on farming as the principal source of household income. Participation in local school activities was very low within the '*Winding Down*', '*Hobby*' and '*Persisting*' trajectories (0%, 17%, 12.5%, respectively). However, the same three groups were, proportionally, more strongly engaged with local church activities. Farms within these trajectories were characterized as generally later in the life cycle, often with grown children living elsewhere. In contrast, the '*Assisted Growth*' and '*Forced Down*' trajectories both displayed higher percentages involved in school activities (46% and 67% respectively), a pattern consistent with a generally younger life cycle stage and higher likelihood of having school-aged children at home.

Table 13.2 Farm trajectory and family involvement in community activities.

Activity	Total sample	Farm focused	Assisted growth	Persisting	Hobby	Forced down	Winding down
Sports (%)	37 (52.9)	13 (56.5)	7 (63.6)	4 (50.0)	3 (50.0)	6 (66.7)	4 (30.8)
School (%)	18 (25.7)	5 (21.7)	5 (45.5)	1 (12.5)	1 (16.7)	6 (66.7)	0
Church (%)	33 (47.1)	9 (39.1)	8 (72.7)	6 (75.0)	2 (33.3)	3 (33.3)	5 (38.5)
Organization (%)	40 (57.1)	11 (47.8)	9 (81.8)	7 (87.5)	3 (50.0)	2 (22.2)	8 (61.5)
Ag. related (%)	21 (30.0)	6 (26.1)	5 (45.5)	4 (50.0)	1 (16.7)	1 (11.1)	2 (15.4)
Non-Ag. related (%)	19 (27.1)	5 (21.7)	4 (36.4)	3 (37.5)	2 (33.3)	1 (11.1)	6 (46.2)

Note: Numbers indicate the involvement of at least one family member in the specific activity. Total responses do not sum to the total number of respondents within each column (e.g. total sample and trajectories), as respondents were able to identify more than one form of participation.

In considering changing levels of interaction, there was some speculative indication that farmers themselves (apart from the household) within certain trajectories were becoming more integrally involved in community-based activities, while other farmers were conscious of a decline in the frequency of these interactions. Over half of the farmers (56%) found to be operating in a '*Forced Down*' trajectory reported that they were becoming more involved in community activities, whereas the most precipitous declines in the frequency and quality of participation were cited by farmers in the two trajectories associated with farm expansion: '*Farm Focused*' and '*Assisted Growth*'.

When asked to comment on reasons for increasing or decreasing involvement in their community there was rarely a clear attribution to a single cause, which would permit easy attribution to business versus household factors. However, family characteristics did emerge as the more important element, with just over half the respondents (53%) citing this factor as the main reason determining why they were involved in various community-related activities. While household considerations seemed to predominate, there was at least one finding of note concerning the role of the farm business in mediating community involvement. Among those younger farmers who indicated a decline in their level of participation in community activities, the demands of the farm business were strongly implicated. It is noteworthy that a number of farmers in the '*Farm Focused*' and '*Assisted Growth*' trajectories who attributed their community involvement to household considerations (children), also indicated that the farm business acted as a counterforce to stronger personal participation: the suggestion being they were obliged to interact in support of children's activities but had elected to let some or all of their own activities subside.

The geography of purchasing

Recent literature has noted a growing tendency for farmers to 'shop around' as a strategy for reducing costs in periods of financial stress (Wilson, 1994). In general terms, the respondents indicated strong economic ties to their local community. However, the analysis differentiated farm business purchases and farm household purchasing to check for possible differences in shopping behaviour. The findings suggested a tendency for most farmers, regardless of farm business trajectory, to purchase farm inputs, or obtain farm services in Lucknow, while household products, especially more expensive consumer goods, were sought first in Goderich. The exception in household purchasing was grocery and banking where farmers were evenly split between the two communities. This pattern is not surprising given the economic profile of both places: 'Goderich is really built up now, more than ever, got malls and the bigger stores; you can get everything you need there. Not like before. But I still buy my feed in Lucknow ...they know me.' In a smaller number of cases, however, these changes in Goderich were less welcome: 'Goderich is a tourist town not a farming town. Goderich doesn't offer anything for farmers any more...they ruined it with all those malls.'

Further analysis of *changes* in the geography of purchasing revealed considerable loyalty to the local community. A majority of farmers stated that they had always shopped in their local community (n = 50, 71%), while the rest indicated that they shopped locally more now than they did 5 years earlier (n = 20, 29%). When asked to provide reasons for local shopping, specifically whether or not it was a decision made in principle to shop locally or if other factors contributed to their tendency to shop in the community, there were a variety of responses (Table 13.3).

Table 13.3 Farms by trajectory and priority to shop locally.

Priority	Total sample	Farm focused	Assisted growth	Persisting	Hobby	Forced down	Winding down
Without a doubt	29	6	5	3	2	3	10
(%)	(41.4)	(26.1)	(45.5)	(37.5)	(33.3)	(33.3)	(76.9)
Most of the time	18	6	3	4	2	3	0
(%)	(25.7)	(26.1)	(27.3)	(50.0)	(33.3)	(33.3)	
Not necessarily	23	11	3	1	2	3	3
(%)	(32.9)	(47.8)	(27.3)	(12.5)	(33.3)	(33.3)	(23.1)

Among those who indicated that shopping locally was a priority, there was a clear sense of commitment to supporting local enterprises as a means of ensuring their continued existence. For those who indicated a general desire to shop locally, but a willingness to look elsewhere, the predominant reasons related to costs and product selection – especially in the case of higher order goods such as farm machinery, furniture and appliances, etc. Yet even among these individuals, there was a stated practice of allowing local sellers to match prices before a purchase was made outside the area. In fact for some respondents, shopping around is more an exercise in determining the fair market value of goods and services than of real purchasing. Those few respondents who indicated they did not make local shopping a priority, indicated that their decisions to shop in the region were based strictly on the availability of a satisfactory range of goods at acceptable prices.

In examining shopping behaviour across the trajectories, there was little evidence of any systematic purchasing bias. However, some subtle distinctions were apparent. Farmers in the '*Winding Down*' group appeared to exhibit a greater economic loyalty to the community which was based on a stated desire to support local business, but also attributable to the effects of habitual behaviour given that members of this group were older with long established ties to the local community. Respondents in the '*Assisted Growth*', '*Persisting*', and '*Forced Down*' trajectories were more likely to shop around for better prices within the local community as well as elsewhere. Such tendencies are not surprising given the fact that all three groups rely on other sources of income to support the farm business, suggesting that cost saving in farm operation is a high priority.

In contrast, respondents in the '*Farm Focused*' trajectory displayed a stronger tendency to shop around within the region and a greater willingness to go further away in their purchasing – especially for farm machinery and supplies. Nearly half of this group expressed an intention to increase shopping outside the local area suggesting a somewhat tenuous economic allegiance to their local community. The potential importance of this tendency is noteworthy given the relatively high proportion of local farmers who were found to be within this development trajectory and the likelihood that such farms, because of their size and heavy reliance on purchased inputs, will constitute the most active consumers of high value farm business equipment and supplies.

Perceptions of community support for family farming

The previous two themes reflect farmers' more tangible interaction with their community. This component of the analysis explored how farmers perceive change in their community and how they believe it is affecting them. Such information aids in understanding specific community-oriented dependencies and sensitivities as they relate to various trajectories of farm business development. Two related dimensions of this issue were explored: farmers' perceptions of how well the community has supported their routine farm business needs, and whether changes in the community have hindered or altered development of the farm – from the perspective of the farm operator.

The majority of family farmers (69%) had few complaints about local support for farming when measured in terms of access to farm services and input distributors. In this sense, the majority of the sample was 'extremely positive' about how well the community met their farm business needs (Table 13.4). Beyond the mere presence of farm-related economic functions and services, many of the respondents spoke in complementary terms about the range and quality of services provided locally. Indications were that local farm service providers enjoyed a good reputation for being credible, capable and competitive.

Among those farmers who believed community support was more 'mixed' (23%), they acknowledged some positive attributes, but were troubled by shortcomings in certain areas. Foremost among the reasons for this sentiment was the changing level of service provided by local banks, which many stated as being downsized and understaffed and lacking in personalized service. Additional complaints related to the growing need to obtain banking information outside the community and the loss of a once-valued personal relationship with the bank: 'Banking is getting worse…they have a different accountant every three months … as soon as you get used to someone they go and change them.'

Table 13.4 Farms by trajectory and perception of community support for farming.

Perception	Total sample	Farm focused	Assisted growth	Persisting	Hobby	Forced down	Winding down
Extremely positive (%)	48 (68.6)	18 (78.3)	4 (36.4)	5 (62.5)	5 (83.3)	6 (66.7)	10 (76.9)
Mixed (%)	16 (22.9)	4 (17.4)	4 (36.4)	3 (37.5)	0	3 (33.3)	2 (15.4)
Largely negative (%)	6 (8.6)	1 (4.4)	3 (27.3)	0	1 (16.7)	0	1 (7.7)

While the above paints a largely favourable picture concerning local community support to farming, it was apparent that many farmers interpreted the issue first in terms of access to traditional services (equipment sellers, financial institutions, veterinary, etc.) in support of the farm business. In a significant number of cases, this interpretation was as far as respondents were inclined to go in their thinking about community support for family farming. This view was most strongly the case for farmers found to be in the '*Farm Focused*' development trajectory – the group most heavily and exclusively committed to farming as a livelihood. However, there were other views on community support that warrant consideration.

A number of respondents noted the importance of adequate employment opportunities in the community as a means of supporting the development and/or persistence of the farm. A number of farmers explained that, in their opinion, off-farm income was essential for the continued survival of not just their farm, but family farming more generally. Such beliefs were found to be more heavily concentrated among farmers in the categories of '*Assisted Growth*' and '*Persisting*'. In more than one case, there were strong suggestions that the absence of such forms of support would place farmers on a track towards continual expansion and formal linkages with agri-business: a decision that many farmers have taken quite willingly, but one that did not sit well with this particular group.

Predictably, the issue of expansion and industrialization arose within this broader consideration of community support for agriculture as a substantial number of respondents felt that the prospects for future community support were caught up in the evolving nature of family farming. Particular concern was expressed (most strongly by farmers within the category of '*Winding Down*') about the effect of continuing farm enlargement and consolidation on community support in all its forms. A number of farmers believed that dwindling numbers of farmers and the erosion of commitment to local shopping would spell the demise of various local businesses. This potential threat was perceived in both economic and social terms. In the words of one participant:

> 'There used to be 15 farms, each with 100 acres around here. Now you need 1000 acres to survive and there are only six or seven farms. Once you lose the small family farms you lose the community. The schools are going now, the churches will be next.'

Interpreting the implications of farm business diversity for community linkages

This chapter has argued that contemporary developments in agriculture are altering how and why farmers and farm families look to their rural community for necessary support (in many forms), and may be affecting how farmers perceive their role in local communities. Though geographically situated in one study region only, this preliminary analysis suggests that differences are building between farms pursuing rigorous expansion and intensification, and those farms that are more economically diversified – sometimes of choice and sometimes of necessity. While participation, purchasing and positive perceptions of support for farming were strong across the sample in North Huron County, farmers in the expansionist trajectories were found to be constrained in their ability to partake in social activities and organizations, somewhat more strategic in their sourcing of farm inputs with less commitment to local shopping, and more 'transactionally' oriented in their appreciation for the role of community in supporting the farm. Conversely, the findings indicated a somewhat stronger recognition among farms in the non-expanding trajectories of other sorts of linkages with communities – most notably employment. Hence, the availability of jobs in town emerges as a critical dependency for many farmers in this region, and surely others.

Analysis of contemporary trends in rural society, and evidence from research such as that reported here, suggests the existence of two broad forms of change affecting agriculture - community interactions in Ontario and elsewhere: integrative and divisive (Joseph *et al.*, 2001; Smithers and Joseph, 1999). The former suggests complementarity and a sense of a shared future, while the latter suggests the opposite and takes the form of mutual neglect or even conflict. An example of *integrative change* is represented by the emergence of new employment opportunities, most often in rural settlements, but also within agriculture itself. For example, the higher labour requirements of organic farming may result in job creation in this part of the farm sector. *Divisive change* is well reflected in local community concerns regarding certain development trends in agriculture such as the introduction of large-scale animal confinement facilities in the countryside and by the threats (real or perceived) implicit in calls by non-farm residents for heightened controls on farming practices. However, the degree to which many changes will create complementarity, individually and cumulatively, or conflict defies easy generalization; it may in fact be determined by local circumstances and differentiated on the basis of certain types of farm development. While additional case examples are needed in Ontario to explore more strategically some of the themes and specific farm business trajectories arising in this preliminary, there is value also in having regard for other regions where these processes are visible. In particular, there are opportunities to synergize insights in Ontario with other analogue regions where the processes and outcomes of rural change well advanced as a means of informing a more general understanding of contemporary farm–community linkages and effective strategies for managing change.

Acknowledgement

The financial assistance of the Social Sciences and Humanities Research Council of Canada and the Ontario Ministry of Agriculture and Food is gratefully acknowledged.

References

Berry, W. E. (1977) *The Unsettling of America: Culture and Agriculture*. Sierra Club, New York.

Bird, E. A., Bultena, G. L. and Gardner, J. C. (1995) *Planting the Future: Developing an Agriculture that Sustains Land and Community*. Iowa State University Press, Ames.

Bowler, I. R. (1992) The industrialization of agriculture. In: Bowler, I. R. (ed.) *The Geography of Agriculture in Developed Market Economies*. John Wiley and Sons, New York, pp.7-29.

Bowler, I. R., Clark, G., Crockett, A., Ilbery, B. and Shaw, A. (1996) The development of alternative farm enterprises: A study of family labour farms in the Northern Pennines of England. *Journal of Rural Studies*. 12, 285-95.

Bryant, C. (1992) Farming in the Urban Fringe. In Bowler, I. R. (ed.) *The Geography of Agriculture in Developed Market Economies*. John Wiley and Sons, New York, 275-304.

Buttel, F. H. (1997) Some observations on agro-food change and the future of agricultural sustainability movements. In: Goodman, D. and Watts, M. (eds) *Globalising Food: Agrarian Questions and Global Restructuring*. Routledge, London, 344-65.

Caldwell, W. J. (1998) Land-use planning, the environment, and siting intensive livestock facilities in the 21st Century. *Journal of Soil and Water Conservation*, 53, 102-6.

Coombes, B. and Campbell, H. (1996) 'Pluriactivity in (and beyond?) a regulationist crisis. *New Zealand Geographer*, 52, 11-17.

DeLind, L. B. (1995) The state, hog hotels, and the 'Right to Farm': A curious relationship. *Agriculture and Human Values*, 12, 34-44.

Durrenburger, E. P. and Thu, K. (1996) The expansion of large-scale hog farming in Iowa: The applicability of Goldschmidt's findings fifty years later. *Human Organization*, 55, 409-15.

Edmond, H. and Crabtree, J. (1994) 'Regional variation in Scottish pluriactivity: the socio-economic context for different types of non-farming activity. *Scottish Geographical Magazine*, 110, 76-84.

Evans, N. J. and Ilbery, B. W. (1993) The pluriactivity, part-time farming, and farm diversification debate. *Environment and Planning A*, 25, 945-59.

Everitt J. and Annis, R. (1992) The sustainability of prairie rural communities. In: Bowler, I. R., Bryant, C. and Nellis, M. D. (eds.) *Contemporary Rural Systems in Transition*. CAB International, Wallingford, pp.213-222.

Fuller, A. M. (1985) The development of farming and farm life in Ontario. In: Fuller, A. M. (ed.) *Farming and the Rural Community in Ontario*. Foundation for Rural Living, Toronto, pp.1-46.

Fuller, A. (1990) From part-time farming to pluriactivity: a decade of change in rural Europe. *Journal of Rural Studies*, 6, 361-73.

Furuseth, O. J. (1997) Restructuring of hog farming in North Carolina: Explosion and implosion. *Professional Geographer*, 49, 391-403.

Gertler, M. (1999) Sustainable communities and sustainable agriculture on the Prairies. In: Pierce, J. T. and Dale, A. (eds.) *Communities, Development and Sustainability Across Canada*. Simon Fraser University Press, Burnaby, pp.121-139.

Goldschmidt, W. (1978) *As You Sow: Three Studies in the Social Consequences of Agri-business*. Osmun and Company, New Jersey.

Green, G. D. (1985) Large-scale farming and quality of life in rural communities, further specification of the Goldschmidt Hypothesis. *Rural Sociology*, 50, 262-74.

Gregor, H. F. (1982) Large-scale farming as a cultural dilemma in US rural development and the role of capital. *Geoforum*, 13, 1-10.

Grey, M. (2000) The industrial food stream and its alternatives in the USA: An introduction. *Human Organization*, 59(2), 143-150.

Ilbery, B., Bowler, I. (1998) From agricultural productivism to post-productivism. In: Ilbery, B. (ed.) *The Geography of Rural Change*. Longman, Essex, pp.57-84.

Ilbery, B. W., Healey, M., Higginbottom, J. and Noon, D. (1996) Agricultural adjustment and business diversification by farm households. *Geography*, 81, 301-10.

Ilbery, B. W., Healey, M., and Higginbottom, J. (1997) On and off-farm business diversification by farm households in England. In: Ilbery, B., Chiotti, Q. and Rickard, T. (eds.) *Agricultural*

Restructuring and Sustainability: A Geographical Perspective. CAB International, Wallingford, pp.135-51.

Joseph, A. E. (1999) *Toward an understanding of the interrelated dynamics of change in agriculture and rural communities.* Population Studies Centre Discussion Papers No. 32, The University of Waikato, Hamilton.

Joseph, A. E., Lidgard, J. M. and Bedford, R. (2001) Dealing with ambiguity: on the interdependence of change in agriculture and rural communities. *New Zealand Geographer,* 57, 16-26.

Lobao, L. M., Schulman, M. D. and Swanson, L. E. (1993) Still going: recent debates on the Goldschmidt Hypothesis. *Rural Sociology,* 58, 277-88.

Marsden, T. (1998) Economic perspectives., In: Ilbery, B. (ed.) *The Geography of Rural Change.* Addison Wesley Longman, London, pp.13-30.

McCoy, M. A. and Filson, G. C. (1996) Working off the farm: impacts on quality of life. *Social Indicators Research,* 37, 149-163.

Moran, W., Blunden, G. and Greenwood, J. (1993) The role of family farming in agrarian change. *Progress in Human Geography,* 17, 22-42.

National Research Council (1989) *Alternative Agriculture.* National Academy Press, Washington.

Penfold, G., Matthews, G. J., Flaming, D. and Brown, D. (1989) Right-to-Farm in Canada. *Plan Canada,* 29, 47-55.

Pierce, J. T. (1990) *The Food Resource.* Longman, Harlow.

Pierce, J. T. (1994) Toward the reconstruction of agriculture: Paths of change and adjustment. *Professional Geographer,* 46, 178-90.

Ramsey, D. and Smit, B. (2001) Impacts of changes in the flue-cured tobacco sector on farmers in Ontario, Canada. *Applied Geography,* 21: 347-368.

Robinson, G. M. (1997) Greening and globalizing: Agriculture in the New Times. In: Ilbery, B.W., Chiotti, Q. and Rickard, T. (eds.) *Agricultural Restructuring and Sustainability: A Geographical Perspective.* CAB International, Wallingford, pp.41-53.

Shucksmith, D. M. and Smith, R. (1991) Farm household strategies and pluriactivity in upland Scotland. *Journal of Agricultural Economics,* 42, 340-353.

Sim, R.A. (1988) *Land and Community: Crisis in Canada's Countryside.* University of Guelph, Guelph.

Smithers, J. A. and Joseph, A. E. (1999) Agriculture and rural community change in Ontario. In: Ogilvie, J., Smithers, J. and Wall, E. (eds.) *Sustaining Agriculture in the 21st Century.* Proceedings of the Fourth Biennial Meeting North American Chapter, International Farming Systems Association, University of Guelph, Guelph, pp.265-74.

Surgeoner, G.A. and Dalrymple, J.R. (1993) Improved efficiencies in Ontario pork production. *Highlights of Agricultural Research in Ontario.* (September), 8-12.

Thu, K. M. and Durrenburger, E. P. (1998) Introduction. In: Thu, K. M. and Durrenberger, E. P. (eds.) *Pigs, Profits, and Rural Communities.* State University of New York Press, Albany, pp.1-20.

Troughton, M. J. (1992) The restructuring of Canadian agriculture. In: Bowler, I. R., Bryant, C. R. and Nellis, M. D. (eds.) *Contemporary Rural Systems in Transition.* CAB International, Wallingford, pp.29-42.

Troughton, M. J. (1997) Scale change, discontinuity and polarization in Canadian farm-based rural systems. In: Ilbery, B., Chiotti, Q. and Rickard, T. (eds.) *Agricultural Restructuring and Sustainability: A Geographical Perspective.* CAB International, Wallingford, pp.279-292.

Van Den Bor, W., Bryden, J. M. and Fuller, A. M. (1997) *Rethinking rural human resource management.* Wageningen Agricultural University, Netherlands.

Ward, N., Lowe, P., Seymour, S. and Clark, J. (1995) Rural restructuring and the regulation of farm pollution. *Environment and Planning A,* 27, 1193-1211.

Whatmore, S., Munton, R., Little, J. and Marsden, T. (1987) Toward a typology of farm business development in advanced capitalist agriculture. *Sociologia Ruralis,* 27, 21-37.

Wilson, O. J. (1994) They changed the rules: farm family response to agricultural deregulation in Southland, New Zealand. *New Zealand Geographer,* 50, 3-13.

Wilson, O. J. (1996) Rural restructuring and agriculture-rural economy linkages: A New Zealand study. *Journal of Rural Studies,* 11, 417-31.

Chapter 14

Engagement with the land: redemption of the rural residence fantasy?

Kirsten Valentine Cadieux

Can rural dwelling facilitate a sustainable relationship with both local and global environments?

Many assessments of the contribution of rural amenity migrants to rural sustainability have been rather bleak, often couched in terms of pessimistic opportunism. Although many who move to rural places to consume an amenitized rural lifestyle contribute significantly to the maintenance of these rural landscapes, the landscape that has resulted from their residential, rather than productive, use of rural areas poses some difficulties for environmental planners (Gertler, 1994; Troughton, 1995). For planners and residents interested in facilitating the negotiation of rural sustainability, it is important to identify ways in which rural in-migration supports, obstructs, or transforms the sustainability of the rural landscape. We could improve our attempts to navigate the contentious relationship between the increasing popularity of rural residence and the widespread condemnation of 'sprawl' by better understanding rural residential practices that contribute to the balance of ecological and human processes of the rural landscape.

The amenitization of post-productivist agriculture provides a parallel to rural residence. Post-productivist agriculture and rural residence share location, symbolic content, and even a vocabulary of debate, especially relating to countryside change. The popularity of rural living and boutique agriculture share a sense of connection to nature that is apparent as a concern for the environment and a desire to incorporate 'nature' into daily life. This concern is manifested in green living environments, a pleasant commute, and organic food.

However, these manifestations of a connection to nature also tend to share a certain superficiality. The problem with this aesthetic superficiality is elaborated by Philip Krang (2002) in his use of Marx's (1906) idea of 'commodity fetishism' (a belief in seemingly magic, unworked-for appearances of desired ends) to account for what has happened to expectations of both agriculture and rural residence. The fetishization process has involved the packaging of a series of ideas into representative symbols that are consumed as a proxy for thinking about, or participating in, those processes which the symbols have come to represent. As connection to nature has become increasingly valued, and as the demand rises for consumables which promise this connection, terms such as 'rural' and 'organic' have become highly valued symbols of desirable outcomes such as good health and environmental sustainability. However, the supply of organic food has been increased by the very industrial techniques against which 'organic' initially defined itself. Acting as a proxy for the complex set of practices and ideology embodied in the term, 'organic' has become available as a consumable, promising good health and environmental sustainability, while obscuring access to practice and ideology. The term 'organic' thus transforms agriculture, pushing the boundaries of industrial agriculture to include organic production, while being itself transformed, as a term, into something more accessible by purchase than practice.

Vigorous discourse about these transformations in both mainstream and alternative media has helped to maintain awareness of, and engagement in, the political processes of naming and marketing, especially as the meaning of organic standards has been negotiated. Authors who point out the social construction of the natural, such as Michael Pollan (2001, 2002a, 2002b) and William Cronon (1995), have played an important role in these negotiations in the USA, particularly in their *New York Times Magazine* stories. In his colourful cover stories, Pollan criticizes industrial organic production, challenging, although at the same time sympathizing with, the easy option of consuming 'certified environmentalism'; an environmentalism someone else will do for you, and that allows your resources to be produced without tarnishing your idea of nature. Cronon's (1995) critique of privileging wilderness over everyday landscapes highlights a similar obstacle to what might otherwise redeem the rural residence fantasy: by encouraging conceptions of nature and the countryside that do not include human work and residence, we prevent ourselves from coming to terms with the processes which support everyday life. Consequently, we also cultivate blindness to the environmental degradations upon which we depend but from which we alienate ourselves by offloading them to what Lawrence Summers infamously called the 'underpolluted' countries (Mokiber and Weissman, 1998).

As geographers, and often vocal participants in the discourse on sprawl, it is important that we, like Pollan, help to keep this discourse open against the tendency to polarize the debate into pro-market and anti-sprawl 'sides'. We should thus remain open to an understanding of the desires of the millions of homebuyers who have made the countryside residential. Just as industrial organic agriculture provides a paradoxical means to achieving a desirable future for farming, rural amenity residents provide a paradoxical way of maintaining the rural landscape by stretching the definition of sustainability to suit as many parties as might possibly support it.

Who wants what from the countryside?

Many of the planning policies shaping the countryside were designed to protect productive agricultural land from residential erosion through ribbon development or severances (see Boyle and Halfacree, 1998 for a current collection of essays on issues of migration into rural areas). As open space and ecological conservation have been added to the goals of farmland protection, it has become clear that the role of rural planning is to protect rural residents from their own bad impulses (Bunce, 1994; Howett, 1997).

Although urban boundaries, prevention of severance, and other measures to stave off the urbanization of the countryside have indeed protected some very scenic countrysides in the UK, Canada and the USA, the construction of a negative connotation for *residence* in rural areas has had some unfortunate effects. By creating planning mechanisms to protect rural residents from the negative aspects of their own urbanization (or the fragmentation of the countryside) that support a polarized discourse of urban (a place where most people should work and live) versus rural (a place where some people farm and other people visit), we seem to be undermining the potential for getting involved in the processes which give rural areas qualities of 'rural' or 'natural.' In cultivating cultural discomfort with these processes, we have seemingly condemned much new rural residence to a stereotypically urban alienation from the landscape.

Some of the problems with the construction of this alienation and with the loss of the processes required to maintain the rural landscape are well illustrated by Fraser and Lewis' (2002) BBC documentary *The European Union Had a Farm...EUEUO* Examining the bureaucratic paradoxes of the Common Agricultural Policy (CAP), Lewis demonstrates that rather than integrating agriculture into society, a cultural commitment to agricultural subsidy can instead create alienation from the food source. Juxtaposing several member countries' ideologies of countryside (and hence goals for the CAP) with agricultural practices in states hoping to join the European Union (and CAP), Lewis highlights a paradox shared with the urbanizing development of the rural landscape driven by urban to rural migration. While countries such as Spain are exploiting both the CAP and their environments (as well as their imported labour) by intensively cultivating vast monoculture crops such as olives and greenhouse crops, new EU states such as Poland will need to restructure their agricultural systems to reflect the EU's admittedly less successful (but more lucrative) systems. Despite Fraser and Lewis' argument that Polish agricultural practice currently reflects what countries like Germany and France see as the goal of the CAP (i.e. a sustainable, but rustic countryside), sweeping, standardized solutions are used to respond to complex problems brought to rural landscapes by modernity. In this way, processes we admire (and should be involved in) are replaced with a bureaucracy that will legislate these processes into being, ignoring geography and the importance of place specificity. As geographers, planners and environmental designers, this is often also our complaint about residential sprawl (Arendt, 1994; Relph, 1976). Applying Fraser and Lewis' critique of the bureaucratized response to agricultural sprawl to *residential* cases provides us with the opportunity to examine what both the CAP and much rural land use planning strive to protect. Although the obvious answer might be 'agriculture', or 'the countryside', I argue that it is not just an agricultural countryside, but an *aesthetic* of agriculture and the natural environment that is being maintained. In the film, Fraser sits with some British farmers on a hillside, overlooking a classic patchwork of fields and hedgerows, and listens to them grouse about the fact that it is the *look* of farming that the CAP preserves for Britain; they have been made park managers despite their desire to work the land.

How do we engage cultural narratives of the rural landscape?

The agricultural working of the land is not, however, considered particularly valuable. Consequently, many of the advances we are making in terms of environmental protection and ecological integrity, both in management and cultural practice, are illusory. They are subsidized by intensive structural intervention or by fulfilment of our resource needs through the exploitation of remote places. In *The Illusion of Preservation*, Berlik *et al.* (2002) call for the local production of natural resources, arguing that our current heavy dependence on less developed nations for natural resources exists largely for cultural reasons: despite using a disproportionate amount of global natural resources, countries like the UK, Canada, and the USA extract proportionately little of their own raw materials, in part because of 'domestic attitudes and policies intended to protect the environment' (2002, p.1). This same desire for a protected environment drives residential densities down and the amount of land in residential use up, leaving us with a large population of people living on land that we think we are treating well, and which is subsidized by a constant inflow of resources from unknown places where environmental policies and practices are, at best, unknown. Those who do find out from whence their resources come, and how, and who attempt to use local land productively, meet stiff resistance from those who wish to protect the local environment against exploitative extractive use: a pernicious cycle. As environmental researchers concerned with rural geography, we might be able to address both the problems of agriculture and residential land use by disrupting the fetishized expectation that the countryside can be both the locus of self-congratulatory ecological restraint and also the magic source of residents' material needs. In order to do this, we can look for the similarity between what people find desirable in rustic Polish agriculture and the appeal of the apparently natural environment of the residential countryside. We can also draw out the relationship between narratives and knowledge of sustainable forms of productive land use and the somewhat impotent environmental concern of the people who are rushing to rural residence.

The cultural imagination of the environment of the rural landscape, and of what people do there, provides us with opportunities to intervene in the consumer economy's escalating cycle of pushing the undesirable effects of productive land use farther from home, especially in the context of bridging between local knowledge and practice and incoming innovation and well-intentioned ideology. In-migrants from urban areas, like affluent member states of the CAP, enjoy the resources and knowledge to increase standards of living – of Polish farmers or long-time rural residents. Likewise, those who wish to preserve the rural landscape often have the resources and the institutional capacity to carry out projects related to their cultural sense of a sustainable relationship with the environment. These conservationists often lack practical or local knowledge about how to enact sustainable projects, or the cultural will to transcend fetishized ideas of nature in order to face openly the relationship between places of resource consumption and production. Consequently, despite the eagerness of many in-migrants to learn about the rural landscape and to share their ideology of environmental sustainability, difficulties in bridging local (often productive) traditions of land use and incoming knowledge of globalization and environmentalism seem not only to limit the integration of the best of both communities' ideas of land use, but also to steer the management of the rural landscape into a parody of both agriculture and sustainable development. Rather than deepen superficial ideas of sustainability or stretch traditional practices of productive land use to emphasize sustainability, 'preservation' of the rural landscape encourages illusory ideas of both the natural environment and processes of resource production.

Over the past six years, I have conducted interviews with rural residents and would-be rural residents in and around Toronto and Boston. Drawing on this evidence and a review of similar studies in places where rural residence is considered desirable (such studies have become increasingly common, but have been around, describing quite similar things for some time: e.g. Alanen, 1995; Blakely, 1984; Bunce and Walker, 1993; Carvalho, 1974; Coppack, 1988; Crump, 1993; Davis *et al.*, 1994; Donahue, 1999; Dorst, 1989; Environics, 2000; Firey, 1946; Gordon, 2003; Meinig, 1979; Patel, 1980; Pindell, 1995; Punter, 1974; Swaffield, 1998; Troughton, 1983; Walker, 1987; Warren, 1916), I have observed that the cultural narratives available to prospective rural residents seem to undermine the potential for these residents to work towards a sustainable inhabitance of the environment. This is true not only in the rural environment, the 'countryside', but also in those environments characterized by heightened contact with 'nature', environments that are increasingly important in the definition of rural, especially in North America. 'Residential tourism' and 'leisure environmentalism' are two important themes of rural residence that have emerged from this analysis. Using these two concepts, I hope to illustrate the utopian potential of the compulsion towards country living. In doing so, I suggest some ways that we might open the discourse on environmental planning to include sustainable relationships that allow for, and support, the carrying out of good intentions towards the environment.

Residential tourism

Rural residence: the residential countryside

Perhaps for the same reasons that CAP countries and other affluent nations subsidize their farmers, urbanites in affluent nations covet, and consequently subsidize, rural living. While urbanization increases on a global scale, demand for rural amenity living increases in Canada and the USA as the single largest growing residential land use type. According to American figures, however, growth of population in non-metropolitan counties occurs most in counties adjacent to counties with metropolitan status (Gordon, 2003; Platt and Macinko, 1983). Although this adjacency could evoke images of ever spreading suburbanization, long-standing evidence suggests that exurban, non-metropolitan growth is qualitatively different, and perhaps separate, from suburbanization.

In addition to the detached single family house in a park like setting, the exurban morphology of the residential countryside emphasizes high levels of privacy with intimacy with nature and a sense of the countryside (Crump, 2003; Gordon, 2003). Despite considerable variety in the morphology of exurban landscapes and residential styles, much non-metropolitan growth conforms closely to a characteristic exurban pattern that I describe as the 'residential countryside'. This 'exurbia' (Spectorsky, 1955) has also been called the 'third metropolitan ring', 'the second ring of suburbs', the 'urban-rural' or 'rural-urban' 'fringe' (Coppack *et al.*, 1988), 'urban field' or 'urban district' (Coppack *et al.*, 1988; Wells, 1904), 'areas of counterurbanization' (Mitchell, 2004), 'rurban areas' (Firey, 1946), 'ruburbia' and 'ruburbs' (Hayden, 2004; Marx, 1991). The fact that many of these terms betray the same urban-centred perspective, apparent in 'non'-metropolitan and 'ex'urban, is not surprising. Several studies, especially in the USA and Canada, suggest that rural residence is largely driven by centrifugal forces from cities, evocatively implied by the term *sprawl*. Sprawl literature suggests that residents from major urban centres are fleeing negative qualities they perceive in the city and are moving not so much to the country but, rather, to someplace or *any*place, else (Homer-Dixon, 2003; Lewis, 1983).

However, given the availability and diversity of suburbs, an explanation based solely on urban flight does not help us understand the growing popularity of rural residence (Crump, 2003; Gordon, 2003). Rather, the perceived amenities of rural residence, such as closeness to nature, the wholesomeness of the countryside, privacy and aesthetic of open space, actively draw people out from the city to seek their residential fortune in the wilds of rure. Participating in a sort of residential tourism, in-migrating rural residents seek a consumable landscape experience. Two defining aspects of the residential countryside appear to make it particularly consumable for in-migrants in the residential countryside outside of Toronto and Boston: an aestheticized narrative of participation in a productive landscape, and an emphasis on the importance and uniqueness of the local.

The aestheticized productive landscape

Spectorsky (1955) explained the pull of the rural landscape by suggesting that those who pioneered the first major exurban settlement of the residential countryside were media types and image makers for America. Describing the plight of an advertizing executive who decides he must have an extensive orchard while working on an apple-sauce campaign, Spectorsky pokes fun at rural in-migrants who buy every possible accessory for their productive ventures before realizing, like the apple-sauce advertizer, that they had been sucked in by their own pastoral images. Their interest in the productive landscape was only as deep as the symbolic fulfilment it advertised.

Still, the landscape aesthetic of rural residence relies heavily on a suggestion of an invisibly worked land. Raymond Williams (1973) argued that by making labour invisible in the landscape, landowners emphasize their own status and control over the amenity use and products of the landscape. The conception of the countryside as 'unworked nature' over the aesthetic of 'a worked landscape' has been used by landowners to alienate those who made a living from the land, prior to enclosure. In a contemporary parallel of this 'aesthetic of enclosure,' the invisibility of those who work the land has been achieved by conceptualising nature as an agent of complexity, ecosystems and environmental harmony: too inviolable, too perfect for interventions such as planning and design.

Along with historians such as Williams and Cronon, design practitioners such as Martha Schwartz criticize this embrace of nature as a misguided environmentalism that leads to an abdication of environmental responsibility (Schwartz, 2003). Having shaped the environment so far, can we naturalize a particular aesthetic of ecologically pleasing rural landscape and abandon it to its own devices in the midst of an environment highly structured by the human economy? Today's watered-down, middle-class version of the rural landscape of nature retreat is underlain by a tension that evokes an aspect of the conflict over enclosure, a conflict between a desire to interact with the working landscape and a practical and ideological alienation from it. This perspective has produced an amenitized version of the productive landscape, a residential countryside in which residents live but do not work in landscapes shaped by productive land uses such as farming, fishing and forestry. Prior to residential development, these productive land uses maintained the environment in its desirable form (open fields or pastures, accessible waterways and attractive forests) but they are available to urban in-migrants only through purchase of a rural residence, which becomes less rural through its residents' aversion to productive land uses. Thus, the residential countryside is undermined by the paradox of productivity predicted by Williams: the residential countryside retains its country aesthetic only as long as it is not too lived in.

Residential tourism

Residential tourism refers to the concept that rural residency allows a form of permanent, or semi-permanent tourism. This link is more explicit where recreational landscapes have paved the garden path for urban residents to follow. Toronto's 'cottage country' and the summer vacation towns around Boston are now staple landscapes of these two cities' exurban residential fringes. But even if the commuter paradise is not a full-time version of the leisure lifestyle, the growth of exurbs testifies to the increasing importance of an imagery of habitation in a recreational landscape. The leisure quality of this landscape lies both in the holiday quality of pleasurable landscape experience and also in the emphasis on this pleasurable quality in the enjoyment of the worked landscape by those who do not work it.

As people move in to enjoy the experience of the rural landscape, the story of those who have made it rural by working the land is increasingly written out of the text of landscape experience. Rural attractions often feature historical and natural elements, and the narrative of escape from the rush of urban, working life overwrites or romanticizes the work of the modern rural. When we use the adjective 'working' to describe farms, does this indicate that the popular imagination of farms suggests that otherwise they are recreational or 'leisure' farms? Or does it mean that they exist only in stories, or histories or, alternatively, that working farms are places for other people and not a usual place for the sort of person that might visit? Despite the potential beneficial exposure to the working landscape that visitor or holiday farms offer visitors, their construction of the rural as a temporary pastoral idyll exacerbates the alienation of the aestheticized rural landscape which bars rural residents – especially new rural residents – from participating in the processes that make the landscape rural. Instead, residential landowners mow lawns, plant trees, or garden – practices which suggest the difficulty of knowing how to live in a rural context that suggests productive land use (which doesn't go along with residential use) or nature (which doesn't want human interference), and difficult practices with which to negotiate the sustainability of the rural landscape.

Engagement as critical neo-agrarianism

Working farms can present opportunities for rural residents to learn about the processes that shape their environment. Brian Donahue (1999) presents an agenda of local agriculture as a means for refamiliarizing the inhabitants of the residential countryside with the processes of utility that once shaped their home landscapes and continue to meet their daily needs. By explicating one community's plan for providing opportunities for engagement with these processes, Donahue engages Cronon's (1995) challenge to disrupt an ethic of habitation based on consumption rather than participation. Donahue's community farming and forestry fleshes out the narrative of the humanly shaped landscape for its current residents, providing them with some of the knowledge, hands-on experience and tools necessary to reshape it to their needs within an ethic of sustainability.

Donahue's book illuminates another aspect of the bureaucratic dilemma discussed earlier in light of the Common Agricultural Policy. In answering the call to participatory, ecologically and culturally sound planning (Arendt, 1994; McHarg, 1969), Donahue brings to the foreground some of the assumptions we make about living in the landscape. By enabling residents explicitly to change the landscape, residents and planners can think about the differences between *looking at* and *participating in* an environment. He suggests that residents have the right and responsibility not just to preserve some remnant of the past (traditionally the realm of public activism in the planning process) but, even more so, to shape the present. This 'engagement' approach disrupts a fundamental distrust of residents' agency and intentions that is commonly manifested in planning policy, while pointing to appropriate contexts for rural residents to shape the environment. Despite many good intentions to preserve the rural character of areas under development, the degree to which standardization of the landscape seems inevitable to many commentators reveals the extent to which bureaucratized decision making (by both the state and the 'market') has come to replace personal and community engagement. Most subdivisional development certainly does not provide residents with a real chance to participate in the shaping of their landscape (Relph, 1981) or to understand how their actions can relate to those that have shaped the region into something they experience or enjoy.

Contrasting sharply with the alienation of the aestheticized landscape, a philosophy of enablement by engagement seems well suited to providing an outlet for the good intentions of rural in-migrants. It provides a connection between the abstracted environment (towards which everyone seems to have good intentions) and the mundane everyday behaviours which so affect this environment. Residents are given purchase on the behavioural details of their surroundings and experience the way the environment has come to be what it is. This engagement could help to bridge the disjunctures in planning policy and land use practice that result in the landscape being turned into something that almost no one wants.

This path of local engagement is not without its pitfalls. Despite its critical radicalism, much of the literature on agrarianism (even residential agrarianism) is heavily coloured by romantic ideas of the countryside and nature, with an alarming, often libertarian rhetoric of a return to a more 'authentic' experience. This denies the degree to which the residential countryside is deeply seated in a highly industrialized web of communications technology and the global economy. Hence, this echo of the call to the garden suburbs, or the bourgeois utopias, seems to lack critical self-awareness, especially in regard to the social and environmental problems that have tended to accompany unconsidered urban flight and the romanticization of the rural landscape (Blum, 2003; Donahue, 2002; Fishman, 1987; Williams, 1973).

Despite the danger of escapist narratives of the pastoral, the deeply rooted aesthetic of amenitized productivity has embedded in it the seeds of a more critical participation in a less romanticized rural landscape. Although I am loathe to endorse the sort of 'country' advertizing used to sell housing and cars to the upper middle classes (and, in fact, suspect that this very commodification provides disturbing insight into the problems of the relationship between capitalism and the rural brand of personally redemptive escape), the results of my interviews in Ontario and New England suggest that the easy availability of narratives of the countryside (for example, children's stories) may, in fact, help those rural in-migrants with intentions to engage in productive landscape processes successfully to envision themselves as part of these processes. However, few children's stories tell you how to take on the aestheticization of systems of power and production (or to farm). Consequently, the efficacy of our cultural narratives for participating in the rural landscape needs to be better understood.

Dwelling in the local

The residential countryside is not only characterized by an aesthetic of amenitized productivity. The lure of the *local* quality of rural residence contributes to the enticement offered to residents of our increasingly globalized metropolitan areas. As their growth integrates them into conurbations such as Ontario's Golden Horseshoe and the north-eastern Megalopolis, cities like Toronto and Boston join trends of urban amalgamation and increased global connectivity (Gottman, 1961; Punter, 1974; Rosenau, 2003). The value of the local, and of *place*, has risen in contrast to the perception of urban homogeneity and placelessness evinced by these mega cities (Lippard, 1997; Relph, 1976). James Rosenau posits that increases in localization are driven in part by the pressures of globalization (2003), that when we are reminded so frequently of our connections with a mind-boggling array of places, it should not be surprising that we try to wrap our locality around us protectively.

In the case of rural residence, the heightened valuation of the local presents a complementary paradox to the valuation of the amenitized productive landscape. Traditional qualities of the Arcadian pastoral gain a new irony in the context of the global relationships linking rural in-migrants almost instantaneously to the rest of the world via telecommunications, the Internet, and world wide overnight door-to-door courier services. As knowledge economies enable more people to live without the necessity of geographical proximity to particular places, trends in rural residence suggest that many people attempt to escape global placelessness by seeking the particularity of the rural. Not only does the repetition of any such trend cause cumulatively Odum's (1982) ecological 'tyranny of small decisions', but the particular expectations of *localness* brought to rural residence by so many people without local ties causes a peculiar dislocation in the meaning of dwelling.

Is it realistic to expect that a move to a place with a heightened sense of the local will confer upon residents the knowledge about how to live in such a place (which, if it is so local, must have particular local ways to live)? Rural sociologist Patrick Jobes (2000) concentrates on the unrealistic expectations that migrants attempt to realize and the destructive influence of these attempts on the places and communities through which they move. Jobes' major insight is that people tend to try to make places into what they imagine them to be, rather than to explore what places are, or to work (perhaps more realistically) with the material that's there. His book ends on a bitter note: most of his interviewees were disillusioned, and four out of five moved on, fairly rapidly, to other communities that they expected to match their original expectations of a rural community. Worse, Jobes charges these transients with leaving their temporary communities somewhat destabilized from their efforts to mould them into the residential countryside: a place suitable for residential tourism, but perhaps not for everyday living.

The community of Bozeman exposes the touristic quality of the residential countryside because, unlike so many of the other rural areas in North American that are experiencing growth due to in-migration, it has no larger metropolitan adjacency. In-migrants were often unable to reconcile an idea of *working* in their residential Edens, and the sorts of jobs that were available in their Arcadia did not appear to be part of the rural residence fantasy. Jobes' conclusions are sobering, in part because they not only highlight illusory qualities of the rural residence fantasy, but demonstrate the durability of these illusions: of the 80% of in-migrants to Bozeman to re-migrate (most within a year), only a small percentage came to recognize that their residential disillusionments had to do with their expectations, rather than the deficiencies of the place. Most moved on to inflict their rural residence fantasies on new communities.

This phenomenon has been reflected in the exurbs of Boston and Toronto, albeit not so dramatically, as these cities provide considerable employment opportunity. Most dissatisfaction seems to relate quite directly to an ideology of the rural landscape or rural community: fundamentally, life in the country is not as different or as ideal, as it should be. One of the most interesting differences I found in a survey of Toronto area exurbanites existed between residents of rural estate subdivisions and those who lived on lots that they have developed themselves, often severances from farms. Those on self-developed residential lots seemed much more able to take part in discourses about the environment, at several levels: in general, they seemed to do more and more varied things with their properties, were more articulate about their goals and projects with their properties, and were more likely to be involved in local governance, both regarding particular issues such as land development and also taking roles as local leaders in office (Cadieux, 2001). While residence in a subdivision did not preclude residents from these activities, subdivision residents spoke much more about the restrictive rather than helpful role of the cultural norms of their neighbourhoods. In contrast to the individual lot residents, who on several occasions mentioned their neighbours as key factors in their own education of how to live in a more sustainable way, subdivision residents complained about the pressure to use pesticides to control dandelions and to keep up with the Jones, and expressed frustration with an aesthetic that placed pressure on their projects for their residential environments. Residents who wished to plant gardens, especially, described problems of finding a suitable place in the lawn-covered subdivision where no one would be able to see their garden interrupting the lawn.

This lawn, in the end, came to represent for many residents this paradox of the undermining of the local. They complained about bringing urban ways of thinking about how to manage land into what was supposed to be their pastoral idyll, and the ubiquity of the lawn aesthetic provided at least some of them with an insight into the increasingly globalized manifestation of what was supposed to be a haven of the local.

'Think globally, act locally': the dilemma of scale in environmental engagement in an age of global landscape processes

The paradox of global applicability is one that infuses the commodification of the local: as locality itself becomes valuable, is its valuation based on an increasingly universal paradigm? The influence of globalized ideology on everyday activity gains a particular significance in the realm of the environment. As concepts, *global* and *environment* inform each other in terms of the encompassing way in which they connect almost every aspect of our experience. Even in the popular discourse on public policy, 'earth summits' suggest ultimate global coordination of human interaction with the 'environment'.

Daily activities obviously engage both local and global. We are here, in a place, but we act at a distance: at conceptual and abstract spheres (trading stocks, absorbing media), in spheres of delegated agency (global governance, economics, war), and, perhaps most obviously, in spheres of consumption, eating pears from Italy and Argentina, citrus from Morocco and California and wearing clothing from the Northern Mariana Islands. But as we think about environment and economy at increasingly complex global scales, is it harder to think about local ones, or at least, about the connections between our locality and global processes. Many of the processes are so complex that perhaps we can understand them only at the most symbolic level.

Leisure environmentalism: undermining work with play

Environmentalism: abstract global stewardship, or commodified relationship with nature?

One of the problems of country living, so obvious as to have become a sort of metonym for the whole ironic package, is the coincidence of the need for a Sports Utility Vehicle (SUV) to round out the environmental experience of rural residence. Several excellent analyses have pointed to the expression of class pretension through both landscape and vehicular choice (Smith, 1993). SUVs and rural residences share a certain degree of privacy, insulation from the hoi polloi, and bulky prestige. SUVs seem to share qualities of bulk, unwieldiness, and expense with rural residences, as well as their paradoxical status in relation to engagement: both SUVs and rural residence might bring one into the woods, but rather than facilitate interaction with the trees, they act as shields from the experience one has set out to achieve. What I find compelling about the metaphor, however, is the difficulty of avoiding the irony of the commonly bewailed environmental impact of an instrument so widely advertised for its capacity to bring the consumer into the heart of nature!

One of the problems underlying this insulative experience of environment has emerged from the popularization of environmentalism. Not only has the dumbing down of environmental campaigns to t-shirt slogans confused people about what recycling has to do with global warming (or local markets with organic agriculture), it has also generated an entire field of marketing, promising environmental sanctity for purchase. Environmentalism by purchase is promoted by many corporations who may be practising more environmentally friendly production, but who are mostly promoting a version of guilt-free consumption – you can consume more, because we are easing your conscience by recycling, or using 'natural' products. Although many of these businesses are run with good intentions and a rhetoric of personal engagement in both global and local environmental issues, these corporations have a vested interest not only in increasing consumption, but also, in their role as middlemen between consumers and environmental issues, in promoting a sense of engagement with these environmental issues that can be resolved through spending money.

This consumer version of environmentalism not only serves to fetishize environmentalism, but also tends to define environmentalism as spatially diffuse, a globalized version of responsibility towards the land. Stores such as Whole Foods Markets target areas of particularly ripe growth in rural residence, suggesting connection between environmentalism by purchase and nature by purchase. This sort of amenitizing of the environment and environmentalism is not all bad – it has certainly increased the visibility of environmental issues in the mainstream and offered increased access to many forms of enjoying the environment. However, the primacy of consumption and enjoyment of nature and the environment hearkens back to Cronon's concern over our cultural sense of habitation in the landscape: given the scale of the challenges facing the sustainability of our environmental uses, might slickly marketed eco-friendly consumption be another way in which we undermine a sense of our own landscapes as related to the natural resources we consume?

In a way, the transformation of rural residence into residential tourism also acts as a shield for residents against an engaged experience of the processes of the rural landscape. The primacy of play over work in the conceptualization of 'nature' as a place of leisure suggests that we must address consumer environmentalism as another form of consumption in our efforts to reinvent our relationships with the productive landscape.

Conclusion

Inversion of rural decline: a new definition for 'ruralization' in the North American context

Does the enthusiasm of urban-to-rural migrants for their new home-places suggest a potential for an inversion of Clout's (1972) 'ruralization', which he defined, rather negatively, as the decline of rural areas? The ascendance of a knowledge economy over the production economy suggests approaches to sustaining knowledge of a productive landscape. By providing access to knowledge of place and place-based processes for in-migrants to rural areas, we could help cultivate an interest in and maintain a higher level of engagement in qualities of rurality.

As affluent urbanites of the knowledge economy spread out into the rural landscape, we are given an opportunity to consider what aspects of the rural landscape attract residents and to connect visibly these aspects with the larger ecological and cultural systems that sustain the rural landscape. Ed Risse (quoted in Gordon, 2003) questions the assumption that sustainable planning requires everyone to live in the urban core, suggesting instead that small settlements in rural areas, referred to as 'Alpha Communities', can be structured to be environmentally, socially, and economically sustainable, in part by making sure that residents understand and pay the full costs of their urban lifestyle. Like the community farm in Donahue's exurban town, Alpha Communities highlight the relationship between sustainability and urban settlement, making the ways in which rural in-migration supports or obstructs the sustainability of the rural landscape more visible, and consequently, easier to do something about.

Exurban hybridity: residential localism and global consumption, or, how to harness aesthetic ends for pragmatic means

If we want to highlight aspects of rural residence that contribute to or detract from the balance of ecological and human processes of the rural landscape, we need to foster a public discourse about the relationship between rural residential localism and global consumption. This must include more than the narrow, and somewhat polemical, focus of the sprawl debate. Engagement in the processes of the rural landscape, or in any environmentalism, reveals that the tension we tend to represent between poles of global and local is more accurately constructed as a hybrid experience of scale. Although we resist this hybridity in places, it has enabled the smallness of rural residential settings through the largeness of global information and economic networks. A critical focus on everyday engagement with the environment (both local *and* global) should not only support the carrying out of good intentions towards the environment by rural in-migrants, but also disrupt the image of country living as an escape from the global metropolis and as a rural arcadia.

This more critical view of the rural landscape might encourage rural lifestyle seekers to assess the goals they associate with rural residence, and to attempt to seat them in the most appropriate locale (Warren, 1993). Many studies have demonstrated a range of goals associated with rural residence. Enabling closer examination of the rural residence fantasy might increase the potential for more selective settlement of the rural environment, based upon the considered desires of those who wish to participate in the rural idyll. The potential for positive change brought by in-migrants to rural landscapes lies in their investment in the place characteristics of exurban residential countrysides and forests. This enthusiasm can be harnessed to encourage involvement in what actually does go on in the rural landscape and in the processes that make it rural.

Narratives of the new rural residential landscape suggest a reconciliation of amenity and productive models of land use in the residential countryside and forest through increased engagement with issues of agriculture, social justice, and resource use in the largely post-productive rural landscape. Although increased land prices, the attractiveness of urban job markets and the restructuring of the agricultural economy have undermined the so-called viability of farming in the areas most experiencing in-migration, agriculture remains one of the most easily accessible ways of participating in the shaping narrative of the rural landscape. Rural in-migrants who participate in productive land uses, especially those who balance their productivity with an interest in environmental sustainability, help to challenge the illusions both of sustainability without effort (and elsewhere) and also of an effortless rural idyll. In this context, types of farming that have been dismissed as 'hobby' or 'marginal' seem particularly significant and warrant further academic attention (see Brooks, 2000; Jobes, 2000; Punter, 1974 as dismissive examples; see Bell, 1996 and DuPuis, 1996 for critical analyses relating to this phenomenon).

If popular trends towards low density living continue to tell us that people want to live in rural landscapes, we, as geographers, environmental planners and educators, should listen to both rural in-migrants and to our own discomfort with the transformations that sprawl has brought to the rural landscape. Our task is to reconcile the desire to consume the countryside with a sustainable future. The residential yard, the place for the enactment of the rural residence fantasy, is a decent place to start.

References

Alanen, A. R. (1995) Back to the land: immigrants and image-makers in the Lake Superior region, 1865 to 1930. In: Thompson, G. F. (ed.) *Landscape in America*. University of Texas Press, Austin, pp.111-141.

Arendt, R. (1994) *Rural by Design: Maintaining Small Town Character*. Planners Press, Chicago.

Bell, M. (1996) Stone age New England: a geology of morals. In: DuPuis, E. M. and Vandergeest, P. (eds.) *Creating the Countryside*. Temple University Press, Philadelphia, pp.29-64.

Berlik, M. M., Kittredge, D. B. and Foster, D. R. (2002) *The Illusion of Preservation: a Global Environmental Argument for the Local Production of Natural Resources*. Harvard

University, Massachusetts.
Blakely, E. J. (1984) The new people in the woods. In: Bradley, G. A. (ed.) *Land Use and Forest Resources in a Changing Environment: The Urban/Forest Interface.* Washington University Press, Seattle, pp.141-148.
Blum, A. (2003) Exurbia, airworld, and the necessary coexistence of the local and remote. American Society of Environmental History Annual Meeting, Providence.
Boyle, P. and Halfacree, K. (eds.) (1998) *Migration into Rural Areas.* John Wiley & Sons, New York.
Brooks, D. (2000) *Bobos in Paradise: The New Upper Class and How They Got There.* Simon & Schuster, New York.
Bunce, M. (1994) *The Countryside Ideal: Anglo-American Images of Landscape.* Routledge, London.
Bunce, M. and Walker, G. (1993) The transformation of rural life: the case of Toronto's countryside. In: Bowler, I. (ed.) *Contemporary Rural Systems in Transition, Vol. 2.* CAB International, Wallingford, UK, pp.44-61.
Cadieux, K. V. (2001) Imagining exurbia: narratives of land use in the residential countryside. MA thesis, University of Toronto, Toronto, Ontario.
Carvalho, M. E. (1974) The nature of demand for exurbia living. Winnipeg Region Study. Report prepared by the second year students of the Department of City Planning, University of Manitoba.
Clout, H. (1972) *Rural Geography: An Introductory Survey.* Pergamon Press, Oxford.
Coppack, P. (1988) Reflections on the role of amenity in the urban field. *Geografiska Annaler B* 70, 353-61.
Coppack, P. M., Russwurm, L. H. and Bryant, C. R. (eds.) (1988) *Essays on Canadian Urban Process and Form III: The Urban Field.* Department of Geography, University of Waterloo, Waterloo, Ontario.
Cronon, W. (1995) The trouble with wilderness; or getting back to the wrong nature. In: Cronon, W. (ed.) *Uncommon Ground: Toward Reinventing Nature.* W.W. Norton & Co., New York, pp.69-90.
Crump, J. R. (2003) Finding a place in the country: exurban and suburban development in Sonoma County, California. *Environment and Behavior* 35, 187-202.
Davis, J., Nelson, A. and Dueker, K. (1994) The new 'burbs: The exurbs and their implications for planning policy. *Journal of the American Planning Association* 60, 45-59.
Donahue, B. (1999) *Reclaiming the Commons: Community Farms and Forests in a New England Town.* Yale University Press, New Haven, Connecticut.
Donahue, B. (2002) The resettling of America. In: Wirzba, N. (ed.) *The Essential Agrarian Reader: the Future of Culture, Community, and the Land.* Lexington University Press, Kentucky, pp.34-51.
Dorst, J. D. (1989) *The Written Suburb: An American site, an Ethnographic Dilemma.* University of Pennsylvania Press, Philadelphia.
DuPuis, E. M. (1996) In the name of nature: ecology, marginality, and rural land use planning during the New Deal. In: DuPuis, E. M. and Vandergeest, P (eds.) *Creating the Countryside: The Politics of Rural and Environmental Discourse.* Temple University Press, Philadephia, pp.99-134.
Environics (2000) *Survey of Farmers, Ranchers and Rural Landowners: Attitudes and Behaviours Regarding Land Stewardship.* Environics Research Group.
Firey, W. (1946) Ecological considerations in planning for rurban fringes. *American Sociological Review* 11, 411-423.
Fishman, R. (1987) *Bourgeois Utopias:The Rise and Fall of Suburbia.* Basic Books, New York.
Fraser, N and Lewis, B. (2002) *The European Union Had A Farm...E.U.E.U.O.* First broadcast 1 April 2003, BBC, London; distributed by TV 2 WORLD, Denmark..
Gertler, M. (1994) Rural communities and the challenge of sustainability. In: Bryden, J. (ed.) *Toward Sustainable Rural Communities.* University School of Rural Planning and Development, University of Guelph, Guelph, Ontario, pp.67-78.
Gordon, A. (2003) The three-acre lot versus the town center: where do Americans want to live? *The Next American City* 1. http://www.americancity.org/article.php?id_article=81, (19/01/05)
Gottmann, J. (1961) *Megalopolis: The Urbanized Northeastern Seaboard of the USA.* Twentieth Century Fund, New York.

Hayden, D. and Wark, J. (2004) *A Field Guide to Sprawl*. W.W. Norton, New York.
Homer-Dixon, T. (2003) The ingenuity gap, keynote speech. International Federation of Landscape Architects Annual Meeting, Calgary.
Howett, C. M. (1997) Where the one-eyed man is king: The tyranny of visual and formalist values in evaluating landscapes. In: Groth, P. and Bressi, T. W. (eds.) *Understanding Ordinary Landscapes*. Yale University Press, New Haven, pp.85-98.
Jobes, P. C. (2000) *Moving Nearer to Heaven: the Illusions and Disillusions of Migrants to Scenic Rural Places*. Praeger, Westport.
Krang, P. (2002) Worlds of consumption. In: Daniels, P., Bradshaw, M., Shaw, D. and Sidaway, J. (eds.) *Human Geography: Issues for the 21st Century*, Pearson Education, Harlow, pp.
Lewis, P. (1983) The galactic metropolis. In: Platt, R. H. and Macinko, G. (eds.) *Beyond the Urban Fringe*. University of Minnesota Press, Minneapolis, pp.23-49.
Lippard, L. R. (1997) *The Lure of the Local: Senses of Place in a Multicentered Society*. New Press, New York.
Marx, K. (1906) *Capital*. Modern Library, New York.
Marx, L. (1991) The American ideology of space. In: Wrede, S. and Adams, W. H. (eds.) *Denatured Visions: Landscape and Culture in the Twentieth Century*. Museum of Modern Art, New York.
McHarg, I. (1969) *Design with Nature*. Doubleday/Natural History Press, New York.
Meinig, D. (1979) Symbolic landscapes. In: Meinig, D. (ed.) *Interpretation of Ordinary Landscapes*. Oxford Press, New York, pp.164-188.
Mitchell, C. J. A. (2004) Making sense of counterurbanization. *Journal of Rural Studies* 20, 15-34.
Mokhiber, R. and Weissman, R. (1998) Memo misfire: World Bank 'spoof' memo on toxic waste holds more irony than laughs. *San Francisco Bay Guardian*, 26 May. http://www.sfbg.com/focus/43.html (19/01/05).
Odum, W. E. (1982) Environmental degradation and the tyranny of small decisions. *BioScience* 32, 728-729.
Patel, D. I. (1980) *Exurbs: Urban Residential Development in the Countryside*. University Press of America, Washington, DC.
Pindell, T. (1995) *A Good Place to Live: America's Last Migration*. H. Holt and Co., New York.
Platt, R. H. and Macinko, G. (1983) *Beyond the Urban Fringe: Land Use Issues of Non-metropolitan America*. University of Minnesota Press, Minneapolis.
Pollan, M. (2001) Naturally. *New York Times Magazine,* 13 May.
Pollan, M. (2002a) This steer's life: the highly unnatural journey of No. 534 from calf to steak. *New York Times Magazine*, 31 March.
Pollan, M. (2002b) An animal's place. *New York Times Magazine*, 10 November.
Punter, J. V. (1974) Urbanites in the countryside: case studies of the impact of exurban development on the landscape in the Toronto-centred region 1954-1971. PhD thesis, Department of Geography, University of Toronto, Toronto, Ontario.
Relph, E. C. (1976) *Place and Placelessness*. Pion, London.
Relph, E. C. (1981) *Rational Landscapes and Humanistic Geography*. Croom Helm, London.
Rosenau, J. N. (2003) *Distant Proximities: Dynamics Beyond Globalization*. Princeton University Press, Princeton.
Schwartz, M. (2003) Keynote speech. International Federation of Landscape Architects Annual Meeting, Calgary.
Smith, J. (1993) The lie that blinds: destabilizing the text of landscape. In: Duncan, J. and Ley, D. (eds.) *Place/Culture/Representation*. Routledge, New York, pp.78-92.
Spectorsky, A. C. (1955) *The Exurbanites*. J.B. Lippincott Co., New York.
Swaffield, S. R. and Fairweather, J. R. (1998) In search of Arcadia: the persistence of the rural idyll in New Zealand rural subdivisions. *Journal of Environmental Planning and Management* 41, 111-127.
Troughton, M. J. (1983) Exurban development and land transformation: the case of Ontario. *Landscape Planning* 10, 335-354.
Troughton, M. J. (1995) Presidential address: rural Canada and Canadian rural geography – an appraisal. *The Canadian Geographer* 39, 290-305.
Walker, G. (1987) *An Invaded Countryside: Structures of Life on the Toronto Fringe*. Atkinson

College, York University, Downsview, Ontario.

Warren, G. F. (1916) Some suggestions for city persons who desire to farm. In: Carver, T. N. (ed.) *Selected Readings in Rural Economics*. Ginn and Co., Boston.

Warren, S. (1993) 'This heaven gives me migraines': the problems and promise of landscapes of leisure. In: Duncan, J. and Ley, D. (eds.) *Place/Culture/Representation*, Routledge, New York, pp.173-186.

Wells, H. G. (1904) *Anticipations and Other Papers*. T. Fisher Unwin Ltd., London.

Williams, R. (1973) *The Country and the City*. Oxford University Press, New York.

Chapter 15

Mammoth Cave National Park and rural economic development

Katie Algeo

Introduction

National parks play a dual role in the conservation and development of rural areas. Although intended to protect and preserve natural areas, national parks promote certain types of development through creation of tourist facilities, services and supplementary recreational experiences. Economic growth through development of tourism infrastructure remains a common goal of community planners. Tourism is perceived as a clean industry with the potential to expand the local business base, generate jobs, increase tax revenues and capture part of an external revenue stream that will then multiply within the local community. Economic impact analysis using input-output models to gauge the indirect effects of tourist spending in various economic sectors generally supports this development strategy (e.g. Cordell *et al.*, 1992; Donnelly *et al.*, 1998). The National Park Service's Money Generation Model, used to estimate direct and indirect effects of tourist spending in the vicinity of 74 parks nationwide (Stynes and Sun, 2003), is no exception. However, the benefits that tourism brings in terms of economic growth and prosperity can often be outweighed by some of the more negative implications of the activity, such as the loss of land base, the predominately low-waged, seasonal jobs, increased traffic congestion, higher costs of living, and escalating real estate costs and taxes.

This chapter investigates the effect of national park designation on economic development using the Mammouth Cave region of Kentucky as a case study area. Tourism has been a prominent component of development efforts in and around Mammoth Cave since the early nineteenth century, and tourist enterprises have played a major role in shaping the local economy and landscape. Early national park boosters made extravagant predictions of the economic prosperity that would follow federal investment in infrastructure, federal management and the cache of national park designation. This chapter critically assesses the influence of the designation of the Mammoth Cave region as a national park on the trajectory of economic development in surrounding rural counties. Using a combination of primary and secondary sources, including censuses, local newspapers and National Park Service archives, the historical development of tourism infrastructure in the Mammoth Cave region is traced and selected indicators of economic well-being are examined to assess the long-term success of tourism as an economic development strategy in the cave region. While these issues cannot be answered fully in this chapter, it appears that benefits of the tourism economy are distributed unevenly and that significant segments of the cave region's populace have not been 'directly and profitably affected', despite the 'great influx'. Tourism has failed to create a robust economy in the Kentucky cave region, although gateway tourism continues to be a preferred type of economic development.

Tourism in the Mammouth Cave region

The bulk of the Mammouth Cave National Park (214 km^2) lies within Edmonson County and occupies about one-quarter of that county's land area (see Fig. 15.1). Small portions of the park lie in Hart and Barren Counties, but these counties contain the primary gateway communities. Seven towns lie within 16 km of the national park and have the potential to serve as gateways for park-bound tourists, but three, Cave City, Park City and Horse Cave, have captured most of that function. The communities are all small, ranging in size from Oakland, with 200 residents, to Horse Cave, with 2200 residents. Most also act as service centres for the surrounding rural areas, so are not wholly reliant on tourism. Brownsville and Munfordville are county seats, so are the locus of government services, which is a role that has tended to overshadow tourism in these two towns. Hart and Edmonson Counties are predominantly rural and their respective gateway communities, Horse Cave and Brownsville, are their largest population centres. Barren County differs from the other two in having a more diverse economy and larger manufacturing sector centred on its county seat of Glasgow, a town of 13,000.

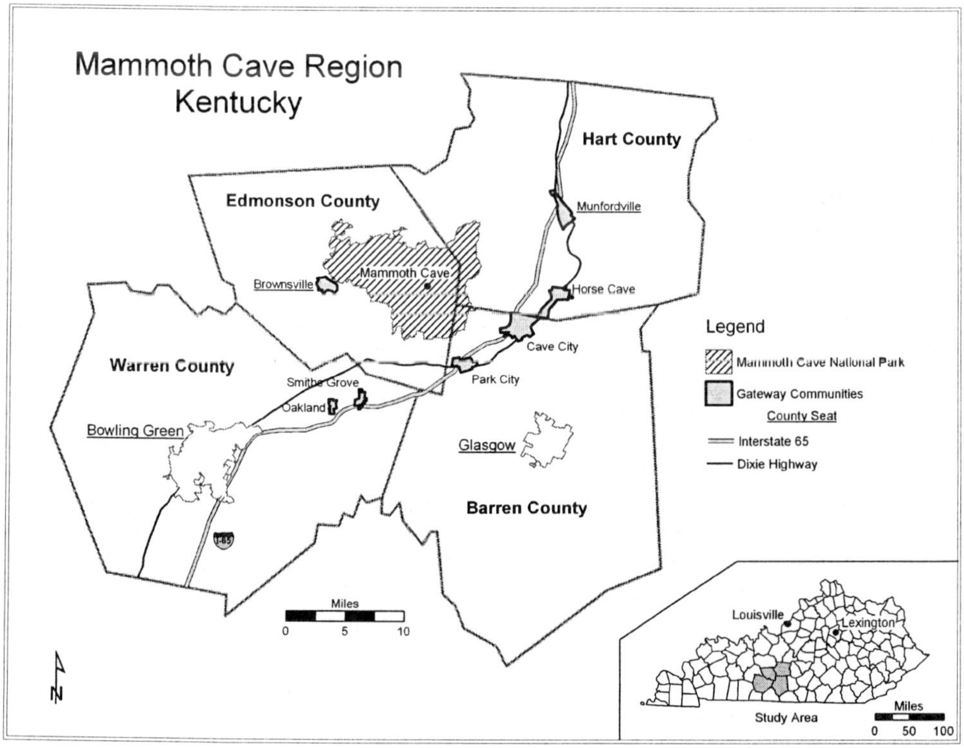

Fig. 15.1 Mammoth Cave National Park and the principal gateway communities.

When boosters in the 1920s proposed the Mammoth Cave region of Kentucky as the locale for one of the first national parks in the eastern USA, the argument they made to the local populace was grounded in the economics of tourism development. A mid-latitude location placed the cave within reach of eastern and mid-western population centres that were major source regions for tourists, while expanding transportation infrastructure made the cave accessible by highway, railroad and river. The result predicted was:

> 'Every type of business in Kentucky, every individual, every industry will be directly and profitably affected by the great influx. Every visitor must spend as he goes. The permanent resident, there to serve him in a business way, derives the benefit of the tourist's need. Many of these casual visitors... will remain as permanent residents, investing their savings in homes and businesses and contributing their labors to community growth. Thus will Kentucky's population be increased and her wealth augmented. It is predicted by students of the subject, that with the establishment and improvement of the Mammoth Cave National Park, there will be ultimately attracted to the State an annual tourist traffic of a hundred million dollars and that in time this great sum will be exceeded.' (MCNPA, 1927, p.20).

These extravagant predictions were based on the assumption that Mammoth Cave would become a more popular tourist destination under federal aegis. The national park designation carries status and would draw attention and visitors to the cave. Mammoth Cave would be made a more appealing and accessible destination by federal investment in infrastructure, such as accommodations and roads. Federal management would ensure consistency of tourist services, and federal ownership would halt the pilfering of stalactites and other cave formations for sale as souvenirs.

All of these assumptions have proven true, and by 2001, Mammoth Cave National Park was Kentucky's most popular tourist attraction, with an estimated 1.8 million visitors whose collective US$94 million in expenditures nearly reached the US$100 million mark glibly predicted by early boosters (Stynes and Sun, 2003, p.14). Direct tourist spending contributes an estimated US$25 million to personal income in the Mammoth Cave region and is credited with creating 1881 jobs. When economic multipliers are taken into account (the indirect effects of businesses that sell goods and services to tourist enterprises), the total impact amounts to US$33 million in personal income and 2499 jobs (Stynes and Sun, 2003, p.17). Moreover, of the 74 parks studied, Mammoth Cave ranked tenth in visitor spending (an average US$50.10 per visit), which is an indication of the potential revenue stream that local tourism enterprizes might tap.

Such visitation and expenditure figures are impressive. Indeed, the results of economic impact analysis are intended to impress. According to the authors of the national park study, 'the primary use of... economic impact analysis to date within the National Park Service has been for public relations' (Stynes and Sun, 2003, p.23). With rather more subtlety and science than the boosters of the 1920s, the Park Service emphasized the positive aspects of the intertwining of economic relations between national parks and gateway communities. However, shortcomings in economic impact analysis, including 'incomplete or outdated secondary data, small unrepresentative survey samples, and problems in computing the multiplier effects of a multifaceted industry like tourism' make it difficult to judge the reliability of such models (Achana and O'Leary, 2000, p.72). Also, the extent to which a region benefits from indirect tourist spending is partially dependent on the level of diversification of the economy (Donnelly *et al.*, 1998). Regions that possess a range of businesses to supply more of the inputs to tourism-related businesses experience higher multipliers from tourist spending than regions that must import those goods.

Even within a region, economic benefits of tourist spending are not homogeneously distributed. Gateway communities, towns that serve as entrance points for a tourist destination and generally support tourism through provision of goods and services such as accommodations, restaurants and supplementary recreational experiences, are nodes of development within their rural hinterlands. Moreover, site, situation and human agency result in different development trajectories for communities that serve as gateways to the same attraction. Thus, an important task remains in the synchronic and comparative study of the national park hinterland to contextualize rather abstract, disembodied statistics. In short, a key question addressed in this chapter is what *difference* does an estimated US$94 million in annual tourist expenditures make in the communities that fringe the national park, and what have been the cumulative effects of cave tourism on these places?

Cave tourism and the origins of Mammoth Cave National Park

Patterns of tourism to Mammoth Cave established during the nineteenth century have had a lasting influence on tourism development subsequent to the creation of the national park in 1941. Unlike the seminal national parks in the western USA, which were in little populated regions that lacked significant tourist infrastructure before achieving national park status, and unlike the settled, but untouristed, Great Smokies, which was granted national park status at the same time as Mammoth Cave, the Kentucky cave region was well populated and had a long history of tourism at the time of its designation as a national park (Algeo, 2004). Tourism began in the second decade of the nineteenth century, with the first known description of the cave published in a newspaper in 1810. Six years later, a lengthy account describing the cave's passages and a mummified Native American taken out of the cave drew significant attention through a series of reprintings during the next decade in the USA and England (Sears, 1989, p.32). At that time, the cave was an industrial site used for the extraction of saltpeter, one of the main ingredients in gunpowder. Mining, which had been highly profitably during the war of 1812, ceased shortly after the end of that conflict.

The cave then changed hands several times before being acquired in 1839 by gentleman-physician John Croghan. After a failed experiment in using the cave as a sanatorium for tuberculosis patients, Croghan concentrated on developing it as a tourist attraction, investing heavily in infrastructure. He expanded the dogtrot-style log cabin that had previously served as a hotel into a substantial building that included among amenities to attract elite tourists, a ballroom, billiard room and veranda for strolling. He installed running water, which was gravity fed through hand hewn oak pipes from a nearby spring, and built roads to the Louisville and Nashville turnpike at points north of Cave City and south of Park City (Brucker and Watson, 1976, 169-70). Then, as now, the Mammoth Cave operation was a source of employment for local residents, who worked as cave guides, hotel staff, and on contract jobs, including renovating the hotel, building roads, making charcoal and cutting timber. Other families grew food and prepared jams, hams, butter and other sundry food items to be sold to the hotel. Arguably, in this age of greater regional self-sufficiency, economic multipliers of tourist spending would have been higher than they are today when most concessions are owned and financed by outside capital and therefore contribute to leakage of tourist dollars as profits and spending are directed outside the region.

For the first half of the nineteenth century, tourism at Mammoth Cave was the provenance of a social and cultural elite who had the time and money to indulge a taste for the scenic tourism made fashionable by the Romantic movement or who had sufficient education to reflect the growing national interest in science and natural history. Successive discoveries of new passageways and chambers kept Mammoth Cave in the media limelight. The location became de rigueur viewing both for visiting Europeans and the upper-class Americans who emulated trans-Atlantic patterns of tourism as a hallmark of cultivation. The majestic dimensions of the cave and the frequent increments to its known extent placed Mammoth Cave in the company of other American sites, such as Niagara Falls, that became early tourist destinations because they represented something uniquely American.

Before his death, Croghan laid the groundwork for Mammoth Cave's metamorphosis to an American icon by encouraging visiting scientists, artists and literary figures to write about their experiences in the cave, enhancing the cave's currency within American culture. Ralph Waldo Emerson, for example, credits Mammoth Cave's Star Chamber with inspiring him to write the poem 'Illusions'. Much publicity came from a travelling exhibit of a type known as a moving panorama. The panorama, a popular entertainment in the middle nineteenth century, featured a long roll of canvas, as much as 800 m in length, on which were painted a series of scenes arranged in sequence to suggest movement through a landscape. The display was accompanied by music and narration, and represented a form of moving picture that foreshadowed popular fascination with the cinema. Artist George Brewer's panorama *Natural Wonders of America,* which used Mammoth Cave as the headline attraction, acquainted thousands with the cave during its American and European tours from 1848 to 1851 and helped establish it as the model of a sublime landscape (Arrington, 1965a, b). By 1850, Mammoth Cave was receiving as many as 300 tourists a day during the high season of August and September and had annual visitor numbers between 30,000 and 40,000 (Arrington, 1965b, p.166).

Transportation and gateway communities

Transportation routes have had a major influence on which communities would serve as gateways to Mammoth Cave since the inception of cave tourism. Successive improvements in transportation technology made Mammoth Cave easier to access and reinforced movement along the Louisville to Nashville transect, encouraging growth in communities along the old turnpike. In the early nineteenth century, Cave City and Park City, which then went by the name of Three Forks, flourished as transfer points between regional and local transportation lines. The Louisville and Bowling Green stagecoach, which by 1834 ran six days a week, stopped in both towns, and travellers could get a local stage to Mammoth Cave (Coleman, 1936, pp.76-77). Bell's Tavern, the stage stop at Park City/Three Forks, became quite famous during the ante-bellum period, and its guest book included the names of prominent politicians and literary figures (King, 1874). Early travel accounts cite abundant board and a locally distilled peach brandy among the tavern's pleasures. No other stop-over in Cave City had the name recognition or reputation of Bell's Tavern. Thus, in the initial period of cave tourism, Park City was the leading gateway community.

The Louisville and Nashville Railroad (L&N), completed in 1859, followed much the same route as the turnpike, with stations in each of the cave region towns along the turnpike. In the decades following the Civil War, the L&N expanded its rail network across the trans-Appalachian South, providing passenger service from Atlanta, New Orleans, St. Louis, and Cincinnati. For the railroad, Mammoth Cave was an attraction that drew visitors, and hence passengers, year-round. The L&N and Mammoth Cave management offered an early form of package tour that made the cave accessible and affordable to the growing middle class and positioned it to become a mass tourist destination. By 1886, the future

Park City had gained two spur lines that added to its importance as a rail hub. One led to the Barren County seat of Glasgow, and the other, built by the privately financed Mammoth Cave Railroad Company, led directly to Mammoth Cave. Around the rail station flourished a commercial district that included the Mentz Hotel, whose namesake served for a time as the president of the Mammoth Cave Railroad. That rail line operated until 1931, by which time maintenance problems, compounded by the popularity of the car, made it obsolete.

The car freed tourists from prescribed transportation routes in ways that realigned commercial development within the cave region. The Dixie Highway, built between 1915 and 1927 by patching together existing roads, was one of the early national roads that, by virtue of being signposted and mapped, encouraged wide-ranging automobile tourism. On its route between northern Michigan and southern Florida, the Dixie Highway followed the same Louisville to Bowling Green axis used by the earlier turnpike and the railroad (see Fig. 15.1), placing Mammoth Cave within easy reach of a growing stream of tourists. Motels, gas stations and restaurants were built along the roadside to serve travellers. Automobile tourists, more mobile within the cave region than train travellers, enhanced opportunities for show caves to compete with Mammoth Cave. Tours had been offered at other caves in the latter half of the nineteenth century, but their success had usually been limited by problems of attracting tourists to out-of-the-way places down bad roads. Intense competition during the 1920s to find and open new caves culminated in the death of Floyd Collins, whose demise while trapped in Sand Cave became a national media event (Murray and Brucker, 1979). Vandalism of cave formations by competitors during this period was one of the factors that lent support to the movement to nationalize Mammoth Cave. Mammoth Cave National Park was authorized in 1926, although land acquisition proceeded slowly, and the park was not formally dedicated until 1941.

If automobile tourism contributed directly to increased visitor numbers at Mammoth Cave and indirectly to the establishment of the national park, subsequent developments in automobile travel complicated gateway communities' attempts to serve travellers. Interstate 65, the Kentucky leg of which was completed in 1969, greatly increased automobile traffic through the cave region, but more significantly, altered the geometry of gateway communities along the highway. The routing of the interstate 1 to 3 km west of the Dixie Highway shifted the tourist stream away from the historic downtowns of these communities. With the interstate between the towns and the national park, cave-bound tourists no longer routinely passed through the towns as they did when traversing the Dixie Highway. Each of the towns along I-65 has bifurcated, with a new node of economic development growing up around the exit ramp, draining economic activity from the traditional downtowns.

Tourism infrastructure

No greater contrast in development trajectories is evident than that between Park City and Cave City, which is a surprising contrast given the similar situations of these towns. Transportation routes made both early gateways to Mammoth Cave, and today their interstate exits are the two officially signposted gateways to the national park. The gulf in tourism infrastructure and economic development that separates them, however, points to the need to understand intraregional variations in the impact of the tourism economy.

In the late nineteenth and early twentieth centuries, Park City seemed poised for success as a gateway community. It had the direct rail line to Mammoth Cave, a spur line to Glasgow, and was the main transfer point for cave-bound L&N passengers. Although the renowned Bell's Tavern had burned, its ruins made an interesting excursion for tourists staying at the adjacent Mentz Hotel. The reasons that Park City failed to capitalize on its early competitive advantages are not clear, but from the 1930s onward, as tourists shifted from trains to private automobiles and as the Mammoth Cave National Park Association acquired the land for the new national park, Park City stagnated. The town's name change from Glasgow Junction to Park City a few years prior signalled residents' hopes for linking their economic future to that of the park. In 2004, despite offering the shortest route into the park, the Park City I-65 interchange is nearly devoid of tourism enterprises and services, with only a single gas station at its exit ramp. National Park Service control of a narrow corridor of land designed to maintain a pristine approach to the park prohibits commercial development on the western side of the interchange. The proximity of downtown Park City to the interchange (at just under a kilometre, it is the closest of the communities to the interstate) offers potential for tourism development, but in 2004, Park City was a dying town. A downtown tour revealed little commercial activity, and residents travelled elsewhere for most goods and services. The former commercial district along the railroad was abandoned, with the surviving buildings empty and in disrepair. While the well-maintained homes along town's central streets evoked a pleasant image of small town USA, the backstreets revealed a different Kentucky, in which decaying trailers and tumble-down houses displayed the human toll of missed opportunity. Park City leaders hoped to revive tourism in their town by preserving the ruins of Bell's Tavern, making it the centrepiece of a historic district, creating a bike path connecting the town to the national park, and building a tourist welcome centre at the interstate, but all of these initiatives are still in the planning phases, with funding uncertain.

Fig. 15.2 Franchise development at the Cave City exit of Interstate 65.

Cave City, in contrast, has been the locus of tourism-related development during the late twentieth century, and its interchange is now the epitome of the tourist strip. Six national hotel chains, eight fast-food restaurant franchises and four major gas stations cluster at the interchange (see Fig. 15.2). Supplementary recreational enterprizes for tourists, including theme parks, souvenir shops, trail riding, miniature golf, a wax museum, game rooms and other show caves, are concentrated along Highway 70, stretching from the interchange to the park border. Much of this development is incidental to cave tourism in that it is of a generic type of entertainment that could be found anywhere, playing on themes of adventure, danger, mystery and the exotic, rather than caves or Kentucky. These businesses were developed by a mix of local and outside capital. One of the larger recreational complexes, Kentucky Action Park, has been pieced together over several decades by a former tenant farmer who consolidated a series of businesses as their original developers pulled out. The newest attraction along cluttered Highway 70 is Dinosaur World, recently opened by a Swedish entrepreneur who operates a similar theme park in Plant City, Florida (see Fig. 15.3). Residences are interspersed with businesses along Highway 70, and the large number of tourists who drive this route afford opportunities for individual entrepreneurship, such as selling souvenir geodes or wood for campfires. While much of the tourism development is confined to the periphery of Cave City, the old tourist route along the Dixie Highway remains a vibrant commercial district. Motor court hotels, including Wigwam Village built in the 1930s, continue to serve tourists. Manufacturers of concrete yard ornaments, a regional specialty using locally available limestone, cater to tourists and retailers buying in bulk. With a variety of businesses serving local residents, Cave City has been the most successful of the gateway communities in maintaining a healthy and diverse local economy.

Fig. 15.3 Will Dinosaur World and the Cave City entertainment complex devour the landscape's aesthetic value?

The other communities close to the national park lack improved highways into the park and the resulting lower tourist flows have limited their gateway functions. Of the remaining four interstate communities, only Horse Cave, with the American Cave Museum, Kentucky Down Under, and a professional repertory theatre, has attractions that specifically target cave tourists. Brownsville is the community that has been most disappointed in terms of realizing benefits from Mammoth Cave tourism. The county seat of Edmonson County and located directly adjacent to the park, it would seem to have potential for gateway development, except that Mammoth Cave and most of the infrastructure that draws tourists is located in the eastern half of the park farthest from Brownsville. In the planning stages for I-65, community leaders in Brownsville argued for a route that would take the interstate north of the national park, placing the town at an interchange. As it is, Brownsville serves only as an entry point to the back country, via a ferry across the Green River. The national park spatially entraps Brownsville, excluding development in much of its immediate hinterland. To extract compensation, the county passed a tax on all recreational ticket sales in 1997, including Mammoth Cave tours that would provide the bulk of the anticipated revenue. The ticket concessionaire contested the tax at the time, but the judge who ruled in favour of the county declared by way of justification, 'the Mammoth Cave National Park is a national asset but (has) provided little in the way of economic benefit for Edmonson County' (Dortch, quoted in Overstreet, 1999). The park presents other barriers to development in Brownsville, notably by prohibiting truck traffic on park roads, a point of contention for businesses in Brownsville that would like to take commercial deliveries via Highway 70.

Smiths Grove, a tightly nucleated village of 780 residents, experienced its heyday during the late nineteenth and early twentieth centuries when passenger traffic on the L&N railroad sparked commercial growth. It was the transfer point to Chalybeate Springs resort, just south of the present national park, where health seekers would take the mineral waters, but it also had a thriving commercial district with two banks, three private schools, a hardware store, a grocery, a hotel, a hospital and a newspaper. Most of these businesses faded with the cessation of passenger rail traffic and the increasing pull of Bowling Green, a burgeoning city to the south. The past decade has seen low-key, but successful tourism development centred on high-end antique stores that have filled empty commercial buildings on Main Street. The town's I-65 interchange has the usual assortment of businesses to serve through travellers, but the town neither cultivates an identity linked to the cave region nor serves as a gateway to the national park. While town businesses benefit from interstate traffic, some of which is related to cave tourism, Mammoth Cave tourism has not otherwise impinged notably on Smiths Grove. The tiny community of Oakland, a combination farm hamlet and dormitory community for Bowling Green, is even further removed from the impacts of cave tourism.

Rather than promoting commercial development in Oakland and Smiths Grove, the national park is being used by residents of those communities to fend off unwanted industrial development. The Kentucky Trimodal Transpark is a proposed 16 km^2 industrial park located 13 km south of Mammoth Cave that has been disputed at lengthy local hearings and in several court cases. Although community residents have a variety of reasons for opposing the industrial park, including preserving prime farmland, maintaining the agricultural landscape, protecting two historic communities founded by free Blacks, and a simple NIMBY response, the issue that has garnered the most widespread opposition to the development has been concern over maintaining environmental quality in the national park. The park has been designated both a World Heritage Site and an International Biosphere Reserve by the United Nations Educational, Scientific, and Cultural Organization in recognition of the region's unique cultural and natural landscapes. In a karst region where most water movement is underground and does not follow surface drainage patterns, however, protection of the park area alone may be insufficient to maintain endangered species' habitat. Concern about impacts of groundwater pollution from manufacturing and transportation facilities planned for the industrial park has been a focal point for opposition to its development. Oakland and Smiths Grove, then, are communities where the recreational tourism engendered by the national park may well be less important to promoting commercial development than the conservation function of the park is in suppressing undesired development.

Regional economic structure

Impacts of cave tourism may also be assessed by examining the regional economic structure and indicators of economic well-being. In this respect, employment structure, level of manufacturing activity, inter-county commuting, income and poverty were examined. In most of the following measures, Hart and Edmonson counties, which are predominantly rural, show similarities, while Barren County, with the larger population centre of Glasgow, has a more diverse economy. Each of the three counties is more agricultural than the state as a whole. While 2% of the state's workforce is employed in agriculture, in Edmonson, Barren, and Hart Counties, the figures are 3%, 4% and 8%, respectively, and tobacco and beef cattle are the chief production systems (KCED, 1999, p.68). In Edmonson County, agriculture remains a more important source of livelihood than manufacturing.

The economic census gathers statistics on businesses and employment, but the importance of the tourist economy in these figures is difficult to gauge because the North American Industry Classification System (NAICS) used by the US Census Bureau defines categories, such as retail, that include both tourism- and non-tourism-related businesses. However, businesses in two categories, NAICS 71 (arts, entertainment, and recreation) and NAICS 72 (accommodation and food services), have a strong tourism component, so are useful in a statewide comparison of the relative importance of tourism in local economies. Figure 15.4 maps the number of employees in businesses categorized as NAICS 71 or 72 as a percentage of each county's total workforce. This spatial distribution suggests that, relative to other parts of the state, tourism employment is not a particularly large sector for the cave region. In Edmonson and Barren Counties, employment in accommodation, food services, arts, entertainment and recreation, numbers less than 5% of the workforce. With the exception of three counties in mountainous eastern Kentucky, the best predictor of employment in these sectors statewide appears to be location at the juncture of two interstates rather than individual tourist attractions.

Relating job numbers within a county to the size of the workforce does not reveal the entire employment picture because workers may live in one county but commute to work in another. Figure 15.5 shows the percentage of each county's workforce that works outside the county. The hinterlands of all Kentucky's larger cities display high cross-county commuting rates, but within south central Kentucky, Edmonson County stands out. Inadequacies in the county's employment situation are suggested by the fact that 52% of the workforce has sought employment elsewhere (US Census Bureau, 2000a). The bulk of these commuters work in the neighboring city of Bowling Green, a situation that led to thoroughly rural Edmonson County being designated part of the Bowling Green Metropolitan Statistical Area (MSA) in 2003. That city's leaders sought the joint designation after the 2000 Bowling Green census fell just shy of the 50,000 population needed for MSA status on its own, a designation that would make it eligible for higher levels of community development funding.

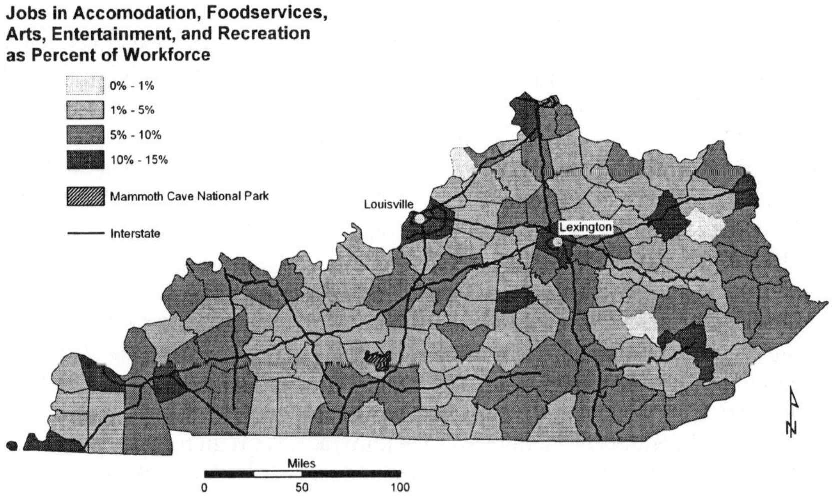

Fig. 15.4 Tourism employment by county, Kentucky, 1997. (Source: US Census Bureau, 1997).

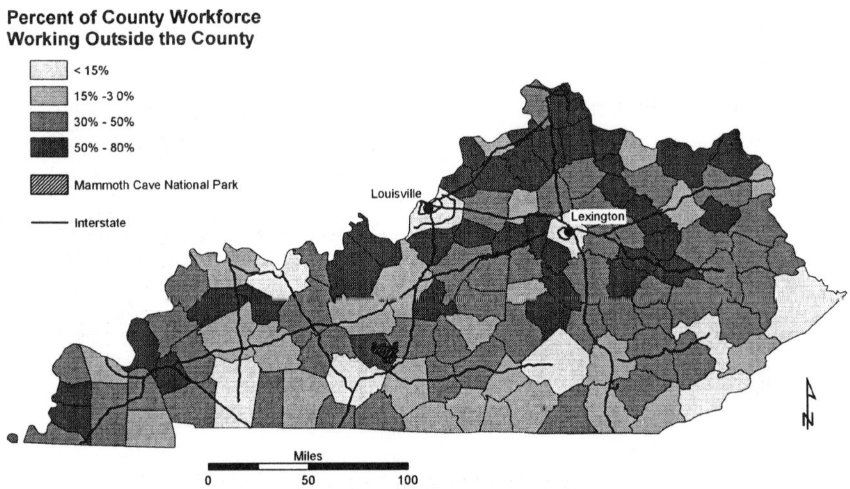

Fig. 15.5 Intercounty commuting rate by county, Kentucky, 2000. (Source: US Census Bureau 2000a)

Edmonson County has virtually no employment in the manufacturing sector, and state and local government make up the largest share of the county's non agricultural employment, accounting for 35% of jobs (US Census Bureau, 1997). Barren and Hart counties have stronger manufacturing sectors, with businesses generally reliant on locally extracted resources, particularly stone and timber. The region is in a limestone belt circling the western coal field, and the same underlying geology that produced the region's extensive cave network is the source of stone for ready-mix concrete and concrete products, including garden ornaments, septic tanks and burial vaults. Forested hills unsuitable for agriculture provide the raw material for rough and milled lumber, cabinetry and reproduction furniture. Two small apparel manufacturers and several specialty machining and printing shops supplement the region's manufacturing enterprizes. The manufacturing situation suggests that tourist dollars are, in general, not being recirculated as capital for local development, and commuting figures suggest that cave tourism is not generating adequate jobs in the immediate surrounding region.

Poverty and income indicators are also not encouraging. The proportion of residents in Edmonson and Hart County living in poverty is well above the state rate of 18% and the national figure of 14% (see Fig. 15.6). Although poverty is not as dire as in the Appalachian mountain region of eastern Kentucky, 20% of Edmonson County's population and 25% of Hart County's population are faring poorly under the current economic regime (KCED, 1999, p.60). A map of income levels within the cave region (see Fig. 15.7) shows a marked urban/rural dichotomy. Per capita income is highest in the block groups around the larger urban centres. Most of the region surrounding the national park has a per capita income below the state figure of $18,093, and the two rural block groups with the lowest per capita income are within 10 km of the park.

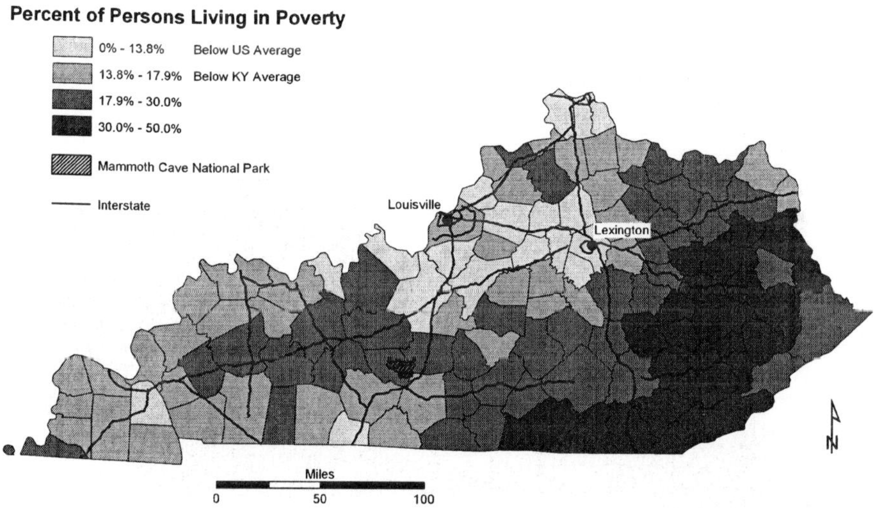

Fig. 15.6. Poverty rate by county, Kentucky, 1997. (Source: KCED, 1999)

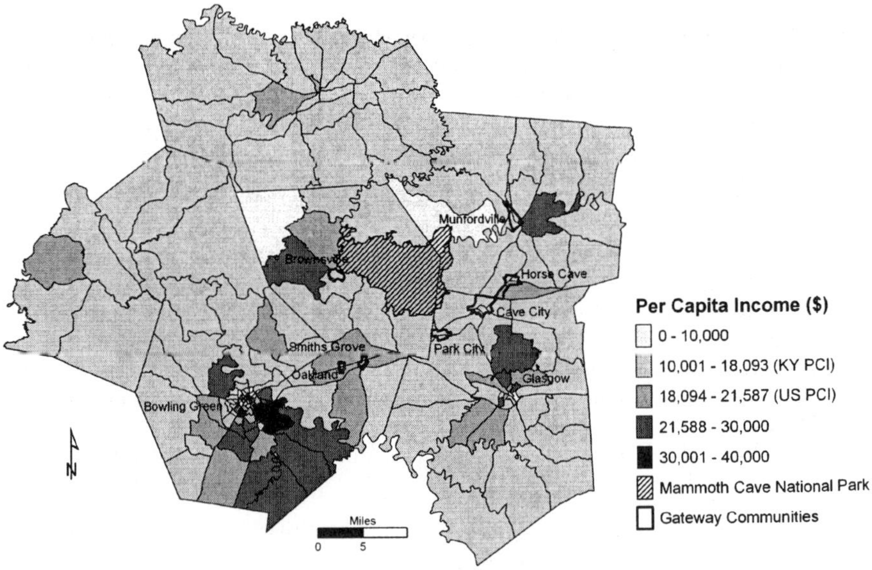

Fig. 15.7. Per capita income by block group, Mammoth Cave Region, 2000. (Source: US Census Bureau, 2000b)

Conclusions

Despite the millions in tourism revenues generated annually by Mammoth Cave National Park, the cave region faces significant economic and social challenges in eradicating poverty, providing local jobs and preserving scenic landscapes. Failures of the tourist economy are most notable in Edmonson County, where the bulk of the national park is located. Here, low incomes, high rates of poverty and cross-county commuting, and limited manufacturing and retailing activity indicate that cave tourism has not led to a robust, diverse economy. One might argue that the situation could well be worse without the national park, which makes contributions to the local economy. The evidence certainly seems to suggest the need to look beyond tourism for development strategies, especially given that Brownsville is unlikely to be a high volume entrance to the park. Gateway tourism development continues to be embraced by community and business leaders in the region, however. Such development has created a vibrant business strip in Cave City, although at the expense of landscape aesthetics. The desire for controlled development of a type compatible with existing land uses is indicated by the vociferous rejection of an industrial park by many residents of Oakland and Smiths Grove. The national park has been successful in promoting a conservation ethic within the cave region, and environmental concerns about the effects of development contribute to a preference for tourism over other kinds of development.

References

Achana, F.T. and O'Leary. J.T. (2000) The transboundary relationship between national parks and adjacent communities. In: Machlis, G.E. and Field, D.R. (eds.) *National Parks and Rural Development: Practice and Policy in the USA*. Island Press, Washington DC, pp.67-87.

Algeo, K. (2004) Mammoth Cave and the making of place. *Southeastern Geographer* 44, 27-47.

Arrington, J.E. (1965a) George Brewer's moving panoramas of the Mammoth Cave and other natural wonders of America, part I. *Filson Club Historical Quarterly* 39, 22-45.

Arrington, J.E. (1965b) George Brewer's moving panoramas of the Mammoth Cave and other natural wonders of America, part II. *Filson Club Historical Quarterly* 39, 151-169.

Brucker, R.W. and Watson, R.A. (1976) *The Longest Cave*. Alfred A. Knopf, New York.

Coleman, J.W. (1936) *Stage-Coach Days in the Bluegrass: Being an Account of Stage-Coach Travel and Tavern Days in Lexington and Central Kentucky 1800-1900*. The Standard Press, Louisville.

Cordell, H.K. Bergstrom, J.C. and Watson, A.E. (1992) Economic growth and interdependence effects of state park visitation in local and state economics. *Journal of Leisure Research* 24, 253-268.

Donnelly, M.P. Vaske, J.J. DeRuiter, D.S. and Loomis, J.B. (1998) Economic impacts of state parks: Effect of park visitation, park facilities, and county economic diversification. *Journal of Park and Recreation Administration* 16, 57-72.

Kentucky Cabinet for Economic Development (KCED) (1999) *Kentucky Deskbook of Economic Statistics*. Kentucky Cabinet for Economic Development, Frankfort, Kentucky.

King, R.H. (1874) Bell's Tavern: A reminiscence of ante-bellum days in Kentucky. *Appleton's Journal* 12, 434-436.

Mammoth Cave National Park Association (MCNPA) (1927) *Now How! A National Park in Kentucky*. Mammoth Cave National Park Association, Louisville.

Murray, R.K. and Brucker, R.W. (1979) *Trapped! The Story of Floyd Collins*. The University Press of Kentucky, Lexington.

Overstreet, M.J. (1999) Edmonson recreation tax ruled valid. *Bowling Green Daily News*, 23 Sept.

Sears, J.F. (1989) *Sacred Places: American Tourist Attractions in the Nineteenth Century*. Oxford University Press, Oxford.

Stynes, D.J. and Sun, Y. (2003) Economic impacts of national park visitor spending on gateway communities, systemwide estimates for 2001. [Online] Unpublished report prepared for the National Park Service. Available at http:www.prr.msu.edu/mgm2/NPSSystem2001.pdf [Accessed 5 May 2003].

US Census Bureau. (1997) *Economic Census*. [Online database] Available at http://www.census.gov/epcd/www/econ97.html [Accessed 20 May 2003].

US Census Bureau. (2000a) 'P26, Place of Work for Workers 16 Years and Over' In: *Census of Population and Housing*, Summary File 3 [Online database] Available at http://factfinder.census.gov/servlet/BasicFactsServlet [Accessed 23 May 2003].

US Census Bureau. (2000b) 'P82, Per Capita Income in 1999' In *Census of Population and Housing*, Summary File 3 [Online database] Available at http://factfinder.census.gov/servlet/BasicFactsServlet [Accessed 23 May 2003].

Chapter 16

Assessing variation in rural America's housing stock: case studies from growing and declining areas

Holly R. Barcus

> 'The ideal image of a thriving small town is one of well-kept, single-family homes, a community school, two or three churches and a bustling main street. Housing is a key element. The visual predominance of single-family, detached, owner-occupied houses set on individual lots embodies the national values of private property rights and the American dream of homeownership.' (Ziebarth, 2000, p.18)

Introduction

Rural America is changing. Economic livelihoods long associated with rural places are in decline and the characteristics of the idyllic pastoral village with church and community at the core is evolving. Global economic shifts produce changes in population composition and shifting migration streams affect rural communities unevenly. Housing, as a physical manifestation of community health and well-being, is a central issue in rural communities, which realize that their housing stock may no longer meet the needs of their changing population. Different population changes result in a variety of housing issues, including shortages in growth communities and declining housing quality in places experiencing population decline. As population grows, housing demand increases and with it so do housing costs, often squeezing out lower income households. Many of these households may choose to relocate to less desirable or less accessible areas nearby in search of less expensive housing (Gober *et al.*, 1993; Fitchen, 1994). On the other hand, in rural areas with declining populations, housing quality is a greater issue than availability. Rural housing markets may contain static housing stock as out-migrants leave behind dwellings once able to house a previously larger population. Decreasing demand for housing reduces the catalysts for any new construction in the market. It also diminishes the incentives for the investments necessary to maintain the quality of the existing housing stock. This pattern results in a static stock of housing units that deteriorate over time. The positive side of this phenomenon is that the declining population and static housing stock tends to improve housing affordability as demand decreases. Too little demand and too much supply result in falling prices that can make housing more affordable. The negative side involves ensuring that the stock fits the reduced demand. The housing stock then ages and deteriorates providing less housing service over time. In either scenario, lower income households are likely to experience the most difficulty securing decent, affordable dwelling units.

In light of these changes and conditions, it is important to examine how housing is redistributed among residents when communities grow or decline, precisely because it allows local officials and residents to institute programmes or seek resources to pre-empt the displacement of vulnerable populations. Assessing residential shifts in the distribution of vulnerable populations, such as the elderly and poor, enhances our understanding of local level implications of larger-scale economic changes.

The objectives of this chapter are twofold. First is to evaluate the differential spatial patterns of housing quality and affordability in two different types of communities: those with positive population growth between 1980 and 2000 and those that declined in total population in this period. That is, does population growth or decline lead to a greater clustering of people vulnerable to housing displacement or a clustering of low quality housing units? Second, do these changes vary between growth and decline counties or are patterns of change similar across counties regardless of similar growth (or decline) trajectories? This chapter explores these issuses in six economically diverse counties located in three major rural regions: Appalachia, the Great Plains and the Upper Great Lakes. Each region has spatially extensive rural areas historically dependent on natural resource extraction. Following this introduction, this chapter reviews the changing fortunes of rural communities and the importance of housing when counties experience significant demographic and economic changes. Sections three and four outline salient features of the study areas, data and methods while section five presents results from the analyis. Finally, section six offers concluding remarks and directions for future research.

Changing fortunes of rural communities

Rural economic restructuring changes rural communities economically and demographically. Castle (1998) suggests that population and manufacturing changes are accompanied by 'generic industrialization' which transforms the countryside directly by changing rural economic activity and indirectly by affecting urban and global systems. This process of industrialization carries with it fundamental changes in rural economies. Increasing specialization threatens the survival of traditional farming operations and environmental quality and places additional stress on rural social services (Castle, 1998). Barkely (1995) notes that the emergence of a new economic environment in rural areas is characterized by greater global competition; an accelerated shift from primary and secondary industries to services; new product organizations and technologies; and industrial restructuring. These changes may be accompanied by a shift in decision making from local control to outside forces, such as global economies and markets and job growth in the form of low wage and low skill manufacturing and service jobs (Castle, 1998; White, 1998). One example of such change is the emergence of large agricultural processing plants and their related industries that offer low skill, low wage work, and often attract immigrants willing to accept poor working conditions and wages (Fink, 1998). Many of these new immigrants are culturally and linguistically different from local residents in such rural communities creating a dual society, both ethnically and economically.

Immigrants are just one group moving to rural areas. Numerous researchers have described the population shifts in rural areas over the past three decades (Gober *et al.*, 1993; Morrison and Wheeler, 1976; Nelson, 1997; Rogerson, 1984; Rundzitis, 1999; Vias, 1999; Watkins, 1990). These shifts are commonly described as the rural renaissance, bust and rebound periods (Johnson and Beale, 1999; Fuguitt and Beale, 1996). During the 1970s and

1990s, internal migration favoured rural areas resulting in population growth. A brief return to 'normalcy' occurred during the 1980s.

While rural counties generally saw decline during the 1980s, the majority experienced growth between 1990 and 1996 (Johnson and Beale, 1999). Migration from metropolitan areas and foreign countries generated much of this growth with a net gain of 1.8 million persons from 1990-1996. Studies of this recent rural growth suggest that migrants move for many reasons including preferences for areas with significant environmental amenities, rural lifestyles and increased flexibility in employment opportunities generated by improvements in transportation and communication technologies (Nelson, 1997; Rundzitis, 1999; Vias, 1999). Return migration and retiree migrants are also significant contributors to rural growth in some areas (Watkins, 1990; White, 1983). On a regional scale, however, shifts in migration vary in their intensity. While many rural regions gained population, others such as the Great Plains, Western Corn Belt and Mississippi Delta, Upper Great Lakes and Appalachia have all experienced prolonged decline interspersed with brief periods of geographically concentrated growth. Changes in economic structure and varying rates of net migration suggest the diversity and complexity of population changes in rural areas. In addition, migration to rural areas declined since 1996 reaching an all time low in 2000 (Cromartie, 2002) re-emphasizing the need to understand the varied outcomes of this change for different rural communities.

The in-migration of different age groups has varying implications based on the major economic activity of the area, particularly in areas with small populations (Fuguitt and Heaton, 1995). Growth of a particular migrant group, such as retirees, generates job growth through increased service demands but it also affects the age structure and needs of the local population. An influx of migrants who are demographically dissimilar to long-term residents also creates changes in local politics and the relative importance of local decisions for such services as schools, infrastructure and new development (Lewis, 1998).

Internal migration such as the out-migration of young people from rural areas, and the in-migration of retirees and certain new immigrant groups changes the demographic profile of rural communities. The movement of young migrants from rural to urban areas leaves an older age structure, higher aged-dependency ratios and lower youth-dependency ratios (Rogerson, 1984). Retiree in-migrants often seek high amenity destinations, bringing substantial non-employment income sources (such as Social Security and retirement funds) with them and may drive up the cost of housing for indigenous, yet often less well-off, elderly residents (HAC, 2000). Hispanic immigrants are also a rapidly growing segment of rural communities (Kandel and Cromartie, 2004) and tend to have lower incomes and inadequate housing in rural areas, often at twice the rate of other rural households (HAC, 2000). Zavala County, Texas provides a good example of disparity between housing conditions in the many Hispanic colonies and those of other, non-Hispanic residents, although recent measures are improving the poor housing conditions of Hispanic residents (HAC, 2001). The rural poor also struggle to maintain adequate housing – a problem complicated by increased rural growth. Among certain rural populations, such as the poor, adequate housing is elusive, spurring frequent residential moves in search of affordable and safe options (Fitchen, 1995). High levels of residential mobility are common among the rural poor as they struggle to find adequate and safe housing on limited income (Fitchen, 1994).

Regarding immigration, the source regions have changed during the past 30 years. Since World War II, non-European migration to the US has increased with the largest proportions of immigrants now arriving from Mexico, Asia and the Americas (Massey, 1995). Communities with expanding agricultural processing and manufacturing activities often attract these new migrants who may be significantly different ethnically, linguistically and economically from current residents (Fink, 1998). Hispanic populations, in particular, have continued to grow as a proportion of the rural workforce (Newman, 2004) reinforcing these patterns of change.

Whether resulting from internal or international migration, compositional changes of rural populations alter demands for housing. As discovered by Ziebarth *et al.* (1997) in a study of 589 midwestern communities, the primary housing issues reported by local residents and leaders as problematic include limited availability of affordable housing, particularly for groups such as the elderly and single parent households; a shortage of rental units; and lack of choice in housing types (i.e. most units were single family detached). Residents also identified the quality of existing housing stock, discrimination and lack of housing opportunities for persons with special needs.

Rural areas experiencing economic growth often develop affordable housing shortages (Swenson and Otto, 1997). Further, in the pursuit of economic development and population growth, rural communities may seek alternative development policies to limit the building of 'undesirable' housing such as mobile home parks, and multifamily rental units. The limitations put on building such housing effectively reduce the availability of affordable housing options for both newcomers to the rural area as well as lower income residents (Ziebarth, 2000). Rural economic development officials are often frustrated by the delayed response of the housing market to rural community growth, fearing the delay will restrict future potential growth. In a study of Sedona, Arizona, a rapidly growing community, Gober *et al.* (1993) found that most people working in Sedona could not afford to live in the community and therefore substituted long-distance commuting from more remote areas in order to find affordable housing. Sedona provides a good example of the effects that significant economic growth can have on local residents. In Hendersonville, North Carolina, a rural Appalachian county in western North Carolina, a dual housing market was created between recent affluent elderly in-migrants and younger, less affluent groups (Rowles and Watkins, 1993). In both cases, the in-migration of groups economically and socially dissimilar to local residents resulted in unfavourable housing outcomes and local tension.

More research is needed to understand the dynamics of rural housing distribution. In his presidential address to the Association of American Geographers in 1984, John Adams called on geographers to explore housing activity as both a social and physical feature of human landscapes. Implicit in his address was the importance of assessing housing from multiple perspectives and different spatial scales (Adams, 1984). Viewing housing issues from a spatial perspective begins to highlight potential investment and policy strategies for communities faced with growing problems. Further, while decline of housing stock is frequently noted in distressed rural communities, the sub-county patterns of decline have not been studied. Other researchers who were trying to more accurately represent rural space have used alternative spatial conceptualizations of rural communities. For example, Smailles *et al.* (2002) use net density to examine settlement changes and classifying non-metropolitan communities in Australia. Such local level expressions of changing economic circumstances are important to maintaining or improving community vitality, particularly in the face of rapid population growth or decline.

Three rural regions: Great Plains, Appalachia, Upper Great Lakes

Great Plains

The changes and concerns discussed above suggest how varied and complex both population fluctuations and economic changes are in rural areas. In particular, different regions in the US may experience different changes and the resulting consequences. The Great Plains region is one example. It has been described as a region of persistent population loss and declining economic opportunity (Albrecht, 1993) and exploited (White, 1998). It is largely dependent upon natural resources, in particular groundwater in the High Plains (Kromm *et al.*, 1992). The Dakotas, northern Kansas and north Texas are characterized by persistent population decline (Zelinsky, 1962); a problem compounded by selective out-migration thereby intensifying the concentration of the service-dependent elderly (Rathge *et al.*, 1998). Zelinsky (1962) categorized most of the Great Plains counties as having experienced a period of population growth followed by decline.

Changes in Great Plains counties have not been uniform. Agricultural decline and industrial restructuring brought significant change to agricultural communities, with those continuing to depend on agriculture experiencing extensive population loss. This region gained manufacturing jobs since 1969 with most of this gain from growth in the meat packing industries, an industry that has geographically concentrated plant locations within 12 counties of three states (McGranahan, 1999). Large plant sizes, and a high percentage of both minority workers and employees with low levels of education, make this a unique type of manufacturing operation. Seward County, a growth county in southeast Kansas, is one such example of this type of growth. Box Butte County in northeast Nebraska, on the other hand, is a county in which the dominant employment sectors historically, the railroad, ranching and farming, remain important today. While Seward grew significantly over the past 20 years, Box Butte's population peaked in 1980 due to increased railroad activity but has been falling since that time.

Appalachia

In the past two decades, the Appalachian region has diversified its economy and seen growing differentiation among its three sub-regions: Northern, Central and Southern Appalachia. Today, the Southern sub-region boasts economic growth and experiences a steady climb out of poverty with lower unemployment rates than most other US regions in both metropolitan and non-metropolitan areas (Isserman, 1996). In terms of economic gains, Northern Appalachia outperforms Central Appalachia but falls short of the progress made by the Southern sub-region. Central Appalachia still lags behind both the Northern and Southern sub-regions with the highest levels of poverty and unemployment (Koebel and Renneckar, 2002). Two counties were chosen from Central Appalachia in eastern Kentucky: Leslie County and Lawrence County. While both counties produce coal, Lawrence County produces far less than most of the coal-field counties, while Leslie County, located adjacent to Harlan County produced more than 9 million tons of coal in 2000, a continuing increase in production, despite declines in total coal-related employment. Unlike the other counties in this analysis, both counties are categorized as persistent poverty counties by the US Department of Agriculture (ERS, 2004).

Upper Great Lakes

The Upper Great Lakes region is also quite diverse. Historically, fur trading, logging and mining all, in turn, dominated the region's economy. Like other rural regions, counties in the Upper Great Lakes experienced boom and bust periods of population growth (Hart, 1984), with most county populations peaking between 1920 and 1960 (Zelinsky, 1962). Between 1999 and 2001, however, the Midwest was the only region in the country to maintain positive population growth through in-migration (Cromartie, 2002). This is attributed to both the desirability of natural amenity areas such as the Upper Great Lakes and relocation of high-tech manufacturing and services industries. Bayfield County,

Wisconsin and Gogebic County, Michigan are both located on the south shore of Lake Superior and are considered non-metropolitan recreation counties (Johnson and Beale, 2002), although neither is considered a high amenity county by the USDA Natural Amenity Scale (McGranahan and Beale, 2002).

Data and methods

Data and scale of analysis

Data for this analysis come from several sources: US Bureau of Census decennial Census reports, local public planning documents, and informal interviews with local non-profits and government officials. Decennial Census reports provide the basic demographic, economic and housing variables at the census tract level (US Bureau of Census, 2001a, 2001b). Census tracts are the primary unit of geographic analysis and are an appropriate scale for assessing rural areas because they provide a unit of analysis that is consistently defined across time and geography and maintain a sample size large enough to minimize sampling error (Cromartie and Swanson, 1996). Tract level data also facilitate the search for local level shifts in housing and income distribution that can be compared over time.

The spatial distribution of housing in any locale is a function of individual preferences and restrictions and the practices of local governing bodies. Local government agencies plan for new housing developments, seek funding for low-income housing, and infrastructure rehabilitation. These agencies play a direct role in facilitating housing development or rehabilitation in a jurisdiction. Local non-profit housing agencies also play a role in providing alternative strategies for homeownership within a community. The focus of the analysis below is on changes in the geographic concentration of low quality housing and vulnerable populations.

To assess the changes, I selected a number of key variables other scholars have identified as important in measuring housing stock. The housing and population variables include housing cost burden, homeownership, lack of complete plumbing, poverty, Hispanic, minority and persons over 65 years of age. As a rule of thumb, households paying 30% or more of gross income for housing are considered cost burdened (Basic Laws of Housing and Community Development, 1991). This means that households are paying too much for housing and may have trouble meeting other basic needs. Previous studies (Mikesell and Wallace, 2000; Whitner, 2000; Ziebarth *et al.*, 1997) highlight the importance of these variables and suggest several relationships examined below. First, that homeownership rates in rural communities are much greater than in most metro areas. Secondly, rural households are more likely to have lower incomes, making affordable housing more difficult to obtain; and they are more likely to find housing unsuitable to their needs (Ziebarth *et al.*, 1997). Thirdly, that housing poverty is greater for ethnic and racial minorities in rural areas (Mikesell and Wallace, 2000; Whitner, 2000). These findings influence the variables selected for this study.

Methods

Location Quotients and the Coefficient of Localization are used to assess the relative distribution of populations by cost burden, homeownership, lack of plumbing and population variables (poverty, elderly, minority and Hispanic). Prior studies of rural housing have not utilized Location Quotients or the Coefficient of Localization to examine patterns of change at the census tract level. Location quotients allow for the comparison of the concentration of sub-group populations to the concentrations of the same groups for larger areas (Plane and Rogerson, 1994). The equation is as follows (using cost burden as an example):

$$LQ_i = \frac{S_i / S^*}{P_i / P^*}$$

S_i = Total population of sub-group in the census tract (1)
S^* = Total population sub-group in the county
P_i = Total population of census tract
P^* = Total population of county

Location Quotients yield a value between 0 and infinity, where 1.0 describes a census tract with the same percentage of Hispanic (for example) as that of the county as a whole. Values less than 1.0 indicate a lower proportion of population than the county and values greater than 1.0 indicate a greater proportion of a particular group in the census tract than at the county level (Plane and Rogerson, 1994). Location Quotients are calculated for 'minority', 'Hispanic', 'cost burdened', 'owner', 'lack of complete plumbing', 'elderly' and 'poverty' for 1990 and 2000 for each county. Changes in levels of concentration are assessed across the two time periods. Coefficients of Localization (CL) are used to compare an overall level of concentration across counties. The CL values range from 0 to 1 with 0 indicating that the distribution of an activity is evenly distributed over the areas and a value of 1 indicating complete concentration (Burt and Barber, 1996). This measure produces a single value and is useful for comparing changes in concentration through time. Here, the Coefficient of Localization is used to compare changes in clustering between 1990 and 2000 for the population and housing variables.

Results

Growth counties: population change and economic shifts, 1980-2000

Economic change and population growth or decline are often parallel processes in rural communities. Building on the extant literature of rural change and population growth, I next investigate the demographic changes brought on by economic shifts in six rural communities and the implications of these changes for housing allocation among different socio-economic groups. The housing landscape provides an intimate view of the health and well-being of a community, clearly revealing distress or prosperity through both subtle and more obvious visual cues (Jakle and Wilson, 1992). Three rural growth counties, Bayfield, Wisconsin, Seward, Kansas and Lawrence, Kentucky, and three rural decline counties, Leslie, Kentucky, Gogebic, Michigan and Box Butte, Nebraska, were chosen to reflect different rural regions of the US that are experiencing different types of rural growth (see Fig. 16.1).

Bayfield County is a non-metropolitan county located in the northeastern corner of Wisconsin on the south shore of Lake Superior. It is located adjacent to the metropolitan county of Superior, Wisconsin. The nearest large cities include Superior, Wisconsin (130 km), Duluth, Minnesota (137 km) and Ashland, Wisconsin (37 km). In 1997, Bayfield was chosen by Alan Solomon, Chicago News Tribune, as the 'Best Little Town in the Midwest'. The town of Bayfield is also known as the gateway to the Apostle Islands National Lakeshore (Bayfield County Economic Development Corporation, 2003).

Historically, the county was dominated by farming and natural resource extraction but today retail sales generated by a new tourist economy prevail. The economies of many of the county's communities are strongly influenced by the Chequamagon National Forest through both logging and tourism. Other communities, such as Barnes, are primarily farming communities. The county's population is relatively dispersed in many small to medium-sized communities.

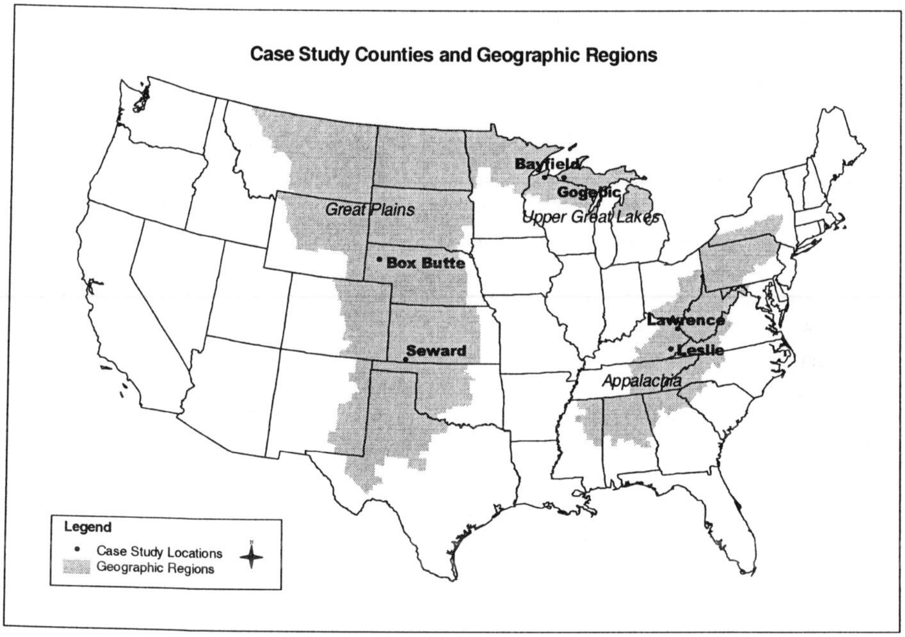

Fig. 16.1 Location of case study counties and generalized regions.

Bayfield County experienced an 8.6% increase in population between 1980 and 2000 (Table 16.1), the lowest of the three growth counties. Bayfield's growth reflects its increasing desirability as an amenity destination, offering rolling agricultural land and deep woods on the interior and hundreds of miles of lakeshore on the its northern border. Second homes have proliferated with about 50% of all residential units occupied year-round. Many once permanent residences have been converted to seasonal homes. The economy shifted from a relatively mixed economy in 1980, with education and government services, manufacturing and retail trade each capturing about 12% of employment, to one which included more tourism-related activities b 2000 (Fig. 16.2). The largest age group currently is 45-54 years (16.3%) followed closely by 35-44 year age group (15.8%). Both youth and elderly age groups declined in Bayfield between 1980 and 2000. Housing units in Bayfield are mixed with many older, well-maintained units and typical rural infill associated with mobile homes. Also, in growth areas of the county, renovated older structures are common, as is new development.

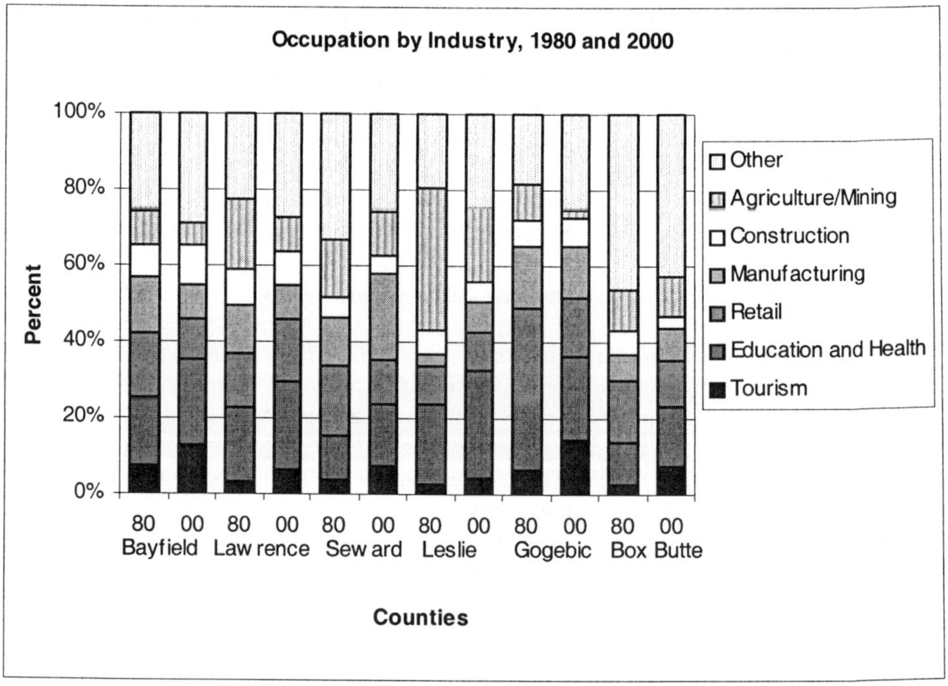

Fig. 16.2 Percentage of total employment by industry for each county, 1980 and 2000.

Lawrence County is located in the far eastern section of Kentucky along the Kentucky-West Virginia border and in the core of the Appalachian Region. The nearest cities include Huntington, WV (53 km) and Ashland, KY (50 km). The construction of the Yatesville Lake Dam, completed in 1989, initiated 'a small renaissance in Lawrence County of people visiting the area to take advantage of the recreational opportunities created by the lake, and an increase in the number of people moving to the county' (Charles, 2001, p.9). The lake draws both recreational users as well as individuals willing to commute to jobs in Ashland and Huntington while living around the lake. The largest age group (14.9% of population) is 35-44 years (Table 16.1). Between 1980 and 2000, both youth and old age groups declined as a perentage of the total population, however, the absolute number of elderly in the county increased by 5.8%.

Lawrence County housing units differ from those of the other two growth counties in that there are many older, poorly maintained or dilapidated units, possibly a reflection of growing poverty and cost burden. Accompanying a 10.3% increase in population, Lawrence County experienced an increase in the percentage of persons in poverty and the percentage of households experiencing cost burden jumped from 15.3% to 27.7% over two decades. In addition to new construction and mobile home infill, the primary stock of the county appears in much worse condition than that of the other two growth communities.

While coal has been mined in Lawrence County both currently and historically, this is no longer a primary economic activity. Coal production peaked in the late 1970s and remained high until the early 1980s (Fig. 16.2). At its peak, Lawrence County mines produced 1.5 million tons of coal (KY Coal Council, 2003), but by 2000 the mines produced less than 700,000 tons of coal, employed only 78 people (1.4% of labour force) and generated US$2.7 million in wages (3.8% of county total wages).

Seward County is located in the southwest corner of Kansas in the heart of the Great Plains. The city of Liberal is positioned at the crossroads of US Highways 54 and 83. Although several small towns are spread throughout the county, Liberal is both the county seat and the centre of population concentration with 87% of the county population residing within the city limits. Seward County, in contrast to Lawrence and Bayfield counties, has grown more than 30% in 20 years (Table 16.1). The population profile changed dramatically as well. In 1980, Seward County was 87% white with a moderate-sized farm population. Today, Seward is 42% Hispanic with a large manufacturing base and shrinking employment in traditional farming (Fig 16.2). Seward County attracts first and second generation Hispanic immigrants who are often originally employed in the local beef processing plant. National Beef currently employs 39% of the working age population. Many of the original Hispanic migrants have opened small businesses and become socially and economically integrated into the community. Also, in contrast to Bayfield and Lawrence counties, the largest age group (15.7% of the population) is 25-34 years of age. Seward County saw a 35% and 36% increase in both youth and aged groups, respectively. Housing units in this county include older, generally well-kept units, new rental units and new moderate- to high-income developments.

In terms of housing, all three growth counties increased the number of housing units available over the 20-year time period. Differences in growth types are most evident when we look at the housing indicators (Table 16.1). Lawrence and Bayfield Counties, both increasing in population partially because of their amenity values, experienced a large increase in vacant units largely attributed to increasing numbers of seasonal homes. Seward County, on the other hand, added 1320 housing units and saw the percentage of vacant units decline. Crowding declined in all three counties. Cost burden also increased in all three counties and poverty increased in both Seward and Lawrence. This general overview suggests that despite growth in all three counties, changes in housing status and vulnerable populations differ across counties, indicating that different types of growth have different outcomes.

Decline counties: population and economic shifts, 1980-2000

Box Butte County, Nebraska is located in the northwestern corner of the state and named for its box-shaped buttes. The main city, Alliance, was incorporated in 1888. Population in Alliance peaked in 1980 with increased railroad activity but has declined since that time (Table 16.1).

Between 1980 and 2000, the population declined by 11.2%. During this period the Hispanic population grew by 80.9% to 930 residents. Youth population declined only slightly from 29.7% of population to 28.1%, while elderly population increased from 12.5% to 14.5%. The railroad industry is the primary employer in Box Butte, although farming and ranching remain important components of the economy (Fig. 16.2). Tourism related employment also grew between 1980 and 2000.

Gogebic County, Michigan lies along the south shore of Lake Superior on Michigan's Upper Peninsula about 100 miles from Bayfield County. The county historically was one of the last areas of the iron range to be mined, with mining operations peaking in the 1920s at 7 million tons annually. The last mine, Peterson mine, closed in 1966. Today, the three towns, Wakefield, Bessemer and Ironwood are working to promote outdoor recreation opportunities and tourism. Location on Lake Superior and 315 named lakes allows the county to focus on water sports in the summer and skiing and snowmobiling in the winter. The county experienced an 11.8% population loss between 1980 and 2000. It follows the expected pattern of declining youth populations and increasing elderly populations. Tourism employment grew from 2.7% to 7.4% between 1980 and 2000.

Table 16.1 County-level measures of housing quality and population characteristics, 1980-2000.

	Growth					
	Bayfield		Lawrence		Seward	
	1980	2000	1980	2000	1980	2000
Total population	13822	15013	14121	15569	17071	22510
Under 18	28.9	24.7	32.2	25.3	31.3	32
Over 64	17.2	16.4	13	12.4	8.6	8.9
Minority	6.8	11.5	0.4	0.5	12.9	34.6
Hispanic	0.3	0.6	0.8	0.2	9.2	42.1
Housing units	9642	11640	5223	7040	6707	8027
Vacant	8.6	46.7	10.7	15.4	8.7	7.6
Owner occupied	80.6	82.6	72	78.1	68.2	64.6
Crowded units	38.5	2.4	6.8	2.8	26.6	11.4
Poverty	13.8	12.4	29.5	30.8	9.6	31.2
Cost burden	18.4	23.2	15.3	27.7	20.1	24.2
Lacks complete plumbing	5.8	11.2	20.1	6.3	9.1	0.7
	Decline					
	Box Butte		Gogebic		Leslie	
	1980	2000	1980	2000	1980	2000
Total population	13696	12158	19686	17370	14882	12401
Under 18	29.7	28.1	24.6	20.4	36.5	24.6
Over 64	12.5	14.5	20.8	22.6	7.7	11.5
Minority	4.7	2.0	2.2	1.2	0.2	0.8
Hispanic	3.8	7.7	0.3	0.9	1.1	0.6
Housing units	5548	5488	10112	10839	4856	5502
Vacant	9.2	12.9	25.1	31.5	5.9	11.2
Crowded units	69.5	70.1	81.5	78.8	70.2	82.1
Poverty	19.1	4.0	15.6	1.2	65.0	2.7
Cost burden	7.0	10.4	11.2	13.8	34.1	32.2
Lacks complete plumbing	7.1	5.7	7.9	5.0	5.7	3.8

Leslie County, a coal county in eastern Kentucky, was created in 1878; it is located in the Eastern Coalfield Region of the state. While coal production fluctuated over the past century, production steadily increased to 9.3 million tons in 1997. Most of this is from underground mining with only 2 million tons from surface mines (KY Coal Council, 2003). Despite increased production, mining employment has declined steadily since the 1980s. The most important employment shift in this county is the decline in mining employment from 34.2% of employment to just 19.4% in 2000. Despite this decline, mining wages still

Table 16. 2 Coefficient of localization values for key variables by county, 1990 and 2000.

County	Minority		Hispanic		Owner		Cost burden		Inadequate plumbing		Poverty		Elderly	
GROWTH	1990	2000	1990	2000	1990	2000	1990	2000	1990	2000	1990	2000	1990	2000
Bayfield	0.57	0.54	0.34	0.25	0.15	0.13	0.13	0.11	0.22	0.23	0.14	0.11	0.06	0.09
Lawrence	0.27	0.16	0.18	0.38	0.02	0.03	0.19	0.16	0.22	0.34	0.13	0.06	0.05	0.05
Seward	0.33	0.19	0.20	0.20	0.08	0.07	0.04	0.10	0.78	0.53	0.18	0.33	0.15	0.18
DECLINE	1990	2000	1990	2000	1990	2000	1990	2000	1990	2000	1990	2000	1990	2000
Box Butte	0.32	0.30	0.31	0.24	0.01	0.01	0.08	0.09	0.61	0.45	0.18	0.18	0.05	0.06
Gogebic	0.62	0.51	0.52	0.22	0.09	0.07	0.13	0.11	0.40	0.44	0.06	0.11	0.05	0.06
Leslie	0.27	0.07	0.26	0.26	0.01	0.01	0.04	0.06	0.06	0.16	0.06	0.06	0.05	0.07

account for 58.4% of all county wages. Between 1980 and 2000, the population declined by 16.7%. Not surprisingly, youth population declined from 36.5% in 1980 to 24.6% in 2000 while the elderly population grew from 7.7% to 11.5%.

The three declining counties are remarkably similar in economic structure and change. Between 1980 and 2000, all three saw declines in agriculture and/or mining, especially Leslie County, as well as reductions in wholesale and retail trade employment (Fig. 16.2). Increases are found in tourism, health and education, and construction employment. Gogebic County is turning towards tourism with new shops catering to mountain bikers, skiers and other outdoor recreation enthusiasts. In Box Butte, the dominant employment in both 1990 and 2000 remains transportation and warehousing; this employment remained strong. Gogebic also experienced an increase in employment related to tourism activities. Leslie County saw increases in manufacturing and health and education as well as slight increases in tourism. Common to all three counties is the decline of the original resource base and shift to tourism and service economies. In these population-loss counties the housing indicators are more consistent. Declines are seen in units lacking complete plumbing, cost burden and crowding while the proportion of owner occupied units increased or stayed about the same. In Box Butte and Gogebic counties poverty rates increased, while in Leslie they decreased. Youth populations declined and elderly population increased in all three places. The indicators suggest that the population may be more stressed economically but housing quality indicators improved supporting the idea that, at least initially, when more units become available access to better quality housing increases for vulnerable population groups.

Sub-county level changes in housing quality and vulnerable populations, 1990-2000

Growth counties

The Coefficient of Localization assesses the relative concentration of each indicator at the county level. As noted above, a score of 0 indicates an even distribution, while a score of 1 indicates complete concentration. In general, for the six counties the levels of concentration are low or declining with most of the housing indicators fairly evenly distributed across census tracts (Table 16.2). Notable exceptions include increasing concentration of cost burdened households and poverty in Seward County. Also in Seward County, units lacking complete plumbing were highly concentrated in 1990 and are still concentrated, although the index value declined, in 2000. Despite the increase in Hispanic population, there is no indication that this population group increased in concentration levels over the two time periods. In Bayfield County, minority concentration was relatively high in both 1990 and 2000, an observation primarily related to the location of the Red Wing Native American reservation. Levels of Hispanic concentration are also notable, although not high. In Lawrence County, Hispanic populations are also concentrated and the level of concentration is increasing. In addition, units lacking plumbing are increasingly concentrated in Lawrence County.

The stable concentration of Hispanics in Seward, but an increase in poverty and cost burden concentrations, suggests that migrants eventually are integrated into the existing and growing neighbourhoods. In both Lawrence and Bayfield Counties, there are increases in the concentration of units without complete plumbing. These concentration values, particularly for structural characteristics, may reflect an improvement in housing conditions

Table 16.3 Census tract location quotients for key variables by county, 1990 and 2000.

County	Census tract	Minority		Hispanic		Owner		Cost burden		Inadequate plumbing		Poverty		Elderly	
GROWTH		1990	2000	1990	2000	1990	2000	1990	2000	1990	2000	1990	2000	1990	2000
Bayfield	9601	4.15	3.99	1.84	2.23	1.14	1.03	1.28	1.24	0.44	0.95	1.55	1.51	0.73	0.71
	9602	0.16	0.24	0.42	1.18	0.89	0.84	0.53	0.30	2.05	2.55	1.37	0.94	1.13	0.97
	9603	0.64	0.58	1.78	0.80	1.63	1.61	1.19	1.04	0.18	0.43	0.71	0.62	0.96	0.88
	9604*	0.19	0.31	0.39	0.49	1.13	1.13	0.74	0.89	1.28	0.94	0.84	1.01	1.04	1.12
	9606*	0.01	0.04	0.36	0.74	0.54	0.58	1.24	1.41	0.94	0.56	0.86	1.06	1.27	1.36
Lawrence	9901	0.44	1.00	0.28	1.24	0.95	0.95	1.45	1.36	0.48	0.24	0.73	0.92	1.11	1.13
	9902	0.06	1.49	0.23	0.00	1.01	0.94	0.23	0.64	1.75	2.26	1.15	1.00	1.04	0.99
	9903	0.21	1.98	0.29	0.00	0.99	1.01	0.64	0.94	1.44	1.25	1.56	1.50	1.02	0.99
	9904	0.14	0.43	0.10	2.73	1.00	1.02	0.81	0.60	1.44	1.76	1.22	0.90	0.86	0.78
	9905	0.15	0.66	0.10	0.00	1.12	1.15	0.82	0.79	0.97	0.87	0.90	0.99	0.85	0.93
Seward	9656	0.40	0.57	0.48	0.78	1.05	1.14	0.69	0.35	0.00	2.56	1.08	0.50	0.75	0.87
	9657	0.33	0.41	0.42	0.34	1.27	1.13	1.00	1.07	0.00	0.67	0.61	0.41	1.40	1.76
	9658	0.50	0.83	0.61	0.76	1.10	1.10	0.96	0.90	4.28	0.67	0.57	0.58	1.34	1.20
	9659	0.88	1.24	1.09	1.23	0.88	0.87	1.13	1.29	0.00	8.45	0.93	0.95	0.77	0.72
	9660	2.61	1.58	2.07	1.63	0.77	0.85	1.02	1.00	0.00	0.34	1.84	1.29	0.74	0.61
DECLINE		1990	2000	1990	2000	1990	2000	1990	2000	1990	2000	1990	2000	1990	2000
Box Butte	9511	0.13	0.43	0.29	0.49	0.98	1.04	0.6	1.0	4.0	3.1	0.8	0.9	1.0	1.1
	9512	0.67	0.65	0.61	0.69	1.03	0.99	1.0	0.8	0.0	0.0	0.6	0.6	1.1	1.1
	9513	1.84	1.72	1.82	1.66	0.99	0.98	1.2	1.2	0.5	0.9	1.5	1.5	0.9	0.8
Gogebic	9501	8.34	5.36	1.57	2.36	0.46	0.56	0.9	1.0	1.9	1.8	0.9	1.0	0.70	0.61
	9502	0.29	0.32	0.00	0.13	1.00	1.08	0.8	0.8	0.8	0.7	1.0	0.8	1.01	1.20
	9503	0.25	0.53	0.64	0.75	1.09	1.08	1.0	0.7	0.1	0.3	1.1	0.8	1.07	0.97
	9504	0.23	0.38	0.00	0.77	1.11	1.08	0.6	0.9	0.8	3.5	0.8	0.6	0.98	1.04
	9505	0.24	0.48	0.00	1.05	1.24	1.14	1.2	1.0	0.4	0.2	1.3	1.4	1.07	1.02
	9506	0.49	0.39	3.29	0.84	1.20	1.11	1.5	1.5	0.2	0.1	1.0	1.2	1.15	1.14
	9507	0.34	0.43	0.69	1.41	0.90	0.97	0.5	0.9	2.8	2.0	0.7	0.8	0.80	0.91
Leslie	9801	1.56	1.16	1.12	2.16	1.00	0.99	1.06	1.08	0.90	0.82	0.96	0.95	1.1	1.2
	9802	0.99	0.93	0.00	0.00	1.04	1.01	1.02	1.09	0.95	1.60	1.22	1.21	0.9	0.9
	9803	0.00	0.81	1.78	3.38	0.95	1.00	0.87	0.79	1.23	0.73	0.86	0.88	0.9	0.9

in a limited number of areas (for example, new construction) in which some areas benefit, or it may reflect a lack of investment, or growth, in specific areas of a county.

The Coefficient of Localization provides a broad assessment of the level of concentration within a county, but it does not indicate where within a county that concentration is occurring. To assess sub-county level changes in housing indicators, I use Location Quotients for census tracts for 1990 and 2000, focusing specifically on indicators that increased in concentration over the study years. Table 16.3 lists all Location Quotients for all variables for 1990 and 2000. In Bayfield County, high concentrations of housing units without complete plumbing are located in tract 9602 (see Fig. 16.3). This is not the tract that has higher concentrations of Hispanic or Native American population, but it is an area along the shoreline that saw a 6.4% increase in population between 1990 and 2000. In Lawrence County, the tracts with the greatest concentration of units lacking plumbing are 9901, 9902 and 9903. All are located the furthest from both the county seat of Louisa and a major state highway. These three tracts also border the county's lake and in 1990 had the largest proportion of vacant housing units. In 2000, 9904 experienced the greatest growth, 17.8% increase since 1990. Finally, in Seward County, concentrations of poverty and cost burden are present in tracts 9660 and 9659, respectively, in 1990 and 2000. Tract 9660 has the largest concentrations of Hispanic and minority populations, poverty households, and lowest concentrations of owner occupied housing units. It appears that despite the lack of concentration of Hispanic households at the county level certain Hispanic households are economically concentrated. These two tracts also experienced 17.6% and 17.8% population growth.

Decline counties

Looking at the Coefficient of Localization (Table 16.2), minority and, to a lesser degree, Hispanic populations, are concentrated in all three counties, although Leslie County saw a sharp decline in the concentration of minority populations while Gogebic remained fairly high. Poverty, cost burden and homeowner indicators were both widely dispersed and stable between 1990 and 2000 in all three counties. Given the moderate levels of poverty in all three counties, it seems that all share equally in economic distress. Housing units lacking complete plumbing are highly concentrated in Box Butte and Gogebic Counties.

At the sub-county level, Location Quotients indicate different patterns of distribution across the key variables (Table 16.3). For Hispanic and minority households, Box Butte has a slight concentration of both groups in tract 9513 as does Leslie (Hispanic only) in tract 9803. Gogebic County has a fairly high concentration (5.36%) of minorities in tract 9501 where an increasing concentration of Hispanic households is found. This finding mirrors evidence of increasing residential segregation among Hispanics across the USA (Kandel and Cromartie, 2004). This tract, 9501, also has the lowest concentration of owner occupied units and the largest (albeit not very large) concentration of units without complete plumbing. Tract 9505 and 9506 have the largest concentrations of owner households and cost burden (9506) and poverty (9505) households. These two tracts also have a slightly greater concentration of elderly populations. In Leslie County, units lacking plumbing and poverty households are concentrated in tract 9802.

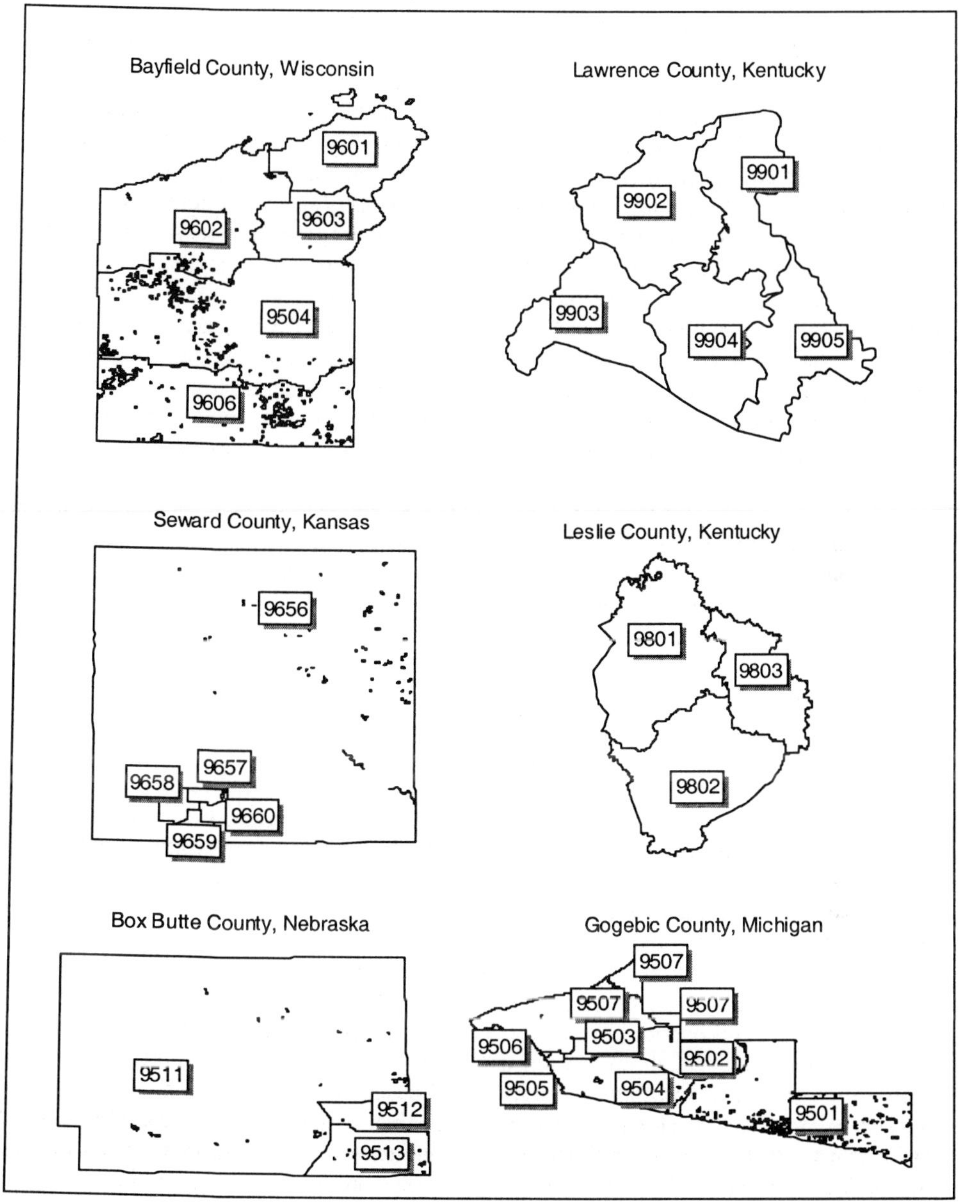

Fig. 16.3 Census 2000 tract boundaries for each case study county.

The three decline counties are remarkably similar. Despite population declines and employment loss, all have witnessed a decline in crowded and cost burdened units and units lacking complete plumbing. These findings suggest an improvement in housing conditions overall. Poverty increased in Box Butte and Gogebic and homeownership declined. In all three counties the number of vacant units increased, in Leslie by as much as 115%. It appears that while the aged population and poverty households increased, fewer of the households are living in low quality units – at least by traditional measures. This condition may be the result of improvements made to deteriorating units or a reshuffling of population to better quality units.

Discussion and further research

The six counties examined above represent different economic types, relative rurality and different types of growth or decline. Despite these contrasts and the differences in regional locations viz. Great Plains, Appalachia and Upper Great Lakes, similarities in the spatial distribution of key housing indicators are relatively consistent. For example, in all six counties, minority and Hispanic populations tend to be concentrated in one or two tracts as are units lacking plumbing, although the patterns are not necessarily the same. In growth counties, high concentrations of minority and Hispanic populations are coincident with high poverty concentrations. Also worth noting is that with the exception of Gogebic County, homeownership is not only very high but spread evenly throughout the counties, regardless of growth or decline conditions.

Units lacking plumbing, in general, declined over the 20-year time period yet remain concentrated particularly in the most rural portions of each county. What this finding suggests is that growth and investment in rural counties are likely taking place in the towns and cities of these counties. City/town governments may be more progressive in addressing housing issues than county governments.

Cost burdened households tend to be widespread with no concentration patterns in either growth or decline counties. Poverty households also tend to have low levels of concentration, however, they are concentrated in the same areas as minority and Hispanic populations when those populations are moderate in size. See for example, Seward, Bayfield, Box Butte and Gogebic Counties. Of interest is that in growth counties, homeowner and cost burden are mixed; Bayfield has high concentrations of both cost burden and homeowner households, Lawrence has high concentrations of homeowner, but in different places than the concentration of cost burden, and in Seward the distribution of cost burden and homeownership is the same. In decline counties, homeownership and cost burden are in similar concentrations, that is, either high or the same.

The question needs to be raised: what do these results mean for rural housing in the USA? As housing quality in general has improved in the USA over the past 50 years, indicators such as lack of complete plumbing are becoming less useful in pinpointing housing issues in rural areas. Also as conditions continue to improve, there is a need for more relevant information about housing quality in rural areas. For example, measures of structural housing quality such as those collected in the American Housing Survey (AHS) would be very useful. Unfortunately, while the AHS collects this information for rural areas, it lacks geographic specificity below the regional level so is not usable for local level analysis.

It seems that new tools need to be developed to assist rural communities in assessing the quality of their housing stock (Ziebarth *et al.*, 1997). Some rural communities embark on housing assessments, often hiring expensive consulting firms who essentially compile US Census data – frequently not only out-of-date by the time the study is finished, but also, as mentioned above, dependent upon variables that are no longer well-suited for an assessment of contemporary housing. Box Butte hired the local economic development office to conduct the assessment but again, it was primarily census-based. One community, the City of Liberal in Seward County, established a 1% sales tax in 1996 to be divided between city beautification projects and housing infrastructure. This money allowed Liberal to hire a full-time Housing and Beautification Director and implement a comprehensive strategy for improving older deteriorated housing units and building new, low-moderate income units. The implementation of the tax was coupled with increased housing inspections and a focus on improving the housing stock in Liberal, where 87% of the county population resides. For a county with such concentrated population, this policy appears to work well since the resources generated by the tax are spent in a concentrated area. Counties with more dispersed populations might have a difficult time stretching such dollars countywide.

Bayfield and Lawrence Counties seem to take a 'hands-off' attitude towards housing development. Interviews with local residents point out a different pattern. A growing group of long-term in-town residents in Bayfield Township have chosen to sell their in-town residences and move just outside township boundaries. Since real estate prices have almost tripled in the last decade, a home that sold for $50,000 in 1990 often sells for up to $150,000 in 2002, local taxes have increased proportionally. The combination of large profits (from selling the original unit) and lowered taxes (being outside the township boundaries) means that many residents who want to stay in the area can now afford to build their own, generally retirement homes, without really leaving the community. Some residents still maintain local jobs. During the summer of 2002, a visual survey of one older neighbourhood in Bayfield Township revealed eight For Sale signs in a four-block area. Most, but not all, of these homes were owned by retirement-age individuals. It seems that for some, the skyrocketing housing market is allowing people to maximize the retirement benefit of homeownership.

For younger families in rural areas, the rapid increase in housing prices has two outcomes. First, younger families that cannot afford to live in town must find housing elsewhere or, second, those who own homes in town, but rely on unreliable service-sector employment often simply do not invest in housing maintenance. The resulting built landscape is one of contradictions. New homes and renovated old-homes exude the financial success of part-time residents or the lucky few with professional wages. These homes are interspersed with deteriorating older homes and infill modular homes of those living on the financial fringe.

This analysis suggests that the type of economic and demographic change occurring in rural America produces different housing outcomes for growth counties, but very similar outcomes for decline counties. Managing growth is important; proactive counties mitigate negative growth effects for vulnerable populations. For decline counties, intervention is also fundamental to minimizing negative housing outcomes. Housing professionals in the decline counties in this study have an acute awareness of the vulnerable populations and housing issues and seem to be taking specific steps towards accommodating needs. Census tract level analysis more clearly depicts rural change, changes often masked at the county level. At a broader regional or national scale, questions about the relationships between housing, population and economic change at the census tract level merit much further investigation. Many questions remain. First, what are the best tools for studying housing stock and access to housing in rural communities? Second, can current federal programmes for rural areas be tailored to reflect economic changes in these communities or is greater intervention necessary (Kravitz and Collings, 1986). Third, what role do not-for-profit organizations play and are they more active in distressed or growth communities? While demographic and economic change will continue in rural America, much greater inquiry into how these changes are linked to quality of life, natural amenities (Deller and Tsai, 2001) and different distributions of jobs and housing is needed (Gober *et al.*, 1993).

Acknowledgment

The Research and Creative Productions Committee at Morehead State University supported this research through an internal university research grant.

References

Adams, J. (1984) The meaning of housing in America. *Annals of the Association of American Geographers* 74, 515-526.

Albrecht, D.E. (1993) The renewal of population loss in the non-metropolitan Great Plains. *Rural Sociology* 58, 233-246.

Barkley, D. (1995) The economics of change in rural America. *American Journal of Agricultural Economics* 77, 1252-1258.

Basic Laws on Housing and Community Development (1991) *H.R. CP No.102-120, 102d Cong., 1st Session,* Washington DC.

Bayfield County Economic Development Corporation. (2003) *Community Profiles* Bayfield County Economic Development Corporation. [Accessed 15th February 2003] Available from http://www.bayfieldwwyp.com

Burt, J. and Barber, G. (1996) *Elementary Statistics for Geographers, 2nd edition.* The Guildford Press, New York.

Castle, E. (1998) A conceptual framework for the study of rural places. *American Journal of Agricultural Economics* 80, 621-631.

Charles, J. (2001) Lawrence County growing and expanding with further development projects. *Big Sandy News,* 10 October, 8-9.

Deller, S.C. and Tsai, T.H. (2001) The role of amenities and quality of life in rural economic growth. *American Journal of Agricultural Economics* 83, 352-356.

Cromartie, J.B. (2002) Nonmetro migration continues downward trend. *Rural America* 17, 70-73.

Cromartie, J.B. and Swanson, L.L. (1996) Census tracts more precisely define rural populations and areas. *Rural Development Perspectives* 11, 31-39.

ERS (Economic Research Service) (2004) *Rural Poverty at a Glance.* Rural Development *Research Report No.100.* USDA, Washington, DC.

Fink, D. (1998) *Cutting into the Meat Packing Line: Workers and Change in the Rural Midwest.* University of North Carolina Press, Chapel Hill.

Fitchen, J. (1994) Residential mobility among the rural poor. *Rural Sociology* 59, 416-436.

Fitchen, J. (1995) Why rural poverty is growing worse: similar causes in diverse settings. In: Castle, E. (ed.) *The Changing American Countryside: Rural People and Places.* University Press of Kansas, Lawrence, pp.247-267.

Fuguitt, G. and Beale, C. (1996) Recent trends in non-metropolitan migration: toward a new turnaround? *Growth and Change* 27, 156-174.

Fuguitt, G. and Heaton, T. (1995) The impact of migration on the nonmetropolitan population age structure, 1960-1990. *Population Research and Policy Review* 14, 215-232.

Gober, P. McHugh, K. and Leclerc, D. (1993) Job-rich but housing poor: the dilemma of a western amenity town. *Professional Geographer* 45, 12-20.

Hart, J.F. (1984) Population change in the Upper Great Lakes states. *Annals of the Association of American Geographers* 74, 221-243.

HAC (Housing Assistance Council) (2000) *Why Housing Matters: HAC's Report on the State of the Nation's Rural Housing.* Housing Assistance Council. [Accessed 15th August 2002] Available from www.ruralhome.org/pubs/hsganalysis/snrh/why/whyhousingmatters.pdf

HAC (Housing Assistance Council) (2001) *Taking stock of rural people, poverty and housing at the turn of the 21st century* Housing Assistance Council. [Accessed 20th December 2001]. Available from http://www.ruralhome.org/pubs/ hsganalysis/ts2000/index.htm

Isserman, A.M. (1996) *Socio-economic Review of Appalachia: Appalachia then and now: An Update of 'The Realities of Deprivation' Reported to the President in 1964 Report to the*

Appalachian Regional Commission. Appalachian Regional Commission [Accessed 17th October 2002.] Available from http://www.arc.gov/images/ reports/thenandnow/then.pdf

Jakle, J. and Wilson, D. (1992) *Derelict Landscapes: the Wasting of America's Built Environment.* Rowman and Littlefield, Maryland.

Johnson, K. and Beale, C. (1999) The continuing population rebound in non-metro America. *Rural Development Perspectives* 13, 2-10.

Johnson, K. and Beale, C. (2002) Nonmetro recreation counties: their identification and rapid growth. *Rural America* 17, 12-20.

Kandel, W. and Cromartie, J. (2004) *New Patterns of Hispanic Settlement in Rural America. Rural Development Research Report No. 99.* USDA, Economic Research Service, Washington, DC.

Kentucky Coal Council and Kentucky Coal Association. (2003) *Kentucky Coal Facts.* Kentucky Coal Association [Accessed 7th January 2003]. Available from www.coaleducation.org.

Koebel, C.T. and Renneckar, P. (2002). *Socioeconomic and Housing Trends in Central Appalachia: Recent Changes in the Appalachian Counties of Kentucky, Tennessee, Virginia, and West Virginia.* Center for Housing Research, Virginia Tech. [Accessed February 2003.] Available from www.arch.vt.edu/caus/research/ vchr/pdfreports/appalachiareport.pdf

Kravitz, L. and Collings, A. (1984) Rural housing policy in America: problems and solutions. In: Bratt, R.G. Nartman, C. and Meyerson, A. (eds.) *Critical Perspectives on Housing.* Temple University Press, Philadelphia, pp.325-334.

Kromm, D.E. and White, S.E. (1992) *Groundwater Exploitation in the High Plains.* University Press of Kansas, Lawrence.

Lewis, G. (1998) Rural migration and demographic change. In: Ilbery, B. (ed.) *The Geography of Rural Change.* Addison Wesley Longman, London, pp.131-160.

Massey, D. (1995) The new immigrant and ethnicity in the USA. *Population and Development Review* 21, 631-652.

McGranahan, D.A. (1999) Can manufacturing reverse rural Great Plains depopulation? *Rural Development Perspectives* 13, 35-45.

McGranahan, D.A. and Beale, C.A. (2002) Understanding rural population loss. *Rural America* 17, 2-11.

Mikesell, J. and Wallace, G. (2000) Unique housing challenges face rural America and its low income workers. *Rural Conditions and Trends* 11, 75-79.

Morrison, P.A. and Wheeler, J.P. (1976) Rural renaissance in America? *Population Bulletin* 31, 1-27.

Nelson, P. (1997) Migration, sources of income, and community change in the nonmetropolitan northwest. *Professional Geographer* 49, 419-430.

Newman, C. (2004) Growth of Hispanics in rural workforce. *Amber Waves* 2, 10.

Plane, D. and Rogerson, P. (1994) *The Geographical Analysis of Population with Applications to Planning and Business.* John Wiley & Sons, New York.

Rathge, R. and Highman, P. (1998) Population change in the Great Plains: a history of prolonged decline. *Rural Development Perspectives* 13, 19-26.

Rogerson, P.A. (1984) The demographic consequences of metropolitan population deconcentration in the US. *Professional Geographer* 36, 307-314.

Rowles, G.D. and Watkins, J.F. (1993) Elderly migration and development in small communities. *Growth and Change* 24, 509-38.

Rundzitis, G. (1999) Amenities increasingly draw people to the rural west. *Rural Development Perspectives* 14, 9-13

Smailles, P.J. Argent, N. and Griffin, T.L.C. (2002) Rural population density: its impact on social and demographic aspects of rural communities. *Journal of Rural Studies* 18, 385-404.

Swenson, D. and Otto, D.M. (1997) Rural housing market response to non-farm employment growth: an analysis of housing development patterns. *International Journal of Public Administration* 20, 1579-1598.

US Bureau of Census. (2001a) *1990 Decennial Census.* US Bureau of Census. [Accessed September 2001] Available from http://www.factfinder.census.gov

US Bureau of Census. (2001b) *2000 Decennial Census.* US Bureau of Census [Accessed September 2001.] Available from http://www.factfinder.census.gov

Vias, A. (1999) Jobs follow people in the rural Rocky Mountain west. *Rural Development Perspectives* 14, 14-23.

Watkins, J.F. (1990) Appalachian elderly migration patterns and implications. *Research on ageing* 12, 409-429.

White, S.E. (1983) Return migration to Appalachian Kentucky: an atypical case of non-metropolitan migration reversal. *Rural Sociology* 48, 471-491.

White, S.E. (1998) Migration trends in the Kansas Ogallala region and the internal colonial dependency model. *Rural Sociology* 63, 253-271.

Whitner, L.A. (2000) Housing poverty in rural areas greater for racial and ethnic minorities. *Rural America* 15, 2-8.

Zelinsky, W. (1962) Changes in the geographic patterns of rural population in the USA: 1790-1960. *Geographical Review* 52, 492-524.

Ziebarth, A. Prochaska-Cue, K. and Shrewsbury, B. (1997) Growth and locational impacts for housing in small communities. *Rural Sociology* 62, 111-125.

Ziebarth, A. (2000) Local housing policy: the small-town myth and economic development. *Rural America* 15, 18-23.

Chapter 17

The geography of housing needs of low income persons in rural Canada

David Bruce

Introduction

This chapter describes the geography of housing needs among low income persons living in rural Canada. Through a case study approach, it examines the geography of these housing needs across different types of rural communities in Canada, focusing on the different types of low income households in different types of rural communities, their differing housing situations, and the differing capacity at the local level to develop housing solutions.

Halseth and Rosenberg (1995) suggest that there is less understanding of housing issues and markets in a rural Canadian context, primarily because of our collective use of universal measures of quality or general characteristics. They argue that rural housing is more complex than we suspect. For example, they note that the rapidly ageing nature of the stock itself may require public policy intervention to rehabilitate and preserve it for future use. They also note that when rural cottage properties are converted to year-round use, infrastructure (water and sewerage) and public service (health, education, transportation, etc.) challenges are posed. Clearly, then, any study of rural housing issues must take into account the variation among the site specific characteristics found in rural communities.

An immediate challenge is defining rural. The Organization for Economic Cooperation and Development (OECD) approach to defining rural is based on population density:

> 'OECD predominantly rural regions refer to individuals living in census divisions with more than 50 percent of the population living in OECD rural [communities with a population density of less than 150 persons per square km] communities. This includes all census divisions without a major city '. (du Plessis *et al.*, 2001, p.6)

Based on this definition, approximately 31.4% of the Canadian population in 1996 lived in predominantly rural regions (du Plessis *et al.*, 2001).

An important defining feature which distinguishes one rural and small town community from another is its distance from a major city. Rural communities and small towns are generally found within one of three types of predominantly rural regions in Canada: rural metro-adjacent regions, rural non-metro-adjacent regions, and rural northern and remote regions. In 1996, 15% of the total Canadian population lived in rural metro-adjacent regions; 14% lived in rural non-metro-adjacent regions and 2% in rural northern and remote regions (Beshiri and Bollman, 2001).

The OECD definition is most helpful when attempting to understand regional issues such as economic development or labour market issues. It is also appropriate for looking at housing issues: the nature of service delivery by the public and non-profit sectors to people with low incomes tends to be on a regional basis. Housing markets in individual small communities tend to be limited. One often has to look at several adjacent communities to find a full range of housing options and choices within the private and public sectors.

The chapter begins with a discussion of the methodology employed in the research. It then moves to a definition of low income and assesses the presence of low income households in rural communities. Their various housing situations are then discussed. The barriers and opportunities facing rural communities in attempts to address low income housing needs are examined. The chapter concludes with a discussion of policy issues and ideas which should be considered in any attempt to address housing needs of low income persons in rural areas.

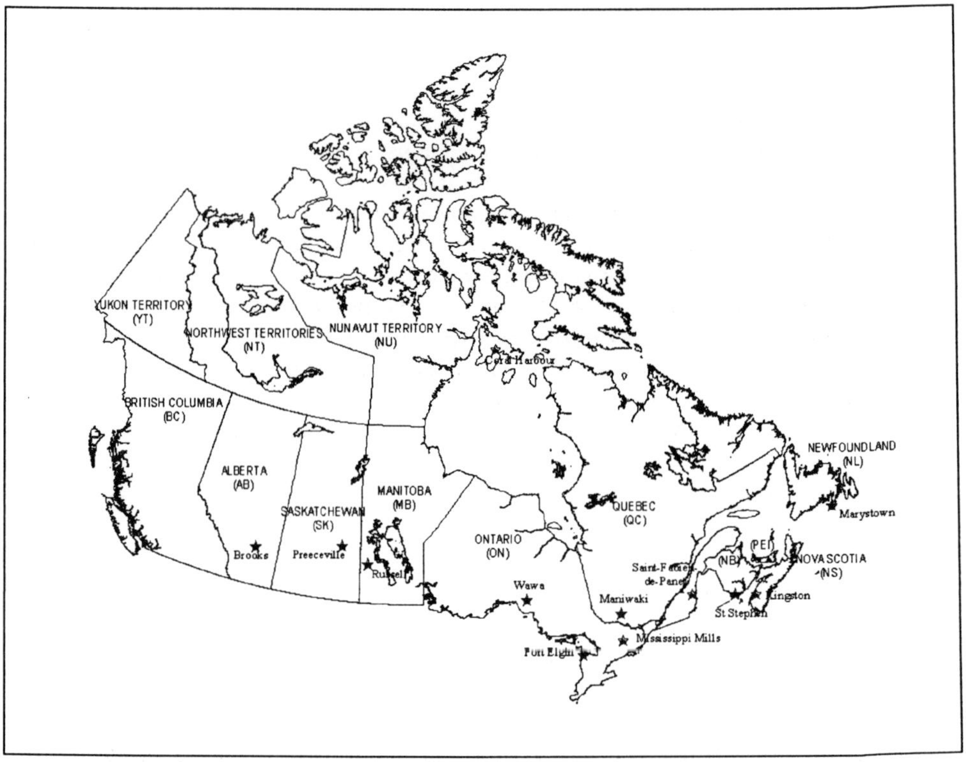

Fig 17.1 Geographic location of the 12 case study sites across Canada.

Methodology

The OECD definition of predominantly rural regions was employed to define rural regions for the purpose of selecting 12 case study communities (where the specific community within a rural region is less than 10,000 in population) for detailed investigation. This permitted a wide-ranging exploration of literature on the subject. It also permitted a review of Census data exploring the relationship between housing, income and households in rural areas, using Canada Mortgage and Housing Corporations (CMHC) Housing in Canada database.

A working typology of rural communities was developed to help frame the discussion about low income households and their housing situations. The typology is based on a mix of variables which extends from the basics of population size and distance from an urban centre to include geography, economic base and demographics. Table 17.1 provides a summary of the elements used to discuss and describe different low income housing situations across different types of rural communities, and identifies the selected case study communities which fall into each category.

A two-step sampling method was used. First, several different predominantly rural regions across Canada were identified. Second, a single municipality within each region was selected for case study. Figure 17.1 shows the geographic location of the 12 case study sites across Canada.

Key informant interviewee selection criteria

A set of 10 to 15 key informant interviews, supplemented with a review of local documents, were completed for each case study. A generic priority list of interviewees was developed from individuals and organizations with knowledge about local housing and related issues. Individuals were invited for an interview if they were in the community or provided local housing service for a particular community. This list was supplemented by a substitute list of individuals and representatives of organizations. Individuals were chosen on a local basis, to replace a refused interview, an unavailable interviewee, or non-existent individuals or organizations, from the priority list.

The priority list, to be interviewed in all communities, included:

- municipal staff or municipal councillor;
- social workers or municipal social service departments;
- economic development officers;
- health care officials, community health workers, directors of the regional health districts;
- housing authority managers;
- real estate agents;
- non-profit housing corporations;
- central housing registries;
- landlords/property managers;
- directors of municipal homes for the aged;
- operators of emergency shelters/transitional housing.

The substitute list, to be interviewed in some communities, as appropriate, included:

- presidents of seniors' associations;
- Royal Canadian Mounted Police (RCMP) or municipal police officers;
- school division superintendents;
- agricultural development officers;
- community planners;
- Human Resources Development Canada/local employment centres;
- support agencies;
- Residential Rehabilitation Assistance Program (RRAP) administrators;
- friendship centres.

Table 17.1 Typology of rural communities.

Type	Geography	Economy	Demographics	Case study communities
Growing communities	located close to urban centres, usually along major highways	fuelled by new economic activity; service centre for other areas	broad mix of population cohorts; more single young males; usually 5,000-9,999 population	Brooks AB; Kingston NS
Stable or Slow Growth Communities	usually within 1 hour of large urban centre	service centre for other areas; presence of social services is important	broad mix of population cohorts; usually 2,500-4,999 population	Port Elgin, ON; Saint Fabien-de-Panet, PQ; St. Stephen, NB
Declining Communities	usually more than 2 hours from large urban centre	slow erosion of economy through business closure; may have been, or still is, a single industry community; seasonal activities may be important	youth out-migration; older population; usually fewer than 2,500 population	Maniwaki, PQ; Marystown, NF; Wawa, ON
Dormitory communities	located near large urban centre	retail and services dominate	young adults and families; semi-retired professionals	Mississippi Mills, ON
Retirement Communities	reasonably close to a large urban centre; may serve as a regional population centre	retail and services; health care sector	high elderly population	Russell, MB; Preeceville, SK
Northern Communities	usually isolated and far distance from markets	resource and service oriented; large gap between high and low paying jobs,limited expansion options	younger population; mostly Aboriginal	Coral Harbour, NU

Defining core housing need

CMHC (1991 and 1994) set out the basic approach to the definition and quantification of housing need, generally referred to as the core housing need methodology. This approach has been accepted in Canada at both the federal and provincial level and has been applied to a variety of statistical sources. It covers three aspects:

- An *adequate* dwelling does not, in the assessment of the occupants, require major repairs or lack (hot or cold) running water and has a full bathroom.
- *The shelter cost* for an *affordable* dwelling must be less than 30% of household pre-tax income. Shelter cost for owners includes mortgage principal and interest, property taxes, condominium charges (if applicable) and utility payments (water, gas and electricity). For renters, the shelter cost includes rent and utilities if the latter are not included in the rent.
- A *suitable* dwelling has enough bedrooms to provide the household with the required amount of space and privacy, taking into account the age and gender of the members using the National Occupancy Standards (NOS) as the measure of space requirements.

To be in core housing need, a household must fall below one or more of the three housing standards of adequacy, affordability or suitability and lack the financial means to access accommodation that would meet adequacy and suitability standards. In most rural places it is difficult to apply the core housing need model because even when households have a sufficient income to be able to afford to move themselves out of a housing problem, the supply of vacant housing in the marketplace may be non-existent.

Low income households in rural Canada

Defining a low income situation is very difficult, and varies from community to community. There are a variety of potential indicators of low income situations, described below.

When low income cut-off (LICO) categories are used, as defined by Statistics Canada, there are fewer low income families in rural areas (10% of households) than in urban centres (16%-18%). However, rural areas have the highest proportion of households with low incomes based on the low income measure, with a rate of 15%, compared to 12% for urban centres with populations of 500,000 or more (Rupnik *et al.*, 2001, p.15).

The incidence of low income, as reported by the Census, is calculated based on income, household size and expenses of the household. It is reported for individuals, for economic families and for the population as a whole living in private households. The incidence across the 12 case study communities ranged from 10% to 25% among economic families, and between 21% and 57% among individuals living alone.

Data collected at the local level on food bank usage and social housing waiting lists varied significantly in terms of its availability and completeness. This was also the case when we asked informants about their impressions of the incomes required to make ends meet in their community. Most had no solid basis for their observations.

All of these indicators were used to construct a typology of low income households. This typology consists of the working poor, single parent families, single persons, seniors, youth, the disabled and Aboriginal persons living off-reserve. Local informants also were challenged to describe what they felt were the absolute numbers and relative composition of the low income typology in their respective communities. This was due to varying degrees of knowledge, lack of contact with low income households and the mix of potential indicators that were used. The relative presence of each of the seven types of low income households found within the six types of rural communities is discussed below.

Working poor families

These households generally have lower levels of education and employable skills, and are one of the largest components of the low income typology in most of the case study communities. In communities where the service sector, tourism and primary industries are major employers, these households tend to be more prominent. They also are prominent in other communities like Kingston, Port Elgin and Mississippi Mills, where there is a large employer paying high wages to a large segment of the workforce, which in turn keeps housing prices relatively higher and thus more expensive for lower income workers. In several of the case study communities it was noted that some working poor families choose to reside in a more rural area outside of a town or village, where land and housing costs are cheaper, or where they may have inherited a family property.

Single parent households

These households were also found to be a large component of the low income typology in most of the case study communities. Many of these households are dependent on social assistance for income, and often occupy social housing units. These households were especially prevalent in Coral Harbour, Marystown, Kingston, Mississippi Mills, Port Elgin, St. Stephen and Wawa.

Seniors

Low income seniors are generally of two types. The first are those who may have been in a relatively low income situation for most of their life. They now find themselves on fixed incomes, living in rental or social housing. They may be single or two-person households. The second are single seniors, usually widowed females, living alone in an older home they own, with a low or modest income. Some of these seniors may have been in a low income situation for most of their lives, while others may have lower incomes now than in their past. These seniors often have challenges associated with the costs of maintaining their homes, and covering the costs of property taxes, high utility bills, and maintenance and repair. Senior households of both types were prominent in Kingston, Preeceville, Russell and Saint-Fabien-de-Panet.

Single persons

These households vary significantly in their circumstances, but are generally young adults with limited education and employable skills. They are more prominent in absolute numbers in communities dependent on natural resources, and thus they tend to form a larger share of the low income group in those types of communities. This includes Coral Harbour, Maniwaki, Marystown, Saint-Fabien-de-Panet and Wawa, and to a lesser extent, Mississippi Mills, Preeceville and Russell.

Single youth

Because there are few economic or housing options for youth when they leave their parental home in a rural area, they often leave the community immediately for larger urban centres. They tend not to be a visible or large part of the low income typology in rural communities. However, there was evidence in most of the case study communities of a small number of single youth accessing food banks, or receiving some form of social assistance.

Disabled persons

This includes both physically and mentally disabled persons. The general observation from the case study communities is that they are not a large component of the low income situation, primarily because of families providing care in their homes, or because of the presence of a well-managed group home. Those without local families or with specific challenges not served by local agencies often move from the community to larger urban centres. Informants in some of the case study communities, such as Kingston, suggested that the number of disabled persons in rural communities is on the rise due to a number of factors, including deinstitutionalization of mentally disabled persons, and an ageing population with more physical limitations. In most of the case study communities there appeared to be limited knowledge about the number of low income disabled persons and their housing needs.

Aboriginal people

Four of the case study communities had some element of an Aboriginal population. Maniwaki, Preeceville and Russell are each located near a First Nation reserve. There is some indication that Aboriginal people from these nearby reserves move into these communities and some of them are in low income situations. Coral Harbour has a large Inuit population. They tend to have lower levels of education and do not always qualify for the higher paying government and management jobs typically held by southerners.

Summary of low income household types

The types of low income households one would find in different types of rural communities vary significantly. In *growing communities* (Brooks, Kingston), the working poor, single parents, and seniors tend to be the most prominent or visible among low income households. They typically feel the brunt of rising housing costs in these communities. Low income households of all types are typically found in both stable (Port Elgin, Saint-Fabien-de-Panet, St. Stephen) and in *declining communities* (Maniwaki, Marystown). In dormitory communities (Mississippi Mills), where higher income households are moving into the community, the result is that the working poor and single parents typically find themselves in low income situations, and are more prevalent than other low income household types. In retirement communities (Russell, Preeceville), one typically finds seniors and single persons as the most visible among low income households. In northern communities (Coral Harbour) there are Aboriginals, single parents, the working poor and single persons prevalent among low income households. Single youth and disabled persons were generally not prevalent among the low income populations observed.

Housing situations of low income households

Different low income groups may find themselves in a variety of different housing situations, facing problems such as affordability, crowding, poor state of repair of their dwellings, and at-risk or absolute homelessness. These are discussed in more detail below.

Affordability issues

The statistical review showed that 15% of rural households faced affordability problems, as measured by the core housing need model, in 1996 (Table 17.2). Most of these are in core housing need, indicating that their incomes are too low for them to solve their housing problems. Senior-led households, non-family households (such as people living alone, or unattached individuals living together) and renters were more likely to have affordability problems compared to other rural households (Table 17.3). This was confirmed in most case study communities.

Table 17.4 examines incomes, shelter cost and shelter cost-to-income ratios (STIRs) for rural households with a comparison to the situation in 1991. In 1996, the household incomes of rural households in need were less than one-third of households not in need. The difference in their shelter cost was less than CA$8. This small difference in shelter cost combined with a large difference in income led to a major difference in the average percentage of income going towards shelter. Households in core housing need spent 42% of their income on shelter compared to only 14% for households not in need. Average STIRs also increased from 1991 to 1996 for all households in need, but remained constant for households that were not in core housing need. It is clear that incomes among low income households are not keeping pace with rising shelter cost. This situation was confirmed in the case study communities.

Table 17.2 Urban and rural households in core housing need due to affordability problems, 1996. (Source: CMHC's Housing in Canada Database, based on 1996 Census).

	Total	Not in need	In need
Below affordability only	12.3%	3.2%	9.1%
Below affordability and adequacy	2.0%	0.4%	1.6%
Below affordability and suitability	0.6%	0.2%	0.4%
Below affordability, adequacy and suitability	0.1%	0%	0.1%
TOTAL	15.0%	3.8%	11.2%

Table 17.3 Incidence of core housing need among rural households, by type and tenure, 1996. (Source: CMHC's Housing in Canada Database, based on 1996 Census).

	Total	Own	Rent
All households	14.2%	10.3%	30.2%
Senior households	14.8%	10.9%	34.7%
Non-senior households	14.0%	10.1%	29.0%
Family households	11.2%	8.3%	26.3%
Non-family households	27.3%	22.5%	34.3%

Table 17.4 Average income, shelter cost and shelter cost-to-income ratios for rural households by need status, 1991-1996. (Source: CMHC's Housing in Canada Database, based on 1996 Census).

		1996			1991		
		Average income	Average shelter cost	Average STIR	Average income	Average shelter cost	Average STIR
All	Total	$43,160	$495	18%	$39,718	$426	17%
	Not in need	$47,794	$495	14%	$43,066	$432	14%
	In need	$15,156	$487	42%	$12,872	$378	40%
Owner	Total	$46,257	$504	16%	$42,255	$434	15%
	Not in need	$49,727	$504	13%	$44,791	$439	13%
	In need	$16,171	$493	40%	$13,759	$373	37%
Renter	Total	$30,297	$458	26%	$29,417	$395	23%
	Not in need	$37,481	$446	17%	$34,719	$397	17%
	In need	$13,716	$479	46%	$11,589	$386	44%

Adequacy and suitability issues

The statistical review showed that 11.3% of rural households have an adequacy problem, but only 4.4% are among those in core need. Suitability is a smaller problem, with 4.1% of rural households being crowded and only 0.9% being among those who are in core need. The need for repair was evident in some case study communities, especially in St. Stephen, Kingston, Maniwaki and Wawa. Crowding problems were apparent only in relatively isolated cases within the case study communities; however, there was evidence of crowding in Brooks and in Coral Harbour in particular.

Heating and utility costs

Lower income households (of all types) tended to live in older housing units which were poorly insulated and had high heating and utility costs. This was particularly evident for low income renters in St. Stephen, Marystown, Kingston, Saint-Fabien-de-Panet, Maniwaki, Port Elgin and Wawa. These high operating costs are particularly acute in northern Canada (Coral Harbour).

Rental housing

Most of the case study communities experienced a lack of new rental housing construction. The result is very little rental housing choice characterized by low vacancies, poor conditions, and high operating costs. The exceptions were in communities where the economy is growing (Brooks, Mississippi Mills), and in communities where a modest number of new units aimed at higher income seniors, usually with extended services and amenities, were being built (Kingston, St. Stephen, Russell, Port Elgin). Social assistance rates created distortions in small rental markets: in some provinces, the rates paid were too low to make ends meet (Ontario, New Brunswick), and in other provinces (Nova Scotia), where there is a shelter component to social assistance, market rents are usually set at that rate by landlords, making it more expensive for other low income households, particularly the working poor.

Ownership housing

This is the main tenure form in rural areas and small towns even for low income households. On a relative basis, a smaller percentage of rural homeowners are in core need - 10% compared to 30% of rural renters (CMHC, 2000). Ownership is often the only tenure choice, especially when rental housing is either unavailable due to low vacancies, or is at least as expensive as owning. Many low income households have problems saving for a down-payment to actually purchase a home. In Brooks, Kingston, Mississippi Mills and Port Elgin, where average incomes are relatively high, average house prices tend to be high and beyond the reach of low income households. In Coral Harbour, where the construction costs (and therefore purchase prices) are exceptionally high, those with low incomes have almost no hope of owning a home. Other low income homeowners have problems affording the costs of maintaining their homes (property taxes, utility bills, and maintenance and repair). This was evident in Saint-Fabien-de-Panet, Port Elgin, St. Stephen, Kingston and Preeceville.

Low income off-reserve Aboriginal households

The incidence of need among rural Aboriginal off-reserve households in 1996 was much higher (28.8%) than among other households living in rural areas, including 17.8% with a housing affordability problem, 21.6% with an adequacy problem, and 18% with a suitability problem (CMHC, 2000). Although this situation was an improvement from the early 1990s, off-reserve Aboriginal households were twice as likely to be in core housing need and seven times more likely to be below multiple housing standards when compared to other rural households. In Coral Harbour there were many housing problems associated with the high costs of construction and utilities coupled with the lack of employment opportunities. This leaves Aboriginal households with major affordability problems, which are often addressed by doubling up in crowded situations.

Comparing the income, shelter cost and shelter cost-to-income ratio of rural Aboriginal off-reserve households by need status, the situation is much the same as among the general rural population. The average income of households in need was less than 40% of that of households not in need. At the same time the average shelter payment of households in need was 84% of that not in need (CA$411 compared to CA$487). Consequently, households in need paid more than twice the proportion of their income on shelter than those that were not in need - 34% compared to 14% (CMHC, 2000). Rural Aboriginal off-reserve households are also more likely than the general rural population to be in core housing need based on adequacy or suitability problems: 11.9% are in core housing need with an adequacy problem; 9.5% with a suitability problem (CMHC, 2000).

Rural homelessness

There is a growing literature in North America on the issue of homelessness in rural areas. The literature is primarily in the form of case studies and community plans (Callaghan 1999; Housing Assistance Council, 2001; National Rural Health Association, 1996) with some also considering the issue in a larger theoretical context (Fitchen 1991; Wright *et al.*,

1998). Collectively, the literature identifies a number of critical elements which are important for understanding the issue of rural homelessness:

- In a rural context homelessness is often more about relative homelessness; this is a population which is vulnerable to becoming homeless (because of low incomes, lack of security in tenure, or poor quality housing), or who are less visible because they live in squalid conditions, double or triple up with family and friends, are transient from one location to another within a community or from community to community, or because they do not have access to services that one might find in an urban setting. The Housing Assistance Council 'suggest[s] that homelessness [in a rural context] be defined as a continuum ranging from a complete lack of shelter at one end to severely inadequate housing conditions at the other' (2001).
- The problem is really one of poverty - rural households with little or no income are vulnerable to becoming homeless.
- There are usually no emergency shelters for homeless singles, youth or families, which means that many rural homeless migrate to urban centres in search of services and solutions.
- There is an increasing reliance of individuals and families on motel accommodation.
- Low income tenants in rural areas are often caught in an affordability squeeze and they have few other housing choices.
- The low density of population and services in rural areas means that poor rural people require a car to get around, which erodes their ability to afford decent housing.
- Rural homelessness is more often caused by economic changes beyond the control of individuals and households; the impact of the loss of energy-related, manufacturing, and farm-related jobs and their replacement, if any, by low-paying and often part-time service sector employment is noted in several reports; when someone loses a job in a rural community, they have few options other than to relocate to another community to find a job, and this is often not a choice people want to make.
- Rural homelessness is less likely to be a result of mental illness than it is in urban settings.
- Fitchen (1991) notes that rural gentrification, whereby people from neighbouring urban areas relocate to obtain cheaper housing or vacationers build second homes (both of which result in rising housing costs for local residents) contributes to the increasing problem of rural homelessness.
- Whole, intact families are more likely to be the face of rural homelessness than are single persons (as is usually the case in urban centres); however, family conflict and domestic violence can also result in the break-up of rural families leaving women and children with a housing problem (Housing Assistance Council 2001; Pauktuutit Inuit Women's Association, undated).

There was very little evidence of absolute homelessness in the case studies. In most of the case study communities there may be only one or two households or individuals who could be described as being absolutely homeless without a permanent place to call home. They would typically be living outside of a small town or community, in a very rural location, perhaps in a tent for part of the year, or in some other modified structure (such as a converted bus or camper).

Summary of low income housing situations in rural communities

The housing situations of low income households one would find in different types of rural communities vary significantly. In growing communities, higher housing costs and declining vacancy rates are problems faced by low income households. In both stable and in declining communities, high operating costs and older housing in relatively poorer condition, are more common problems. Also, there may be declining vacancy rates in stable communities due to the lack of new construction. In dormitory communities, where higher income households are moving into the community, the result is that the working poor and single parents typically find themselves in low income situations, and they also face rising costs and fewer housing choices. In retirement communities, seniors may face rising housing costs as new units are built. In northern communities where construction and operating costs are much higher, there are major affordability and crowding problems, especially among Aboriginal households.

Barriers and opportunities

There are a variety of barriers to addressing the housing needs of low income households beyond the fact that they have low incomes, including: high building costs (especially the costs of materials and shipping); the NIMBY (not in my backyard) syndrome which makes it difficult to site and build affordable, below market housing; limited economic options, economic uncertainty, and lack of viable housing markets; limited economic return on rental housing (which makes new construction by the private sector almost impossible); lack of an extended community services infrastructure; poor social conditions (including lack of life skills among social assistance recipients); lack of community leadership (housing needs are not on the local political agenda); and restrictive public policies and regulations (which provide disincentives for new affordable housing construction).

Despite the fact that there are many barriers to addressing the housing needs of low income households in rural communities, the collective experience of the 12 case studies reveals that there are many potential opportunities to address some of the needs: introducing new municipal planning and land development practices; having proactive community leadership; converting buildings to housing; responding to market demand from seniors; integrating housing and services for seniors, disabled people, and others; embracing manufactured housing; and continuing research and development on building technology for the north (Tables 17.5 and 17.6).

Discussion

The development of strategies to address housing needs in rural communities and small towns must be considered in concert with broader community development activities. The investigation in case study communities clearly revealed that some communities are better prepared to address needs. In these cases, the community adopted broader approaches to the issue, through the establishment of community housing or social agencies (Brooks, Mississippi Mills), developing a population growth or retention strategy (Saint-Fabien-de-Panet), making use of community assets and resources (Saint-Fabien-de-Panet, Port Elgin, St. Stephen), and linking social development and social or health services with the provision of housing (Mississippi Mills, Port Elgin, Russell,

Preeceville). Other communities recognized the need to put in place a more coordinated social development strategy to boost people's life skills so that they would be better housing occupants and would have a greater chance of obtaining meaningful employment (St. Stephen, Coral Harbour).

Table 17.5 Summary of low income households and their housing situations in rural communities (part 1).

Type	Barriers	Housing Situations	Opportunities
Growing communities	NIMBY; need for community services; need for community leadership; social assistance rates	low or falling rental vacancy rates; rising costs	supply of affordable land; proactive community leadership; market demand from seniors
Stable or slow growth communities	NIMBY; limited land for development; economic uncertainty; poor social conditions; lack of viable housing market; need for community services social assistance rates	low or falling rental vacancy rates; poor quality units; high heating and operating costs	supply of affordable land; land development; housing policy; conversion of non-residential buildings; proactive community leadership; market demand from seniors; manufactured housing; integrating housing and services
Declining communities	lack of construction sector; limited land for development limited economic options; poor social conditions; lack of viable housing market; need for community services; need for community leadership; social assistance rates	poor quality units; high heating and operating costs;	supply of affordable land; land assembly/ sub-division; proactive community leadership

Table 17.6 Summary of low income households and their housing situations in rural communities (part 2).

Type	Barriers	Housing situations	Opportunities
Dormitory communities	NIMBY; need for community services; social assistance rates	low or falling rental vacancy rates; rising housing costs	proactive community leadership; integrating housing and services;
Retirement communities	lack of construction sector; lack of viable housing market	rising costs for newer units; some older units with high utility costs;	supply of affordable land land assembly/ subdivision; proactive community leadership; market demand from seniors; manufactured housing; integrating housing and services
Northern communities	high building costs; limited economic options; poor social conditions; lack of viable housing market; social assistance rates	high utility costs; high construction costs; limited or no private sector housing provision; affordability problems due to lack of jobs among local Aboriginal population; crowding among extended families	building technology; research; integration with economic development strategies

The housing needs of low income households are also, in part, an income problem. Many rural communities have limited economic options (Coral Harbour, Marystown, Wawa, Saint-Fabien-de-Panet, Maniwaki, Preeceville), which reduce the potential residents have to obtain sustained, year-round employment capable of providing an income sufficient to pay for their housing. Thus, it is critical that housing development strategies be explicitly linked to broader community economic development strategies which will create jobs and income for lower income households.

Conclusions

The housing needs of low income persons in rural Canada are quite diverse and vary from place to place across the country. One finds seniors, working poor families, single parent households, some single youth and young adults, some disabled persons, and some Aboriginal persons represented in the typology of low income persons in rural Canada. The relative presence of one or more of these groups among local low income populations varies from place to place dependent upon the local circumstances, as do the housing situations they face, and the barriers and opportunities associated with addressing their housing needs.

Housing development generally, and rental markets more specifically, tend to be problematic in most rural communities and small towns. From a demand perspective, those with relatively high or even modest incomes can afford homeownership. Thus renters are often characterized as being low income, on social assistance, or transient. From a supply side, then, it is difficult for rental owners and investors to earn sufficient income for reinvestment purposes, and this leads in part to a decline in the quality of the stock.

This context poses real challenges for low income households looking for ways to solve their own housing problems. The implicit assumption of the core housing need model is that, with sufficient income, a household can resolve its housing difficulty. In an urban setting with a viable rental market, this assumption is probably reasonable. In a rural setting, supply can be constrained so that, even with financial resources, other options may not exist. The lack of housing supply in rural areas must be borne in mind when assessing housing need in rural areas.

Federal, provincial, and territorial governments must explore ways to increase the supply of affordable housing in rural communities. A recently announced (2003) affordable housing supply programme (cost-shared between the two levels of government) holds some promise, but it is too early to know if any new funds will be directed to rural communities.

To facilitate an improved policy framework for addressing rural housing needs, there are at least three related research topics which require further investigation. First, we need to know more about rental investment and rental demand in a rural context. The demand thresholds for rental housing are limited; one vacancy in a ten-unit apartment building can be problematic from a cash-flow and profit perspective. Research is needed to assess risks and develop strategies for reducing those risks. Related to this issue is the challenge associated with small numbers of households in different income groups (low income, working poor, social assistance, transient labour, professionals) each with specific rental housing needs. We need more information about how to develop private sector mixed income rental developments and how they are managed financially and socially.

Second, there is a need to explore other indicators of housing need. In environments where housing markets are small or non-existent, households in core housing need have more difficulty in addressing their housing needs. In order to establish benchmarks for core housing need in non-market areas, specific measuring criteria and indicators of housing need should be created. The calculations used to determine if a household is in core need, for example, depends on average market rents as one factor. However, in many rural communities (often non-market) there is little or no supply for making the basis for this calculation, or for those in core need to turn to in an attempt to address their housing needs. These indicators (the income required to afford average market rents for adequate and suitable accommodation in the community without exceeding 30% of income) are thus meaningless. There is also the fact that the 30% guideline is often unrealistic given high costs of other basic necessities because of transportation costs. There may also be cultural differences in what represents crowding and adequacy. Hence our traditional indicators of need may not be relevant. Research is required to establish new ways to measure core housing needs in many rural contexts.

Third, we need to know more about the impact of the ageing population. Seniors are retiring to small towns near large urban centres, or other 'central places', where access to (health) services is readily available. We need to know more about the needs and resulting innovations associated with housing an ageing population, especially among those with lower incomes. Related to this is the depressed income growth among seniors. The predictions of the late 1980s and early 1990s centred on a growing elderly population with higher incomes reshaping housing markets may no longer hold. Given the changes in the global economy and smaller growth in retirement incomes (particularly for seniors who have worked in primary sector occupations), we need to know more about how this has and will affect demands for housing and services. Finally, in small towns there is a need to examine how to structure a full range of housing options (from independent to assisted to enriched to nursing home care) to serve a broad regional territory.

Acknowledgments

This chapter is based on work completed by the author and several colleagues on a project entitled Housing Needs of Low Income People Living in Rural Areas, for the Canada Mortgage and Housing Corporation (CMHC Project # 6620-49). Permission has been granted by CMHC Project Officer Anna Lenk to use research materials prepared for that report. The author acknowledges the intellectual and research contribution of Tom Carter, Jino Distasio, Ed Starr and Paddy Fuller.

References

Beshiri, R. and Bollman, R. (2001) *Population Structure and Change in Predominantly Rural Regions. Rural and Small Town Canada Analysis Bulletin Volume 2, Number 2.* Statistics Canada, Ottawa.

Callaghan, M. (1999) *Understanding Homelessness in Muskoka.* District of Muskoka, Ontario.

Canada Mortgage and Housing Corporation (CMHC) (1991) *Core Housing Need in Canada.* Canada Mortgage and Housing Corporation, Ottawa.

Canada Mortgage and Housing Corporation (CMHC) (1994) *Technical Guide to Understanding 1991 Core Housing Need Databases and Estimates.* Canada Mortgage and Housing Corporation, Ottawa

Canada Mortgage and Housing Corporation (CMHC) 2000 *Housing in Canada: Electronic Data Series Database.* Canada Mortgage and Housing Corporation, Ottawa.

du Plessis, V., Beshiri, R., Clemenson, H. and Bollman, R. (2001) *Definitions of Rural and Small Town Canada Analysis Bulletin Volume 3 Number 3.* Statistics Canada, Ottawa.

Fitchen, J. (1991) Homelessness in rural places: perspectives from upstate New York. *Urban Anthropology* 20, 177-210.

Halseth, G. and Rosenberg, M. (1995) Complexity in the rural Canadian housing landscape, *Canadian Geographer* 39, 336-352.

Housing Assistance Council (2001) Rural Homelessness: The Problem [Accessed 15th December 2001] Available from http://www.ruralhome.org/pubs/infoshts/rhomeles.htm.

Marshall, J. and Bollman, R, (1999) *Rural and Urban Household Expenditure Patterns for 1996 Rural and Small Town Canada Analysis Bulletin Volume 1 Number 4.* Statistics Canada, Ottawa.

National Rural Health Association (1996) *The Rural Homeless: America's Lost Population.* National Rural Health Association, Kansas City.

Pauktuutit Inuit Women's Association (undated) *Inuit Women: The Housing Crisis and Violence.* Canada Mortgage and Housing Corporation, Ottawa.

Rupnik, C., Thompson-James, M. and Bollman, R (2001) *Measuring Economic Well-Being of Rural Canadians Using Income Indicators Rural and Small Town Canada Analysis Bulletin Volume 2 Number 5.* Statistics Canada, Ottawa.

Wright, J., Rubin, B. and Devine, J. (1998) *Beside the Golden Door: Policy, Politics, and the Homeless.* Aldine de Gruyter, New York.

Chapter 18

Change in rural North Carolina: the new Latino presence

Owen J. Furuseth

Introduction

The USA 2000 Census of population presented evidence of a dramatic demographic shift during the latter portion of the twentieth century. Between 1990-2000, Hispanics/Latinos had emerged as the largest ethnic/race group in the country. Indeed, during this time, Hispanics were the fastest growing US minority group, increasing from 22.4 million to 35.3 million persons. This represented an impressive 58% increase. As a result, Hispanics eclipsed African-Americans as the largest minority group in the country, 12.5% to 12.3%. Pushed by high birth and immigration rates, the Latino population jumped another 4.7% from 2000 to 2001, and comprised 13% of the USA population.

Beyond the absolute growth in Latino-Americans, the 2000 census revealed significant shifts in the national and community level scale of Latino geography. In general, the Latino population growth shifted away from the conventional immigrant settlement norms. Three geographical changes were important in these shifts. First, at the national scale, the highest rates of Latino population increase were states and parts of the nation that were formerly non-Latino. Latinos have been traditionally concentrated in the southwestern states; however, the 1990s evidenced a decentralization of Latinos populations throughout the US. Second, the data revealed a declining importance associated with large gateway cities to new Hispanic immigrant streams. Formerly, Latinos, like other immigrant groups, tended to arrive in the US at selected gateway cities and subsequently were concentrated in these places. The 2000 data revealed that the settlement appeal and subsequent long-term residence of Latinos in these urban areas greatly declined during the past decade. The data revealed a growing desirability for suburban areas versus central cities as the preferred residential locations for Latino-Americans. Taken together, each of these findings suggests that if contemporary Latino settlement trends continue, a reshaping of American ethnic geography is underway.

Among US regions caught up in the Latino settlement transformation, the South is experiencing the most significant impacts. For most of the past 100 years, the region was passed over by non-Anglo immigrants. While there was limited international migration, the process was incremental and the low numbers of immigrants were easily assimilated. Within this framework, societal and cultural stresses are rooted in race. Historically, social and economic relations in the American South have been constructed along racial divides: black and white. But beginning in the late 1980s, the collective consciousness of southerners was blurred by unprecedented growth in domestic and foreign-born immigration into the region. Southeast and South Asians arrived in large numbers, but the largest growth in new southerners was Latino, with the largest single group being Mexicans (Murphy *et al.*, 2001). The 2000 Census provided powerful evidence that a fundamental demographic change is underway. The Southeast region has emerged as the second largest home of Latinos in the USA, with 69 million Hispanics living in 13 states and the District of Columbia. Across the south-eastern states, the rate of Latino population increase was over 100% in every case, except Florida, the state with the largest Latino population in 1990.

Latino migration to North Carolina

Among these states, North Carolina experienced the highest rate of Hispanic/Latino population increase, a 386% jump between 1990 and 2000. In 1990, there were only 76,726 Hispanics in the state, but by 2000 North Carolina was home to 372,964 Hispanics. (US Census Bureau, 2002). The most recent census estimates for 1 April 2000 to 1 July 2003, offer evidence of continued rapid growth, with 87,741 new Hispanic residents, a 23.1% increase (US Census Bureau, 2004). The official numbers evidence a dramatic increase, but most knowledgeable analysts view these data as undercounting Latinos, particularly the illegal and underclass segment of the population. Indeed, census undercount research carried out in other states if applied to North Carolina suggest that the number of Latinos in the state could be two to three times the official enumeration. The result would be a Latino headcount of between 750,000 to 1.2 million persons.

Until now, the only comprehensive analysis of recent state-wide Latino settlement in North Carolina was completed by Karen Johnson-Webb and James Johnson in 1996. Their research, completed without the benefit of the 2000 Census, used commercial population estimates for the post-1990 censual period and offered an examination of North Carolina's incipient Latino population growth. Based upon county scale analysis of Latino population distributions for 1980, 1990 and 1995, their findings were that contemporary Latino settlement was focused in two types of places: (i) counties with large US military complexes (Onslow, Cumberland and Craven counties) and (ii) 'metropolitan or urban crescent' counties (Wake, Durham, Orange, Chatham, Guilford, Forsyth, Cabarrus and Mecklenburg) positioned along the Interstate Highway 85, which arcs through the Piedmont region of the state. Moreover, using these data, Webb-Johnson and Johnson (1996) posited that future Latino population growth is likely to be increasingly concentrated in North Carolina's rapidly growing metropolitan, urban areas. These communities exhibit diverse economic structures that would support a wide array of job skills, including traditional manual labour as well as skilled labour requiring language and technical expertise.

The findings of Johnson-Webb and Johnson present a strong counterpoint to traditional thinking about North Carolina's historically small Latino population. In the past, the Latino presence was mainly rural based and closely associated with agriculture. North Carolina is an important agricultural state with a heavy reliance on immigrant workers for farm labour. Currently, the state ranks fifth in the number of migrant farm workers (Dale *et al.*, 2001), with Latinos, especially Mexican workers, comprising more than 90% of the work force. In particular, North Carolina's large-scale poultry and hog (pig) farms and the labour intensive fruit, Christmas tree and tobacco farming operations have production regimes that create a year round as well as seasonal demand for agrarian skilled labour.

In similar fashion, agricultural processing facilities located near production districts are attractive employers for immigrants. The strong relationships between poultry, hog and beef processing industries and Latino immigrants is widely documented (Griffith, 1993; Stull *et al.*, 1995; Guthey, 2001). The sustainability and growth of agricultural processing facilities, or other blue collar manufacturing jobs in rural counties raises questions surrounding an urban-centred trend for Hispanic population expansion in North Carolina.

In the context of the growing Latino presence in North Carolina, the geocoded county level data from the 2000 Census permit answers to the questions surrounding where North Carolina's newest residents are choosing to settle and, in a related fashion, what the impacts these new residents on local economic, social and political relations.

A cartographic examination of Latino settlement patterns in 1990 compared with the 2000 census findings offers some support for the inferences contained in the Webb-Johnson and Johnson research, but also informs alternative findings with significant rural impacts (see Figs 18.1 and 18.2). A review of the county scale Latino population in 1990 finds a highly concentrated Spanish speaking population. (Fig. 18.1). Across most of the state, but particularly in the mountainous western portion and the northeastern Coastal Plain counties, Latinos were virtually absent. Consider that among the 100 counties, 22 had less than one-half of one percent Latino residents in 1990. As a consequence, counties with fewer than 50 Latino residents in 1990 were not unusual.

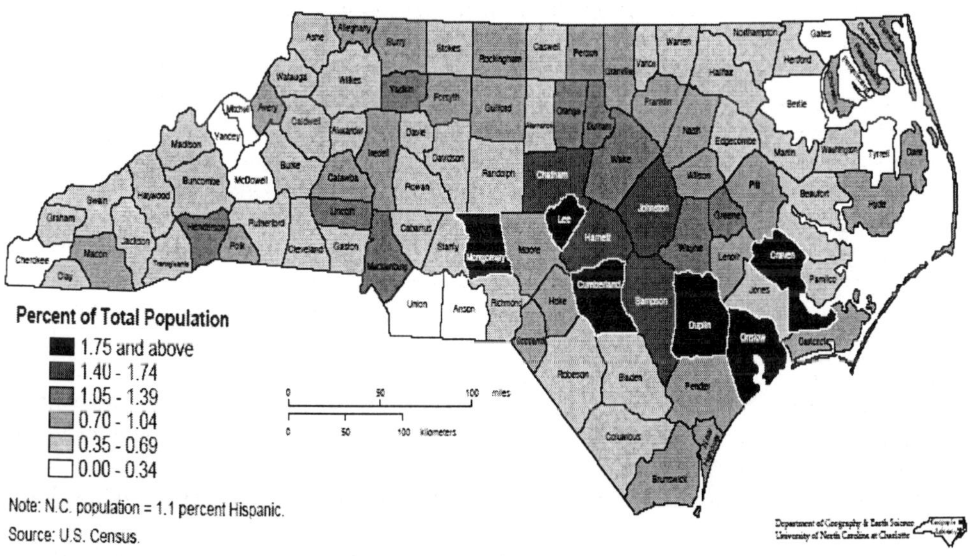

Fig 18.1 Distribution of Hispanic population, 1990.

The largest concentration of Latinos resided in the western Coastal Plain in eastern North Carolina and eastern section of the Piedmont region. The ten counties in this group were home to 29,237 Latinos, or 38.7% of the state-wide total. While some of North Carolina's leading agricultural production districts are included within this group, the single most critical underlying factor explaining Latino residency is the concentration of USA military facilities. Cumberland, Onslow and Craven counties have large US Army, Air

Force and Marines Corps training bases. In 1990, these three counties contained 30.2% of North Carolina's Latinos. While the absolute numbers of Latinos is clear, their residency is not a product of conventional settlement processes. Rather, it is structured around the military lifestyle. In practice, following enlistment in the USA military, troops are assigned to facilities where they may reside for a few months or years. At the end of their 'tour of duty', they are reassigned to other duties, often at military posts in other regions of the USA or abroad. Residency is temporary. In fact few of the military residents rotating through North Carolina military facilities return to permanent residence.

During the Reagan and George W. Bush era military build-up, the numbers of military stationed on bases in eastern North Carolina greatly increased. This, in turn, translated in larger numbers of Latino soldiers and air personnel moving through the state.

The second significant concentration of Latinos was located in six metropolitan urban counties linked by the Interstate highway 85. These economically prosperous and rapidly growing counties were the residence for 20,411 Latinos or 26.2% of the state-wide totals. These were the counties that Webb-Johnson and Johnson envisioned as the future settlement destination for Latino North Carolinians. Taken together, the two geographic clusters encompassed included 65.3% of all Latinos.

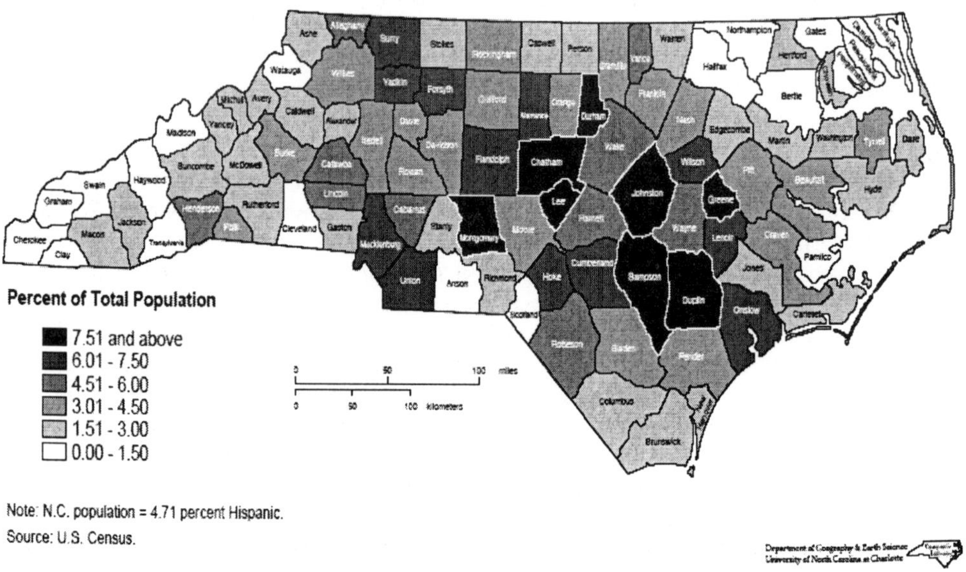

Fig 18.2 Distribution of Hispanic population, 2000.

Changes in Latino settlement 1990–2000

A review of the 2000 Latino settlement geography shows the areal effects of the explosive growth in Latin population change (Fig. 18.2). In general sense, the numbers and proportions of Latinos in the two 1990 core areas have increased dramatically and spread to surrounding counties. The geographical concentration of Latinos in the coastal plain and eastern Piedmont has expanded to the south and west. In absolute terms, the counties absorbing the largest number of Latino residents include heavily agricultural counties (Duplin, Sampson, Johnston, Robeson and Union); the largest Interstate 85 urban counties

(Mecklenburg, Wake, Forsythe, Durham and Guilford); and exurban counties with strong manufacturing oriented economies (Alamance, Cabarrus, Catawba and Randolph).

Interestingly, in the context of a growing and expanding Latino influx into North Carolina, large sections of the state remain bypassed. Most of these communities are rural but they are quite disparate. Broadly speaking they are the mountain and foothill counties in the west; the outer Coastal Plain counties, generally the north-eastern region; and the northern tier of counties along the state boundary with Virginia. In a number of cases, the increase in Latino residents in these counties was negligible during the decade.

A review of the migration literature and complementary empirical work on contemporary Latino settlement geography suggests several alternative patterns may be evidenced. Based upon the important role that social relationships play in providing facilitating and fostering Latino movement into Anglo settings, chain migration processes may be important determinants for the new settlement geography. Thus, communities with existing Latino population in 1990, or rural communities that have been traditional job sites for migratory Latino labour would be expected to have gained large Latino populations during the past decade.

In a related way, counties with large-scale agriculturally oriented economies may also have been popular settlement destinations for Latino migrants. Among Mexican migrants to the USA, home residence type is a powerful steering variable. That is to say, Mexican migrants from urban areas tend to move to urban areas in the USA, while Mexicans from rural areas of Mexico gravitate to rural settings in the USA. Social processes play a role in this outcome, but the employment history and the skill set of the migrant are more important factors. Mexicans originally from urban areas are better educated and therefore equipped to compete most effectively in the urban job market, while those from rural areas find it easier to transfer their farming skills to North American agricultural settings.

An alternative traditional labour market model, around which Johnson-Webb and Johnson constructed their settlement forecast, posits that Latino population growth is largely driven by economic opportunities. In the context of North Carolina this means urban counties with large diverse labour markets would be attractive targets for Latino migrants. Within these communities, skilled as well as unskilled Latinos would find numerous opportunities in a large-scale labour market. In particular, the market for domestic workers in urban business districts and affluent suburban settings increases the opportunities for women, especially single women, to move to the state. On a national level, Latinos are greatly under-represented among Hispanic immigrants, e.g. 55.9% male, 44.1% female. In North Carolina, the gender margin is wider. Currently, the Latino population is 59.8% male and 40.1% female.

Finally, a related economically based migration framework may exist around place-based prosperity, exclusive of rural or urban characteristics. Within this context, counties experiencing high growth rates, and enjoying high incomes and lower rates of poverty would be attractive migration destinations because of their diverse and robust economies. Across North Carolina there are a number of counties, cutting across rural and urban settlement characteristics that match well with this prosperity-oriented template.

Regression analyses of 1990–2000 migration flows

In order to examine the saliency of these alternative settlement models, linear regression analyses were carried out. Separate multiple regression models were run for each of the four models and following tests for multicollinearity, a cumulative model was developed. The dependent variable used in the analyses was the percentage of Latino county residents in 2000. Based upon model specifications, ten independent variables likely to shape the settlement process were employed. The input data were derived from the 2000 Census of population and the 1997 Census of Agriculture.

A review of the four regression model results reveals significant variation in the strength of association (Tables 18.1 and 18.2). The most effective predictor of Latinos settlement geography was the model structured around antecedent Latino population and the importance of commercial agriculture. Labelled the 'Hispanic Settlement-Agricultural Model', this emphasis of social reproduction factors explained 50% of the variation of Latino settlement (Table 18.1).

The next most useful model was centred on agricultural characteristics ('Agricultural Model'). Using commercial agriculture values, farm-related attributes and rural population, the model was able to explain almost 37% of the variation in Latino settlement geography.

Table 18.1 Rural-centric regression models for explaining Latino settlement in North Carolina counties, 1990–2000.

	Hispanic Settlement – Agricultural Model	
Variable	Beta	Standard Error
Hispanic population, 1990	0.475	0.259
Market value of agricultural products, 1997	0.430	0.000
R^2 adjusted	0.500	

	Agricultural Model	
Variable	Beta	Standard Error
Commercial agriculture value, 1997	0.000	0.486
Number of farms, 1997	0.001	−0.153
Number of farmers, 1997*	-	-
Percentage rural population, 2000	0.007	−0.262
R^2 adjusted	0.373	

* Dropped for multicollinearity.

Table 18.2 Urban and Growth Oriented models for explaining Latino settlement in North Carolina counties, 1990–2000.

	Urban Opportunities Model	
Variable	Beta	Standard Error
County population, 2000	0.000	0.087
Percentage urban population, 2000	0.012	0.143
Percentage population change, 1990–2000	0.023	0.297
R^2 adjusted	0.137	

	Place-based Prosperity Model	
Variable	Beta	Standard Error
Median household income, 2000	0.000	0.148
Percentage population change, 1990-2000	0.030	0.249
Percentage persons in poverty, 2000*	-	-
R^2 adjusted	0.115	

*Dropped for multicollinearity.

Both of these most useful models tended to exhibit rural-centric elements. In contrast, the remaining models were more urban oriented, and both performed at approximately the same level. Neither approached to the same degree of precision as the former models (Table 18.2). The 'Urban Opportunity Model' and the 'Placed-based Prosperity Model' each explained less than one half of the variation explained by the rural oriented models. The former model was developed to represent urban orientation and community growth characteristics. The latter model was constructed around measures of community wealth and growth.

Following up on the individual models, the seven variables that were not significantly inter-correlated were used in a step-wise composite multiple regression analysis (Table 18.3). Dropped from the model for multicollinearity was the percentage rural population variable. The Cp statistic was used to select the most efficient model. The final model explained approximately 60% of the variation in Latino settlement geography. This cumulative model included two agricultural attributes, two economic-demographic variables, and the antecedent Latino settlement. Omitted from the model were two urban variables. Significantly, the stepwise process included the antecedent Latino settlement first, followed by commercial agricultural income and median income. These three variables were dominant in providing the explanatory power to the model.

Table 18.3 Cumulative model for explaining Latino settlement in North Carolina counties, 1990–2000.

Step	Variable	Standard Error	R^2 Adjusted	Increases in R^2
1	Hispanic population, 1990	0.242	0.328	-
2	Market value of agricultural products, 1997	0.0002	0.500	0.172
3	Median household income, 2000	0.0004	0.579	0.079
4	Percentage population change, 1990–2000	0.020	0.591	0.012
5	Number of farms, 1997	0.0006	0.596	0.005

Montgomery County NC: case study for rural Latino migration

Among the rural counties in North Carolina experiencing the new Latino presence, Montgomery County is illustrative. Located in the southern Piedmont, this county is situated beyond the urbanized edge of the three Piedmont metropolitan regions. Faced with the challenges of economic restructuring as well as locational disadvantage, Montgomery County has been slowly losing ground to other rural Piedmont communities. In 2000, the county's population was 26,822. During the last two census periods, the county experienced only modest population growth. Indeed, during the 1990-2000 period, the rate of population increase for the decade was only 14.1%; or 66% of the state-wide average. This increase was among the lowest in the Piedmont region.

Geographically, the county is divided along east-west lines (Fig. 18.3). This east-west bifurcation is structured around physical characteristics, but is also translated into pronounced economic and social differences. The western portion of the county is dominated by the rugged and heavily forested terrain made up of Monadnocks, resistant rocks from the early Appalachian Mountains. Resembling a patch of the Appalachian foothills in the Piedmont physiographic province, the west of the county is racially more white than the rest of the county with an economy dominated by logging and timber related business. East of Little River, the county landscape is coastal plain. The colloquial term 'sandhills' is applied to a landscape dominated by pine woods and flat, well-drained sandy soils. The economy of the east county is mixture of large scale commercial agriculture, fruits, tobacco, and cotton and low wage manufacturing. The small eastern towns of Star, Briscoe and Candor are dominated by a mix of small and large textile mills and furniture plants. Racially the east is far more diverse, African-American and whites have traditionally lived here in roughly equal numbers.

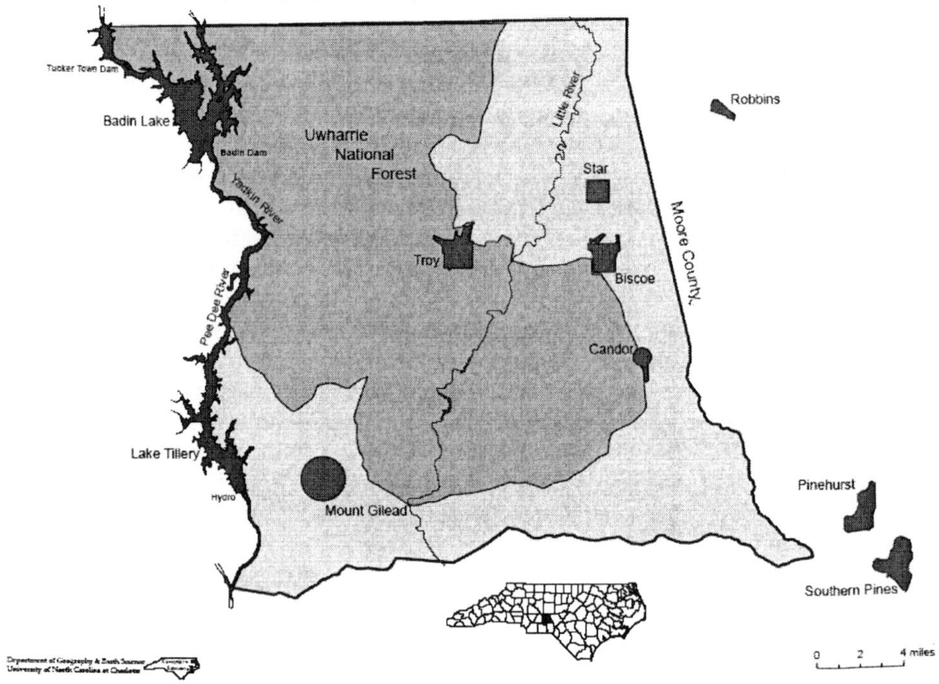

Fig. 18.3 Montgomery County.

The decade-long restructuring in the traditional low wage manufacturing sectors, textiles and furniture, that are the core employment base for many North Carolina rural counties, including Montgomery, has translated into significant community disruption. Nonetheless, manufacturing remains the centre of the century's economy. A total of 51% of the work zone is employed in the manufacturing sector. Weaving and knitting textiles and furniture manufacturing and milling are the dominant employers. The average manufacturing weekly earnings (second quarter, 2001) are US$449.43, compared to a state-wide average of US$719.21. Viewed in economic terms, Montgomery County is increasingly marginalized when compared with other parts of the wealthy Piedmont region. Beyond low wage rates, the movement of low skill manufacturing jobs off shore has been devastating. The most recent, annual unemployment rate for 2002 was 8.9%. The latest monthly unemployment figures (April, 2003) are a record 10.5%, 4.1 points higher than the state-wide rate. According to the 2000 Census 15.4% of county residents are living in poverty (3.1% higher than the state-wide figure), with median household income at US$32,903, or just 84% of the state-wide median.

Traditionally, the racial and ethnic composition of Montgomery County has closely mirrored that of the state. White persons make up 69.1% of the county and African-Americans 21.8%. In turn, social relations were constructed along these lines. But, beginning in the late 1980s, Latinos began moving to Montgomery County. The 1990 Census recorded 556 Hispanic/Latino Montgomery County residents. This was 2.4% of the county population, making it the fourth highest Latino proportion in North Carolina. In 2000, 2279 persons or 10.4% of Montgomery residents were Hispanic. Again this is the fourth highest proportion of Latino's residents in the state. During a time when the absolute growth in the county population was slightly less than 3,500, Latinos comprised 78.5% of the new resident stream.

The most recent census estimates document a continuation of the demographic trends of the last decade. Between 2000–2003, Montgomery County's population increase was nearly flat, 1.8% (US Census Bureau, 2004). In contrast, new Hispanic residents grew by a robust 20.2%.

Montgomery County's Latino immigration is propelled by Mexicans moving into the county from other parts of the USA as well as international migration. Over 91% of Latinos identify themselves as Mexican, with fewer than 1% originating from any other single Latin American nation. These new residents are predominantly male (56.6%) and young. Most are either of working age or less than 18. The proportion of Latino families grew significantly between 1990-2000, reflecting a maturing of the immigration stream into the county.

Geographically, the influx of Latinos is almost exclusively restricted to the eastern portion of the county. Within the five census tracts in the county, the two western tracts have only 32 Latinos out of more than 8000 residents, whereas the two eastern tracts have 2,334 Latinos, or 15% of the total population in the eastern area. The eastern county towns have become increasingly multiethnic. Biscoe, the largest town with 1700 residents, is 23% Latino Mexican. Candor, population 825, is 25% Mexican and 2% other Latino. Finally, Star is 9.2% Latino.

Latino migration stories

While the census data provide quantitative insights into the rapid influx of Latinos into Montgomery County, a deeper understanding of the migration histories and individual choices made by migrants can only be pieced together through qualitative data sources. In this research, the author made several trips to Montgomery County to talk with Anglo and Latino residents and community leaders. While all the conversations were useful and offered particular insights, discussion with Father Ricardo Sanchez, the parish priest at Our Lady of the Americas church in Biscoe was the most important. In an immigrant community with strong religious and cultural foundations centred around the church, Father Sanchez, a Mexican born priest, is perhaps the most important leader in Montgomery County's Latino community as the parish is overwhelmingly Mexican and, of the 400 households in the parish registry, all but seven are Spanish speaking (Sanchez, 2003).

The emergence of east Montgomery County as a Latino migration destination is best explained by three sets of factors operating at varying geographic scales. On a national and international level, the persistent economic turmoil in Mexico has fuelled a modern culture within this Latin nation that fosters and sustains immigration to the USA. In turn, evolving national policies in the USA towards Mexican emigration have alternatively spurred both legal and illegal immigrant streams. Most analysts agree that the most important policy adjustment affecting the new Latino geography was the passage of the Immigration Reform and Control Act of 1986 (IRCA) (Hernandez-Leon and Zuniga, 2000; Durand *et al.*, 2000). The IRCA granted potential amnesty to an estimated 2+ million Mexicans living illegally in the USA. With legalization of residence and the spectre of deportation removed, Mexican-Americans could for the first time move around the country freely. Indeed, Hernandez-Leon and Zuniga (2000) state that IRCA was the single most important cause for the new Mexican diaspora. However, because the legalization affects of the IRCA were phased in between 1987 and 1989, the geographical impacts of the legislation, the declining importance of gateway places and the emergence of the South as a Latino migration destination, did not surface until the 1990 Census.

The second factor informing an understanding of local Latino settlement in the area, is the historic employment opportunities in Montgomery County. The earliest Latino presence in eastern Montgomery County was seasonal farm labour in the 1970s and 1980s. Many of these workers arrived under the H-2 programme. This programme allowed labour contractors to provide farmers with temporary foreign workers during the peak labour demand period of harvest. In theory, H-2 workers returned home following the harvest season. Some did, but others remained in the agricultural work force. Latino (Mexican) workers from Florida and Texas were a small, but persistent segment of the workers who moved through the Sandhills region picking peaches and cropping tobacco. This is a pattern that has been widely evidenced in other parts of the USA during many harvest cycles. The young men and some cases whole clans from single rural villages would travel together. The reasons for this process were varied. Network recruitment as well as social relations between workers played roles in these groupings. According to local informants, in some instances the temporary agricultural employment was transformed in permanent farm-based employment, which is tending fields and orchards after or before the high labour season. Montgomery County's low cost of living, tolerance to outsiders, comfortable rural environment, and the steady jobs made the area an attractive settlement option. Following the gender related migration patterns observed elsewhere, these earliest Latino settlers were primarily working-aged men.

As is the case in other rural communities, Latino labour was soon absorbed into the local industrial labour market. Latinos developed a reputation among local manufacturing employers as hard working and reliable. Local textile and furniture factories, and a large poultry processing plant in nearby Robbins (Moore County) needed a stable and docile labour force. In particular, the Latinos were willing to take on the most undesirable type of industrial labour. Often these jobs are viewed by local whites and African-Americans as inferior and/or undesirable. Thus, they have a higher turnover and a problematic for employers. Today, for example, over 50% of the floor workers at the Citation Foundry in Biscoe, a job characterized as hot, dirty and dangerous, are Latinos.

Because of the tolerance to poor working conditions, less trained rural Mexican migrants are drawn to these jobs. *Campasinos* used to working long hours for low wages, are not discouraged by the tedium and/or discomfort working in a textile weaving room or poultry processing line. Indeed, one recent estimate from Iowa is that among Mexican migrants working in a meat packing plant, 60% to 70% came from agricultural backgrounds (Griffith, 1993).

These *trailblazer* Latinos encouraged friends and family to move from other parts of the USA and Mexico to Montgomery County. Take together, the implementation of IRCA, the existing immigrant community, and the strength of the industrial sector in the mid 1990s made Montgomery County an attractive residential destination for Latinos no longer restricted to traditional gateway states and cities.

Griffith and Kissam (1995) have analysed the Latino penetration of the low wage food processing sectors in the rural south and refer to this accelerating Latino migration into the industrial sector in terms of a 'colonization' process. In the first stage, employed Latinos help find jobs at their workplaces for family and hometown friends. Subsequently, the growing network of workers improves their status within the work place, helping management meet labour needs. And, finally, reciprocity within the growing Latino population allows individual workers to 'move up' as well as protect each other.

The final factor revolves around the social processes that fuel immigration destinations. Perhaps surprisingly to some given the racist reputation of the rural south, eastern Montgomery County does not present an anti-immigrant environment. The existence of an incipient Latino community, low key rural lifestyle, the surplus housing and business building environment, and a strong labour market for low skill occupations within community distance of Montgomery County has fuelled the reputation of the county as a good place to move to.

Attraction of rurality

According to Father Ricardo Sanchez (2003), the most important factors attracting Mexicans to eastern Montgomery County are, in order: the availability of steady jobs; the receptivity of the community to Mexican migrants; the 'high quality of life', which for his parishioners revolves around issues of low crime and educational opportunities; and 'the countryside'. According to Father Ricardo, 'at least 80%' of his Mexican parishioners are from rural Mexico and they regard a rural place as an inviting place to live. This is especially important to Mexicans relocating from southern California ('LA') and urban areas in Florida and Texas.

From discussions with Father Ricardo and a review of the parish planning documents, the migratory trajectories of the Mexican immigrants into Montgomery County can be constructed. Approximately half of the parishioners moved to Montgomery County from other parts of the USA. California and Florida, followed by Texas, were the primary source areas in the 1990s. The influx of Californians, especially from urban southern California, is an important new migration source during the past 4 years.

Among the Mexican immigrants a majority of newcomers are from Guanajuato State, with Veracruz providing the second largest numbers followed by Michoacan and San Luis Potosi. Chihuahua State and Districto Federal were minor contributors to Montgomery County's Mexican migrants. Among the Mexicans moving to the county from other parts of the USA, most had family ties to these states. Thus, a cousin or sister formerly living in Los Angeles or Tampa, but originally from Guanajuato State, may move to eastern Montgomery County at the encouragement of relatives.

Migration impacts on the community

The maturation of the Latino presence in Montgomery County is evidenced by the explosive growth in Spanish speaking children in the Montgomery County public school system and the major inroads by Latinos into the small business sector. During the initial trailblazer phase of Latino migration, single men without families were the demographic profile of the newcomers. By the early 1990s, the stream of Mexican migrants was becoming more balanced, with wives and children joining their husbands/fathers; and intact families relocating to Montgomery County.

The rapidity of this process and the effects on the school system are remarkable. Consider, in the 1994-1995 school year there was no need for a Spanish language 'English as a Second Language' (ESL) teacher in Montgomery County. But as of the 2002-2003 school year, 24.3% of the entire school system are Spanish speakers (Reynolds, 2003). With one exception all of these children attend schools in the east of the county. Most importantly, school planners expect the influx of Latino students to continue to grow. For example, at Candor Elementary in the 2002-2003 school year, 84% of the newly enrolled kindergarten or first grade students were Spanish speakers. At the same school, the 307 Spanish speaking students include 34.5% who were born in Mexico, but 53.4% were born in North Carolina. These birthplace data confirm that growing Mexican families are settling and putting down roots in Montgomery County.

The integration and delivery of services to Latino students in Montgomery County is one of the most easily documented successes of the new migration. School officials report strong parental support and interest in student educational outcomes. Latino students are reportedly well adjusted and behaved. There are few incidents of documented social or ethnic conflict. Most importantly, the end of the year tests results for 2002-2003, show that Latino students as a group out performed their white and African-American peers.

The recent growth in the Latino business sector is also verification of permanence in the migrant settlement. Throughout the towns of Biscoe and Candor, a growing number of small Spanish language, Mexican oriented grocery stores, clothing and jewellery stores, beauty shops, restaurants, and even a CPA, insurance, and immigration advice service have opened. Two bus lines specializing in transit to Texas and Mexico serve these towns. The Latino businesses are modestly capitalizing and occupying space in formerly vacant structures. An empty service station is converted to a Tienda (grocery) or Mexican restaurant (Oscar, 2003). The Latino businesses fill the vacant and often derelict void left by departed Anglo businesses. Indeed, local community leaders are pleased and supportive of Latino entrepreneurship, according to the Mayor of Biscoe (Blake, 2003).

The availability of Mexican made products and services meet the demand created by the immigrant community, while helping foster entrepreneurship within the Mexican community. In turn, the growth of Latino business interests represents tangible inroads by migrants towards greater economic opportunity.

One measure of the newness of the Latino presence in Montgomery County is the pattern of housing tenure. Economic security and commitment to long-term residency are reflected in home ownership. Across Montgomery County home-ownership, combined from 'stick built' homes and mobile homes, is quite high. Nearly, 77% of Montgomery residents live in homes that they own. Among Latinos, however, the rate of ownership is only 43.8%. Economic barriers clearly factor into this low figure, but the recent emergence of Latino families as a significant presence in the community is also likely important.

The housing tenure data provide a starting point for examining the overall economic status of Montgomery County's new Latino community. As a group, the Latino migrants are far more economically and socially disadvantaged than their black and white neighbours. Nearly one-third of all Latinos had 1999 incomes below the official poverty level; and among Latin children the rate was 34%. Not surprisingly, the census data revealed that 53.7% of Latino households had incomes below the county median household income of US$32,903. In a related way, 26% of Latino adults were identified as having employment restricting disabilities in 2000.

A major challenge for these new North Carolinians is language. Among Latinos aged 18 to 64 only 24.6% claim to speak English 'very well'. In contrast, 21.7% claim not to speak English (i.e. 'not at all') and another 40.8% characterize their English proficiency as 'not well'. Even among children, the group best equipped to speak English, 27.9% fall into the 'not well' or 'not at all' categories. Without language skills Latinos are severely disadvantaged in their range of employment opportunities, their capacity to establish social relations outside the Latino community, and live normal lifestyles in an English speaking community.

Conclusion and policy implications

The 1990s marked a new era for Latino settlement in the USA. Spurred by social connectedness, new political freeness, widened employment opportunities, and heightened sensitivity to quality of life issues, millions of Latinos moved to parts of the nation that were until recently beyond their residential reach. The speed and scale of this non-traditional process is such that few scholars have examined it.

Regionally, the South has emerged as new target for intra-national and international Latino migrants. A number of southern cities and their suburbs have been labelled the new gateway cities owing to a growing concentration of new Latino residents during the past decade. However, rural communities in the South are also experiencing significant Latino immigration and, indeed, rural areas may feel greater impacts because the influx of newcomers is proportionately greater.

The data for North Carolina show that Latino immigration in the last ten years has continued to be guided by social processes as well as new, evolving market place opportunities. The existence of even small Latino communities is a powerful magnet for new immigrants. For Latinos moving from rural places or those with ties to rural lifestyles, the presence of large-scale commercial agriculture or low skill manufacturing jobs are attractors. The combination of trailblazer migrants and job opportunities has a powerful synergistic effect, as shown in the Montgomery County case. Place prosperity alone is not an effective predictor for Latino population growth.

In the short term, Latino newcomers provide challenges for rural southern communities. The provision of services and the day-to-day adjustments to cultural differences can create social tension. But over the mid to long term, the influx of Latino newcomers may be an important new ingredient for sustaining community in marginalized rural areas. In a period when rural population growth is flat or modestly increasing, Latino migration infuses critical labour and capital into bypassed rural communities. While Latinos present assimilative challenges, they hold cultural values quite similar to rural whites and African-Americans. The importance of family, community, faith and individual initiative are core values that resonate across all three groups. They provide entrepreneurial talents and offer capital for local business investment.

References

Blake, J. (2003) Interview by author. Mayor of Town of Biscoe, Biscoe, NC, June 2003.

Cravey, A. (1997) Latino labour and poultry production in rural North Carolina. *South-eastern Geographer* 27, 295-300.

Durand, J., Massey, D.S. and Charvet, F. (2000) The changing geography of Mexican immigration to the USA: 1910-1996. *Social Science Quarterly*. 81, 1-15.

Griffith, D. (1983) *Jones's Minimal, Low Wage Labour in the USA*. State University of New York Press, New York.

Griffith, D. and Kissam, E. (1995) *Working Poor. Farmworkers in the USA*. Temple University Press, Philadelphia.

Guthey, G. (2001) Mexican places in southern spaces: globalization, work and daily life in and around the North Georgia poultry industry. In Murphy, A.D., Blanchard, C. and Hill, J.A. (eds.) *Latino Workers in the Contemporary South*. University of Georgia Press, Athens, pp.57-67.

Hernández-Léon, R. and Zúñiga, V. (2000) Making carpet by the mile: the emergence of a Mexican immigrant community in an industrial region of the US historic South. *Social Science Quarterly* 81, 49-66.

James, F.J. Romine, J.A. and Zwanzig (1998) The effects of immigration on urban communities. *Cityscape* 3, 171-192.
Johnson-Webb, K.D. and Johnson, J.H. (1990) North Carolina communities in transition: the Hispanic influx. *The North Carolina Geographer* 5, pp.21-40.
Murphy, A.D., Blanchard, C. and Hill, C.A. (eds.) (2001) *Latino Workers in the Contemporary South.* University of Georgia Press, Athens.
Oscar (2003) Interview by author. Tienda Owner, Biscoe, NC, June 11.
Reynolds, T. (2003) Interviews by author. Director of Curriculum and Instruction, Montgomery County Schools, Troy, NC, June 2 and 11.
Sanchez, Fr. R. (2003) Interviews by author. Parish Priest, Our Lady of the Americas Church, Biscoe, NC, June 2 and 11.
Stull, D.D., Broadway, M.J., and Griffith, D. (eds.) (1995) *Any Way You Cut It: Meat Processing and Small Town America.* University of Kansas Press, Lawrence.
US Census Bureau (2002) *Census Fact Finder.* US Census Bureau (accessed 2002). Available from http://www.census.gov
US Census Bureau (2004) Table 3. Annual Estimates of the Population by Sex, Race, and Hispanics or Latino Origin for North Carolina: April 1, 2000 to July 1, 2003. US Census Bureau, Washington DC.
Zúñiga, V. and Hernández-Léon, R. (2001) A new destination for an old migration: origins trajectories, and labour market incorporation of Latinos in Dalton, Georgia. In: Murphy, A.D., Blanchard, C. and Hill, J.A. (eds.) *Latino Workers in the Contemporary South.* University of Georgia Press, Athens, pp.126-136.

Chapter 19

Finding the 'region' in rural regional governance

Ann K. Deakin

Introduction

Both geographers and regional governance advocates in the USA have neglected to consider the concept of 'region' within the context of regional governance. Researchers in the UK have, on the other hand, questioned the appropriateness of regions as units of governance, and, to this end, have stated the difficulty of recognizing, understanding and defining regions (see, for example, Jones, 1998; MacLeod, 1998; Marden, 1997). A similar examination has yet to occur among geographers in the USA. Further, practitioners and advocates of regional governance have failed to recognize that 'region' is not necessarily synonymous with pre-existing administrative units, such as counties. Nor have they considered the applicability of a typically urban solution to a rural setting. The goal of this chapter is to inform the regional governance approach through the new regional geography and the analytical and integrative potential of GIS (Geographic Information Science).

Governance refers to a style of governing where the boundaries between and within public and private sectors are blurred (Stoker, 1996, cited in Goodwin, 1998). The shift from government to governance is associated with the decline of the federal government's direct management and sponsorship of social and economic projects (MacLeod and Goodwin, 1999). The region that is typically staked out for regional governance typically coincides with county boundaries. For example, cities and counties share some services or perhaps even consolidate in an effort to curb such economic and social ills such as urban sprawl, racial segregation and high taxes, among others.

The promulgation of the region as a solution to the economic and social ills of urban centres in the USA has been ongoing for some time. For example, in 1962 the city of Nashville, Tennessee was consolidated with surrounding Davidson County. The resulting Metro government consolidated the functions performed by the city and the county, with some compromises to appease residents living outside the former city limits of Nashville (Rusk, 1995). Although there are not many unified governments, there are many instances where cities and counties have decided to share some services in an effort to cut costs and ultimately improve the economic situation of the 'region' and enable participation in the world economy.

Regional governance in the USA has evolved as primarily a remedy to *urban* problems and is often referred to as '*urban* regional governance'. The project described in this chapter looks at a predominantly *rural* area, Chautauqua County, which has decided to try the same remedy that has, as Blair *et al.* (1996) observe, yet to be proven in an urban context. On one hand, Chautauqua County's political and business leaders consider it to be on the leading edge with its efforts towards regionalizing governance in a rural context. On the other hand, it may be taking a risk applying an unproven urban remedy to rural problems. Numerous scholars from a variety of disciplines have written about the differences between rural and urban (see, for example, Berry *et al.*, 2000; Fitchen, 1991; Galston and Baehler, 1995; Reynolds, 1995). Yet those differences are being ignored or minimized as urban solutions are applied to rural problems.

While policy makers and the academicians who advise them are ardently pursuing regional governance, this change in the structure of local government does have critics. In spite of the need for more thoughtful consideration, proponents of regional governance continue to forge ahead without altering their course. Blair *et al.* (1996) caution that such restructuring is without 'strong empirical support and should not be considered panaceas for central city decline' (p.345). Deas and Ward (2000) observe that regionally based economic development efforts in North West England have exacerbated the tension between the city of Manchester and the surrounding North West region due in no small part to a lack of regional consciousness. Similarly, such restructuring should not be assumed to be a panacea for troubled rural areas either.

In the context of rural Britain, Goodwin (1998) speculates that the rhetoric surrounding governance in general, 'stress[es] notions of inclusion and empowerment, but given the paternalistic and 'non-political' traditions dominant in many rural areas, there may be a tendency to involve only key actors in the new structures of governance and marginalize the wider community' (p.11). In an urban setting, MacLeod and Goodwin (1999) have similarly observed that London's shift from government to governance may compromise its ability to function on behalf of most of its residents. Furthermore, the evidence worldwide is that the human trend is towards devolution (World Bank, 2000) rather than mergers of places and removal of boundaries. Within the USA, for example, there are secession movements afoot in the cities of Los Angeles, Miami and New York (*The Economist* 26 February 2000, pp.40-42). Newman and Paasi (1998) observe the institutionalization of boundary studies as evidence that boundaries, and the separation they imply, are more, rather than less, important.

Hirst and Thompson's (1995) assertion regarding the staying power of state boundaries could be applied to the local or administrative level, as well. For the average person, one's municipality is also, according to Hirst and Thompson (1995), a 'community of fate'. The globalization of the economy and all that it entails should not be assumed to be a death knell for those communities (Hirst and Thompson, 1995). Newman and Paasi (1998, p.197) observe that administrative boundaries, 'have a far greater impact on daily behavioural patterns of most individuals than do national and international boundaries' and that, 'most of life's functions take place within the context of local boundaries'. Administrative or local boundaries would appear to be undergoing a change in meaning rather than disappearing altogether.

Similarly, Huntington's (1996) 'Who are you?' in terms of cultural identity at the state level is perhaps more appropriately asked at the local level due to globalization. If governance is only for economic development, then by all means boundaries for wealth tend towards larger territories. If, however, we do not want to lose the cultural identity to which we cling then we must not erase the boundaries that make smaller territories possible nor should we preclude the potential for new boundaries to evolve.

The reason for being concerned with the accurate use of region as applied to governance is that there is a danger of marginalizing the wider community. Referring to a county as a region dismisses the potentially vast diversity and regions that exist within that county. A failure to recognize the regional differences that may exist within a county means that everyone is painted with the same brush whether it is appropriate or not. This is of particular concern in Chautauqua County (which will be described in more detail below), where potential for regional variation is high. One example of this potential is the major watershed divide that separates the county. In Chautauqua County and counties like it forcing residents from potentially different regions within the same county to don the same regional identity may affect their level of comfort and belonging to a particular region.

Identity with regions is a very important and emotional attachment. People thrive on comparing their part of geographic space, regardless of scale, to the geographic space of others: northern Californians versus southern Californians and Easterners versus Midwesterners, for example. Association with geographic space is a vital part of a person's identity, in general, and, therefore, not to be dismissed. The purpose of this project then is not to determine whether regional governance works or not or whether it is good or bad. Rather, the purpose of the project is to provide evidence that caution, consideration and sensitivity may be in order if regional governance is to be implemented at the county level. That evidence will consist of an exploration of a variety of data sets to determine the regions coincident with the county.

Regional governance

The use of the word 'regional' in 'regional governance' implies that governance will occur at the level of a region. For geographers, a region is typically defined as an area distinguished by one or more unique characteristics or a larger-sized territory that encompasses many places. All or most of these places share similar attributes in comparison with the attributes of places elsewhere. Regions can occur at many different scales using many different attributes. There is, for example, the world region referred to as Western Europe and the region of the USA referred to as the Midwest. How does this type of region translate to the region of regional governance? In practice, it really does not. In the USA, governance at the regional level means governance at the county level. Technically, some geographers might refer to a county as a formal region, which is characterized by some legal and administrative characteristics that may distinguish one county from another to varying degrees. Most geographers would, however, argue that a county is merely, to use Entrikin's (1994) words, 'a handy container' rather than a region.

A region represents a process that is 'continually reproduced and transformed in and through the practices of individuals and institutions at a variety of spatial levels' (MacLeod, 1998, p.833), where those practices reflect cultural and historical identities. In addition, the idea of a region is also tied to the physical environment and is not rigid or closed (Allen *et al.*, 1998). Adapting Paasi's (1996, p.xviii) definition of the reconstructed regional geography, regions can be differentiated by 'the social construction of [their respective] territorial communities and consciousness and identities on different spatial scales'. So, in comparing a region to a county, the former might be argued as emerging through predominantly internal forces while the latter emerges through predominantly external forces.

Despite these inherent differences between administrative units such as counties and regions, regional governance advocates continue to use region to mean county. David Rusk, the former mayor of Albuquerque and arguably the foremost advocate of regional governance, justifies this regional or county approach as the way to address urban sprawl and social injustice (Rusk, 1995). For Rusk, consolidating the city, or at least some of its services, with the surrounding suburbs up to the county boundary means the following: (i) access to low-skilled jobs can improve, (ii) fiscal equity can be achieved and (iii) greater economic integration can be promoted. By increasing the size of the areal unit, the concentrations of social ills are less apparent as they are diluted across a larger area.

In a predominantly rural context, the same reasoning is being applied. Small cities, villages, towns, or school districts that have the greatest concentrations of poor people within a rural county, for example, could be consolidated with that county in an effort to share the wealth with those places needing it most. Again, this consolidation can mean sharing a few services or even just a vision for growth management to the elimination of boundaries. The latter, in most cases, requires action at the state legislature level.

Ironically, while the proponents of regional governance decry the need for so many local governments, they unwittingly reinforce these seemingly arbitrary political territories by assuming that the county territory is somehow more logical merely because it is bigger. In Jane Jacobs's (1961, p.410) words: 'A region…is an area safely larger than the last one to whose problems we found no solution.' Bruce Katz (2000, p.2), the editor of a recent collection of essays on regional governance, states that 'the term *region* is certainly ill-defined in popular usage' (Katz, original emphasis). Just as the 'Standard Region' may be 'assuming an increasingly hollow ring' as the basis for administration in the UK (MacLeod, 1998, p.835), so too might the 'County' in the USA.

There is tremendous historic, geographic, social, political, cultural and economic diversity within the boundaries of a county. Counties were demarcated as 'handy containers' for administration, while regions are not (Murphy, 1991). Counties were, by and large, imposed on the landscape, perhaps dictated by geography, perhaps not. It is not the purpose of this study to assess whether regions are more or less 'natural' than counties, nor is the purpose to claim that there are any existing criteria capable of classifying regions such that they would coincide perfectly (Bourdieu, 1993). This study will, however, explore the way regions are produced in comparison to counties as this has implications for governance. This exploration will include applying the criteria others have used to determine the characteristics that are most emblematic of a particular region rather than attempting to search for an all-inclusive set of regional characteristics. The nature of regions and efforts to explain them are discussed further below.

The new regional geography

Of what use can the new regional geography and a GIS environment be in the context of regional governance? First, what is the new regional geography, particularly when compared to 'old' or 'traditional' regional geography? In general, regional geography has meant the description of areas distinguished by one or more unique characteristics. Dissatisfaction with this paradigm of spatial separatism was evidenced in the debates on the new regional geography occurring in the 1980s (Paasi, 1991). Thus, more recently, regional geography has become the analysis of environmental and human factors that produce territories with distinctive landscapes and cultural attributes. The perspectives on this new regional geography have scrutinized the human factors and can, according to Gilbert (1988), be distilled to three approaches:

1. Focus on the spatial organization of social processes that are associated with the political-economic basis of regions.
2. Focus on the region as a setting for social interaction.
3. Focus on the region as the cultural relationship between the residents and the land.

Anssi Paasi's (1991) perspective builds on a combination of these three, strengthens the role of culture and adds the historical constitution of regions. Thus, in order to arrive at a reconstruction of the new regional geography, Paasi, in a sense, deconstructs regions to find the factors that have been missed. For Paasi (1996), then, the new regional geography should also focus on the historical process of region construction, and 'the social construction of territorial communities and consciousness and identities on different spatial scales' (p.xviii). It is assumed that the process is ongoing and the boundaries may be blurred. Murphy (1991, p.28), not unlike Paasi, is concerned that the new regional geography is not sufficiently exploring the nature and significance of the region chosen for study. The region must be seen as an integral part of the study, not merely as a framework for the study.

Paasi (1996) has applied his approach to the new regional geography in an effort to understand the contested form of the Finnish-Russian border and its impact on local and national identity. MacLeod (1998), describing Paasi's framework as a historical and spatially sensitive mode of analysis, uses it to conceptualize modern Scotland as a distinct territorial unit and understand its distinctive institutional expression for governing economic development. Murphy (1991) provides an example in the ethnolinguistic controversy in Belgium. He argues that it is imperative to consider the evolution of Brussels as a regional unit in order to understand the nature and significance of language group identity and interaction there. Murphy further states that it is 'only by probing the nature and significance of the spatial framework within which interactions and mediations [such as governance] occur can we hope to capture the inextricable interrelationships between society and space' (p.32). Murphy's examination must also occur if an area is to be adequately engaged in the process or merely seen as a setting for governance.

In addition to the environmental factors, it is Paasi's approach to the human factors of the new regional geography that are used to guide this study. In summary, the new regional geography, as it is applied here, is defined as the analysis of environmental and human factors that produce territories with distinctive landscapes and cultural attributes. The environmental factors include terrain, weather, vegetation, and drainage, and the human factors include historical, political-economic, transport and cultural factors.

Chautauqua County, New York

The geographic laboratory for this study will be Chautauqua County, a predominantly rural county in the southwest corner of New York (see Fig. 19.1). Chautauqua County covers an area of 2800 sq km, is comprised of 44 municipalities, and has a population of approximately 140,000 people. This county was chosen because of its intriguing characteristics within the context of regional geography and regional governance. Chautauqua County borders Pennsylvania on two sides and includes part of the Cattaraugus Seneca Nation of Indians Reservation. Its northern boundary coincides with the Erie lakeshore, and, from the northeast corner, it is only an hour's drive to the Canadian border and Niagara Falls.

Like many counties, Chautauqua County is considered part of a number of 'regions,' depending on the context. Chautauqua County belongs, for example, to at least four different 'regions'. At the national level, it is considered part of the Appalachian region. Within Appalachia, it is part of the sub-region called Southern Tier West, which refers to the sub-region's location relative to the rest of New York. Within New York, Chautauqua County is part of the region referred to casually as Western New York, in an effort to distinguish it from the parts of the State that are better known, such as New York City and

Upstate New York. According to the Institute for Local Governance and Regional Growth at SUNY (State University of New York) at Buffalo, Chautauqua County is part of the Buffalo-Niagara Region, which includes eight contiguous western New York counties and two regional municipalities in southern Ontario.

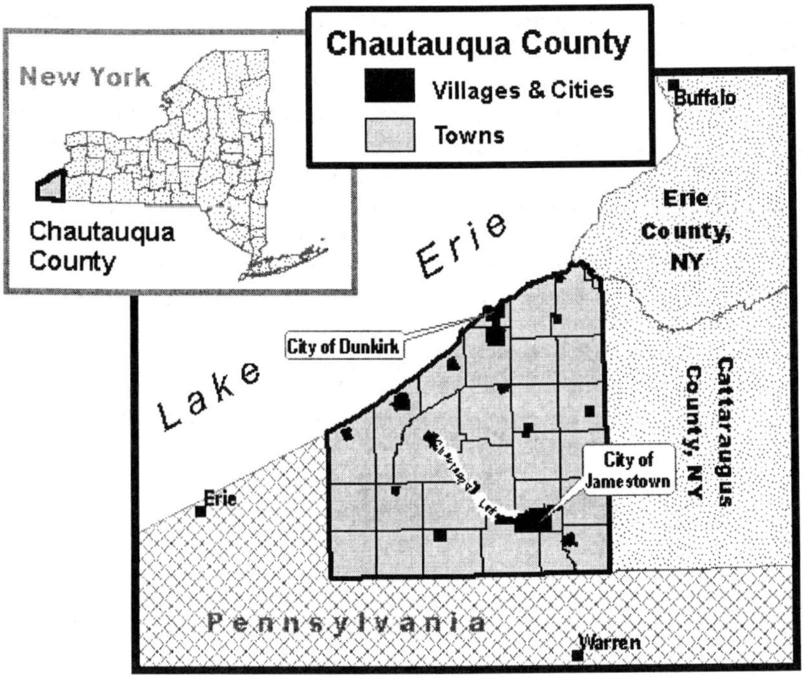

Fig 19.1 Chautauqua County, New York.

Within Chautauqua County, many residents distinguish between the 'North County' and the 'South County'. Indeed, until recently there were two County Chambers of Commerce: one for the 'North County' and one for the 'South County'. In the ongoing movement to regionalize government to the county level, the Chambers decided to merge and there is now one County Chamber of Commerce; its single office is located in the City of Dunkirk, the former North County Chamber of Commerce office. (It is too early to assess what impact this may have in terms of the Chamber's ability to represent and appeal to businesses countywide.) The County's Community College has both a North and a South Campus. County summits and public meetings are typically duplicated to appeal to both South and North County residents. There is no documented boundary between the North and South County and no effort to date has been made to locate such a boundary. This boundary 'decision' has been left up to the businesses and students, in terms of what businesses participated in the North versus the South County Chambers of Commerce or what students attend the North versus the South County Community College Campus. Thus, the boundary is undoubtedly a fuzzy area, or perhaps a region itself, that depends, by and large, on the perceptions of those making the 'decision'. It is hypothesized that this perceptual divide may, in reality, coincide with the Erie Escarpment, a major watershed boundary. It is anticipated that this perceptual partitioning of the county may become more clear and measurable as a result of this study.

In addition to its rural character and potential for regional variation, the current political and economic condition of Chautauqua County also makes it a fertile test site. As stated earlier, political and business leaders in the County have been greatly influenced by the governance approach in general and regionalizing governance in particular. With regard to the former, the county legislature has adopted performance-based budgeting due primarily to its success in Sunnyvale, California. Other counties in New York are watching with great interest and some admiration as the process moves forward both rapidly and smoothly. Efforts at regionalizing governance in the form of shared services, the consolidation of school districts, and municipalities is making very little progress, due in no small part to local interests within the County. A recent effort to consolidate two school districts in the county, for example, was voted down by those who would be most affected by it – the residents. Sharing services has met with slightly more success. Thus far, the suggested merger of village and town governments has met with very stiff resistance.

Although Chautauqua County is the particular area of study in this chapter, it is anticipated that the procedures followed here could be followed elsewhere. The overall framework used to determine the region or regions in evidence in Chautauqua County could be applied to other similar geographic areas and perhaps even at different scales where a regional governance approach is under consideration. Chautauqua County, in spite of its unique qualities, is very much like other parts of the USA. It is a rural area actively pursuing strategies such as regional governance as a remedy for economic decline.

Methods

The method used to identify the region or regions that coincide with Chautauqua County is, by and large exploratory. Listed below are the data that will be explored and analysed in a GIS environment for the purpose of this study.

- terrain
- drainage
- agriculture
- ecoregions and subregions (Bailey *et al.*, 1994)
- questionnaire responses
- school districts
- ancestry
- transportation, roads
- evolution of administrative boundaries since Holland Land Company Purchase

Perhaps the most difficult aspect of Paasi's new regional geography approach to incorporate is that of local identity. As Paasi (1996) observes, regional identity is not a topic that is routinely discussed in casual conversation by most people. Rather it is typically expressed through everyday activities and affiliations. Further, Bourdieu (1993, p.220) stresses that regional identity is not easily 'objectified' because in social practice it is an act of:

> 'perception and appreciation, of cognition and recognition, in which agents invest their interests and their presuppositions, and of *objectified representations*, in things (emblems, flags, badges, etc.) or acts, self-interested strategies of symbolic manipulation which aim at determining the (mental) representation that other people may form of these properties and their bearers' (original emphasis).

A proxy for trying to understand identity within Chautauqua County was tested at the outset of what will eventually be a long-term study. Personal interviews were conducted with residents asking questions about their travel behaviour in and around the county. These interviews include questions regarding their town or village of residence, length of residence, where they shop, where they socialized, where they seek medical attention, among others. Twenty residents from across the county were interviewed. The residents agreed to volunteer 30 minutes of their time to answer the questions and consider the maps. Appointments were scheduled with the participants in advance and the interview conducted at a location that was known to and comfortable for the participants. For example, many of the interviews were conducted in the participants' homes. As the study progresses, it is anticipated that focus groups may be formed throughout the county to discuss the specific notion of regional identity. Prior to participating in the personal interview, participants read a description of the study and an informed consent document. They were asked to provide written informed consent for their participation. A brief survey asking for basic demographic information was also administered immediately following the personal interview session.

The personal interview data were geocoded by residential street address and added to the human and environmental data in the GIS environment. An overlay analysis was used to detect the region or regions that emerged. The region or regions were interpreted using the new regional geography. These data were also analysed using exploratory and descriptive statistics.

Initial results

Even a cursory visual inspection of the physical geography of Chautauqua County suggests that two regions coincide with it (see Fig. 19.2). Water flowing down the north side of the Erie Escarpment finds its way to the St. Lawrence River, while water flowing down the south side connects with the Allegheny-Ohio-Mississippi River system. While the escarpment is a minor landform in comparison to other 'divides,' such as the Rocky Mountains, it is significant enough to be a climatic, ecological, and agricultural boundary, marking the transitional zone between the Allegheny Plateau and the Great Lakes Basin, or in the terms of Bailey *et al.*'s (1994) ecoregions, the Western Glaciated Allegheny Plateau Section and the Erie and Ontario Lake Plain Section.

The Erie Escarpment rises from approximately 260 m to about 400 m within 2.7 km. At first the slope is gradual at about 3%, but then quickly increases to 13% at about 300 m. To the north of the Erie Escarpment is the Lake Erie Plain, which is a lowland belt that runs in a southwest to northeast direction along the shore of Lake Erie. This plain rises from an elevation of 174 m at the wave-cut shale bluffs of the Lake Erie shoreline to about 260 m at the base of the Erie Escarpment approximately 4.2 km inland. The Lake Erie Plain is famous agriculturally for its fruit crops; it is the greatest grape producing region in the US outside of California (Baldwin, 2001). The grape vineyards stop abruptly as the Escarpment becomes steeper. Once the Allegheny Plateau is reached, the agriculture shifts abruptly to Christmas tree plantations and small dairy farms (see Fig. 19.2).

The Allegheny Plateau is characterized by highlands covering most of New York south of Interstate Highway 90 (the New York State Thruway) and west of the Hudson River, northern and western Pennsylvania, West Virginia, and eastern Ontario. It constitutes the northern half of the Appalachian Plateau. Locally, the Allegheny Plateau is described as an area of 'rolling valleys' that have been cut by erosion. The Wisconsin Glacier covered much of this area, creating a number of glacial landscape features, such as kettles and kames, drumlins, eskers and moraines, among others. Chautauqua Lake is a glacial feature that is not unlike the Finger Lakes of Central New York.

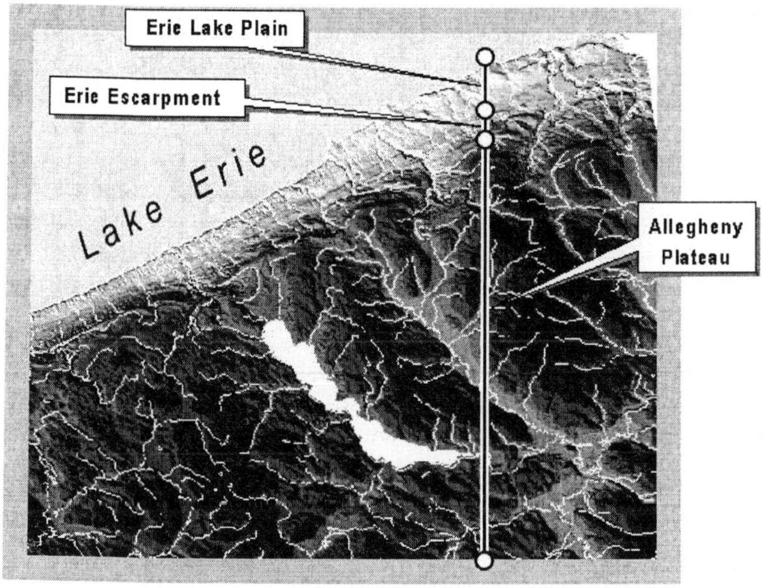

Fig. 19.2 Topography and drainage of Chautauqua County, New York.

Although predominantly rural, Chautauqua County does have two small cities, Dunkirk (2000 population: 13,131), which is north of the Escarpment, and Jamestown (2000 population: 31,730), which is south of the Escarpment at the southern tip of Chautauqua Lake. Both of these cities are too small to provide much in the way of urban amenities such as can be experienced in Buffalo and Erie, Pennsylvania. Buffalo (2000 population of Buffalo-Niagara Falls Metropolitan Statistical Area: 1,170,111) is approximately 80.5 km to the northeast, and Erie (2000 population of Erie: 103,173) is approximately 80.5 km to the southwest. Thus, the pull factors of both of these larger cities undoubtedly play a role in the formation of regions and regional identity in the form of residents' travel behaviour within Chautauqua County.

More data, particularly in the form of personal interview or focus group data, should be captured to state with any certainty what region or regions might coincide with Chautauqua County, there are some suggestive results. The Erie Escarpment clearly divides the County into two regions in terms of its physical geography. Are those regions evident in terms of human geography? The partitioning of Chautauqua County into municipalities occurred without much consideration of the natural environment. The municipal boundaries roughly following the systematic rectangular survey design of the township grid, which was used widely by the federal government as new territories were opened for development following the Revolution. The deviations that evolved from the original survey conducted by Joseph Ellicott on behalf of the Holland Land Company show some consideration for the 'irregular' physical environment. Note, for example, the shared borders of the municipalities of Westfield, Chautauqua and Portland, which follow the Erie Escarpment and the gorge carved by the Chautauqua Creek.

Chautauqua County was carved out of the Holland Land Purchase and opened to settlement in 1800. Unfortunately for the Holland Land Company, land sales were far from brisk. Thus, the Company financed land surveys to shape the wilderness 'into an orderly geometry of townships and lots; company villages were established as centres of economic and social activity; company-constructed roads [would connect] settlers with one another and the outside world' (Wyckoff, 1988, p.2). This 'orderly geometry' was, of course, representative of the abstraction of space begun in the New World by European colonists when presented with 'a vast, distant, unknown, and novel space' (Sack, 1986, p.131).

When settlement of the Holland Land Purchase began, the Purchase area was part of Ontario County. In 1802, the New York State Legislature agreed to a proposal that created Genesee County, which coincided with the boundaries of the Purchase area. In 1808, Genesee County was divided into four counties, the southwestern portion of the purchase became Chautauqua County to shorten the distance residents needed to travel to the county seat to conduct business and give settlers more control over local affairs and revenues. Thus, at least in the case of Chautauqua County, the county as a defined area has appeared through the efforts of external forces acting, to some extent, in the interests of local residents.

In terms of transportation routes, the Lake Plain enjoys the most ease of transportation. Interstate 90, which connects Boston with Seattle, parallels the Lake Erie shore as do US Routes 5 and 20, which are two-lane highways. It is, however, more difficult to travel north and south. There are a very few north-south routes that directly connect the Lake Plain with the southern portion of the County. New York State (NYS) Highway 60 is one such two-lane route. This is a well-travelled route by those commuting between Jamestown and Dunkirk.

In terms of railroad lines, there are three lines that connect the east to the west coast and also parallel the Erie Lake shore. These lines carry both freight and passenger trains. The towns and villages along the rail lines used to all have active stations, but most are now inactive and house local museums or provide local historical interest. There are two short lines that run southwesterly connecting Buffalo to Gowanda and Gowanda to Jamestown, the Buffalo Southern Railroad (a switching line) and the New York and Lake Erie Railroad/Trillium Railroad (a handling line). These lines are typically only used for freight. Every few years there may be an excursion train that runs on the New York and Lake Erie Railroad/Trillium Railroad for a special event. To travel by air, the closest hub is Buffalo, although Jamestown does offer some connecting service to larger cities.

While the administrative units were, for the most part imposed on the landscape, the actual settlement of the various European nationalities indicates that the physical environment had some influence. The Scandinavians, Swedes in particular, located on the Allegheny Plateau, while the Poles located along on the Lake Plain. The Italians also appear to be found in higher percentages on the Lake Plain, with the exception of the city of Jamestown and the surrounding town of Ellicott. The Hispanic population, which has immigrated primarily from Puerto Rico and Mexico in search of agricultural work, can be found on the Lake Plain in close proximity to the grape vineyards.

An initial inspection of these data are representative of the ethnic landscape of the county as one travels it. There are a number of Polish clubs (Dom Polski, Kosciuszko, First Ward Falcons, Moniuszko) in the city of Dunkirk. Polish surnames are very common in the area. In the southern part of the county, Swedish surnames are common and there are a number of shops that specialize in Scandinavian goods and crafts, such as rosemaling – the Norwegian art of floral painting. Other European nationalities, such as German and British are more ubiquitous throughout the county.

A profile of the interview respondents and their travel patterns

Of the 20 people who participated in the pilot portion of the study, 12 are female and 8 are male. They range in age from 18 to 75 years, with an average age of 35 to 44 years. Their annual household income was between $10,000 to over $100,000, with an average income of $42,000. Seventeen of the respondents are white, one is black, and two are Latino. Eighteen use a private car as their primary means of transportation, while two walk or use public transportation. Fourteen of the respondents were born in Chautauqua County, and the average length of residence is 35 years.

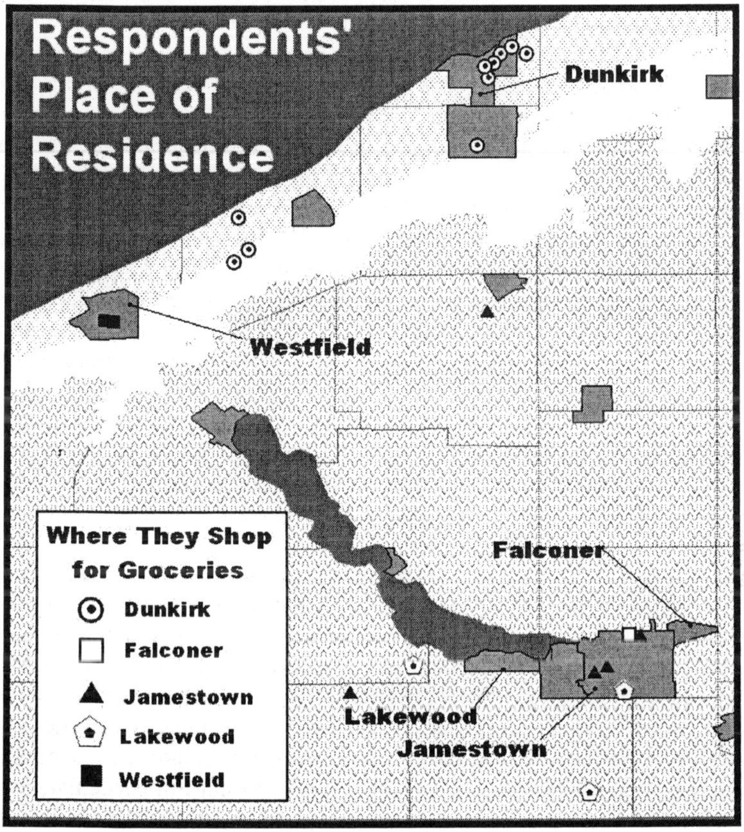

Fig. 19.3 Respondents' place of residence and where they shop for groceries.

In general, the respondents' travel patterns appear to be dictated by physical geography. Of the 20 respondents, eight of the ten who live on the Lake Plain also work on the Lake Plain, one works from home, and the other commutes to the Allegheny Plateau. Eight of the nine who live on the Allegheny Plateau also work in that area and one commutes to the Lake Plain. Similarly, all respondents shop for groceries with the geographic section where they live as can be seen in Fig. 19.3. For example, those that live on the Lake Plain shop for groceries there, too. In terms of large purchases, such as a car, furniture, or refrigerator, the respondents living on the Lake Plain were willing to travel further, while the respondents living on the Allegheny Plateau focused on the city of Jamestown and its suburbs. The Lake Plain respondents chose, by and large, to travel to the cities of Erie, Pennsylvania or Buffalo to make large purchases. One Lake Plain respondent said she would journey to Jamestown to make a large purchase.

Most respondents stated that they socialized regularly with people who live in the same geographic area. One respondent each, from the Lake Plain and the Allegheny Plateau, stated that (s)he socialized with people who live in the other geographic area. Generally most respondents seek entertainment in the form of dining out and going to the cinema within their respective geographic areas. The respondents living in the Lake Plain are willing to journey to either Buffalo or Erie in search of entertainment, while those living on the Plateau appear to have sufficient options in the Jamestown area. What is, perhaps, telling in terms of the division of the county is that in spite of the fact that Jamestown is closer (48 km) to many of the respondents on the Lake Plain, they choose to journey further to either Erie (80.5 km southwest) or Buffalo (80.5 km northeast). Only one respondent on the Plateau said he would go to Erie for entertainment, and he lives in the extreme part of the county, just 2 km from the Pennsylvania state line.

Summary and preliminary conclusions

The purpose of this chapter is to contribute to the present state of knowledge in the areas of regional governance, the new regional geography, and regional identity through an interdisciplinary case study. The project's effort at identification of a region or regions in Chautauqua County responds, in part, to the work done by researchers in the UK to recognize, understand and define regions in the context of regional governance. This aspect of the study also serves as another application of the recent theoretical discussions surrounding the new regional geography. This study also provides regional governance advocates and practitioners with quantifiable data that can be used to implement their approach more sensitively or perhaps even reconfigure it.

It is anticipated that future work will include significantly more personal interviews and/or the use of focus groups. Focus groups may be useful in this study over interviews in order to probe the significance of regional identity without the leading nature of structured questions. Interviews emphasize questions and responses, while focus groups rely on interaction among the participants based on the topic or topics that are supplied by the researcher (Morgan, 1997, p.12). The use of focus groups is also appealing in the minimization of bias (Glynn *et al.*, 1999) because the 'participants talk to one another in their own language, rather than simply reacting to the questions and language of an interviewer' (Conover *et al.*, 1991, p.805).

Thus, the addition of a focus group component to the study would seek to understand regional identity through participants' discussions of more tangible topics such as the following, for example:

- how they would describe a region in general and their region in particular;
- what their criteria are for distinguishing one region from another;
- whether there are any symbols that they consider representative of their region;
- where they conduct their everyday activities;
- where their friends and family live;
- where they feel at home;
- what obstacles discourage them from going from one part of the county to another.

In addition to the use of focus groups to expand this study, the data analysis could move beyond the exploratory to the confirmatory level. For example, the analysis will include a search for spatial clusters of block groups or municipalities that are homogeneous and an identification of the significant contributing variables to the existence of the region(s).

As stated earlier, more data must be gathered to determine with any certainty the approximate location(s) of the region or regions that coincide with Chautauqua County. For now, though, it is evident that the potential for two or more regions exists. At least two regions are defined by physical geography where the Lake Plain transitions to the Allegheny Plateau via the Erie Escarpment. The human geography, with some exceptions, seems to bear this out and perhaps suggest evidence for the existence of more regions. Rural regional governance advocates would, therefore, be remiss in proceeding to 'regionalize' governance to the county level without any consideration for the existence of those regions. To do so is to ignore the rich variations that exist and the identity its residents use to navigate their everyday lives. This is not to suggest that efforts to regionalize rural governance cannot be made. Rather, it is a call to such advocates to look beyond a rural County's fiscal condition in an effort to implement a solution – regional or otherwise – that is sensitive to local concerns and is, therefore, tailored to meet them.

Acknowledgement

The author gratefully acknowledges the assistance of Maclain Nichols (Executive Director, Center for Rural Regional Development and Governance at the SUNY Fredonia).

References

Allen, J., Massey, D. and Cochrane, A. (1998) *Rethinking the Region*, Routledge, London.

Bailey, R.G., Avers, P.E., King, T. and McNab, W.H. (1994) *Ecoregions and Subregions of the USA. Map Prepared for the US Department of Agriculture, Forest Service. US Scale 1:7,500,000; Colored.*. Geological Survey, Washington DC.

Baldwin, M. (2001) *Natural History Atlas to the Chautauqua-Allegheny Region.* Roger Tory Peterson Institute of Natural History, New York.

Barris, N.J. (1990) The formation of Chautauqua County Towns 1804–1918. *Chautauqua Genealogist* 13, 26-30.

Berry, K.A., Markee, N.L., Fowler, N. and Giewat, G.R (2000) Interpreting what is rural and urban for western US counties. *The Professional Geographer* 52, 93-105.

Blair, J.P., Staley, S.R. and Zhang, Z. (1996) The central city elasticity hypothesis: a critical appraisal of Rusk's theory of urban development. *Journal of the American Planning Association* 62, 345-353.

Bourdieu, P. (1993) *Language and Symbolic Power.* Harvard University Press, Cambridge, Massachusetts.

Conover, P., Crew, I. and Searing, D. (1991) The nature of citizenship in the USA and Great Britain: empirical comments on theoretical themes. *Journal of Politics* 53, 800-832.

Deas, I. and Ward, K.G. (2000) From the 'new localism' to the 'new regionalism'? The implications of regional development agencies for city-regional relations. *Political Geography* 19, 273-292.

Entrikin, J.N. (1994) Place and region. *Progress in Human Geography* 18, 227-233.

Fitchen, J.M. (1991) *Endangered Spaces, Enduring Places: Change, Identity, and Survival in Rural America.* Westview Press, Boulder, Colorado.

Galston, W.A. and Baehler, K.J. (1995) *Rural Development in the USA: Connecting Theory,Practice, and Possibilities*. Island Press, Washington D.C.

Gilbert, A. (1988) The new regional geography in English and French-speaking countries. *Progress in Human Geography* 12, 208-228.

Glynn, C.J.; Herbst, S., O'Keefe, G.J. and Shapiro, R.Y. (1999) *Public Opinion*. Westview Press, Boulder.

Goodwin, M. (1998) The governance of rural areas: some emerging research issues and agendas. *Journal of Rural Studies* 14, 5-12.

Hirst, P. and Thompson, G. (1995) Globalization and the future of the nation state. *Economy and Society* 24, 408-442.

Huntington, S.P. (1996) *The Clash of Civilizations and the Remaking of World Order*. Simon and Schuster, New York.

Jacobs, J. (1961) *The Death and Life of Great American Cities*. Random House, New York.

Jones, M. (1998) Restructuring the local state: economic governance or social regulation? *Political Geography* 17, 959-988

Katz, B. (2000) *Reflections on Regionalism*. The Brookings Institution, Washington D.C.

MacLeod, G. (1998) In what sense a region? Place hybridity, symbolic shape, and institutional formation in (post-) modern Scotland. *Political Geography* 17, 833-863.

MacLeod, G. and Goodwin, M. (1999) Space, scale and state strategy: rethinking urban and regional governance. *Progress in Human Geography* 23, 503-527.

Marden, P. (1997) Geographies of dissent: globalization, identity and the nation. *Political Geography* 16, 37-64.

Morgan, D.L. (1997) *Focus Groups as Qualitative Research*. Sage, London.

Murphy, A. (1991) Regions as social constructs: the gap between theory and practice. *Progress in Human Geography* 15, 22-35.

Newman, D. and Paasi, A. (1998) Fences and neighbours in the postmodern world: boundary narratives in political geography. *Progress in Human Geography* 22, 186-207.

Paasi, A. (1986) The institutionalization of regions: a theoretical framework for understanding the emergence of regions and the constitution of regional identity. *Fennia* 164, 105-146.

Paasi, A. (1991) Deconstructing regions: notes on the scales of spatial life. *Environment and Planning A* 23, 239-256.

Paasi, A. (1996) *Territories, Boundaries and Consciousness: The Changing Geographies of the Finnish-Russian Border*. John Wiley & Sons Ltd, Chichester.

Reynolds, D.R. (1995) Rural education: decentering the consolidation debate. In Castle, E. (ed.) *The Changing American Countryside: Rural People and Places*. University of Kansas Press, Kansas, pp.451-480.

Rusk, D. (1995) *Cities without Suburbs*, Johns Hopkins University Press/The Woodrow Wilson Center Press, Baltimore.

Sack, R.D. (1986) *Human Territoriality: its Theory and History*. Cambridge University Press, New York.

Smith, D.M. (1990) Introduction: The sharing and dividing of geographical space. In Chisholm, M. and Smith, D.M. (eds.) *Shared Space, Divided Space: Essays in Conflict and Territorial Organization*. Unwin Hyman, London, pp.1-21.

Stoker, G. (1996) Governance as theory: five propositions. Unpublished manuscript available from author at Department of Government, University of Strathclyde, Glasgow.

World Bank (2000) *Entering the 21st Century: World Development Report 1999/2000*. Oxford University Press, Oxford.

Wycoff, W. (1988) *The Developer's Frontier: the Making of the Western New York Landscape*. Yale University Press, Connecticut.

Chapter 20

Corporate-community relations in the tourism sector: a stakeholder perspective

Alison M. Gill and Peter W. Williams

Introduction

Tourism has become a powerful force in transforming rural places and landscapes, often creating a homogeneity that diminishes the distinctive sense of place of destinations. In particular, globalization has resulted in the emergence of large, often multinational, tourist corporations whose influence on the destinations that they control has been referred to as 'the corporatization of place' (Rothman, 1998). Globalization has also introduced other new actors, including local authorities and non-government organizations (NGOs), in the decision making arena. All have their own agendas for appropriate and efficient development of the regulatory system, thus creating new interactions between the actors in the form of partnerships and collaboration agreements.

Norcliffe (2000) argues that globalization is a contested and sometimes even chaotic process, which needs to be interpreted in the context of the local geographies and histories in which it is embedded. The intent of this paper is to provide a framework for embedding 'place' into an understanding of emerging corporate resort strategies associated with the forces of globalization. The focus is on how corporate responses to an increasingly competitive environment are affecting tourism corporations and their relationships with communities. While a rapidly growing literature exists on these aspects of strategic management from a business management perspective, there has been little attempt to extend these ideas beyond the firm to encompass the geographical frame, although Hayter and Soyez (1996), Patchell (1999) and Hayter and LeHeron (2002) are beginning to explore the theme of corporate environmentalism. The discussion draws upon stakeholder theory as a conceptual framework for considering how a corporation works with, interacts with and shapes the destination within which it is located. The importance of building and maintaining good stakeholder relationships in order to achieve competitive advantage has been posited by numerous authors (e.g. Hillman and Keim, 2001; Jones, 1995; Prahalad and Hamel, 1990; Svendson, 1998). At the community level, participatory approaches to decision making have highlighted the importance of stakeholder involvement and much attention is being paid to the processes involved in such engagement.

A discussion of stakeholder theory first establishes a frame of reference within which to examine corporate-community relations. The case of Whistler, British Columbia is then used to explore the relationship between Intrawest, the main resort corporation, and the community. Some researchers (e.g. Sternberg, 1997; Bannerjee, 2001) argue that there are limitations to stakeholder theory because the definition of stakeholders is too broad to be of practical use to organizations. Nevertheless, in the context of resort communities, the successful integration of corporate and community functions are critical to the sustainability of both the company and the community and a stakeholder perspective provides an appropriate framework within which to understand this relationship.

Conceptual framework

Within the strategic management literature, stakeholder management is generally seen as a component of corporate social responsibility (CSR), a broad multidimensional construct that also includes such notions as social issues participation (Carroll, 1999), corporate environmentalism (Berry and Rondinelli, 1998) and the resource-based view of the firm (Wernerfelt, 1984; Grant, 1991). This chapter draws upon these various literatures to identify a framework that is pertinent to the mountain resort context of this discussion.

Stakeholder theory

Svendsen et al. (2002, p.1) contend '[in] a rapidly globalizing, knowledge-based economy, sources of value creation in business are shifting from tangible assets such as land and equipment, to intangibles such as intellectual and human capital'. It is in this context that companies are conceiving the value of good stakeholder relations. Stakeholders were originally defined by Freeman (1984) as any group or individual who can affect or is affected by the achievement of a firm's objectives. Primary stakeholders include shareholders and investors, employees, resource suppliers, customers and community residents (Clarkson, 1995; Hillman and Keim, 2001). Other definitions also include those individuals and groups who speak for the natural environment (Wheeler and Sillanpää, 1997). Clarkson (1995, p.106) identifies the public stakeholder group as 'the governments and communities that provide infrastructures and markets, whose laws and regulations must be obeyed, and to whom taxes and other obligations may be due'. From this perspective, the firm is viewed as a set of interdependent relationships among stakeholders. A key principle of stakeholder theory is that a corporation is given the licence to operate by virtue of its social contract with stakeholders (Robson and Robson, 1996).

Effective management of these relationships 'can constitute intangible, socially complex resources that may enhance firms' abilities to out-perform competitors in terms of long-term value creation' (Hillman and Keim, 2001, p.127). According to the resource-based view of the firm, firms have distinct capabilities derived from both tangible and intangible assets (Wernerfelt, 1984). When these resources are unique and cannot be replicated by competitors, they produce sustainable competitive advantages (Barney, 1991; Hillman and Keim, 2001). Such resources as reputation, corporate culture, and long-term relationships with suppliers and customers are significant intangible resources that may lead to competitive advantage. Such relationships involve investments of resources and time to build 'webs of interdependencies' (Harrison and St. John, 1996). Collaborative or cooperative approaches with competitors and other firms in geographic locales to support infrastructural investments in communities are relational transactions that build value (Hart, 1995; Jamal and Getz, 1995). Stakeholders act as gatekeepers to resources that firms need (Svendsen *et al.*, 2002). For example, a community has control over the allocation of land-use within the community as well as the establishment of bylaws, including for instance, design standards.

Stakeholder involvement is also integrally linked to the principles of 'sustainability'. There is increasing reference to the 'triple bottom line' incorporating economic, environmental and social responsibility as corporations seek to maintain their market competitiveness in the face of global change and instability (Wheeler and Elkington, 2000). Empirical studies suggest that in some cases, and to varying degrees, such behaviour by firms does indeed lead to enhanced financial performance (Berman *et al*., 1999).

Svendsen *et al.* (2002) identify three ways in which positive stakeholder relationships can be related to competitive advantage. First, there may be a risk to shareholders if stakeholder relations are not well maintained. Public awareness campaigns by advocacy groups may result in damage to reputation and declining sales. A second benefit of good stakeholder relations is the opportunity for innovation from both within the firm and also along the supply chain. A third factor is the need to maintain a good reputation. This factor depends on building trust and credible relationships with stakeholders, or what Fomburn (1996) refers to as 'reputational capital'.

Corporate environmentalism

Hoffman (1997) asserts that although motives for environmental management have emerged from within industry, more substantial change is due to outside forces emanating from changing societal environmental perspectives. Environmental management strategies have evolved through various phases. In the 1960s to 1970s, they were reactive, coping with environmental crises as they occurred and seeking to control damage. This approach was followed in the 1980s by a regulatory phase as firms sought to react to increasing environmental legislation. Since the 1990s, at least some firms have been proactive in anticipating the environmental impacts of operations and taking measures to reduce waste and pollution in advance of regulation (Berry and Rondinelli, 1998).

Presently, proactive corporations are incorporating environmental values as an integral part of corporate cultures and management processes with a growing number of companies auditing and accounting for environmental impacts as a 'second bottom line'. Progressive corporations have now moved beyond compliance with increasingly stringent regulations. They must protect or enhance their ethical images, avoid serious liabilities, satisfy the safety concerns of employees, respond to government regulators and stakeholders, and develop new business opportunities in order to remain competitive in world markets (Berry and Rondinelli, 1998).

A stakeholder model of corporate-community relations

A conceptual model of corporate-community stakeholder relationships embedded in place is presented in Fig. 20.1. Both the corporation and the community have networks of stakeholders that operate within the context of a place that is characterized by a unique set of attributes that reflect the economic, political/regulatory, socio-cultural and biophysical environments. The primary corporate stakeholders include customers, employees, suppliers, competitors and shareholders. With the exception of employees, the majority of these stakeholders are located outside the community, although to varying degrees some may be within the community. The community's primary stakeholders are local government and community leaders, residents, local businesses and NGOs and community groups. With the exception of some NGO groups who may not be community based, most stakeholders can be considered as local. The figure greatly simplifies the complexities of the various interactions between these stakeholder groups, but suggests some possible linkages (dashed lines). For example, employees are in many instances residents. Other residents may be customers or even shareholders. Local businesses may function as suppliers or competitors.

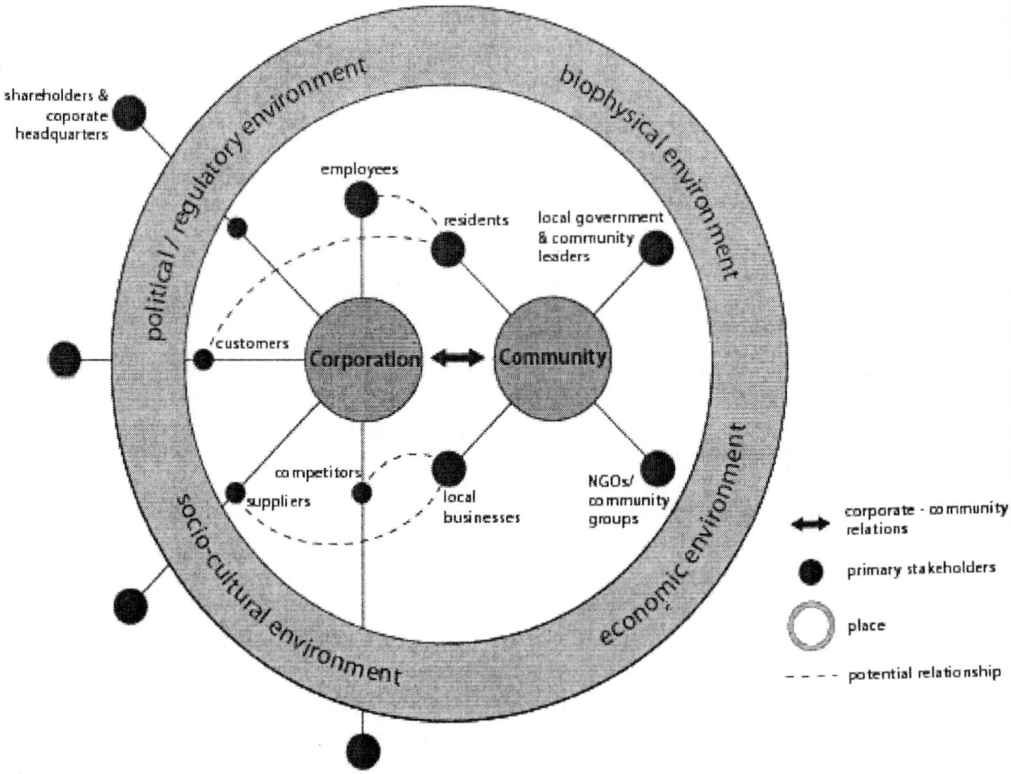

Fig. 20.1 A place-based model of corporate-community stakeholder relationships.

The model presents a generic situation. In the context of mountain resorts, such as discussed in the case study, there are a number of distinct features that characterize resort destinations. First, with respect to the context of place, ski developments are mainly on public lands and there is generally a rigorous set of environmental regulations that must be met in the development and operation of such areas. Second, the resident population is distinctive in that often more than 50% of 'residents' are second homeowners who do not live permanently in the community. Third, in many cases a high proportion of employees are not community residents because of the lack of affordable housing. This feature results in a commuting work force. Finally, and perhaps most importantly, the entire community is commodified as part of the tourism product, thus, there is a high degree of interdependency between the corporation and the community. While the corporation represents the main economic driver upon which the community depends, the corporation is also dependent on the community, not only to provide regulatory approval for its development plans, but also to maintain a high quality environment and service level to complement the corporation's image.

The mountain resort sector

Keller (2000) notes that the globalization process aggravates competition between places and individual destinations. In response, they must develop unique products to differentiate themselves and reposition themselves with the help of new strategies. The predominant strategic changes in mountain resort corporations are the development of corporate environmental strategies (Bhat, 1996), and consolidation and strategic alliances (Crotts *et al.*, 2000). Consolidations are readily apparent in a tourism context where numerous examples of travel, accommodation, food and beverage and resort destination organizations continue to merge vertically and/or horizontally (Crotts *et al.*, 2000; Keller and Bieger, 2001). Debbage (1990) points to the emergence in the USA of what he calls 'tourism oligopoly' whereby the tourism supply industry (e.g. hotels, airlines) is increasingly concentrated in the hands of large transnational corporations. In many of these contexts, this process involves companies entering into inter-firm relationships that are designed to gain the critical mass required for such strategic activities. These activities include: accessing capital and resources; creating economies of scale; building marketing leverage; diversifying risk; resolving seasonal variations in cash flow; developing strategic synergies; and, adapting to changing stakeholder demands (Buhalis, 2000; Bullock, 1998; Poon, 1993). From a resource-based view, the incentives to merge or acquire other businesses are related to recognition that the firm lacks the critical competencies necessary to achieve significant competitive advantage.

During the past decade, mountain resort-based corporations have been consolidating properties at a rapid rate, especially in North America. Rothman (1998) refers to this consolidation as the 'corporatization of place'. Where large corporations dominate resort destinations, their corporate decisions significantly influence the nature and appearance of those places. As Horner (2000, p.13) states 'resorts will evolve at the intersection of capital that is simultaneously local and global, public and private; consequently their form, function and image as marketed will be the outcome of the relative power of the actors representing these sources of capital'.

The North American ski industry is dominated by three publicly traded companies, referred to as 'the Big Three' (Clifford, 2002) - Intrawest Corporation, Vail Resorts Incorporated and American Skiing Company, who together with three private players, Booth Creek Ski Holdings, Boyne Resorts and Aspen Ski Company, control more than 30% of the North American market. The three publicly traded companies emerged in the 1990s and have interests in over 25 ski resorts in North America. Consolidation is motivated by financial market conditions with a fear of competitive pressures in the global market place. A review of prospectuses of the leading mountain resort consolidators suggest that their rationale for mergers and consolidations is largely associated with improving resort operations by integrating investments in on-mountain capital improvements with the development of controlled mountainside real estate, facilities and four-season resort villages. The rationale for such strategies is to increase utilization, lengthen visitor stay by high spending markets and increase the efficiencies and cost effectiveness of marketing, operations and management. As Clifford (2002, p.7) observes with respect to Vail Resorts' approach 'Vail Resorts intends to get its corporate fingers into a lot of businesses in the neighborhood'. Such expansion is bringing a great deal of pressure to mountain towns in which the 'Big Three' operate as corporations and, in their search for profits, are expanding their interest to include retail stores, restaurants, hotels, property management, travel services and tour operations and other businesses. Clifford (2002) argues that skiing is no longer the key to corporate profit but rather serves as an amenity to sell real estate. In so doing, he contends that modern ski resorts are losing what it was that made them special in the first place and are becoming like the 'rest of America'.

Intrawest: the corporation

Intrawest began as a Vancouver-based urban real estate developer involved in not only residential, but also 'festival marketplace' development. Intrawest began operations in Whistler in 1986 with its acquisition of Blackcomb Mountain from the original operator, a Canadian subsidiary of Aspen Skiing Corporation, who had owned the skiing operations since 1980 when the village and the two mountains, Blackcomb and Whistler, became fully operational. In competition with Whistler Mountain, both operations invested major amounts of money in infrastructure that, together with the successful village development, had, by 1991, attained the resort recognition as the number one ski resort in North America (by *Snow Country* readers).

Subsequently, Intrawest shifted its corporate interests out of urban real estate and in a decade, from 1991, focused on acquiring ten mountain resorts in North America. These resorts included Mt Tremblant and Mt Ste Marie, Quebec; Panorama and Whistler Mountain, British Columbia; Blue Mountain, Ontario; Stratton Mountain, Vermont; Copper Mountain, Colorado; Snowshoe, West Virginia; Mammoth Mountain, California and Mountain Creek, New Jersey. Other expansions involved partnerships to build residential units at Keystone, Colorado; Snowmass, Colorado; and Squaw Valley, California. In 1997, Intrawest began trading on the New York and Vancouver stock exchanges. A year later, following the merger with Whistler Mountain and with continued investment in on-slope infrastructure, Whistler became the first North American winter resort to have 2 million skier visits.

The recent Intrawest purchase of a 16% equity interest in Compagnie des Alpes (CDA), the leading European ski operator, reflects its strategy of geographical diversification to spread the risk of such factors as poor snow years or regional economic downturn. Intrawest is developing real estate for CDA at Les Arcs, France, employing its successful model of resort village development. Further diversification is seen in the expansion of Intrawest's interests to warmer destinations including golf resorts in Palm Desert, California and Sandestin, Florida. The company has also bought the Breeze and Max ski-shop chain and a 45% share of Canadian Mountain Holidays, Canada's largest heli-skiing and heli-hiking company. Its intent is to brand itself to appeal to both the cold and warm weather vacation needs of its customers. As such, Intrawest's goal, like those of other mountain resort companies, is to 'move aggressively into real estate development and the conversion of ski areas into high-volume, industrial-scale four-season resorts' (Clifford, 2002, p.44).

Intrawest's key strategy has been to focus on village-centred developments. This expertize is now a marketable commodity. Concern has been expressed about the standardization and homogenization that corporate ownership of resorts may entail (Clifford, 2002; Rothman, 1998). Indeed, the vision for Intrawest's village developments is derived from the successful design of Whistler Village, before Intrawest's involvement in Blackcomb. It is the vision of Eldon Beck, a landscape designer, who envisaged creating pedestrian oriented environments modelled on European alpine villages that would evoke a 'sense of place' (Dorward, 1990). It was Intrawest's initial failure in the late 1980s at creating a successful commercial development at the base of Blackcomb Mountain that led to recognition of the success of the original design of Whistler Village. In turn, they adopted the model as their own and hired Eldon Beck to apply his ideas elsewhere in their corporate properties.

Intrawest position themselves as *innovators in the leisure business* (Intrawest, 2005). In addition to the village-centred resort concept, other innovations include new forms of vacation-ownership (timeshares), new pre-selling techniques for resort real estate and innovative business processes and technology in their central reservations business. Intrawest has also developed a proactive environmental management programme that is managed at a senior level within the company. This programme is best developed at its 'flagship' property in Whistler, British Columbia.

Changing community stakeholder relationships in the evolution of Whistler, BC

Whistler, British Columbia is a new planned resort community located 120 km north of Vancouver that began operations as a ski resort in 1980. The evolution of community-corporate stakeholder relationships in Whistler can be traced through various stages that represent not only changing political, economic and social circumstances within the community, but also evolving corporate responses to market demand and local opportunities and constraints. As Svendsen et al. (2002, p.15) observe '[t]he stage of development of a business can affect the business value of various stakeholder relationships'. The case study is based on over a decade of field-based observations associated with various projects in Whistler. Information is drawn from a variety of sources: published sources, local newspapers, local government documents, corporate reports, graduate theses and both formal and informal interviews with key informants in the community.

The growth machine years

In the 1960s and early 1970s, the community of Alta Lake expanded from a few seasonally occupied summer lodges and cabins to a small, unorganized winter community centred at the base of the early downhill ski development on Whistler Mountain. In 1975, the provincial government passed an Act that initiated the development of Whistler and gave it special status as a 'Resort Municipality'. The following year, a mayor and council were elected and the first Official Community Plan (OCP) was developed. The land-use plan centred on compact resort development and the establishment of a village centre to complement the development of ski infrastructure on two adjoining mountains, Whistler and Blackcomb. Separate companies developed the two mountains with Whistler Mountain developed by a family-owned local company, Whistler Mountain Skiing Company, whereas Blackcomb Mountain was developed by 'outside' interests, a Canadian subsidiary of Aspen Skiing Company.

Gill (2000) has characterized the early phase of development in Whistler (from 1975 to 1990) as the 'growth machine' phase. A key stakeholder during this phase was the management body - initially, the Whistler Village Land Co. Ltd. (WVLC), a wholly owned subsidiary of the Resort Municipality that was created to oversee financing and development of the village centre. All members of council were automatically directors of the company and the mayor was chairman. Thus, the role of local government was integrally bound to land development and several councilors were actively involved in development projects. The provincial government was also a major player as they had provided lands and financing and set up the regulatory system. The provincial government also appointed a member to council to represent their interests. The two ski companies were the largest and most powerful of the developer voices as plans for the village called for small parcels of land to be developed by a diversified group of developers in order to avoid an 'instant town' appearance. There was virtually no community 'voice'. There were a few residents, but all supported development as they welcomed control over the previously uncoordinated settlement growth and anticipated increased value for their homes as well as entrepreneurial opportunity. Essentially, then the early phase of development was a government (local and provincial)/developer relationship in which the government acting as both regulatory authority and arms-length developer worked with private developers to ensure adherence to the conditions of the OCP. Bankruptcy of WVLC during the recession of 1982, just two years after the ski hills operations began, led to the buy-out of the municipal development company by the provincial government, who wanted to protect their major investment in infrastructural development.

With the opening of Whistler Resort in 1980, a new agency was added to the stakeholder mix, the Whistler Resort Association (WRA - more recently named Tourism Whistler). They represented the marketing arm of the resort and all commercial operators, including the two large ski companies who had representatives on the board, paid dues to the Association to conduct coordinated marketing of the resort.

The balance of power within the resort began to change around 1986, when Intrawest bought Blackcomb Ski Corporation and began to exercise its rights to develop residential and commercial real estate at the base of its lifts – a major incentive for their purchase of the company. The rights to development had been granted under the 'lands for lifts' policy whereby developers were offered the incentive of development rights on land in return for the construction of ski infrastructure. This arrangement provided an opportunity for more rapid and lucrative return on capital investment from real estate development rather than the returns from user fees for the lifts. A formula that tied lift capacity to bed-unit development was developed which gave one bed unit for every two-skier days of lift capacity. Real estate development had been delayed on Blackcomb Benchlands because of the planned phased-in development strategy that required concentration in Whistler Village in order to achieve a critical mass of development. The allocation of bed units (or the right to develop land) has been a very important issue in the control of development by the Resort Municipality of Whistler (RMOW). The amount of bed units agreed to under the OCP was set initially at 45,000 but raised to 52,500 in 1989 - where (at least officially) it remains today. Commitment of all bed units was virtually complete by the early 1990s. A belief by many residents in the importance of maintaining the established bed unit limit has strengthened as the community has approached build out.

Development during the late 1980s was fuelled by competition between the lift companies on the two adjacent mountains, Blackcomb and Whistler, as each sought to remain competitive by upgrading lift infrastructure. During this period, Intrawest spent over $40 million on snowmaking and lift capacity as well as many million dollars more on on-hill service facilities. This surge of development in the late 1980s led to the emergence of the first hint of concern about growth by residents, whose numbers had increased from a few hundred in 1980 to around 4000 by the end of the decade.

In summary, during this initial uncontested growth machine phase in Whistler, the needs of the resort's growing resident population were overlooked as all efforts were placed on revenue-generating construction that would guarantee investors a return on their capital (Gill, 2000). Despite specific acknowledgement in the Resort Municipality of Whistler Act of the responsibility of the provincial government to ensure the well-being of the local community, such considerations were ignored in favour of protecting economic investments. Likewise, the local government did not exercise its rights to impose development cost charges to cover the cost of providing employee housing. However, there was virtually no resistance from local residents at this stage as all stood to gain if the resort succeeded and conversely lose if it failed. While local residents continued to see Whistler Mountain as the locals' mountain during this phase, Intrawest's Blackcomb operation gained the financial edge.

The emergence of community as stakeholder

The year 1990 marked the beginning of significant changes in attitude by local residents who began to contest the priorities placed on resort growth and demanded that community issues be addressed. The community was also becoming more organized with the emergence of interest groups who began to raise community awareness and lobby local government. The most influential of these groups was the Association of Whistler Area Residents for the Environment (AWARE), a community-based environmental organization, formed in 1987 (Reed, 1997).

In part, this attitudinal change was related to the changing demographics of the community as it matured. The expansion to a year-round resort brought year-round employment and an increasing number of permanent residents with children. The desires of residents to acquire permanent and affordable housing and appropriate community services were heightened. In 1990, the community had no community or recreation centre and no high school, day care, church or adequate library facilities. Concern over protection of the environment, the provision of affordable housing and improved recreation facilities were also widely voiced priorities. The need for some form of institutional response to the increasing anti-development attitude of local residents led to the 'growth management era' (Gill, 2000).

Growth management entails the active involvement of residents and other stakeholders in the establishment of growth priorities. Since 1994, the RMOW has acted more in line with the defining characteristics of growth management that tie new development to housing, transportation and environmental needs, and engage the public in planning decisions. As part of the growth management approach, the resort has now instituted a comprehensive resort and community monitoring system that collects data on a wide range of economic, environmental and social indicators. The results are communicated to residents in an interactive forum at an annual town hall meeting. The issue of affordable housing is also being tackled, although because the problem has been addressed in a post hoc fashion, the problem is far from being resolved. All developers are now required to either build employee housing to support commercial developments or contribute an equivalent amount to the Whistler Valley Housing Authority that then has the responsibility of building units.

The growth management approach that Whistler adopted was designed to acknowledge the competing forces within the resort community. As indicated in the following statement, the RMOW recognizes very clearly that its mandate lies in understanding the interdependence of the community and the resort business, they state: '[t]he health of our community depends upon the economic viability of the resort. Likewise the continuing success of the resort is possible through the support of this strong, flourishing community' (RMOW, 1999, p.8). To address this issue, the RMOW introduced a paradigm shift from one in which economic competition alone drove the system to one in which economic goals were mediated by the necessity to consider social and community values. Growth management was initially introduced as more or less a buzzword and a compromise that might appeal to both growth and no-growth factions - with business interests focusing on growth and anti-growth interests emphasizing management. However, it has resulted in greater public benefits deriving from development in the form of improved community facilities; an affordable housing policy; a more participatory and accountable form of governance; and, increasingly stronger environmental protection measures.

The reaction to rapid growth and the recognition that limits of development were being reached was reflected in the 1996 election of a pro-community/anti-growth mayor, who was committed to such issues as affordable housing and environmental quality. Indeed, for the first time in the resort's history, a council made up of three no-growth councillors, two neutral members and only one pro-growth advocate marked a distinct change in the style of local governance. A new approach to governance, based on a 'corporate' model (Horner, 2000), was adopted. This approach is in line with the recommendations of Flagestad and Hope (2001) who suggest that winter sports destinations should adopt the strategic management approaches employed by firms. The main mandate has been to engage in longer-term planning initiatives. This change has taken the form of an intensive visioning exercize using widespread public input to establish guidelines for when the development cap has been reached. Collaboration and partnerships were key concepts of this vision (RMOW, 1999). Intrawest has participated as a corporate stakeholder in the on-going evolution of the community's visioning process. Its values, along with those of other community stakeholder interests are reflected in the vision's concerns for environmental stewardship, enhancing the visitor experience, promoting partnerships in new mountain recreation ventures, as well as encouraging the revitalization of Whistler Creek (RMOW, 1999). While these are all points of convergence in terms of Intrawest's values aligning with broader Whistler values, they do not necessarily carry equal weight within the organization. Intrawest's core values are clearly more closely linked to its leading position as a creator of carefully themed and animated mountain villages that eventually translate into real estate development opportunities and sales for the corporation (Williams *et al.*, 2004).

Shortly after the election of the new municipal government, Intrawest took over the Whistler Mountain Skiing Corporation, thus eliminating its local competitor and essentially turning Whistler into a one-company town. The new mountain operation was named 'Whistler-Blackcomb'. This operation, together with the new approach taken by the RMOW, set the stage for a changed relationship between the corporation and the community.

Key events in corporate-community stakeholder engagement

While there has always been interaction and negotiation between the corporation and the RMOW over development issues, two contentious issues that have been matters of extensive public debate among the stakeholder groups stand out.

The first example is that of the development of Whistler Creek by Intrawest. When Intrawest purchased Whistler Mountain, it also acquired development rights to 2500 bed units at the base of Whistler Mountain - on the opposite side of the mountain to the village. Unlike Blackcomb, where the bed unit allocation under the lands for lifts policy had already been built out, Whistler had not exercized its rights to develop its lands. Intrawest saw this situation as an opportunity to construct a new village in which the corporation would have complete control of all commercial and residential properties. At the same time, Intrawest also wanted to commit to the development of all its remaining bed units, although this proposal exceeded the limits set for the new village area. To plan these developments, Intrawest developed a Comprehensive Development Strategy (CDS) that, in addition to the new village, also called for the development of a new residential neighbourhood of mixed market and affordable housing and an exclusive subdivision development of high-end single-family homes. This large development proposal that bundled all the proposals together was an unusual, but strategic, move on the part of Intrawest.

The proposal for the new village at Whistler Creek (later known as Creekside) was developed through a visioning process that included significant input from the municipality and local residents in the 'scripting of place'. While primarily envisioned by the corporation's designers who articulated an identity for the development, that vision was refined through the participation of invited community and municipal representatives in design workshops. This dialogue led to intense discussion concerning the relative values placed on real estate development, protected area networks, and social values espoused by the corporation and the community. The 'vision' for Whistler Creek was one of nostalgia that recalled the early days of skiing in this place – the original base for ski development on Whistler Mountain. The intent was to not compete with its own existing upscale market in the Village, but capture a more local/regional market. While maintaining control of the retail operations, the intent was to develop retail services that would have local appeal, while at the same time adding animation to the new village. However, to make the vision workable from a financial perspective, Intrawest also needed approval to develop a high-end elite subdivision. This view was contrary to the vision and values of the community. Therein lay the core of the debate and negotiations.

In the new spirit of partnership and collaboration that characterized the new corporate model of the municipality, a lengthy negotiation that lasted more than two years was undertaken. The RMOW recognized that, with no more bed units to develop, this opportunity was their last chance to negotiate community benefits from developers. Thus, they had a significant wish list of needs. The highest priority was a new elementary school together with other community amenities such as a daycare centre, parks and recreational trails. The deal that the RMOW struck with Intrawest was widely supported in the community as the municipality had negotiated many amenities. However, several pro-environment councillors (who were also members of AWARE) were vehemently opposed to allowing the development of an expensive elite neighborhood. As Horner (2000, 19) observes '[t]he irony … was that the pro-community faction were willing to forego an unprecedented package of community benefits for the sake of not compromising and not sharing power with the corporation'.

A second example of controversy between the corporation and the community surrounds the 'Emerald Forest deal'. Emerald Forest is a 56-ha parcel of land in the valley bottom that was owned by a development company that had rights to develop large parcel lots (8 ha). However, the RMOW wanted to preserve the land as part of the valley corridor ecosystem. The RMOW could not afford the estimated CA$10 million to buy the property, and several compromise deals with the property owners failed. The solution suggested was to make a deal with a third party. Intrawest agreed to purchase the land (for an undisclosed sum) and return it to the RMOW as parkland in return for 434 bed units and CA$1million in cash. The bed units were to be applied to develop a new hotel on a parking lot near the Blackcomb ski base. This proposal appeared to benefit all three parties involved. However, the RMOW did not have any more bed units available as all 52,500 under the agreed upon cap were allocated. Nevertheless, the deal was hastily agreed to due to a forthcoming local election and caused significant local reaction (Barnett, 1999). While some resident stakeholders were upset that democratic principles had been violated due to an increase in the cap on bed units and inadequate public debate, others believed that Intrawest had exerted its power to acquire extra profit at the expense of the community. The most vocal opponents of the deal were not even residents, but second-homeowners with properties adjacent to the proposed Intrawest development who objected on the basis of anticipated noise and congestion. On the other hand, many residents, including AWARE members, were pleased that the ecological integrity of the valley had been preserved. As Horner (2000) observes, in actual fact Intrawest had not been overly enthusiastic about entering into this arrangement. Although they undoubtedly made money, they were not keen to be involved in this rather minor but controversial development issue because they were in the process of developing the Whistler Creek project which was dependent on gaining community support. One might speculate therefore that they did it in the interests of being 'good citizens'.

Moving towards sustainability

Maintaining environmental quality is a key value for both the community and the corporation. While for the community this value is tied to maintaining a sense of place and protecting the environment for future generations, for the corporation it translates into good business practice. Due to the fact that the development and management of ski areas have recognized environmental impacts, the ski industry in North America has for some time engaged in a variety of environmental management techniques in many cases to meet regulatory requirements and avoid litigation. Intrawest is no exception. As early as 1989, when it engaged in a pubic participatory approach to gain approval for an expansion of its ski area into Garibaldi Provincial Park, the company engaged an environmental expert to assist in presenting the corporation's environmental perspective. By 1993, Intrawest developed an Environmental Management System (EMS) at Blackcomb, although it appears the motivation for this system came from concern at corporate headquarters about litigation (Williams and Todd, 1997). Nevertheless, in line with the corporation's overall strategic approach, the EMS adopted by Intrawest in their on-mountain operations is one perceived by the company as innovative. As they observe 'we are working to design and implement a comprehensive environmental strategy that can be modeled by other resorts' (Intrawest, 2005). The current EMS is guided by the basic principles of the National Ski Areas Association's *Sustainable Slopes: The Environmental Charter for Ski Areas* (National Ski Areas Association, 2000). The EMS template includes aspects of ecosystem management, conservation, waste management, community outreach and environmental education. While the majority of ski operations in North America endorse the *Charter*, compliance is voluntary (Clifford, 2002), and few places have implemented as comprehensive an EMS as the one in Whistler.

Whistler-Blackcomb's Master Plan aims to establish the best recreational designs with the least amount of environmental impact (avoiding wetlands and riparian zones; using natural contours, developing glades and natural trails) and maximizing the visual quality with attention to natural viewscapes. Whistler-Blackcomb's EMS is in line with a similar initiative by the RMOW who established a Whistler Environmental Strategy and subsequently a Comprehensive Sustainability Plan known as *Whistler: Its Our Future*. From the company's perspective, the community's commitment to environmental quality is especially important for their summer visitors as their recreational activities are less concentrated on the mountains than those of winter visitors.

The integration of sustainable environmental practices with those of community stakeholders is an important element of Whistler-Blackcomb's overall objectives. A recent example is their membership as Early Adopter for the Whistler Sustainability Initiative as part of The Natural Step (www.naturalstep.org) – 'an international organization that uses a science-based, systems framework to help organizations and communities understand and move towards sustainability'. The Natural Step (TNS) works with 'progressive' corporations and institutions committed to a sustainable path. TNS focuses on generating solutions and on forming strategic partnerships to help move society towards a sustainable future. TNS forms long-term, working relationships in order to build internal capacity for clients. Transparency, stakeholder involvement and continual learning are cornerstones of the TNS approach. In Whistler, Whistler-Blackcomb have partnered with the Resort Municipality of Whistler; a local ENGO - Whistler Area Residents for the Environment

(AWARE); the Fairmont Hotel; Tourism Whistler (the marketing agency) and a local photography store (Whistler's Foto Source). Begun in September 2001, they are training staff on the Natural Step framework and working with other early adopters to develop sustainability toolkits for small businesses, households and schools. Working with the community and making sustainability a priority within the business they hope to contribute to the goal of becoming a sustainable community.

Other community partnerships include the development of a Habitat Improvement Team (HIT) which involves helping local groups engage in hands-on habitat work for fish, bears, etc. (tools, transportation, instruction, refreshment). Whistler-Blackcomb also has established an Environmental Fund that funds community environmental projects through money raised in staff donations and matched by the corporation.

Environmental education of staff and guests is also seen as critical to the success of the mountains' environmental programme and involves staff training and interpretive programmes for guests. Whistler-Blackcomb also supports bear research and an on-mountain environmental education programme for elementary and high school students from the region. Waste auditing and the 3Rs (reduce; reuse; recycle) are now standard practice.

Whistler-Blackcomb has been successful in reaching a level of environmental management that has attracted a number of awards including the 2003 Golden Eagle Award for overall environmental excellence in the North American ski industry. The mountain has also won Silver Eagle Awards for excellence in stakeholder relations (2002); environmental group relations (2001, 1999); habitat protection (1999); and, environmental education (1998).

Conclusions

In this chapter, the evolving stakeholder relationship between Intrawest and the Resort Municipality of Whistler offers some insights into how communities and corporations are responding to changing conditions of competitiveness. Due to the fragmented but interdependent nature of the tourism industry, collaboration is seen as the key to developing a successful resort destination (Jamal and Getz, 1995). However, the conditions under which this collaboration can successfully occur depend to a considerable extent on the power relations that exist between the key actors (Reed, 1997). In the example given of Intrawest's relationship with the resort community of Whistler, there is evidence of a convergence towards collaboration and partnership as the result of changes both within the corporation and within the community.

As Norcliffe (2000) states the implications of the forces of globalization are place specific. As suggested in the model presented earlier in the chapter, the corporate community relationship is bounded by various attributes of place such as the political/regulatory, biophysical, economic and socio-cultural environments. Thus, while Clifford (2002) and Rothman (1999) both conclude that the corporations that control mountain activities negatively impact mountain communities, this statement would seem to be an over-generalization. While they cite many examples of poor relationships between mountain corporations and communities, the situation in Whistler appears to be less hostile. While, some residents will always be suspicious of the motivations of large corporations, overall there has been a fairly good relationship between the corporation and the community. This situation does not appear to be the case, however, in Intrawest's dealings in the USA. Clifford (2002) highlights numerous tensions between Intrawest and the communities with which it deals in Colorado and California such as Copper Mountain and

Mammoth Lakes, where there appears to be a high level of mistrust in the corporations dealings. As one administrator in Copper Mountain, Colorado observed, the company is perceived as a Canadian company that does not understand US regulations (E. Black, Colorado, personal communication, 2002). The short time span in which Intrawest has acquired many US properties to become the largest of the 'Big Three' is also a cause for unease. Horner (2000) suggests that Intrawest's relatively positive relationship with the community may reflect the fact that, although Intrawest was perceived by some to be (relative to the Whistler Mountain Skiing Corporation) an outside interest with a head office in Vancouver, in fact the CEO and senior management were also Whistler residents.

A further factor that appears to have affected the relationship is the fact that Whistler was a new planned resort community with a well-defined and strict regulatory framework that established certain 'rules of engagement' between the mountain corporations and the community. This engagement included collaborative marketing ventures through the WRA, design regulations and control by the RMOW of the rate and location of development. Thus, from the start the RMOW had considerable control. Further, as Gill (2000) observes, in the early days of development (before Intrawest became involved), the corporation and the community shared a common goal of growth at all costs.

With respect to whether or not Whistler has been, to use Rothman's (1998) words, 'corporatized', is difficult to judge, as there were few original residents and the initial village development created an 'instant town'. No one would disagree that Whistler has, with growth, become more urbanized with a highly developed retail and commercial landscape. Whether one likes such developments or hates them is a matter of personal preference. Unlike some of Intrawest's newly created villages, such as Mt Tremblant, Quebec, Intrawest was not the sole developer in Whistler. However, Eldon Beck's vision has guided the majority of development and, despite the high quality and highly successful nature of his designs, similarities are evident between Whistler and other resorts with which he is associated. The new development at Creekside will be the most corporately planned and controlled element of Intrawest's operation in the community. With the exception of the mountain operations, Intrawest's corporate interests are dispersed throughout the community, but often hidden behind the face of a management company or operator.

In tracing how the corporate-community stakeholder relationships evolved in Whistler, it is evident from residents' letters and editorial commentary in the local newspaper (see, e.g. *The Pique NewsMagazine*, 1997) that the acquisition of Whistler Ski Corporation by Intrawest in 1997 created a fear of the consequences of concentration of power in the hands of a single corporation. Elements of this concentration are evident in the debates surrounding the Emerald Forest deal and the development of Whistler Creekside village. Despite these issues, there does appear to be a convergence of vision between the community and the corporation especially with respect to environmental quality and sustainability. It is the issue of a shared vision between the two entities that is critical to successful collaboration.

Is there a corporate commitment to a stakeholder approach? This question is much more difficult to answer. Some community members have argued that Intrawest's negotiations over both Emerald Forest and the Whistler Creek proposal provide evidence of Intrawest's community mindedness (Horner, 2000). If, one considers the environmental initiatives that Intrawest has engaged in at Whistler, including the more recent commitment to work with other community members in following 'The Natural Step' then the answer might be 'yes'. Further, Clifford (2002, 57) quotes Joe Houssian, CEO of Intrawest, as saying: 'We're in the environment business, we're in the nature business. We can't come into a town, into a mountain valley and change it. Our objective is to put it in a better condition and to provide to the community things it would not otherwise have – churches and schools'. The degree to which an individual accepts this statement will be dependent on their attitudes towards corporate social responsibility.

That Intrawest's competitive strategy is to demonstrate innovation across many aspects of their operation is clearly stated in their corporate website (Intrawest, 2005). This innovation is evidenced in their environmental program at Whistler as they strive to be ahead of other resorts. Joining 'The Natural Step' initiative represents an attempt to join the ranks of innovators – an important selling point of TNS. However, this type of commitment to sustainability initiatives seems limited to the Whistler-Blackcomb context. As the initiators of TNS in Whistler were the RMOW, it appears to be a localized commitment rather than one of the corporate head office. Indeed, mention of environmental issues is virtually lacking from annual reports and the corporation's website (www.intrawest.com). It is clear, from annual reports that as a publicly traded company, responsibility to shareholders financial expectations is the primary driver of decisions (Williams *et al.*, 2004). However, research in stakeholder management does suggest that the intangible resources linked to reputation do positively affect the bottom line (Waddock and Graves 1997).

In conclusion, there are many corporate influences in rural areas such as agribusiness, forestry or mining. Although the corporate tourism industry, because of its dependency on place and the interdependencies of the industry, has perhaps the most to lose if it does not build good relationships with the local community, the need for engaging community stakeholders is of relevance to all sectors. In response to the changed power relations that have resulted from forces of globalization, the necessity of working with local governments and agencies and engaging residents in participatory processes is now a requirement. From a strategic business perspective, stakeholder management is increasingly seen as a corporate response that is argued to result in competitive advantage. As demonstrated in this chapter, such relationships are complex and actually determining the underlying drivers of corporate action, is challenging.

References

Banerjee, S.B. (2001) Managerial perceptions of corporate environmentalism: interpretations from industry and strategic implications for organization. *Journal of Management Studies,* 38, 489-513.

Barnett, B. (1999) Emerald Forest deal, panned and praised. *Pique Newsmagazine,* 27, August, p8.

Barney, J.B. (1991) Firm resources and sustained competitive advantage. *Journal of Management,* 17, 99-120.

Berman, S.L., Wicks, A.C., Kotha, S. and Jones. T.M. (1999) Does stakeholder orientation matter? The relationship between stakeholder management models and firm financial performance. *Academy of Management Journal,* 42, 488-506.

Berry, M.A. and Rondinelli, D.A. (1998) Proactive corporate environmental management: a new industrial revolution. *Academy of Management Executive,* 12 (2), 38-50.

Bhat, V.N. (1996) *The Green Corporation: The Next Competitive Advantage.* Quorum Books: Westport, Connecticut.

Buhalis, D. (2000) Marketing the competitive destination of the future. *Tourism Management,* 21, 97-116.

Bullock, G. K. (1998) The UK package holiday industry. *Management Accounting,* 76(4), 36-38.

Carroll, A. B. (1999) Corporate social responsibility. Evolution of a definitional construct. *Business and Society,* 38, 268-295.

Clarkson, M. (1995) A stakeholder framework for analyzing and evaluating corporate social performance. *Academy of Management Review,* 20, 92-117.

Clifford, H. (2002) *Downhill Slide*. Sierra Club Books, San Francisco.

Crotts, J.C., Buhalis, D. and March, R. (2000) Global alliances in tourism and hospitality management. *International Journal of Hospitality and Tourism Management,* 1, 1-10.

Debbage, K.G. (1990) Oligopoly and the resort cycle in the Bahamas. *Annals of Tourism Research,* 17, 513-527.

Dorward, S. (1990) *Design for Mountain Communities: A Landscape and Architectural Guide.* Van Nostrand Reinhold, New York.

Flagestad, A. and Hope, C.A. (2001) Strategic success in winter sports destinations; a sustainable value creation perspective. *Tourism Management,* 22, 445-461.

Fomburn, C. J. (1996) *Reputation: Realizing Value From the Corporate Image.* Harvard, Boston.

Freeman, R.E. (1984) *Strategic Management: A Stakeholder Approach.* Pitman, Boston.

Gill, A.M. (2000) From growth machine to growth management: the dynamics of resort development in Whistler, British Columbia. *Environment and Planning A,* 32, 1083-1100.

Grant, R. (1991) The resource-based theory of competitive advantage: implications for strategy formulation. *California Management Review,* 33 (3), 114-135.

Harrison, J. and St. John, C. (1996) Managing and partnering with external stakeholders. *Academy of Management Executive,* 10, 46.

Hart, S. (1995) A natural-resource-based view of the firm. *Academy of Management Review,* 20.

Hayter, R. and LeHeron, R.B. (2002) Industrialization, techno-economic paradigms and the environment. In: Hayter, R. and LeHeron, R.B. (eds) *Knowledge, Industry and Environment: Innovations and Institutions in Territorial Perspective.* Ashgate, London, pp.11-30.

Hayter, R. and Soyez, D. (1996) Clearcut issues: German environmental pressure and the British Columbia forest sector. *Geographische Zeitschrift,* 84,143-156.

Hillman, A. and Keim, G. (2001) Shareholder value, stakeholder management and social issues: what's the bottom line? *Strategic Management Journal,* 22, 125-139.

Hoffman, A.J. (1997) *From Heresy to Dogma.* New Lexington Press, San Francisco.

Horner, G. (2000) Mountains of money: the corporate production of Whistler Resort. MA thesis, The University of British Columbia, British Columbia, Canada.

Intrawest (2005) www.intrawest.com (accessed 5 January, 2005).

Jamal, T.B. and Getz, D. (1995) Collaboration theory and community tourism planning. *Annals of Tourism Research,* 22, 186-204.

Jones, T. (1995) Instrumental stakeholder theory: a synthesis of ethics and economics. *Academy of Management Review,* 20, 404-437.

Keller, P. (2000) Globalization and tourism. In: Gartner, W.E. and Lime, D.W. (eds) *Trends in Outdoor Recreation.* CAB International, Wallingford, UK.

Keller, P. and Bieger, T. (eds) (2001) *Tourism Growth and Global Competition.* International Association of Scientific Experts in Tourism. Vol 23, St Gallen, Switzerland.

National Ski Areas Association (2000) *Sustainable Slopes: The Environmental Charter for Ski Areas.* National Ski Areas Association, Lakewood, Colorado.

Norcliffe, G. (2000) Canada in the global economy. *The Canadian Geographer,* 45, 14-30.

Patchell, J. (1999) Creating the Japanese electric vehicle industry; the challenges of uncertainty and cooperation. *Environment and Planning A,* 31, 997-1016.

Poon, A . (1993) *Tourism, Technology and Competitive Strategies.* CAB International, Wallingford, UK.

Prahalad, C. and Hamel, G. (1990) The core competence of the corporation. *Harvard Business Review,* 68 (3), 79-91.

Reed, M.G. (1997) Power relations and community-based tourism planning. *Annals of Tourism Research,* 24, 566-591.

Resort Municipality of Whistler (RMOW) (1999) *Whistler 2002: Charting a Course for the Future.* Resort Municipality of Whistler, Whistler, British Columbia.

Robson, J. and Robson, I. (1996) From shareholders to stakeholders; critical issues for tourism marketers. *Tourism Management,* 17, 553-540.

Rothman, H. (1998) *Devil's Bargain: Tourism in the Twentieth-century American West.* University of Kansas Press, Lawrence, Kansas.

Sternberg, E. (1997) The defects of stakeholder theory. *Corporate Governance: An International Review,* 5, 3-10.

Svendsen, A. (1998) *The Stakeholder Strategy: Profiting from Collaborative Business Relations.* Berrett-Koehler, San Francisco.

Svendsen, A., Boutilier, R.G., Abbott, R. M. and Wheeler, D. (2002) *Measuring the Business Value of Stakeholder Relationships: Part 1.* Center for Innovation Management, Simon Fraser University, Burnaby, British Columbia.

Waddock, S. A. and Graves, S. (1997) The corporate social performance-financial performance link. *Strategic Management Journal,* 19, 303-317.

Wernerfelt, B. (1984) A resource-based view of the firm. *Strategic Management Journal,* 5, 171-180.

Wheeler, D. and Elkington, J. (2000) The end of the corporate environmental report? Or the advent of cybernetic sustainability reporting. *Business Strategy and Environment,* 10, 1-14.

Wheeler, D. and Sillanpää, M. (1997) *The Stakeholder Corporation.* Pitman, London.

Williams, P.W., Gill, A.M. and Chura, N. (2004) Branding mountain destinations: the battle for 'placefulness'. *Tourism Review,* 59, 6-15.

Williams, P.W. and S. Todd. (1997) Toward an environmental management system for ski areas. *Mountain Research and Development,* 17, 75-90.

Chapter 21

Resource town transition: debates after closure

Greg Halseth

Introduction

Rural and small town places across British Columbia's non-metropolitan hinterland are experiencing change. While change has been a fundamental part of this landscape for more than 100 years, the pace of change is accelerating. As a result of social, economic and political restructuring, people and places across rural and small town British Columbia (BC) are pressed to adjust (Reed, 2003; Morford and Kahlke, 2004). With a focus upon Gold River and Tumbler Ridge (Fig. 21.1), this chapter examines local debates following a major industrial closure. The principal finding is that different geographic imaginations of rural places and their contributions to the provincial economy have limited the options for local renewal (see also Hodge, 2004).

British Columbia

British Columbia has been described as a resource periphery (Bradbury, 1987). Hayter (2000, p.35) argues that its 'geographic marginality' is the defining context for the evolution of both its resource economy and settlement system. Much of the industrial base, communication and transportation infrastructure, and non-metropolitan settlement developed during the post Second World War economic boom (Williston and Keller, 1997). Resource revenues were used to develop the province, and the economy remains dependent upon limited manufacture resource commodity exports (Baxter and Ramlo, 2002).

The province's settlement geography is strongly organized by two characteristics. The first is a mountainous topography that restricts the amount of land available for settlement and limits connecting transportation networks to a small number of valley corridors. This pattern of small settlement nodes along narrow transportation corridors means that the province did not develop the extensive rural settlement patterns seen across the rest of western Canada. 'Rural' BC is perhaps better conceptualized as 'non-metropolitan' BC, where the usual rural characteristics of small populations and large distances remain in effect but where the topography has concentrated those populations into small settlements. The topography reinforces a sense of isolation between regions and settlements.

The second settlement geography characteristic is a clear metropolitan/non-metropolitan division (Davis and Hutton, 1989). The metropolitan region of Vancouver and Victoria, located in the southwest corner of the province, is the centre of population, government, and high order business and retail services (Wynn and Oke, 1992). This metropolitan region has dominated the province's settlement structure since it joined Canada in 1871. By 2001, metropolitan Vancouver-Victoria contained 58% of BC's population. This concentration of population and political power is geographically removed from the resource extraction communities and regions which drive the provincial economy. As widely noted of late, this geographic separation has the effect of 'hiding' BC's economic dependence on natural resources, and the plight of hinterland communities and workers involved in these industries, from metropolitan decision makers (Baxter and Ramlo, 2002; Hutton, 2002; Lee, 2003).

Fig. 21.1 British Columbia, Canada.

The non-metropolitan landscape can be characterized by a set of regional centres and a constellation of small resource dependent towns. Many of these settlements have little economic diversification. In concert with expectations derived from staples theory, both public policy and business interests have built and rebuilt a provincial infrastructure designed to extract and export large volumes of minimally processed raw resources and concentrate the benefits from that activity in the metropolitan core (Gunton, 1997; Hayter and Barnes, 1997; Hutton, 1997). The result is an isolated set of small settlements, dependent upon resource management controlled from a metropolitan heartland, and with few local economic alternatives. Restructuring of both resource production and public service provision have meant that the past 20 years have been marked by accelerating instability, change and job losses in many resource communities (Barnes and Hayter, 1994; Halseth and Sullivan, 2002). Indeed, rural and resource industry restructuring is a common story in developed economies (Marsden *et al.*, 1990; Neil *et al.*, 1992; Mackenzie and Norcliffe, 1997). Despite many indicators that economic restructuring is altering BC's resource industry landscape, the dependence of the provincial economy, and the dependency of many rural communities, on resource extraction reinforces a particular rural imagination at both levels.

To illustrate the actions of these rural imaginations, two resource town transition planning exercises are examined. The communities of Gold River and Tumbler Ridge each experienced the closure of a major industrial operation within the past 4 years. The research for Gold River is based on extensive newspaper searches and transition reports prepared for the community. In Tumbler Ridge, additional research was completed that involved interviews, focus groups and surveys, as well as the analysis of transition reports and newspapers.

Institutional players

The last 20 years have marked an increased interest in local control of economic and development practices. Dauncey (1996) suggests that this interest has spawned a wide variety of concepts and acronyms, including macro-economic development, local economic development, community economic development, sustainable development and sustainable communities. Each concept brings its own nuanced interpretation of issues such as power and participation, and its own understanding of the desired goals or outcomes (Campfens, 1997; Maser, 1997). Within this diverse literature there is some commonality of view around the idea that local places should be initiating and generating their 'own solutions to their common economic problems and thereby build long-term community capacity and foster the integration of economic, social and environmental objectives' (McRobie and Ross, 1987, p.1). This shift of capacity to local levels suggests, in BC at least, an important role for the provincial government which has considerable responsibility for development policy. It also suggests that attention to the interactions between local governments and local groups is important.

In community development and revitalization debates, it is clear that some institutional players simply carry more weight. Hayter and Barnes (1997), writing about the restructuring of BC's forest economy, have suggested four key institutional players. These are: the state (principally the provincial government), capital (principally corporate capital), environmental NGOs (many of which are professionalized and multinational), and aboriginal groups (most of which are locally organized).

In bringing the focus of the research to the community redevelopment level, we can redefine some of these key players. As noted above, the BC provincial government manages and controls a number of transition activities and its resource policies can create or limit local development opportunities. A second key player is the local government, which acts for local interests and can drive local activities through its community economic development officer. Local governments often become the official conduits through which senior government aid and programmes flow. A third key player, increasingly important since the 1990s, is local civil society organizations. While a complex concept, this chapter follows Massam's (1996) understanding of civil society organizations as collective groups of local residents situated between the state and capital. They range from loosely organized interest groups to highly organized community groups with a legacy of local activity (Reimer, 1999).

Provincial government

BC's economy has experienced a succession of resource development waves over the years. From the fur trade, to gold rushes, to mining booms, to the explosion of interest in fisheries and forest products, BC's economy is closely tied to its natural resource base. The election in 1952 of W.A.C. Bennett's conservative Social Credit party marked a turning point in resource exploitation. The post Second World War period became a directed enterprize with clear policy goals aimed at 'province building' (Williston and Keller, 1997). As was happening elsewhere in Canada (Himelfarb, 1977), BC's resource endowment was imagined as a foundation for provincial prosperity. In terms of hydroelectric power development, the Bennett government launched a two rivers policy. Extensive damming in the Columbia River basin to mitigate flooding and supply a steady flow of water to US power generators earned immediate development cash. Two hydroelectric dams in the Peace River basin created the power source to expand industrial activity in BC.

To support new resource developments, new towns were needed. Historically, such towns may have come about in a chaotic fashion - but such was an expensive and often wasteful process. Both Canada and British Columbia have long experience with resource-dependent and single-industry towns. 'Company towns', 'planned communities', 'resource towns', and 'instant towns' are found in all provinces (Bowles, 1982; Randall and Ironside, 1996). These towns share an economic orientation on a single industry, their economies are overwhelmingly natural resource-dependent and they suffer fluctuating global demands and prices for their resource commodity. Both Gold River and Tumbler Ridge are planned instant towns, where considerable effort was put into their development to create an attractive and diverse community (Gill, 2002; Halseth and Sullivan 2002). This was generally thought to be a good idea by both government service providers, who hoped to minimize the social costs of isolated living, and by resource companies, who anticipated the benefits of a stable workforce (Bradbury and St. Martin, 1983; Gill, 1991; Halseth, 1999). It was also thought to be a good idea by resource companies who no longer wished to be involved in the running or maintenance of towns or the civic infrastructure (Stein, 1952; Bowles, 1992).

In addition to the sweeping policy changes needed to carry out such large resource developments, the provincial government also sought a mechanism to create these 'instant towns' in remote resource sites (British Columbia, 1998). Amendments to the provincial Municipal Act were:

> 'it is in the public interest to establish a municipality in conjunction with the development of a natural resource ... [the Provincial Government] may, by Letters Patent, incorporate ... any area of land ... into a municipality upon the receipt of a petition from at least five owners of land within the area'.

Thus, resource companies could purchase small blocks of land and petition for incorporation. Through a variety of legislative amendments, often collectively referred to as the 'Instant Town Act', the province now had a tool to create towns in advance of resource developments.

In terms of the province's forest base, large industrial forestry firms were encouraged (Williston and Keller, 1997). The goal was to provide local employment, and thus provincial income tax revenue, as well as spin off economic activity from a growing industrial base (Hayter, 2000). The transition to large firms was seen as a way of stabilizing the flow of provincial revenue and various additional policies, such as the requirement to cut a minimum percentage of their timber allowance each year regardless of market conditions, ensured a base income for the province. In many respects, rural BC was imagined by policy makers as a vacant resource bank ready to be brought into 'productive use'.

Successive BC governments, regardless of political orientation, continue to endorse and reinforce this large industrial resource development policy. In 1972, the social democratic government of Dave Barrett was elected. They had proposed a radical agenda to change BC's resource policies. This agenda included diversifying access to resource tenures for small scale operators, to increase competition, to force out inefficient corporations, and to increase provincial royalties. Their proposed radical agenda was thwarted, however, by the demands of the provincial treasury and resistance from both corporate and union sectors (Wilson, 1997). Preservation of the status quo satisfied all three players. Industry inefficiencies (i.e. small firms, outdated machinery and techniques) identified in the early 1970s, particularly with respect to coastal forestry operations, continue to plague the BC forest industry 30 years later.

The election in 1975 of Bill Bennett's conservative Social Credit government brought little change to the province's resource policies. In fact, the government response to a provincial recession in the early 1980s was to look back to the early 1950s and the use of resource developments to 'jump start' provincial prosperity. The updated version involved the Northeast Coal project. Costing nearly CA$3 billion, it involved the creation of two huge coal mines in north-eastern BC, the creation of Tumbler Ridge as the town to house the coal miners and their families, significant upgrades to the BC Rail and CN Rail transportation networks, and construction of a bulk loading port at Prince Rupert (Gill, 1984). Again, both social democratic and conservative governments looked to rural BC as a resource bank to bolster the provincial economy and provide financial support for the services increasingly concentrated in metropolitan areas.

The return of a social democratic government in 1991 meant another set of governments following a well worn pathway based on a geographic imagination of a rural resource hinterland. Two significant policy directions did not divert this large resource industry dependence (Hayter, 2000). The introduction of a new 'Forest Practices Code' and the development of extensive resource planning processes sought to bring 'peace in the woods', and a soft landing transition to a more sustainable forest economy' (Wilson, 1997, p.83). The Forest Practices Code brought in stringent harvesting and management guidelines while the resource planning processes sought to identify lands for protection as well as lands for development. In many respects, these were clearly aimed at ensuring continuing market access vis-à-vis international environmental boycotts, real or threatened (Hayter, 2003; Gunton, 1997).

A second policy adjustment was in response to the recognition that many regions of the province had been harvesting more timber than was sustainable. To deal with increasing uncertainty around timber supply and annual harvests (and the provincial revenues based on those harvests), the province sought to review the calculation of stumpage rates (Wilson, 1997). This is the levy charged to forest harvesters for the right to cut on Crown land. The forest industry argued that both the terrain and the cost of retooling industrial plants varied across the province and appealed for relief. The policy response was called 'waterbedding'; that is, lowering stumpage rates in one forest region while compensating by raising stumpage rates in another forest region. Again, the provincial dependence on steady resource revenue flows meant that there was no fundamental change to the large resource tenure systems in place since 1950.

The current conservative government of Gordon Campbell is perpetuating the provincial focus upon 'resource revenues at all costs' (British Columbia, 2003, 2004). This includes policy directions to loosen government regulations and allow larger firms enhanced freedom and the potential for greater profitability. It also includes the expansion of raw log exports in order to maintain provincial revenues during the softwood lumber dispute with the USA. Finally, the province's push to open up coal-bed methane exploration in the northeast and to have the oil and gas development moratorium on the northwest coast lifted is clearly driven by the provincial treasury's need to replace proportionally dwindling fishing and forestry revenues as the provincial population continues to grow.

In summary, the province's rich resource endowment has been seen as a base for intensive industrial development since 1950. Following staples theory arguments, little industrial diversification beyond raw commodity processing has occurred. Successive governments have adjusted policy to allow large resource companies to remain competitive and export oriented. This has occurred despite that since 1980 massive restructuring in resource industries has meant job losses and mill closures. Successive governments have used public policy in support of industrial capital (and to a degree organized labour) in order to maintain provincial tax revenue.

Gold River

The district of Gold River is BC's first instant town, incorporated by Letters Patent under the 'instant towns' legislation on 26 August 1965 (British Columbia, 1965). Among the provisions of those Letters Patent were the boundaries of the town, its financing and borrowing privileges, the membership of the interim council, and the requirement to hold civic elections in 1969. Industrial development was protected by provisions which allocated lands for industry and prevented the town council from passing bylaws which would tamper with construction or development on those lands. In Gold River, this industrial base was a pulp and paper mill. From the start, economic diversification was hampered in Gold River, as in most of BC's other 'instant towns', by the relative isolation of the town site.

The town got off to an optimistic start. Porteous (1970, p.322) remarked that these modern instant towns 'could well be the forerunner of the wholesale replacement of the obsolete company town idea by the fully planned and socially responsible instant town concept'. Such appraisals were based on the idea that local responsibilities had passed from the company to the local government.

Despite this shift of responsibilities, the local economy remained dependent upon a single resource and control of the local economy rested outside the town (often outside the country). Significant forest industry restructuring after 1980 generated local fluctuations in employment and mill operations. In the mid-1990s, the mill changed ownership several times and one of the new owners sold the mill's paper making machine to an offshore competitor (Chow, 1995). It was widely felt that the loss of the paper making machine left little hope for the viability of the mill. These concerns turned to reality in 1998 when Bowater Industries announced that the pulp mill was to be closed (Crosspoint and Associates, 2000). For Bowater, the purchase of the Gold River mill had secured a new customer base while its closure had removed a competitor from the market (Hamilton, 1998). With no alternate economic base, Gold River was plunged into crisis. From a population of 1820 in 1998, the estimated population of Gold River in 2002 was 1390 (British Columbia Statistics, 2003).

Recovery strategy

In response to this crisis, a two-stage recovery strategy was initiated. The first part was to address the immediate financial and employment needs of workers. Unemployment insurance, worker transition funding, retraining support and local counselling were all made available immediately after the mill closure (Table 21.1). In addition to accessing standard employment assistance programmes, the crisis programme also initiated a number of transition planning activities. Many of these focused on promoting community dialogue and participation around the changes occurring within the town and its local business environment, and the kind of 'future' which residents might desire.

Table 21.1 Crisis Support Programmes - Gold River.
(Source: Crosspoint and Associates, 2000).

Standard programmes:
Employment Insurance Programme- federal
Forest Worker Transition Programme - provincial
Industrial Adjustment Services - joint federal and provincial
Employment Assistance Service - federal
Community Futures - joint federal and provincial
Crisis planning activities:
Engaging of the Forest Community Transition Secretariat
Establishing the Community Action Network Committee
Formation of the Community Transition Committee
Updating of the local Economic Development Strategy
Funding of the Gold River Futures Society Recreation Development Project
Initiation of a Small Business Transition Study
Conducting the Gold River Futures Forum

As part of the crisis programme, the provincial government also initiated a five-year CA$8 million support package (Crosspoint and Associates, 2000). Included in this funding was CA$1.8 million to pay off approximately 75% of the town's debt (a debt largely associated with loans for a recently built aquatic centre). The provincial funding would also contribute (approximately CA$4.2 million) to the municipal tax base over five years to compensate for lost industrial property taxes. Since property tax revenues are the main source of funding for local governments in BC, the loss of taxes from the closed pulp and paper mill had a significant impact on the town's budget. Additional areas of support were in the provision of key local services (see also Furuseth, 1998 on the importance of local services vis-à-vis rural deprivation). These included education (CA$440,000 to ease the transition to lower enrolment levels), health (CA$360,000 for service adjustments), and children and family services (CA$327,000 for counselling support during the transition period). Provincial funding was also identified for economic development support and planning. Small business support funding (CA$150,000) was identified to assist local businesses adjust to the new economic circumstances, and an economic development planning fund (CA$250,000) was provided to support a strategic economic planning exercise.

In addition to securing basic local services, and providing short-term compensation for lost industrial taxes, moves were also made to secure a longer-term property tax base. Key to this initiative was the disposal of company owned houses (at low prices) to stabilize a residential property tax base.

The strategic economic planning exercise, that included economic diversification, formed the second part of the community recovery strategy. The diversification plan for Gold River appears broad based (Table 21.2), including everything from home-based businesses, to tourism, to aquaculture. It includes an emphasis on traditional forestry activities and expanded resources processing. It also incorporates currently 'fashionable' opportunities like sports training centres and retirement housing. In many respects it is a 'shopping list' that tries to catch all of the possible options without either closing the door on possibilities or clearly identifying areas of strength. Such shopping lists are increasingly common products in economic development consultations in BC's rural and small town places. This lack of focus is problematic, especially given that the consultant's report specially identifies the 'key issue for Gold River is how to establish a strong industrial or business cluster' (Crosspoint and Associates, 2000, p.15). The town has an infrastructure developed to support a pulp and paper mill industrial cluster, and it may take considerable adjustment of that infrastructure to support a different type of economy. Without some sense of direction, investments may not be strategically coordinated or may even be counter-productive.

Table 21.2 Gold River diversification strategy.
(Source: Crosspoint and Associates, 2000)

home-based businesses
small businesses
tourism
mariculture/aquaculture
value-added forestry
retirement industry
arts and crafts
sports training centre
natural resource processing

The provincial government review report on Gold River's community recovery strategy also identifies the key challenge of building collective capacity to adjust to change. Among the indicators identified for assessing this capacity change are 'learning new skills, developing a welcoming attitude towards entrepreneurs, and collaborating for economic development' (Crosspoint and Associates, 2000, p.19). Each of these areas is somewhat *terra incognita* for towns which had relied upon a single large employer to provide for both personal and community well-being.

Despite the breadth of alternatives, there has been limited action in Gold River. The diverse rural 'imaginations' of the different players have created various barriers. Local civil society groups, with their diverse memberships, interests, and capacities, have been active participants in local diversification debates. As a result, there has been a lack of consensus and considerable fragmentation. As noted:

> '[s]ome people are eagerly pursuing the future, while others are about to lose hope for the future. The community as a whole has not yet completed the

> grieving process. The community cannot easily develop and grow until the vast majority have accepted the past and are ready to move forward' (Crosspoint and Associates, 2000, p.19).

New inquiries into the community are often greeted with some type of protest voice. For example, closures to local crab and prawn fisheries, and health concerns among the local First Nations band, had long raised opposition to pulp mill activity in Gold River (The Gazette, 1989). In the more than 4 years since the pulp mill closed, Gold River has seen only limited recovery.

The town's government has embraced the language of change. The Gold River municipal website states that since the closing of the mill, 'Gold River has been focused on rebuilding the community. Many new residents have moved into town bringing with them many new skills and interests. We are in the process of building a new diversified industrial tax base and creating new jobs' (Village of Gold River, 2003). While the local government expresses wide support for diversification there is also a clear emphasis on attracting a new large industrial firm to either restart the mill or invest in a different processing facility.

In early 2003, the excitement in a large employer was rekindled when it was announced that the closed mill had been sold to Muchalat Industries and Green Island Energy Ltd. (Lavoie, 2003). The plan is to rebuild it into an environmentally friendly electrical generation plant. The plant would burn wood waste and produce about 250 megawatts of power. Recent changes in power regulation in BC mean that independent producers can sell power into the provincial grid. Again, the provincial government is using the old strategy of resource development projects (this time around power generation) in the non-metropolitan hinterland as a way to boost provincial revenues and economic investment. One advantage for the plan is that the mill was already connected to the provincial power grid with a 250 megawatt transmission line. Besides returning a large industrial contribution to the town's tax base, the new plant would create between 40 and 60 jobs (at the time of its closure, the pulp and paper mill employed about 360 people). Investors include mostly American and offshore interests.

Tumbler Ridge

Located in northeastern BC, on the eastern slope of the Rocky Mountains, Tumbler Ridge began as a coal mining town during the 1980s (Gill, 2002; Halseth and Sullivan, 2002). Its economic heart was two open pit coal mines, 'Quintette' (operated by Denison Mines) and 'Bullmoose' (operated by Teck Corporation). As part of the Northeast Coal Project, Tumbler Ridge was BC's last great experiment in instant town development. Planning began in 1976 and the provincial government issued Letters Patent to incorporate the town on 9 April 1981. Town construction activity started in earnest in 1982 following the signing of a 14-year deal to ship coal to a consortium of steel mills in Japan. The extensive project development costs were also part of a provincial policy to kick start a recession-mired economy and expand industrial development in the region.

As with Gold River, the Tumbler Ridge Letters Patent begin with a rationale defining the establishment of the town as part of a provincially important resource development project. They go on to identify that the town site was to be 'completed by the time the permanent residents arrive so that there will be a socially cohesive and well-planned community conducive to attracting and retaining a stable work force' (British Columbia, 1981). The early 1980s was a period of economic recession in BC, and Northeast Coal was one of two mega projects initiated by the provincial government to stimulate the economy.

As with W.A.C. Bennett's province building policies, the Bill Bennett government 20 years later again used resource developments to stimulate the economy.

Unfortunately, coal prices fell almost as soon as Tumbler Ridge opened (*Tumbler Ridge Tattler*, 9 January 1984, 14 May 1984). This created concern about the viability of both the mines and the town. A 1990 contract dispute with the Japanese buyers led the Quintette Mine, the larger of the two mines, to file for bankruptcy protection. Under the bankruptcy plan, management of Quintette was transferred from Denison Mines to Teck Corporation. Through the 1990s, falling coal prices meant successive rounds of layoffs in Tumbler Ridge. In March 2000, Teck Corporation announced that the Quintette Mine would close 2 years ahead of schedule. Like in Gold River, years of warning about the likely demise of the local resource industry had not been heeded, and the mine closure announcement signalled a crisis.

The reaction of the provincial energy and mines Minister to the Quintette closure announcement was that he had been 'blindsided'. In fact, his Ministry had been in active negotiations with Teck Corporation on a new rail and shipping deal in order to lower company costs and keep the mine viable. When confronted with the closure announcement, the Minister reported that 'the town was built for one purpose, as a place to house the workers for the two mines, and without the mines, I don't know that you could convert the town to some other use' (Hunter and McInnes, 2000, p.D2).

Local residents and government officials were also 'blindsided'. Quintette had recently signed a 5-year export agreement with its Japanese buyers and people felt the mine would operate until at least 2003. Despite the gloomy provincial Minister's response, town leaders were committed to renewal. The tone in the local newspaper was defiant, arguing that:

> '[d]espite the media messages, and the abandonment of Tumbler Ridge as a disposable [town] by Provincial politicians, there is a strong current of confidence about our future running through town. Many residents are survivors from other mining towns … [they] call Tumbler Ridge home and are not willing to move.' (Tumbler Ridge Community Connections, March 15, 2000, p.1)

Six days after the closure announcement, the local council had created an emergency action plan. As a first stage, they convinced regional, provincial, and federal government offices to support Tumbler Ridge through a transition period. With that support in place, the next step involved the creation of a Revitalization Task Force (Task Force). The Task Force included a broad range of interests (Table 21.3). According to its terms of reference:

> '[t]he purpose of the Tumbler Ridge Revitalization Task Force Committee is to focus on the [town] and its ability to deliver municipal, educational, community health, and social services to a viable and stable community infrastructure. Special emphasis will be put into creating an economic environment that creates economic diversity.' (Tumbler Ridge Revitalization Task Force, 2000, p.1)

Stabilization strategy

Similar to Gold River, immediate assistance was offered to workers with respect to unemployment insurance, retraining, and relocation assistance. The Task Force identified three time horizons for action. In the *short term*, the stock of rental, vacant, and company owned houses was to be sold to create a municipal property tax base. Local services (especially education, health and social) were to be stabilized through Emergency Services Agreements to last for a period of 5 years (Sullivan, 2002). Over the *intermediate term*, population change and municipal operations were to be founded upon an orderly transition plan. In the *long term*, a strategic marketing and diversification plan was to be developed as only a diversified economy would help create a more stable population. The high quality residential, commercial, and industrial service infrastructure of Tumbler Ridge was featured prominently in economic development promotional materials.

One of the first areas of concern was returning the town's housing stock to private households in order to secure a residential property tax base. About 80% of local housing was tied up as rental housing by either mines (Quintette or Bullmoose) or Canada Mortgage and Housing Corporation (CMHC). With the Quintette closure announcement, the town negotiated with CMHC and Quintette to acquire 985 units which were then marketed through the Tumbler Ridge Housing Corporation (Hope, 2001). Houses sold for CA$25,000 and apartments sold for CA$12,000. The housing sale went far better than expected and within 2 years 97% of properties were sold. These homeowners now began to pay property taxes, something which helped pay for municipal services such as sewer and water, road maintenance, and the local recreation/aquatic centre.

Table 21.3 Tumbler Ridge revitalization task force.
(Source: Tumbler Ridge Revitalization Task Force, 2000)

Town	Position held
Victoria	Minister of Energy and Mines
Victoria	Ministry of Energy and Mines
Victoria	Assistant Deputy Minister
Peace River	MLA
Peace River Reg. Dist.	Administrator
Peace River Reg. Dist.	Chairperson
Tumbler Ridge	Mayor
Tumbler Ridge	Administrator
Tumbler Ridge	Councillor
Tumbler Ridge	Councillor
Tumbler Ridge	Councillor
Tumbler Ridge	Councillor
Tumbler Ridge	Councillor
Dawson Creek	Mayor
Dawson Creek	Administrator
Fort St. John	Mayor
Fort St. John	Administrator
Chetwynd	Mayor
Chetwynd	Administrator
School District #59	Chair
School District #59	Administrator
South Peace Health Unit	Chair
Peace Health Unit	Chief Administrative Officer

A second early area of concern was key local services such as education, health and social services (Halseth *et al.*, 2003). Each of these priority areas was the focus of a special Services Agreement to guarantee funding through the transition. With respect to education services, the provincial Ministry of Education was to provide 75% of the pre-Quintette closure funding. The agreement was to last for 5 years, with the goal of providing certainty of access to education for people wishing to remain in, or move to, Tumbler Ridge. In fact, the population transition in Tumbler Ridge was so successful that the number of students has remained relatively stable.

With respect to health, a similar emergency funding package was put in place. Health care funding was especially important for both seniors moving to town and for families with young children wishing to stay in town. The package was sufficient to secure the services of two physicians and a nursing staff. Tumbler Ridge, like most of rural Canada, still faces difficulty in recruiting a nursing staff, but the arrival of a large number of older residents seeking a retirement community has bolstered physician caseloads.

With respect to social services, the challenge was to continue providing a full range of services on a reduced budget. The Ministry of Health had previously funded assessment and referral services, counselling and self-improvement programmes as separate services. These were consolidated into one office, and a social worker was specially added to assist youth going through the transition process. This umbrella office was then supported by the Tumbler Ridge Family Support Society - a local volunteer group with a long history of local service.

A third area of concern was the town's debt. Tumbler Ridge still owed a CA$10.8 million infrastructure debt left over from the original town construction. As with Gold River, town debt could be a significant drag on local diversification. To eliminate the debt, the provincial Municipal Finance Authority provided CA$5.3 million, the provincial government provided CA$3.7 million, and the town provided CA$1.8 million. Tumbler Ridge cleared its debt on 22 December 2000.

As part of the Task Force's long-term priorities, an economic diversification strategy was also developed. Consultants hired to research local economic potential highlighted a diverse range of opportunities. As with Gold River, the diversification strategy for Tumbler Ridge ranged from traditional resource development to the growing sectors of recreation and tourism for remaking Tumbler Ridge into a retirement or educational centre (Table 21.4). While the town's services and infrastructure are ready to accommodate most of the diversification suggestions, many residents and decision makers still look towards big resource sector projects.

Table 21.4 Tumbler Ridge Diversification Strategy. (Source: Sullivan, 2002).

resource development (forestry, mining, oil and gas)
recreation (backcountry, ecotourism)
tourism (wilderness, park-based)
home businesses
retirement centre
educational centre

While the local government professed broad support for diversification, it has been clearly searching for a replacement large resource industry. The town's website states that Tumbler Ridge 'has a strategy to encourage growth and development in every sector of the community. The main areas of focus are natural gas exploration, forestry including sylvaculture, tourism and recreation' (underline in original) (District of Tumbler Ridge website, accessed 3 March 2003). It has formed a local tourism committee, but has refused to release one of the remaining small apartment buildings for use by a tourism or educational cooperative on the argument that a new coal mine proposal by Western Canada Coal could hinge on availability of immediate worker housing. The council is clearly pinning its hopes on a new coal mine. As early as 2000, Western Canadian Coal Corporation was exploring properties between the Quintette and Bullmoose sites (Vancouver Sun, 2001). By 2004, reserves were proven and the development process was well underway (Prince George Citizen, 2004; Tumbler Ridge News, 2004). In addition, Duke Energy is preparing for an oil and gas pipeline and processing plant expansion in the Tumbler Ridge area (National Post, 2003). Both projects build upon the provincial government's resource expansion policies which have 're-opened' the province to development much as has each successive provincial government for the past 50 years.

Both in concert with, and contrary to, the position of the council are arrayed a range of civil society groups. A Bed and Breakfast Association, Tourism Co-operative, Dinosaur Museum Society, together with existing naturalist and environmental groups, are all pushing for small-scale diversification away from resource dependence. While a lack of consensus at the local level continues to limit action, the tourism/dinosaur museum initiative is gathering momentum. In part, the lack of tools, in the form of legislation and grants, in existing provincial policy is restricting options to aggressively move on tourism or other alternatives.

A survey was conducted in Tumbler Ridge one year after the mine closed. One of the questions was about resident opinions for diversifying the town's economy (Table 21.5). In many respects, the table mirrors the consultant driven Economic Diversification Strategy suggestions. A focus upon the top five suggestions highlights the potentially conflicting visions for this small community. Resource development and tourism set the foundation for two very different rural imaginations positing two very different economic development futures.

Table 21.5 Recommendations to Improve the Tumbler Ridge Economy

Recommendations	Respondents
Better shopping/stores	75
Big resource industry	68
Cell phone high speed internet	11
Diversified resource industry	142
Expand post-secondary education	11
Expand small businesses	112
Fix roads/bus service	47
New vision at town hall	29
Re-open mine/new mine	100
Ski hill	26
Tourism	147
Other	17

Source: Halseth and Ryser (2002).

At the close of its first 2 years of community transition, the plans of the Tumbler Ridge Revitalization Task Force have met with considerable success. Following the Quintette closure, many households moved out of town, but this departing population was almost entirely replaced by people attracted to the community and the town's available housing.

Conclusions

In summary, this chapter has evaluated the responses of two small resource dependent towns in non-metropolitan BC to crises precipitated by closure of the main local employer. In both cases, it is clear that deeply held geographic imaginations about a resource commodity exporting hinterland with limited economic diversification is truncating revitalization activities. The provincial government is continuing with a five-decades-old policy to support large-scale resource industries through granting use rights to large industrial capital. Despite successive government changes, the dependence of the provincial treasury on resource revenues has resulted in these governments simply reinforcing the existing structure. The recession of the early 1980s has been seen as a turning point from resource industry expansion to resource industry crisis. Despite this, provincial government policy continues to try to repair a system which has been identified as needing fundamental change.

At the local government level, there is a clear commitment to continued resource extraction and export. In some respects, this meets the geographic imagination of its local constituency which has been raised under a mill or mine regime. As well, replacement of one major company with another is an easy and familiar pathway. Despite this, local governments remain open to alternative revitalization ideas, but provincial policy provides few real opportunities for change.

An increasingly important player in local revitalization debates, local civil society organizations are so diverse as to create an often conflicting landscape which precludes local consensus. Some organizations look for a replacement of resource jobs, with their high wages and union contracts. Other groups argue that a closure crisis provides an ideal opportunity for diversification. This lack of consensus can be a barrier to local capacity building. Limited local government vision hampers activities such as community asset mapping and the aggressive pursuit of alternative economic futures. A limited vision of the rural hinterland as a resource bank is restricting provincial government policy to the point where few diversification incentives or tools exist. Each participant's actions are grounded in their own conceptualization of 'rural'. In British Columbia, the geographic imagination of a metropolitan heartland and a resource dependent hinterland continues to both drive local and provincial debate, and hinder rural and small town diversification.

References

Barnes, T.J. and Hayter, R. (1994) Economic restructuring, local development, and resource towns: Forest communities in coastal British Columbia. *Canadian Journal of Regional Science* XVII, 289-310.

Baxter, D. and Ramlo, A. (2002) *Resource dependency: the spatial origins of British Columbia's economic base.* The Urban Futures Institute, Vancouver, British Columbia.

Bowles, R.T. (1992) Single-industry resource communities in Canada's North. In: Hay, D.A. and Basran, G.S. (eds.) *Rural sociology in Canada.* Oxford University Press, Don Mills, Ontario, pp.63-83.

Bradbury, J. (1987) British Columbia: Metropolis and hinterland in microcosm. In: McCann, L.D. (ed.) *Heartland and hinterland: A Geography of Canada.* Prentice-Hall Canada Inc., Scarborough, Ontario, pp.400-441.

Bradbury, J.H. and St-Martin, I. (1983) Winding down in a Quebec mining town: a case study of Schefferville. *Canadian Geographer* 27, 128-144.

BC Government Caucus (2004) *Rebuilding the Heartlands – What we've accomplished together.* Queen's Printer, Victoria, British Columbia.

British Columbia (2003) *The B.C. Heartlands Economic Strategy – A Plan to Revitalize Our Entire Province.* Queen's Printer, Victoria, British Columbia.

British Columbia (1998) *British Columbia Municipal Act (consolidated).* Ministry of Municipal Affairs, Victoria, British Columbia.

British Columbia (1981) *Letters Patent - Tumbler Ridge.* Ministry of Municipal Affairs, Victoria, British Columbia.

British Columbia (1965). *Letters Patent – Gold River.* Ministry of Municipal Affairs, Victoria, British Columbia.

British Columbia Statistics (2003) *Gold River Community Facts.* British Columbia Statistics [Accessed January 23, 2003]. Available from http://www.bcstats.gov.bc.ca

Campfens, H. (ed.) (1997) *Community Development around the World: Practice, Theory, Research, Training.* University of Toronto Press, Toronto, Ontario.

Chow, W. (1995, August 15) Hope for Gold River mill dies as Fletcher buys equipment. *The Vancouver Sun* Final Edition, D1.

Crosspoint and Associates. (2000) Toward a sustainable future for Gold River: A research design aimed at monitoring and evaluating provincial funding for Gold River. Prepared for Ministry of Community Development, Cooperatives and Volunteers, Community Transition and Corporate Policy Branch, Province of British Columbia, Victoria, BC.

Dauncey, G. (1996) *After the Crash: The Emergence of the Rainbow Economy.* Merlin Press and Green Print, Basingstoke.

Davis, H.C. and Hutton, T.A. (1989) The two economies of British Columbia. *BC Studies* 82, 3-15.

District of Tumbler Ridge (2003) *District of tumbler ridge* [Accessed March 3, 2003]. Available from http://www.district.tumbler-ridge.bc.ca/econ.html

Furuseth, O. (1998) Service provision and social deprivation. In Ilbery, B. (ed.) *The Geography of Rural Change.* Longman Ltd, Essex, pp.233-256.

The Gazette (1989) B.C. pollution forces 'devastating' ban on crab, *The Gazette* November 24, p.B1.

Gill, A.M. (2002) Respecting context in northern resource town planning: the case of Tumbler Ridge. *Western Geography* 12, 113-129.

Gill, A.M. (1991) An evaluation of socially responsive planning in a new resource town. *Social Indicators Research* 24, 177-204.

Gill, A.M. (1984) Resource towns in British Columbia: the development of Tumbler Ridge. In: Johnston, M.E. (ed.) *Geographic Perspectives on the Provincial Norths, vol. 3.* Lakehead University, Centre for Northern Studies, Thunder Bay, Ontario, pp.134-150.

Gunton, T. (1997) Forestry land use and public policy in British Columbia: the dynamics of change. In: Barnes, T.J. and Hayter, R. (eds.) *Troubles in the Rainforest: British Columbia's Forest Economy in Transition. Canadian Western Geographical Series Volume. 33.* Western Geographical Press, Victoria, British Columbia, pp.65-74.

Halseth, G. (1999) Resource town employment: perceptions in small town British Columbia. *Tijdschrift voor Economische en Sociale Geografie* 90, 196-210.

Halseth, G. and Ryser, L. (2002) *Tumbler Ridge Community Transition Survey 2001: Report on New 'Tools' for Community Economic Development.* UNBC Northern Land Use Institute, Prince George, British Columbia.

Halseth, G., Ryser, L, and Sullivan, L. (2003) Service provision as part of resource town transition planning: a case from northern British Columbia. In: Bruce, D. and Lister, G. (eds.) *Opportunities and Actions in the New Rural Economy.* Rural and Small Town Programme, Mount Allison University, Sackville, New Brunswick, pp.29-56.

Halseth, G. and Sullivan, L. (2002) *Building Community in an Instant Town: A Social Geography of Mackenzie and Tumbler Ridge, British Columbia.* UNBC Press, Prince George, British Columbia.

Hamilton, G. (1998) US firm to buy Avenor; Abitibi offer blown away: South Carolina pulp powerhouse Bowater will become one of world's largest producers. *The Vancouver Sun* Final Edition March 10, D1.

Hayter, R. (2003) 'The war in the woods': Port-fordist restructuring, globalization, and the contested remapping of British Columbia's forest economy. *Annals of the Association of American Geographers* 93, 706-729.

Hayter, R. (2000) *Flexible Crossroads: The Restructuring of British Columbia's Forest Economy.* UBC Press, Vancouver, British Columbia.

Hayter R. and Barnes, T.J. (1997) Troubles in the rainforest: British Columbia's forest economy in transition. In: Barnes, T.J. and Hayter, R. (eds.) *Troubles in the Rainforest: British Columbia's Forest Economy in Transition. Canadian Western Geographical Series Volume 33,* Western Geographical Press, Victoria, British Columbia, pp.1-11.

Himelfarb, A. (1977) *The Social Characteristics of One-Industry Towns in Canada: a Background Report. Royal Commission on Corporate Concentration, Study No. 30.* Queen's Printer, Ottawa, Ontario.

Hodge, I. (2004) The economic diversity of rural England: stylized fallacies and uncertain evidence. *Journal of Rural Studies* 20, 263-272.

Hope, M. (2001) Town that wouldn't take a tumble: huge real-estate sale brings new life to B.C. community. *Calgary Herald* Final Edition 16 June, HS04.

Hunter, J. and McInnes, C. (2000) Miller powerless to stop death of Tumbler Ridge: minister said he was blindsided by Teck's decision to close mine. *The Vancouver Sun* 2 March,, D1.

Hutton, T. (2002) British Columbia at the crossroads: new regional development pathways for the 21st century - a review of the literature on regional development processes, issues, and policy implications for BC. Centre for Human Settlements, School of Communitiy & Regional Planning, University of British Columbia. Report prepared for Project 250., Regional Economies Expert Panel, Vancouver, British Columbia.

Hutton, T.A. (1997) Vancouver as a control centre for British Columbia's resource hinterland: aspects of linkage and divergence in a provincial staple economy. In: Barnes, T.J. and Hayter, R. (eds.) *Troubles in the rainforest: British Columbia's Forest Economy in Transition. Canadian Western Geographical Series vol. 33,* Western Geographical Press, Victoria, British Columbia, pp.233-262.

Lavoie, J. (2003) Dying Gold River dares to hope again. *Times-Colonist* Final Edition 4 May, C1.

Lee, M. (2003) *Bleeding the hinterland: A regional analysis of BC's tax and spending cuts.* Canadian Centre of Policy Alternatives - BC office, Vancouver, British Columbia.

Mackenzie, S. and Norcliffe, G. (1997) Restructuring in the Canadian newsprint industry. *Canadian Geographer* 41, 2-6.

Marsden, T., Lowe, P., and Whatmore, S. (eds.) (1990) *Rural Restructuring: Global Processes and their Responses.* David Fulton Publishers, London.

Maser, C. (1997) *Sustainable Community Development: Principles and Concepts.* Delray Beach, FL St. Lucie Press, Delray Beach, Florida.

Massam, B.H. (1996) *An Essay on Civil Society.* Community Economic Development Centre Discussion Paper, Simon Fraser University, Burnaby, British Columbia.

McKeown, L., Davidman, K., McIver, D., and Lasby, D. (2004) *The Rural Charitable Sector Research Initiative – A Portrait of the Nonprofit and Voluntary Sector in Rural Ontario.* Prepared for the Foundation of Rural Living, Canadian Centre for Philanthropy, Toronto, Ontario.

McRobie, G. and Ross, D. (1987) A feasibility study for a Centre for Community Economic Development at Simon Fraser University. Community Economic Development Centre, Simon Fraser University, Vancouver, British Columbia.

Morford, S. and Kahlke, R. (eds.) (2004) *Communities and Natural Resources in Transition: Linking Social Science, Decision Makers, and Practitioners for a Sustainable Future – Forum Summaries and Abstracts.* FORREX – Forest Research Extension Partnership, Kamloops, British Columbia.

National Post (2003) Duke Energy gets go-ahead for pipeline extension. *National Post National Edition* 7 January, 2.

Neil, C., Tykkläinen, M., and Bradbury, J. (1992) *Coping with Closure: An International Comparison of Mine Town Experiences.* Routledge, New York.

Porteous, J.D. (1970) Gold River: An instant town in British Columbia. *Geography* 55, 317-322.

Prince George Citizen (2004) Permit issued for coal mine near Chetwynd. *Prince George Citizen* 11 September, 3.

Randall, J.E. and Ironside, R.G. (1996) Communities on the edge: an economic geography of resource dependent communities in Canada. *Canadian Geographer* 40, 17-35.

Reed, M.G. (2003) Marginality and gender at work in forestry communities in British Columbia, Canada. *Journal of Rural Studies* 19, 373-389.

Reimer, B. (ed.) (1999) *Voluntary Organizations in Rural Canada*. Canadian Rural Revitalization Foundation, Montreal, Quebec.

Statistics Canada (2003) [Accessed 18th May, 2003]. Available from http://geodepot.statcan.ca/Diss/Reference/COGG/.

Stein, C.S. (1952) Planning objectives. Reprinted in *Kitimat Town Site 1962 Report*. District of Kitimat, Kitimat, British Columbia.

Sullivan, L.M. (2002) The geography of community crisis: A case of Tumbler Ridge, BC. MA Thesis, University of Northern British Columbia, Prince George, British Columbia.

The Vancouver Sun (2001, February 1) Tumbler Ridge coal feasible, study says. *The Vancouver Sun 1st February*, D3.

Tumbler Ridge Community Connections (2000). Tumbler Ridge – in for the long haul. *Tumbler Ridge Community Connections* 15 March, 1.

Tumbler Ridge Revitalization Task Force. (2000). *Terms of reference: Structure of the Tumbler Ridge Revitalization Task Force*. District of Tumbler Ridge, Tumbler Ridge, British Columbia.

Tumbler Ridge News. (2004) Western Canada Coal Open House. *Tumbler Ridge News* 9 June, 1.

Tumbler Ridge Tattler (1984) Price a question. *Tumbler Ridge Tattler* 9 January, 1.

Tumbler Ridge Tattler. (1984) Price hikes held off. *Tumbler Ridge Tattle* 14 May, 2.

Village of Gold River (2003) *About Gold River*. Village of Gold Website [accessed February 12, 2003]. Available from http://www.village.goldriver.bc.ca/community/profile/profile.html.

Williston, E. and Keller, B. (1997) *Forests, Power and Policy: the Legacy of Ray Willistion*. Caitlin Press, Prince George, British Columbia.

Wilson, J. (1997) Implementing forest policy change in British Columbia: comparing the experiences of the NDP governments of 1972-75 and 1991-? In: Barnes, T.J. and Hayter, R. (eds.) *Troubles in the Rainforest: British Columbia's Forest Economy in Transition. Canadian Western Geographical Series Voume 33*. Western Geographical Press, Victoria, British Columbia, pp.75-98.

Wynn, G. and Oke, T. (eds.) (1992) *Vancouver and its Region*. UBC Press, Vancouver, British Columbia.

Chapter 22

Narratives of community-based resource management in the American West

Randall K. Wilson

Introduction

In the past decade, collaborative community-based approaches to public land management and planning have proliferated rapidly across the rural western USA (Jones, 1996; Wondolleck and Yaffee, 2000; Brick *et al.*, 2001).[1] Community-based resource management (heretofore CBRM or simply collaboration) refers to efforts to increase local resident participation in public land management discussions in order to improve decision making, cultivate local support for monitoring and implementation and avoid or anticipate land use conflicts. Recent estimates have found collaborative management processes underway in well over 200 rural communities throughout the West (Weber, 2000).

However, given the potential significance of this trend – especially in terms of the spatial restructuring of public land use decision making it implies[2] – little critical attention has been paid to the underlying socio-political contexts within which collaboration has emerged. The dominant explanations posit the rise of collaborative management as a response to a crisis over the use, control and meaning of public lands and resources in the West. Expressed as political gridlock between federal managers and various public land interest groups, the crisis is explained as a byproduct of the failure of top-down environmental regulation to balance the ecological health of public lands with the changing socio-economic realities of rural western communities. Relying upon theoretical concepts drawn from civic republican thought, these accounts point to the structural contradictions of federal governance institutions as leading not only to political stalemate, but to an eroded sense of community belonging in many rural western areas. By providing a public forum for greater local participation in decision making processes, and by encouraging open and constructive dialogue, shared learning, and consensus building among participants, CBRM is thus presented not only as a means of improving management decisions and resolving land use conflicts, but as a vehicle for rediscovering a sense of community belonging and responsibility.

The purpose of this chapter is to rethink these narratives and their implications for collaborative resource management. Following recent developments in political ecology research (Peet and Watts, 1996; Escobar, 1996; Willems-Braun, 1997; Wilson, 1999), this discussion will assess the established conceptual categories that comprise hegemonic explanations for the rise of collaborative conservation.[3] The aim is to better understand the political dynamics underlying these constructs and to trace out the potential political and material consequences that they hold for those engaged in collaborative projects. Viewing the dominant narratives of collaboration in this way reveals that, while these accounts do offer important insights into some of the major forces and trends shaping the emergence of CBRM, they are nonetheless incomplete. That is to say, they fail to critically consider underlying currents of socio-environmental power relations that continually shape and reshape the place-specific contexts within which CBRM efforts occur. At first glance, these narratives may appear to hold little significance either for the actual practice of CBRM or for its broader implications upon public land management and policymaking in the USA. However, as this chapter will demonstrate, these accounts wield effects of their own. Insofar as the dominant narratives situate CBRM projects within specific historical socio-environmental trends and trajectories, they set up a number of problematic assumptions and expectations that hinder the progressive potential of collaborative projects.

The discussion begins with a brief overview and assessment of the major narratives that make up the dominant explanations for the rise of CBRM. These narratives are then critically re-examined with reference to both broader regional socio-political trends and to selected empirical data from a case-study of collaborative planning on the San Juan National Forest in Colorado. Finally, an alternative reading is proposed that exposes the 'on the ground' implications of these narratives while seeking to make explicit the complex and dynamic web of socio-environmental forces through which collaboration has emerged in the American West.

Narratives of collaboration

The dominant explanations[4] for the emergence of CBRM can be broken down into two inter-related sub-narratives, both of which cast collaboration as a needed response to a general crisis in public land use decision making expressed through political gridlock. The first sub-narrative explains this impasse as a historically evolving crisis of federal governance. In contrast, the second narrative points to a crisis of community and citizenship in the rural West as the locus of the problem. The specifics of each narrative are examined in turn.

Sub-narrative I: the crisis of federal governance

West and Brechin (1991) suggest that the primary distinction between public land institutions in the USA and other countries in the world resides in the fact that in the USA, public land management is based on a 'exclusive' rather than and 'inclusive' model. That is to say, historically, public land institutions in the USA were established to de-couple local interests from the use and control of national (i.e. public) resources so as to ensure their most efficient and rational use via state-controlled conservation management strategies. By contrast, in Western Europe and other countries, an inclusive model was adopted. In this model, the participation and interests of local communities were more actively incorporated into public land management and planning processes from the outset.

This presumption provides the starting point for narratives explaining the rise of CBRM as a response to a crisis of public land management and policy making caused by the failure of federal governance. These accounts are couched in a historical framework that traces the evolution of public participation in resource management and policy making over time. In this framework, the inception of federal resource agencies in the progressive era established the 'absolute' authority of the federal government over the use and control of public lands and resources. This authority, in turn, was premised on the notion that only the state could provide the scientific expertise and disinterested perspective necessary to manage these resources in a rational and efficient manner (Pinchot, 1910; Hays, 1959).[5] However, as the constituency for public resources has expanded and diversified over time, the dominance of federal control over land use decisions has gradually given way to ever greater formal levels of public participation, increasingly expressed in terms of local- or community-state collaborations.

Progress along this trajectory can be measured by shifts in public land policy. One of the first examples occurred as a response to the dramatic economic growth and urban expansion of the post-war period that helped to trigger an upsurge in recreational activities on public lands. This expansion of the public land use constituency was reflected in the passage of the Multiple Use and Sustained Yield Act of 1960. Although this law entailed minimal substantive change in actual management practices and priorities, it served to formally acknowledge non-resource extraction activities (i.e. recreation and wilderness preservation) as legitimate uses of public lands (Davis, 1997).

Another major shift occurred in the late 1960s and early 1970s with the rise of modern environmentalism as a mainstream political movement. Sparked by a number of concurrent trends and events in American society (e.g. the civil rights movement, the Vietnam War protests, the Watergate scandal, the publication of *Silent Spring*, etc.), environmental organizations quickly gained in both membership and political clout. Increasingly, they deployed this new found power to question the authority, competency and motivation of federal agencies involved in managing public lands. Demands for greater transparency and an expanded role for the public in the policy making process were reflected in a host of new laws. These included the 1964 Wilderness Act, the 1969 National Environmental Policy Act (NEPA), the 1973 Endangered Species Act (ESA), and the 1976 National Forest Management Act (NFMA). The effect of these policies was two-fold. First, they formalized greater public participation in decision making processes through new mandates such as the creation of environmental impact statements required under NEPA. Agencies were ordered to gather public comments and hold public hearings before initiating new management plans and projects. Second, these laws often contained citizen suit provisions allowing the public to directly challenge agency decisions through the courts. Organizations such as the Natural Resources Defense Fund and Sierra Club Legal Defense Fund (both established in 1971) began to develop their own scientific databases and legal skills in order to exercise these new rights.

The trend towards expanded formal public participation in the policy making process continued in the 1980s and 1990s. The anti regulatory orientation of the Reagan and G. W. Bush Sr. administrations not only reduced political and financial support for federal agencies, but allowed greater public access to government reports and databases (e.g. 1986 Emergency Planning and Community Right to Know Act). As a result, environmental groups became increasingly effective in challenging both the scientific and legal basis of federal resource management decisions.[6]

In the Clinton Era, political support for fiscal conservatism and reductions in the federal bureaucracy continued apace. As exemplified in the Clinton Administration's 'Reinventing Government' initiative and the 104th Republican Congress' 'Contract with America', federal agencies were strongly encouraged to partner with private businesses and organizations in order to carry out their resource stewardship duties. Meanwhile, partly in response to the growing success of environmental organizations in affecting policy, groups opposed to environmental preservation (e.g. the wise use movement and various property rights groups) also began to mobilize. These groups mounted their own scientific-legal challenges to federal agency decisions bolstered by new laws emphasizing the role of cost-benefit analysis in federal management, the legitimacy of regulatory takings claims by property owners, and the need to eliminate unfunded federal mandates upon state and local governments (Echeverria and Eby, 1995).

The tendency towards greater public participation received additional momentum as the field of ecology – and in particular, ecosystem management – gained scientific legitimacy and acceptance by federal agencies (Cortner and Moote, 1999). Managing resources according to an ecosystem framework meant adopting new ways of thinking, new assumptions, and new techniques that often challenged the conventional wisdom and expertise of federal managers. Meanwhile, the holistic orientation of ecosystem management – emphasizing both human and non-human nature as parts of a single integrated system – provided yet another incentive for federal agencies to take seriously the practices, interests and concerns of local communities, and to find ways of integrating these into resource management plans.

In the last stages of this narrative, one finds these various socio-economic, political and scientific trends winding together to produce a crisis in public land management in the mid-1990s. The authority of federal resource agencies eroded dramatically as agency actions and decisions became subject to constant juridical scrutiny by all sides in public land use debates. The combination of new political and economic priorities, management paradigms and unprecedented levels of public input have exacerbated the inability of federal resource management institutions to respond to these challenges and adapt to new roles (Nelson, 1995). The result was political gridlock, leading to rising levels of frustration and uncertainty for local activists, and ultimately, poor management of public lands and resources. In the wake of this crisis, CBRM emerged as the latest evolutionary step in the progression towards greater public participation. By facilitating meaningful and constructive public dialogue among disparate public land use interests (including federal agencies), collaboration offered a means of relieving the tensions and contradictions inherent in rigid federal institutional structures while improving the health and integrity of public lands.

Sub-narrative II: the crisis of community

The second sub-narrative explaining the rise of CBRM in the American West characterizes the crisis in public land decision making as a crisis of community or civil society. The point of departure for this account is the New West socio-economic and demographic transformations that have dramatically reordered the environmental politics of many rural communities during the 1990s (Powers, 1996; Riebsame *et al.*, 1997). More specifically, these transformations reflect the rapid influx of ex-urbanite immigrants into the region, many of whom tend to prioritize the aesthetic and recreational values of public lands. In contrast, for many long-term residents (often termed 'old' or 'traditional westerners') local landscapes are valued primarily for commercial resource extraction and development purposes. As these disparate sets of values increasingly rub up against each other in local communities, the potential for conflict has risen dramatically.

However, according to this narrative, the primary cause of the crisis is not the potential tension or conflict engendered by the New West changes per se. Nor is it purely a matter of failed federal political institutions to accommodate these tensions (although federal governance does play an important role). Rather, the crux of the problem is the inability of rural communities to cope with these changes due to an eroded sense of community belonging, civic responsibility and public trust.

For some commentators (Bellah *et al.*, 1985; Etzioni, 1993) the crisis of community in the rural West is a regional manifestation of a much broader national-level erosion of community values. In this context, the crisis is a product of social and economic globalization, the increased mobilization of the population and industry (facilitated by advances in communications and transportation technologies), and the dual homogenization/commoditization of local cultures and place-based identities as amenity resources for tourism and urban development activities. One result of this combination of social and economic forces is a decline in social capital, which, Putnam (1996, 2000) suggests, is crucial for maintaining healthy civic institutions. According to this view, the lack of engagement in civic affairs leads not only to an erosion of public trust, but weakens community-based norms and networks of interpersonal relations. This process in turn undermines the capacity of citizens and civic institutions to deal constructively with conflict, to work towards compromise, and to seek the common good.

Kemmis (1990, 1995, 2000) offers one of the most influential regional accounts of this crisis in the rural West. He argues that the weakened ability of individuals to act as responsible community members arises from an overdependence on federalist institutions of governance.[7] Rather than communicating and working together to solve local problems, local residents, especially those with diverse and competing interests, have tended to look past each other. Attention is focused on convincing federal arbitrators (e.g. either resource management agencies or the courts) to settle disputes in their favour. This federal dependency, along with its concomitant emphasis on individual rights, has impeded the capacity of citizens to view themselves as having common place-based interests (i.e. as communities of place) rather than as members of free-floating special interest groups decoupled from any specific locational tie (i.e. communities of interest). In other words, while federalism tends to excel in providing incentives for individuals to act defensively, to protect their respective interests or rights, it often fails to facilitate initiatives that might benefit the common good.

Consequently, as local public land interest groups have diversified and grown in strength in recent years, land use disputes have become increasingly locked in stalemate. While each group is capable of defending its own interests by blocking the initiatives put forth by the opposition (including federal agencies), no individual interest group is dominant enough to win outright. In this individualistic and competitive atmosphere, there is little incentive to compromise and seek the common good.

For Kemmis, the answer to this dilemma is a return to Jeffersonian ideals of democracy, as expressed in the notion of civic republicanism (Sandel, 1996). The idea is to devolve regulatory power over public lands and resources to the local level through new community-based forms of governance (including CBRM, see below). Granting residents this responsibility, however, requires a simultaneous re-definition of community citizenship along liberal pluralist lines (e.g. from an inherent right to a moral responsibility that is expressed through a commitment to the common good over individual interests). This move, in turn, presumes that local residents can in fact identify the common good without falling into the trap of interest group politics. One way out according to Kemmis is through the cultivation of 'sense of place'. In this context, sense of place refers to the common set of values held by individuals that result from living in the same place or landscape (again, as communities of place). Insofar as these values reside 'in the ground', they offer an objective and material basis for establishing the common good for all local citizens. In this way, the common *good* becomes common *ground* in both a figurative and material sense.

In this sub-narrative, community-based collaboration is thus presented as one ideal means of putting these notions into practice (Shannon, 1992; Kemmis, 2000; Snow, 2001). By devolving responsibility for public land planning and management to the community level, CBRM affords opportunities for local residents to exercise their civic responsibilities. To the extent that resolving land use disputes through collaboration leads to the cultivation of a shared sense of place, CBRM can contribute in significant ways to the identification a broader community-based understanding of the common good.

Rethinking the rise of CBRM

Each of the two narratives described above offers important insights into the forces and trends shaping the emergence of CBRM. However, at the same time, they each reproduce problematic expectations and assumptions that carry potentially adverse implications for the way that CBRM is understood and put into practice. The most problematic of these involves the conceptualization of place and community, and the way each of these notions are related to state-based resource management institutions. In both narratives, there is a need to rethink these conceptions, and in the process, give greater attention to the dynamic, historical, and place-specific socio-environmental contexts within which collaborative projects occur.

Whither the crisis in federal governance

In the first sub-narrative, this rethink requires taking a closer look at the starting points offered as a basis for the historical trajectory. A critical examination of the notion that the creation of public lands and management agencies initiated a 'clean break' between the interests of the federal government and local residents, finds it problematic on at least three accounts. First, it serves to endorse and perpetuate an entire series of epistemological positions that hinder the implementation of CBRM projects (Wilson, 2000). These include the dualistic assumptions of the modern state as disinterested bearer of scientific rationality (versus subjective/traditional knowledge), the zero-sum division of the western landscape into public land or private property, and the conceptual separation of human from non-human nature (i.e. as objectified resources). Each of these assumptions pose potential barriers for collaborative efforts in which groups of private individuals seek to engage in the management of lands and resources deemed 'public', employing tools (i.e. local knowledge) and spatial frameworks (i.e. ecosystem or watershed-based units) that tend to blur such bipolar distinctions (see also Willems-Braun, 1997).

It bears noting that the dominant narrative *does* point to the fact that the crisis of federal governance is due partly to the problematic 'Progressive Era' assumption that the authority of federal managers is tied to their role as unbiased bearers of scientific expertise.[8] However, this explanation fails to make a distinction between the conceptual assumption and the politics through which it was construed in the first instance. In other words, to paraphrase Latour (1993), federal agencies have never been 'modern' unbiased bearers of scientific expertise. Rather, via the institutionalization and practice of state-based natural resource management, government agencies were (and are) inherently implicated in rendering various regional and place-specific socio-political effects.

This point leads to a second implication of an assumed 'break' between federal agencies and local interests, namely, the way it simultaneously serves to erase or gloss over certain historical socio-environmental relations while endorsing others. More specifically, the narrative fails to acknowledge the selectivity of the state in legitimating certain local interests and activities (e.g. Euro-American residents involved in resource-extraction industries) at the expense of other local interests (e.g. indigenous or Hispano residents involved in non-commercial resource use and/or alternative forms of land tenure). By simplifying the notion of local community to a single uniform entity (from which the state is 'distanced'), this assumption not only ignores the complexity of place-based social relations, but formalizes the unequal rights of access to public lands and resources granted to different social groups. In so doing, it creates a conceptual barrier for collaborative projects seeking to establish alternative (e.g. commons-based) and more egalitarian socio-environmental relations (see Wilson, 1999).

Third and finally, this 'absolute' or pure beginning implies linearity in the historical trajectory enframing the crisis in federal governance. It suggests an aura of progress, that society is going somewhere new with CBRM rather than circling around to revisit the multiple histories and complex socio-environmental politics inherent in different places in the West. The danger in this way of thinking is simply that those interested in initiating or participating in a collaborative process may fail to take account of the historically politicized terrain within which it occurs.

An alternative reading I

Rather than suggesting that the institutionalization of public lands in the USA occurred on a *tabula rasa*, an alternative account makes explicit the politics inherent in the process. It considers the role of the federal government in displacing and dismantling indigenous societies and socio-environmental systems respectively, while facilitating Euro-American settlement in the West. For example, in the case of the San Juan National Forest (SJNF) in Colorado, its creation in 1905 served to both freeze and reorder local socio-environmental relations.[9] It formally reinforced the displacement of Ute peoples from the San Juans, while establishing resource-extraction and development as the most appropriate uses of these lands. This point is illustrated in the original 1903 report (Dubois, 1903, p.22-23) outlining the purpose of the new reserves:

1. Farming around the base of the mountains is dependent on irrigation, and the irrigation water all rises in the mountains. Therefore, to prevent floods in spring and drought in summer the forest on the heads of these streams must be preserved.
2. To ensure a steady supply of timber for developing the mineral resources of the country, the forest in the mining districts should be protected from fire and theft.
3. To prevent the overstocking of the range and total destruction of the forest floor, and to regulate the disputes between sheep and cow men, and between Colorado and New Mexico sheepmen, the summer range should be under Government supervision.
4. To prevent and control the repeated forest fires, reserve management and administration is necessary.
5. To ensure a steady supply of timber to the local markets and prevent the control of the supply from passing into the hands of large companies, a reserve is necessary.

Contra West and Brechin (1991), then, it can be argued that the creation of the SJNF served less to detach local interests from forest use and management, than to reorder local socio-environmental and community-state relations. It actually facilitated the establishment and expansion of several local Euro-American communities. Durango, Pagosa Springs, Dolores, Bay Field and Cortez, all owe a major part of their economic growth and development to the continued availability of resources managed and provided by SJNF officials. Moreover, in some instances, the timber contracts brokered with the national forest service led to the wholesale creation of *new* communities. The logging town of McPhee, Colorado owed its entire existence to such arrangements (O'Rourke, 1980). Despite these origins, local ranchers, loggers and miners at times resented the regulatory intrusion of federal agencies in the San Juans. Nonetheless, on balance, from the early twentieth century to the post-World War II period, forest service managers worked to serve the needs of these local and regional industries (White, 1991).

Resituating the origin of western federal lands and management institutions within their historically and politically dynamic contexts alters the narrative and the expectations it renders. Rather than tracing the rise of collaboration as the culmination of a rather rigid linear trajectory of ever-greater public participation, the emergence of CBRM can be viewed as the latest manifestation of an ongoing dialogue that is about and occurs through local socio-environmental relations of power. At stake is not merely the resolution or amelioration of site-specific conflicts, but negotiations over what constitutes appropriate or legitimate relations between the community and the forest, between local residents and the state (or federal government), and between individual residents who are attempting to redefine the meaning of 'community.'

Whither the crisis of community

The concepts of community and place lie at the centre of the second sub-narrative and provide the focus for an indepth re-examination of the conceptual assumptions underlying this account. Insofar as the cultivation of a sense of place and community belonging serves to reinvigorate individual commitments to the socio-cultural, political-economic and ecological 'good' of local communities (i.e. the common good), these concepts hold enormous potential in facilitating more progressive and egalitarian socio-environmental relations. However, as Young (1990) observes, while the *ideal* of community provides a worthy goal, the way there can be fraught with potentially adverse and often unexamined political implications. Foremost among these are the way such efforts affect issues of community membership.

This point is demonstrated in recent work examining efforts to 'rediscover' a sense of community belonging as a response to the dual forces of globalization and the devolution of state institutions (Staeheli *et al.*, 1997). Staeheli (1997) argues that the extent to which this move requires a redefinition of community membership along liberal pluralist lines – that is to say, from an inherent right to a moral responsibility – it can function to exclude socially and economically marginalized groups. Depending on how civic responsibility is defined, the shift can undermine social service provisions and weaken the citizenship rights and protections upon which these groups rely to meet basic needs. A key dynamic in determining such an outcome is the process through which community and the common good are defined. For collaborative efforts, these concerns speak to the issue of public participation and the way 'relevant' stakeholders are identified and solicited. Pluralism assumes equal and open access to political processes, including community-based collaborative forums. However, as argued above, the historically-politicized place-based terrain upon which CBRM occurs plays a profound role in determining who does and does not have a voice at the collaborative table.

Related to these concerns about community are the assumptions underlying the concept of place, or sense of place. According to the sub-narrative, sense of place offers one means of bringing greater objectivity to the process of redefining community and articulating the common good by providing a material basis for these discussions. The notion that shared ties to a local area can serve as a stimulus for forging new place-based coalitions of interest in political struggles is well documented (Cox and Mair, 1991; Harvey, 1996). However, the forging of these coalitions requires very explicit political negotiation between diverse groups to identify the way their varied interests in local places might in fact align under certain conditions. This position is quite different from suggesting that a basis for a political or moral agenda resides objectively in the landscape (needing only to be discovered). Similar to the assumptions of objectivity underlying modern state-based resource management institutions described above, such essentialist thinking can mask the inherently political process of constructing place identities or meanings as something apolitical, objective, or natural (Harvey, 1996). This logic can allow reactionary or other narrowly-based interests to cast a veil of legitimacy upon their own agendas simply by claiming them as 'common ground'.

An alternative reading II

An alternative reading of the crisis of community begins from the assumption that community belonging and sense of place are the products of dynamic socio-environmental relations. Rather than assuming that place holds objective meanings and values, it is acknowledged that these may differ for different social groups. Hence, while the promise and appeal of rediscovering a sense of place in the face of accelerating globalization helps to explain public support for CBRM, it does so *not* because place offers an objective alternative to political negotiation. Rather, the widespread appeal of CBRM is due precisely to the fact that different (in fact, quite disparate) interest groups hold unique assumptions about what place, community and 'going local' actually mean. This point is exemplified in three of the major groups of public land interests and actors in the West, all of which have given some level of support to CBRM in recent years

The first group consists of federal resource management agencies. As outlined above, the reasons for endorsing collaboration in the 1990s were three fold. These included the growing influence of ecosystem-management, the increasing political and legal challenges to agency decisions and competency, and the impacts of government downsizing including new incentives for entering into public-private partnerships. Insofar as CBRM provided a vehicle for addressing these issues, it began to be integrated (to varying degrees) into agency resource management protocols.

However, a lingering and unresolved issue is the *level* of 'power sharing' or input granted to the public in final decision making processes. Due in part to the challenge, this move poses to the authority and legitimacy of state-based resource stewardship, it is far from clear how a balance might be struck. Aside from the BLM's resource advisory committees and the unique case of the Quincy Library Group,[10] most agencies are reluctant to transfer such power to groups of local residents and, in fact, are legally forbidden to do so under the constraints of the 1972 Federal Advisory Committee Act. In short, for federal agencies, support for 'going local' did not necessarily imply concomitant support for reductions in agency authority. The distance between agency expectations and those held by local residents over what CBRM actually entails in shaping public land use decisions represents a potentially significant barrier to achieving common ground.

For members of the second major group, environmental preservationists, the move towards CBRM holds a different set of potential meanings. The dominant narratives explain support as resulting from the mounting fatigue experienced by grass-roots environmental activists faced with political and bureaucratic gridlock. However, the appeal of community- or place-based approaches to environmental management is rooted in more deeply embedded, historical strands of thought within the environmental movement. For example, bioregionalists and deep ecologists have long warned of the dangers posed by the large-scale commercial development of environmental resources (Sale, 1985; Jackson, 1994; Pena, 1998). Alternatively, they have espoused the importance of 'going local', of acknowledging and cultivating place connectedness or sense of place as a basis for initiating socially progressive and environmentally sustainable forms of land use practices and relationships.

Sense of place in this guise, however, carries a dual meaning. First, the term refers to the cultivation of local ecological knowledge that can be used to guide environmental resource management decisions in sustainable ways (Worster, 1979). In addition to this rather scientific-ecological meaning, sense of place also implies a particular morality, a personal and often emotional attachment and obligation to ensuring the integrity of one's local environment. Sense of place then is expressed not only as local knowledge, but as a commitment to a set of ideals that motivate one to inhabit a place in a sustainable manner.

A third major group consists of those with interests in resource-extraction and development activities on western public lands. Characterized by the wise use, county supremacy, and property rights movements, the endorsement of community-based collaboration in the 1990s by this group often coincided with calls for deregulation and greater local control over public lands and resources (Cawley, 1993). This stance was premised in part on the need for greater local control over public matters (i.e. lands and resources) that had been abused by what were perceived as bloated, incompetent and inefficient federal bureaucracies (Nelson, 1995). For this group, going local implies relief from the regulatory restrictions imposed by environmental protection laws and greater freedoms for individual landowners to use public resources as they see fit. It is also premised on a particular view of community and sense of place.

In this context, rediscovering a sense of place implies a renewed appreciation of the traditional Euro-American socio-economic relations and values that historically shaped many rural western communities (Hage, 1990). It prioritizes the knowledge and experiences of local residents involved in resource-extraction activities as the most appropriate basis for land use decision making.[11] Insofar as collaboration implies greater local influence, which in turn facilitates the continuation of traditional land-based economic activities, it is seen as a means of improving the well-being of both the land and local communities.

It is important to note that each of these groups contains cross-currents of interests and ideologies that deviate from the accounts offered above. For example, within the environmental movement, there are serious concerns about the extent to which the devolution of decision making processes implied by collaboration will result in an erosion of federal environmental protection laws. Since many collaborative efforts occur in rural towns historically tied to ranching, timber and mining industries, CBRM is likely to favour those interests while devaluing or muting the voices of non-local urban constituents of national organizations (McCloskey, 1996). Similar concerns are present in the pro-resource-extraction and development group. The most common expression of dissent occurs through NIMBYist concerns voiced by property owners wishing to protect land values threatened by site-specific development projects. Likewise within many government agencies, those who favour CBRM as an integral part of the evolution of public land management are often at odds with opposing views that see collaboration as undermining the rational scientific-basis of resource management.

Nonetheless, the key point is that although each of these three groups supported CBRM, they did so for very different reasons. Their endorsement hinged (to varying degrees) on group-specific conceptions of place and community belonging, especially on diverse understandings and expectations about what ‘going local’ might actually mean. In the case of the SJNF planning effort, the range of support was evidenced in the diversity of motivations local residents claimed for their participation in the collaborative process.[12]

From this purview, then, the rapid proliferation of CBRM in the 1990s was largely due to the fact that interest groups across the political spectrum each perceived collaboration as a means of furthering their own, at times rather disparate, agendas. A series of scientific, political, economic and ecological forces converged creating a terrain of conditions from which a variety of interests groups came to envision community-based and collaborative approaches as a preferred approach to public land use decision making. While understanding this dynamic helps to explain the widespread enthusiasm for and legitimation of collaboration as a dominant trend in the 1990s, it also raises questions regarding the future viability of CBRM when and if these incompatible expectations are not realized ‘on the ground’.

Conclusions

Massey (1993) suggests that achieving a 'progressive sense of place' necessitates understanding the full complexity and fluidity of places, viewing them as particular 'constellations' of dynamic socio-spatial power relations operating at the full range of spatial scales. From this perspective, attempts to define place identities can be seen as efforts to temporarily and artificially freeze socio-environmental relations at a particular moment according to a particular ideology. Acknowledging these dynamics allows one to engage in a politics of place that might re-shape these meanings and identities along more socially and environmentally progressive lines. In this chapter, the discussion has demonstrated that, like sense of place, CBRM is a similarly complex and politicized process. This observation becomes especially evident as one begins to view collaboration in its place-based social and environmental contexts. In contrast, the dominant explanations for the rise of collaboration in the American West tend to gloss over the politics of place. In fact, the extent to which these narratives successfully establish and reproduce their influence turns, in part, on their ability to represent the complexities of collaboration as highly generalized abstractions. Place-specific socio-environmental relations are translated into conceptual binaries that serve to justify certain norms, values and environmental practices while, at the same time, the narratives themselves are cast as self-evident common sense accounts.

More specifically, the problematic conceptual assumptions contained within these narratives serve to hinder the inclusive community-building goals of collaboration in very real ways. Revealing, anticipation and grappling with these barriers requires a critical reexamination of CBRM processes. In short, realizing the progressive potential of CBRM necessitates *re*capturing the complexity of the socio-environmental contexts and relations through which collaboration has emerged and been put into practice in the rural West. For example, making explicit the fully politicized terrain upon which state-based resource management institutions were originally established helps to explain why certain socio-cultural groups might respond adversely to participation in CBRM projects initiated or facilitated by historically antagonistic interests. Failure to acknowledge these relations may serve to re-enforce detrimental pluralistic assumptions, namely, that those who 'choose' to participate are the only voices that matter since legitimate community membership is defined in terms of civic responsibility (i.e. participation).

Similarly, realizing that CBRM participants do not abandon their respective interests, histories or politicized-selves at the door (along with their hats), but bring them in tow as they negotiate and articulate various visions of the common good, helps to illuminate the diverse assumptions and expectations participants deploy in collaborative discussions. It also underscores the need for facilitators to be sensitive to the unequal levels or forms of socio-political influence wielded by different participants. By ignoring or downplaying these differences, the dominant narratives risk leaving CBRM practitioners and scholars with the false notion that participants are drawn to the table for the sake of compromise when this is simply not the case.

Moreover, the problematic assumption of some objectively-derived common sense of place in collaborative discussions can produce a series of troubling ripple effects. Just as this assumption masks the diversity of place meanings and experiences held by local residents, it can also gloss over important distinctions in other vaguely defined terms. Notions such as 'forest health', 'multiple use,' 'sustainable development', and 'community viability' are often held up as evidence of shared values or common ground when, in fact, they may carry very specific meanings for different groups and individuals. Although participants in collaborative discussions might agree that 'forest health' should be a management priority, they may differ vehemently about what constitutes 'health' and how it should be measured or achieved. Failure to clarify these differences in meaning creates a potential discordance between participant expectations and the actual 'on the ground' management policies resulting from collaborative processes. In such instances, collaboration may *exacerbate* rather than ameliorate local land use conflicts.[13]

In addition to identifying these problematic expectations, this chapter also points to alternative loci for explaining the dynamics underlying collaborative processes. In other words, if consensus is not achieved via collaborative processes, it has little to do with the failure to 'identify' a pre-existing notion of common ground. Understanding the common good as a dynamic product of socio-environmental relations undermines the notion of some autochthonous rendering of common values of place as suggested by Kemmis (1990) and others. Common ground is not discovered but made. Hence, to understand the forces driving collaborative processes one should look to the dynamic and complex socio-environmental contexts and relations upon which and through which they occur. The future potential of CBRM rests heavily upon the extent to which collaborative projects serve to mediate these relations.

Notes

1. CBRM does not always entail a local or community-based component (as evidenced by federal and state efforts to work together to manage whole ecosystems or landscapes). My use of the term collaboration in this chapter refers to community-based management processes involving community/state and intra-community collaboration.
2. This trend refers to the extent to which CBRM entails a spatial shift in the locus of public land use discussions from the national political arena to local community-based forums (see McCloskey, 1996; Wilson, 2000; Brick *et al.*, 2001).
3. This approach is sometimes termed 'neo-Gramscian' insofar as it builds upon Gramsci's (1971) observations that reproducing hegemonic power relations in society requires not only coercive power, but consent. One means though which consent is achieved involves discursive practices that present these relations as 'common sense' or self-evident truths lying beyond the realm of questioning. Post-structuralist discourse analysis serves to deconstruct and draw out the relations of power that are reproduced through such discursive interactions. The goal is to demonstrate how representational practices render very real material and political effects upon social and environmental actors (see McKay, 1994; Hall, 1996; Escobar, 1996).
4. Following the critical theoretical approach informing this chapter, the dominant discourse explaining the rise of CBRM does not derive from a single account. Rather it reflects a compilation of narratives sharing a series of epistemological assumptions that define the discursive and conceptual parameters surrounding an issue, idea or conflict. As Foucault (1980) demonstrated, these parameters are reproduced by all actors engaged in discursive interaction, regardless of their respective political interests or ability to affect relations of coercion and power. The ability of a hegemonic discourse to accommodate diverse political agendas while simultaneously placing distinct limitations on what constitutes legitimate epistemological positions helps to account for its 'benign' appearance and, in turn, its lasting influence on society.
5. In practice, the USDA Forest Service has always maintained close ties with local residents and worked hard to meet local economic needs. The authority of the agency, however, has premised a conceptual separation between the objective scientific ideals of the

Forest Service and subjective interests of local and regional stakeholders.

6. This participation was perhaps most famously demonstrated in the old growth/spotted owl controversy in the Pacific Northwest (see Foster, 1991).

7. For Kemmis, this overdependence is part of an enduring colonialist legacy that continues to characterize the West's relationship with the federal government.

8. Recall that according to the sub-narrative, insofar as the expansion of public participation in public land use decision making challenges or undermines official knowledge, it simultaneously serves to raise questions about federal authority.

9. Initially established as three forest reserves – the San Juan, Uncompaghre and Montezuma Reserves respectively – the SJNF assumed its approximate current boundaries in 1947.

10. The Quincy Library Group gained decision making authority through an act of Congress in 1998.

11. For some commentators, including Hage (1990), the implications of this local knowledge and experience (e.g. 'custom and culture') go even further. In accordance with wise use rhetoric, this knowledge in combination with the continual historical use of the public domain by local residents translates into the right to claim portions of the public domain as the private property of local users.

12. In a survey of participants the most frequently cited reason for attending the SJNF collaborative planning meetings was to 'promote personal views' (Wilson, 2000).

13. The potential for discordance is likely to grow as collaborative efforts in the West are increasingly translated into specific land use practices with measurable social and ecological impacts.

References

Bellah, R., Madsen, R., Sullivan, M., Swidler, A., and S. Tipton. (1985) *Habits of the Heart.* University of California Press, Berkeley.

Brick, P., Snow, D. and S. Van de Wetering. (eds.) (2001) *Across the Great Divide: Explorations in Collaborative Conservation and the American West.* Island Press, Washington D.C.

Cawley, R. (1993) *Federal Land and Western Anger: The Sagebrush Rebellion and Environmental Politics.* The University Press of Kansas, Lawrence.

Cortner, H., and Moote, M. (1999) *The Politics of Ecosystem Management.* Island Press, Washington D.C.

Cox, K. and Mair, A. (1991) From localized social structures to localities as agents. *Environment and Planning A* 23, 197-213.

Davis, C. (ed.) (1997) *Western Public Lands and Environmental Politics.* Westview Press, Boulder.

DuBois, C. (1903) *Report on the Proposed San Juan Forest Reserve, Colorado.* USDA Forest Service, Washington D.C.

Echeverria, J. and Eby, R. (eds.) (1995) *Let the People Judge: Wise Use and the Property Rights Movement.* Island Press, Washington, D.C.

Escobar, A. (1996) Constructing Nature: Elements for a Poststructural Political Ecology. In: R. Peet and M. Watts (eds.). *Liberation Ecologies.* Routledge, London, pp.46-68.

Etzioni, A. (1993) *The Spirit of Community: The Reinvention of American Society.* Touchstone, New York.

Foster, J. B. (1991) Capitalism and the ancient forest. *Monthly Review* 43, 1-16.

Foucault, M. (1980) *Power/Knowledge: Selected Interviews and Writings, 1972-1977.* Pantheon, New York.

Gramsci, A. (1971) *Selections from the Prison Notebooks.* Translated by Q. Hoare and G.N. Smith. International Publishers, New York.

Hage, W. (1990) *Storm Over the Rangelands: Private Rights in Federal Lands, Second edition.* Free enterprize Press, Bellevue.

Hall, S. (1996) *Stuart Hall: Critical Dialogues in Cultural Studies.* In: D. Morley and K-H. Chen (eds) Routledge, New York.
Harvey, D. (1996) *Justice, Nature and the Geography of Difference.* Blackwell, Cambridge.
Hays, S. (1959) *Conservation and the Gospel of Efficiency.* Harvard University Press, Cambridge.
Jackson, W. (1994) *Becoming Native to this Place.* University Press of Kentucky, Lexington.
Jones, L. (1996) Howdy neighbor! As a last resort, westerners start talking to each other. *High Country News* 28, 1, 6, 8.
Kemmis, D. (1990) *Community and the Politics of Place.* University of Oklahoma Press, Norman.
Kemmis, D. (1995) *The Good City and the Good Life.* Houghton Mifflin, New York.
Kemmis, D. (2000) *This Sovereign Land: A New Vision for Governing the West.* Island Press, Washington DC.
Latour, B. (1993) *We Have Never Been Modern.* Harvard University Press, Cambridge.
Massey, D. (1993) Power geometry and a progressive sense of place. In: J. Bird, T. Curtis, T. Putnam, G. Robertson, L. Tickner (eds) *Mapping the Futures: Local Cultures, Global Change.* Routledge, London, pp.59-69.
McCloskey, M. (1996) Collaboration has its limits. *High Country News* 28, 6.
McKay, I. (1994) *The Quest of the Folk: Antimodernism and Cultural Selection in Twentieth-Century Nova Scotia.* McGill-Queen's University Press, Montreal.
Nelson, R. (1995) Public Lands and Private Rights: The Failure of Scientific Management. Rowman and Littlefield, Lanham.
O'Rourke, P. (1980) *Frontier In Transition: A History of southwestern Colorado.* Bureau of Land Management, CO State Office, Denver.
Peet, R., and Watts, M. (Eds) (1996) *Liberation Ecologies: Environment, Development, Social Movements.* Routledge, London.
Pena, D. G. (ed.) (1998) *Chicano Culture, Ecology, Politics: Subversive Kin.* The University of Arizona Press, Tucson.
Pinchot, G. (1910) *The Fight for Conservation.* Doubleday, Page & Co., New York.
Powers, T. (1996) *Lost Landscapes and Failed Economies: The Search for a Value of Place.* Island Press, Washington DC.
Putnam, R. (1996) The strange disappearance of civic America. *American Prospect* 24, 34-48.
Putnam, R. (2000) *Bowling Alone: The Collapse and Revival of American Community.* Simon & Schuster, New York.
Riebsame, W., Robb, J., Limerick, P., and W. Wilkinson. (1997) *Atlas of the New West: Portrait of a Changing Region.* W.W. Norton, New York.
Sale, K. (1985) *Dwellers in the Land: The Bioregional Vision.* Sierra Club Books, San Francisco.
Sandel, M. (1996). *Democracy's Discontent: America in Search of a Public Philosophy.* Belknap Press, Cambridge.
Shannon, M. (1992) Community governance: an enduring institution of democracy. In: Library of Congress, Congressional Research Service (ed.) *Multiple use and sustained yield: changing philosophies for federal land management?* GPO, Washington D.C., pp.219-246
Snow, D. (2001) Coming home: an introduction to collaborative conservation. In: Brick, P. *et al.*, (eds) *Across the Great Divide.* Island Press, Washington DC, pp.1-11.
Staeheli, L. (1997) Citizenship and the search for community. In: Staeheli, L. *et al.*, (eds) *State Devolution in America: Implications for a Diverse Society.* Sage, Thousand Oaks, pp.60-75.
Staeheli, L., Kodras, J., and Flint, C. (eds) (1997) *State Devolution in America: Implications for a Diverse Society.* Sage, Thousand Oaks.
Weber, E. (2000) A new vanguard for the environment: grass-roots ecosystem management as a new environmental movement. *Society and Natural Resources* 13, 237-259
West, P., and Brechin, S. (1991) *Resident Peoples and National Parks.* University of Arizona Press, Tucson.
White, R. (1991) *It's Your Misfortune and None of My Own: A New History of the American West.* University of Oklahoma Press, Norman.
Willems-Braun, B. (1997) Buried epistemologies: the politics of (post)colonial British Columbia. *Annals of the Association of American Geographers* 87, 3-31.

Wilson, R. (1999) 'Placing nature': the politics of collaboration and representation in the struggle for La Sierra in San Luis, Colorado. *Ecumene* 6, 1-28.

Wilson, R. (2000) Public lands, community and the politics of place: exploring collaborative resource management in southern Colorado. Ph.D. Dissertation, The University of Iowa, Iowa City.

Wondolleck, J., and Yaffee, S. (2000) *Making Collaboration Work: Lessons from Innovation in Natural Resource Management.* Island Press, Washington DC.

Worster, D. (1979) *Dust Bowl: The Southern Plains in the 1930s.* Oxford University Press, Oxford.

Young, I.M. (1990) The ideal of community and the politics of difference. In: Nicholson, L. (ed.) *Feminism/Postmodernism.* Routledge, New York, pp.300-323.

Chapter 23

Youth, partnerships and participation? Action research as a methodological approach

Christinc Corcoran

Introduction

The aim of this chapter is to draw together rural research on partnerships and youth. Although there has been interest in both partnerships and youth in rural geography, there has, to date, been little work carried out to combine the two. This discussion addresses that gap by drawing on recent doctoral research on a partnership that had a remit to address the issues and concerns of young people.

The research addressed a number of tensions between partnerships and young people living in rural areas, specifically the problem of combining 'top-down' and 'bottom-up' voices: an integral part of the partnership ideal. Despite rhetoric and intent to represent and integrate young people, the partnership under scrutiny had no direct input from the young. This essential but missing element questioned the integrity of the partnership philosophy and added to concerns identified by those involved in youth studies that the young are indeed powerless in an adult world (see, for example, James, 1990; Rosenbaum, 1993; Valentine, 1997). To overcome and resolve these issues, 'action research' was undertaken to bring the voices of the young back into the partnership.

Although a considerable part of this project involved researching the issues and problems faced by young people in rural areas, it is not the intention to report these findings here (but see Corcoran 2003a, 2003b for details of these). Rather, the emphasis is on tracing the political relationships between young people, the researcher and a government-steered partnership. The remainder of this chapter therefore outlines four elements of this action research project. First, it reviews the background to the debate on youth and partnerships; secondly, it discusses the contextual background of the case-study partnership; thirdly, it examines implications of action research as a methodology for researching the young; and, finally, it evaluates the degree of integration achieved between young people and policy makers by the partnership. It then concludes with a short discussion on the success of action research as a methodology.

Background to the debate

Young people

During the early part of the 1990s concerns were raised that a rich stream of 'other' voices such as the young, elderly and women were missing from rural academic research (Philo, 1992). These voices were rarely heard in local political debate as they had been eclipsed by louder voices belonging to the dominant majority, namely white, heterosexual males of middle class and middle age (Cloke and Little, 1997). Ironically, as Philo (1992) pointed out, research was being carried out both *by* and *on* the same cohort: groups of academic middle-aged, middle-class heterosexual males were carrying out research on groups of middle-age, middle-class heterosexual males.

Debate over the value of a post-modern approach to rural studies sparked an exchange of views that highlighted two approaches to studying 'others': those who advocated listening to 'others' as a means of contributing to our store of knowledge (Philo, 1992); and those who argued that in order to further our understanding what we focus on the power relations surrounding these groups of 'others' (Murdoch and Pratt, 1993). Although the debate has since moved on to consider questions of positionality, social construction and the spatial implications of post-modernism, both of these approaches had merit in putting neglected groups at the centre stage of rural geography.

Young people in rural areas have been the focus of initial, and ongoing, attention in the 'others' debate. They mirror the dichotomy of this discussion in that, on the one hand, they are not often heard (be that in research or in the community at large) and, on the other, they are excluded from political power. They frequently find themselves the object of decisions made for and on their behalf, and over which they have no control. Politically, (that is their ability to engage with decision makers) they are almost totally disempowered, a somewhat contradictory position given the current climate of encouraging 'bottom-up' participation. As a result of this powerlessness they have been largely over-looked by policy makers.

Although there is evidence of increasing geographical interest in the lives of the young, three areas remain neglected:

- the majority of youth studies are focused on urban areas: youth in a *rural* context have been absent;
- with one or two exceptions, studies have been on or about youth, not *from* youth. The voices of the young themselves are still missing;
- in the main studies have focused on issues such as identity, migration and the environment. Research is needed into the rural political arena (with a small 'p') that concerns itself with the experiences of young people and their relationship with the decision making process.

To begin bridging these gaps, it is not only important to focus on the young in rural areas, but to ask them to speak for themselves about the issues that are important to them and, further, to ensure that these concerns are heard by decision making bodies. The ideal vehicle through which to develop such a project is through a focus on partnerships, a form of governance very much in vogue at present.

Partnership

Partnership is a term that is widely used to describe members from the public, private and voluntary sectors joining forces with a view to targeting specific commonly agreed goal(s) and objective(s) and has been broadly defined as:

> 'an arrangement which deliberately draws together the resources of specified partners in order to create a capacity to act with regard to a defined objective or set of objectives.' (Edwards *et al.*, 2000, p.2)

Despite the growing interest and application of partnerships, they are not a particularly recent concept. Imrie and Raco (1999) trace their use from the 1960s; during the 1970s they formed a central plank of government strategy particularly in regeneration projects (Geddes, 1997), and in the 1980s were utilized in the implementation of social policy (Berthoud *et al.*, 1981). In terms of public policy, these partnerships concentrated on economic regeneration with their main targets being specific areas of decline or disadvantage (Taylor, 2000).

By the 1990s, a different emphasis framed policy targets in a discourse of partnership that broadened their remit to include and consider social capital (Taylor, 2000). Over a passage of time, government rhetoric came to recognise that partnerships could transcend 'political' boundaries and acknowledged their contribution to policy coordination. Partnership working recognised that changes, where they are needed, are better seen to come from 'within', and not imposed from 'without' (HMSO, 1995).

A frequently stated aim of partnership working is to bring members of the community into the policy debate and, thereby, to integrate voices from the 'top' and the 'bottom'. Partnerships increase the ability to network and embrace all manner of people from all manner of sources such as members of local government, commerce, the clergy and academia. In this way partnerships have the capability of providing a forum for local people, particularly those often excluded from discussion such as the young and the elderly, to contribute to the decision making process (Snowdon and Slee, 1997). This is particularly relevant in terms of young people as increasingly service provision is being made in partnership between the public and other sectors (Hackett, 1997). In addition, partnerships have the ability to be flexible and cope with the wide range of issues that are a notable facet of work with young people (Garonna and Ryan, 1991).

During the 1990s the growth of partnerships was extensive and they now apply to a range of policy sectors including education and training, housing, crime and policing, health, community care, social services and community regeneration (Geddes, 1997). The following section, however, focuses on the use of partnerships in the development of anti-poverty policies at the level of local government.

Contextual background to the Worcestershire anti-poverty partnership

The impetus for the development of local anti-poverty strategies derived from the UK's New Labour Government's mandate was to address the inequalities that had allegedly emerged under the Conservative Government (Mulgan, 1998). Such was Government's concern with issues of poverty and exclusion that the Prime Minister made this the theme of his first speech after the General Election (Mulgan, 1998):

> 'For 18 years, the poorest people in our country have been forgotten by government. They have been left out of growing prosperity, told that they were not needed, ignored by the Government except for the purpose of blaming them. I want that to change. There will be no forgotten people in the Britain I want to build.' (Social Exclusion Unit, 1997, p.3)

Subsequently the Prime Minister set up the Social Exclusion Unit (SEU) in the Cabinet Office with the remit of attacking social exclusion. In the case of young people, particular emphasis was placed on developing early intervention strategies that would prevent their descent into poverty and social exclusion (SEU, 1997). The SEU recognized that there was no single cause of poverty, issues were frequently interconnected and therefore departments could no longer work in isolation (Halpern, 1998; Mulgan, 1998). In order to develop policies to address these concerns, different government and non-governmental agencies would be required not only to engage in 'joined-up' thinking, but also to engage in 'joined-up' working as well.

It was considered to be bureaucratically easier to move these policies forward at the local level (Mulgan, 1998). Here, the role of local authorities in tackling exclusion and poverty is particularly significant as they have the potential to develop and enable economic improvement programmes and act on behalf of people and areas that are particularly deprived (Balloch and Jones, 1990).

It was in response to this background and rhetoric on poverty and exclusion that Hereford and Worcester County Council[1] (HWCC) initiated a conference in 1996 aimed at encouraging inter-agency collaboration. The general consensus from the subsequent feedback session was that an inter-agency forum should be initiated to investigate an appropriate way forward for various agencies to help reduce poverty. Subsequently, the Worcestershire Anti-Poverty Partnership (WAPP) was established as a facilitative partnership that aimed to develop policies to tackle social exclusion.

Membership of the partnership reflected its policy-based nature as it was comprised largely of representatives from the public sector. Of the original 23 members of the partnership, 16 were from the public sector representing council departments such as Planning, the Environment, Benefits, Health and Welfare as well as members of the County Council secretariat. Of the remaining seven, three were from academia, three from voluntary agencies (a local Citizens Advice Bureau and a local charitable organization helping people on benefits) and one representative from the church. Given that partnerships often include those from private, public and voluntary sectors it was surprising that were no members from the former.

One of the main missions was to develop early intervention strategies aimed at young people. It was this element of their remit that prompted the liaison with University College Worcester. What the anti-poverty partnership wanted was for the University College to field an academic who would become a member of the partnership and, in this capacity, would conduct research on young people and disaffection. The findings of this research would be reported to them and then be incorporated into policy, as would any areas identified that would improve inter-agency working.

Following a period of negotiation between actors at UCW and WAPP, some funding was made available by WAPP to supplement a studentship from UCW to support doctoral research into these issues. Collaborative policy research of this kind is becoming increasingly widespread and as far as both the academic and non-academic organizations are concerned, can be mutually beneficial (Macmillan and Scott, 2003; Barrett *et al.*, 2003). The academic institution has a new source of funding and the collaborative organization, in a climate of rising demand for policy relevant research, has a substantiated and professionally supervised report (but see Banks and MacKian, 2000; Peck, 1999, 2000; and

Pollard *et al*., 2000; Fuller, 1999; Lees, 1999; and Martin, 2001 for a fuller debate on these issues).

This negotiation and funding allowed me to take up a position as a full-time research student in UCW's geography department. This allowed me to be a member of the WAPP yet, at the same time, gave me the independence and resources to research young people in an academic, geographically relevant manner.

Although this opportunity had the potential for a mutually beneficial relationship, the difficulty was that the partnership had no direct input from young people. This challenged the integrity of the partnership whose supposed *raison d'être* is to integrate 'top' and 'bottom' voices and further, one of the aims of the thesis was to gauge the effectiveness of this kind of integrative approach to policy making. Without a direct connection both the partnership and the thesis were left in two halves with no obvious link between partnership and youth. However, this could be overcome by drawing on a participatory approach through the application of action research.

Action research

The most direct way of assessing the effectiveness of integrative approaches to policy making would have studied an established rural partnership that worked more directly with young people and involved them in decision making, such as a youth parish council. In these types of partnerships, young people are involved as partners (rather than the subject of partnerships) and it would have been possible to examine the ways in which young people were empowered and involved in their partnership in a detached, evaluative manner (see Fig. 23.1). Here I would have acted as an 'observer' rather than participant.

However, due to funding agreements, the focus of the work had to be on WAPP, a partnership that did not directly involve young people in its decision making. Consequently, the difficulty was that because the partnership had no direct input from the young, youth and the partnership were unconnected (see Fig. 23.2). Given this situation, it would have been extremely problematic to link research on youth voices to the development of policies by the partnership to address these concerns. This posed a research dilemma: how could I integrate work on voices of young people with the actions of the partnership if the two were not connected? Further, how could the partnership claim to integrate voices from the 'top' and 'bottom' voices unless that gap could be bridged? Action research provided a way of conducting academically relevant work as well as linking the partnership more closely with young people.

Restoring linkages through action research

Action research combines three elements of the research process: *research* into and of the chosen subject area; *participation* with and in those areas of concern being researched; and *action* where some form of social change is brought about. Therefore, it aims to do more than give voice, participant observation and ethnographic approaches are perfectly capable of doing that, but to participate in the research process and bring about social change with and on behalf of the stakeholders. This empowers the researched and democratizes the research process at the same time. The researched group *have* an interest in the research, rather than being *of* interest to it (Cahill, 2004; Hoggart *et al*., 2002).

Participation is achieved by drawing the researcher from the *outside* (see Fig. 23.1) to the *inside* (see Fig. 23.3). In my position as researcher, I was able to establish a link between the young and the members of WAPP that allowed their voices to be heard, and potentially acted upon, by policy makers. Thus, the ability to assess the extent to which these voices were integrated into partnership policy would be met through my active participation, working on behalf of the young.

I occupied this 'middle' position and, as such, was able to research the issues facing young people. This research adopted a qualitative methodology, using interviews and focus groups to give, as much as possible, young people to express their stories, concerns and opinions in their own words and actions (see Corcoran, 2003b). In turn, their voices were accurately reported as qualitative research findings to WAPP on behalf of young people. This position also allowed me to assess whether or not these voices were being integrated into policy strategies. This allowed not only the aims of the thesis to be reconnected, but also enabled the partnership philosophy of bringing voices from the 'bottom' into the arena to be re-established. Furthermore, the three components of action research (*research*, *participation* and *action*) could be incorporated into this process by: considering background to and problematic of research into the young (*research*); engaging with the study group and discovering issues of concern to them (*participation*); and accurately reporting these issues to the partnership with a view to invoking change (*action*).

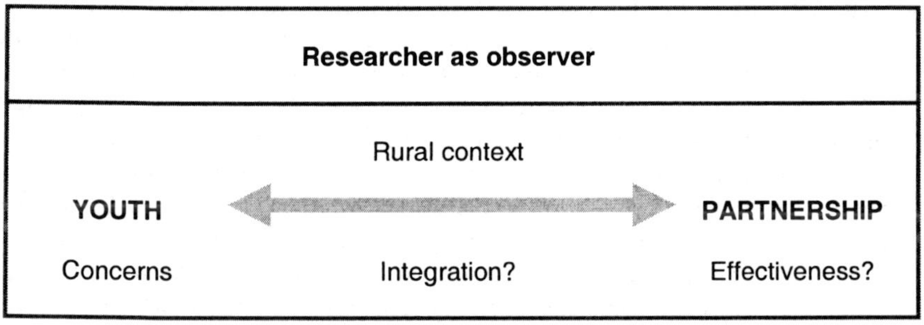

Fig. 23. 1 Ideal thesis plan in diagrammatic form.

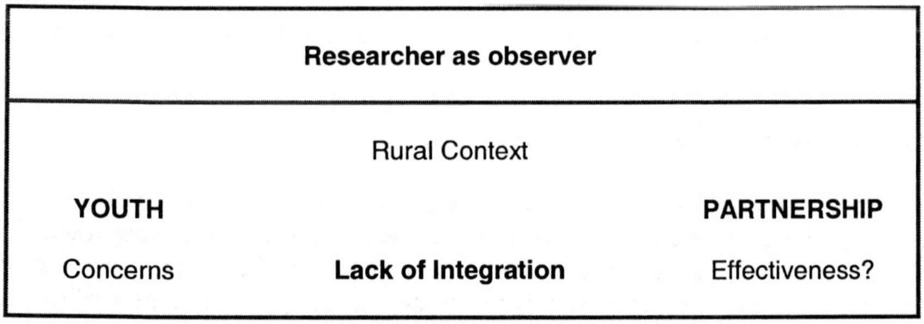

Fig. 23.2 Youth and partnership in isolation.

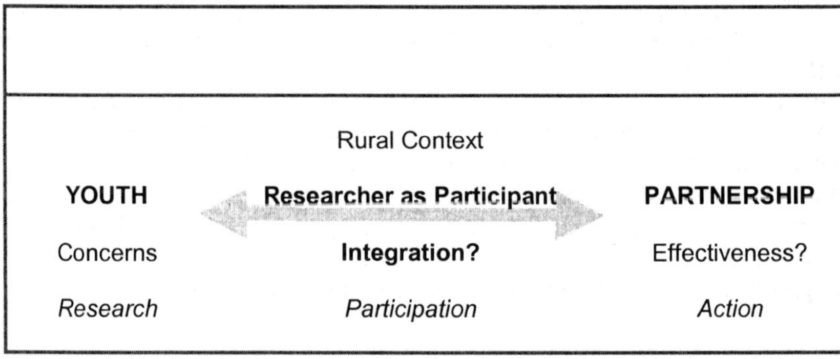

Fig. 23.3 Linkages restored.

However, there are two main concerns with this approach: first, the ability to maintain and distinguish between the role of both researcher and activist (see, for example, Fuller, 1999; Halfacree, 1999; Hubbard, 1999; Maxey, 1999; Stanley and Wise, 1993); second, that action research can leave the researcher in an undefined land of third space.

Researcher versus Activist

Under more objective methodologies, the distinction between researcher and researched is clear and the role that each played is defined: neither one should contaminate the other. Maintaining distance ensures that the researcher's position is never in doubt, but what happens to that position once the line dividing research and action has been crossed? Once the researcher becomes actively involved in agitating on behalf of the researched, no longer are these identities separate: interests can become blurred and the hitherto unquestioned position of the researcher is now potentially open to both conflict and scrutiny. This is diametrically opposed to the traditional research ethos of leaving the 'field' as you found it (Flowerdew and Martin, 1997; Neuman, 1994).

It has been suggested that the problem with continued adherence to traditional approaches, is that it denies the possibility of alternatives and as a result perpetuates academic 'ivory tower' syndrome through a lack of engagement at grass roots level (Hoggart *et al.*, 2002; Robson, 2002). This argument has arisen because historically academics have rigorously defended objectivity but at the expense of recognizing that knowledge can be produced in other ways. As a result, accusations have been made of maintaining a grip on pedagogic authority through reluctance to transcend the boundaries between academia and activism (Kitchin and Hubbard, 1999).

However, as Fuller (1999) points out, while it is essential to maintain academic integrity, over-reliance on deliberative practices can be counter-productive as they can stifle spontaneity and can, unwittingly, contrive to miss opportunities for innovative and reflective research. Action research is a way of countering and addressing these concerns.

Third space

The development of third space has evolved through a perceived inadequacy with spatial metaphors such as 'insider'/'outsider', 'centre'/'margins', 'core'/'periphery' (Valentine, 2001). Soja (1989, 1996) drawing on the work of Lefebvre (1991, 1992), rejects these approaches as reductionist as they fail to take into account a changing society where hybrid identities often require more than one position to be occupied simultaneously (Crang, 1998;

Lewis, 2002). Soja argues that where these intersect, be they real or symbolic spaces of contestation, a third space is created that enables other positions and understandings to emerge (Lewis, 2002).

In terms of action research, the parallel is that the researcher maintains a hybrid identity that simultaneously occupies the position of academic and activist and as such must negotiate a position that satisfies the demands of both roles. The problem is that researchers can fall between two stools and find themselves in an uncomfortable third space between them - neither fully accepted nor fully rejected (Hoggart *et al.*, 2002). However, this need not be overly problematic as it has merit in offering a different view of the research process and the opportunity to develop reflective practices from a unique position.

Personal positionality and reflexivity

Critical researchers appear to be in general agreement that the way to approach action research and overcome concerns of positionality is through due reflex(ct)ion. There are two areas that impacted directly on this piece of work. First, through claims of action research to political democracy; second, with regard to the relationship between adults and the young, how to position oneself in a realm which one has already left and has no prospect of returning.

Democracy and positionality

The claim of action research is that it democratizes the research process by working on behalf of the researched and taking its lead from them. This means that the research could be taken from any number of political standpoints and, in its pure form, action research should follow these through regardless of whether they were morally right or wrong (Hoggart *et al.*, 2002). This is clearly untenable and although the literature is a little thin, common-sense must prevail to prevent abuse and avoid situations that support nefarious activities or threaten academic professionalism.

There was no such problem of association with the young. It was not anticipated that any such moral maze would occur to prevent action research being applied in the democratic fashion in which it is meant, that is with respect for diversity and to empower the disenfranchized (Hoggart *et al.*, 2002).

Positionality and reflexivity

Positionality refers to our own beliefs and experiences that affect the way we make sense of the world and go about researching it, reflexivity is a process of reflection on what we are, what we know and how we come to know it (Valentine, 2001). Reflexive research has been defined as:

> 'self-critical sympathetic introspection and the self-conscious *analytical* scrutiny of the self as a researcher.' (England, 1984, p.82 in Flowerdew and Martin, 1997, p.113, italics in original text)

Challenging the overtly objective approach to research requires detaching and distancing oneself from the subject and, drawing on feminist approaches, geographers have argued the case for a more empathetic and empowering approach characterized by the development of mutual understanding between researcher and researched (Kitchin and Tate, 2000; Robson, 2002). This requires that the researcher accepts the responsibility to question not only the research process, but also themselves and from whence their knowledge is originating. It also questions whether or not one is seeing what one wants to see, or hearing that which one wants to hear (Bailey *et al.*, 1999).

However, even though a reflexive approach serves as a continual memorandum of objectivity to the researcher, it cannot turn back the clock and allow the researcher to see the world through a child's eyes. In which case one could argue that any attempt to research the young is at best a fictionalized account of the researcher's version of the subject's reality (Robson, 2002), or that it is not possible to see the world through *anybody's* eyes rendering all research invalid. But, as Maxey (1999, p.230) points out:

> 'given that academic research is in part the production of our personal experience, morality and politics, does our positionality and privilege preclude us from making meaningful contributions to debates?'

Although there is a certain inevitability surrounding ethical considerations in research, rather than abandoning or ignoring them, researchers should embrace them and use them to solve and avoid future methodological difficulties (James, 1990): a partial understanding is surely better than none at all (Matthews, 2001). Thus when researching vulnerable groups we should be aware of the impact our own history, our norms and values and evaluate the research critically for these hidden agenda (Alderson, 1995).

There are inevitable tensions that will exist in action research as the researcher adopts a number of different roles as researcher, activist and academic, and how to manage these roles and maintain research integrity, requires considerable thought. Therefore, in order to keep the research grounded and focused, I adopted a rather pragmatic approach and devised the following strategy/framework (see Table 23.1) that set out quite clearly the underlying assumptions relating to action research (Robson, 2002). This proved an invaluable tool as it ensured that young people remained the focus of the project and served as a continual reminder to consider and reflect at each stage of the project.

Table 23.1 Underlying research assumptions.

1. Young people's version of reality will be the one adopted.
2. Concerns and issues that they require to be changed are real and genuine.
3. When attempting to report these issues on their behalf, anticipate a certain resistance.
4. Action research requires continual repositioning of the researcher as roles and requirements change demanding due pause and reflection at each stage.

Degrees of integration

The way that young people's voices were drawn into discussion and integrated into policy occurred at two levels. First, they were discussed by members of WAPP. Secondly, WAPP in turn responded to the findings and subsequent discussion of them by reporting to decision makers at higher levels and, ultimately, responsible, for developing policy. The following sections reflect on the way that my research findings influenced these deliberations.

Youth, partnership and integration

If youth issues and action research were to be taken seriously, then a strong commitment was also needed to WAPP. Thus the researcher attended seven meetings out of a total of eight, and contributed on behalf of young people whenever possible. Further, as the project was partly funded by members of WAPP, they had a vested interest in reviewing progress on the project and, consequently, the research and its findings were always allocated time at their meetings. This ensured that young people's issues were continually on the WAPP's agenda.

A noted characteristic of partnerships is their volatility of membership (Edwards *et al.*, 2000). Indeed membership of WAPP frequently changed, with personnel leaving, joining or being replaced. My presence throughout the partnership's life lent some stability and also ensured that WAPP's interest in young people remained continuous. Further, as new faces were continually being introduced, this gave wider opportunities for discussions with a range of actors and agents about their responses to young people's issues and their opinions on the effectiveness of partnership working.

However, where the partnership was most effective was that it was able to schedule opportunities for different members to 'air' their particular cause: its watching brief was, after all, to achieve understanding of issues of exclusion. This was achieved was by giving partner members, or relevant others, the opportunity to present their interests formally to the partnerships. For example, presentations were given on 'The Westlands Rapid Appraisal Report' (a local housing estate targeted for social support); low cost affordable housing; and 'The Libraries on Line Information Service'. In due course, I was allowed to speak about young people's concerns and, in doing so, give them a voice, albeit through an adult 'interpreter', on WAPP by reporting my research findings

The report I presented was comprehensive and covered the majority of issues raised as a result of the study. It was well received and was the subject of positive post-presentation discussion although the content was considerably abbreviated as the meeting was running late. Despite this, the report and discussion were extensively reported in the minutes, which reached a wider audience than those just attending the meeting as it was circulated to all members of the partnership. With one or two small inaccuracies, the minuted discussion was a reasonably faithful account of the meeting and tabled report. In this way the concerns and issues impacting on young people were effectively broadcast and their voices were integrated into the partnership. Given the broad membership base of the partnership these voices were heard by a range of agencies and actors, which has been recognized as an important attribute of a facilitative partnership (Edwards *et al.*, 2000).

Policy integration

Even though youth concerns were widely transmitted among the partners of WAPP, this was no guarantee that change would take place. However, one of the terms of reference for the partnership was to 'influence decision makers' (see Table 23.2). Accordingly, correspondence had been generated between the chair of the partnership and the Head of Policy, which culminated in an invitation for members of the core group to contribute to the County Council Scrutiny Panel on Social Exclusion.

Table 23.2 WAPP's terms of reference of influencing decision makers.

[WAPP aims] to link, coordinate and share local strategies and actions already being carried out in the County in order to share best practice i.e. 'Goal of continual improvement, which is about effectiveness NOT efficiency'; reduce duplication and wastage of resources and to consider the reconfiguration of resources where necessary. This will be achieved through advising on effectiveness measures; advising on interventions such as transport policy, housing provision, child care, benefits advice, etc.
All agencies and local authority departments involved in the core group to define how their services currently impact on those people in poverty and how they interact with one another.

The Social Exclusion Scrutiny Panel was established to examine the Council's response to social exclusion and to consider its future strategy. It was comprised entirely of elected Councillors who lend some legitimacy to the proceedings as they, at least, have some level of accountability to the general public if not young people directly.

Their outlined timetable was in nine stages and it was at Stage 5 that the consultation meetings at which the partnership were to contribute took place. Four members from the partnership were selected to present their area of interest to the Panel although in the event only three took part, representing interests from social exclusion and health, drug action and young people.

I took the opportunity to give a complete overview of the research and present as clearly as possible issues that are of concern to young people. The subsequent minutes to the meeting were a masterpiece of brevity. Although they contained some minor inaccuracies and omissions, they managed to condense the key findings into a manageable sound-bite.

Four more stages remained during which the Panel aimed to deliberate before finalising its report on policy recommendations. The report on young people survived all of these and appeared in the Social Exclusion Scrutiny Panel Report in the Executive Summary:

> 'The quantitative research has not been able to identify the incidence of social exclusion in rural areas. However, qualitative research indicates that younger people in rural areas may be at risk of social exclusion due to the unavailability of transport. The Panel would welcome views on other social exclusion needs in rural areas.' (Worcestershire County Council Social Exclusion Scrutiny Panel Report, 1999, p.4)

as well as in the main body of the report albeit in greatly reduced form:

> **'2.3 Study of Young People and Social Exclusion**
> The Panel received the initial findings of doctoral research being undertaken of the needs of 12 to 16-year-olds in the Wychavon area by Ms. C. Corcoran of the Centre for Rural Studies, University College Worcester. One of the key findings of the research is the problem of transport that often faces young people in rural areas, particularly when families have lower incomes and no alternatives to public transport. These young people find that they are denied access to both friends and facilities in urban areas. This problem is particularly acute for 16 year olds, who are charged adult fares on buses.' (Worcestershire County Council Social Exclusion Scrutiny Panel Report 1999, p.4)

Attention was therefore only given to one element of my research and so a number of issues that it raised were not absorbed into policy, for example: the treatment of young people as a homogeneous rather than heterogeneous group or young people's feelings of lack of consultation. It seems that these issues of attitude and culture and could not be changed with the stroke of a policy pen.

Despite these limitations the partnership *did* achieve one of its aims, which was to influence decision makers. Decision makers *did* hear young people's views and concerns; and one of these concerns (transport) *was* integrated into policy proposals to help eradicate social exclusion. For these reasons and in this instance, the partnership was effective and *did* integrate voices. Once the Social Exclusion Scrutiny Panel admitted youth concerns into its policy, young people had a return on their investment and although never present in person, they had indirectly achieved integration through the medium of the partnership and the application of action research (see Fig 23.4).

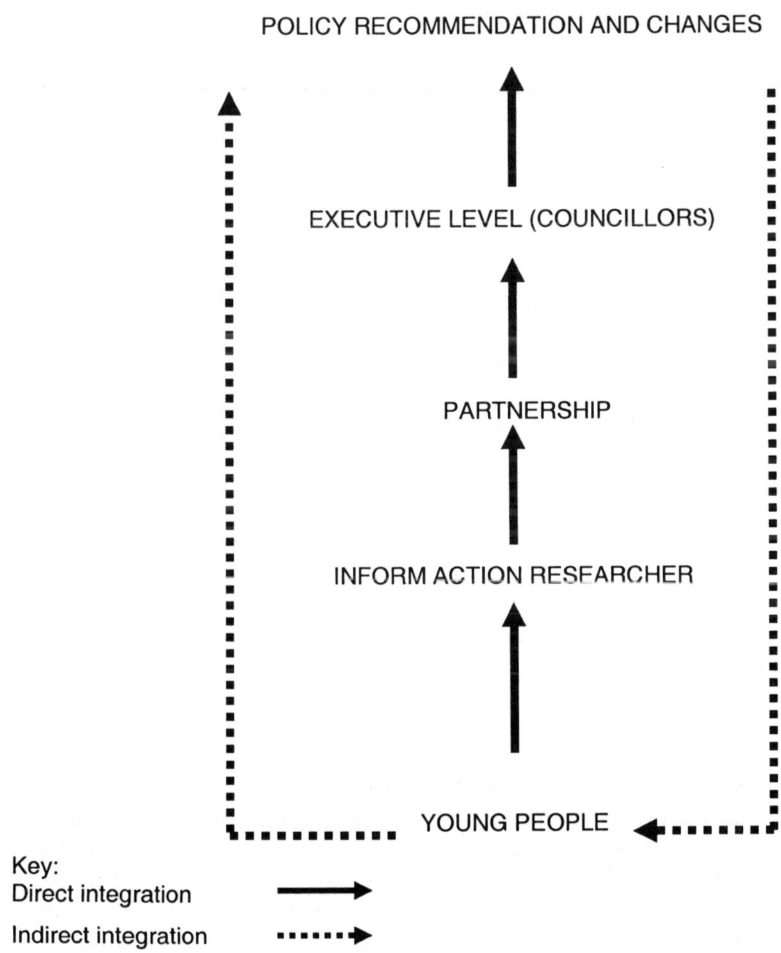

Fig. 23.4 Circuit of integration.

Although this is some measure of success it is worth noting the difficulties encountered when trying to report back to the young people involved in the study. As the research progressed it became increasingly apparent that by the time the reports had been compiled and policy changes initiated the study group would have moved on. This is not an uncommon finding in action research particularly when associated with attempts to introduce policy change, which can be notoriously slow (Pain and Francis, 2003). The transient natures of the young as they change school, migrate with their parents or simply choose not to re-engage with the research compound the difficulty of returning to the original cohort.

Despite these challenges every effort was made to report back to the study group through the schools who had participated in the original research and by using the partnership as a network through which to inform a wider community of young people's concerns. Thus, even though the original cohort may have moved on those following would at least benefit from their input.

Conclusions

Given the disconnection between the partnership and young people, the methodology of action worked well. It drew young people's voices into a partnership that eventually allowed their voices to be heard by other policy makers. The route taken was more circuitous than anticipated: although the partnership was unable to bring about change itself, it provided a platform through which to voice the concerns of young people to a wider audience. As a result, eventually a level of integration was achieved that enabled young people's concerns about transport to be brought to the notice of decision makers. Subsequently, policy recommendations have been made to address these problems.

However, this form of research does pose problems of positionality and reflexivity. Not only does it challenge the traditional role of the researcher by drawing them from the 'outside' to the 'inside' but also, particularly when researching vulnerable groups, requires the researcher to reflect on what they are bringing into the relationship and how this influences the interpretation of the results. This is particularly pertinent when researching young people as it is widely acknowledged that an adult has left the world of the child and cannot see it with their eyes. For this reason, the development of a framework of research assumptions ensured that the study remained grounded and focused on young people.

Finally the most difficult problem to overcome was in reporting back to the study group. The combination of a transient study group and a lengthy policy implementation procedure conspired to problematize returning to the original research cohort. Reporting to the participating schools and ensuring young people's concerns were kept at the forefront of the partnership agenda, overcame this although it is acknowledged that by this stage at least half of the original group had moved on. However, this is still an important element of the action research process and should be built into the initial design. Without that reference to the study group, the research becomes extractive and loses its democratic edge.

Note

1. Hereford and Worcester County Council was the local authority responsible for the county of Hereford and Worcester prior to 1998. In 1998, following local government reorganization, the county was divided into two counties called Herefordshire and

Worcestershire, each with its own county council. The Worcestershire Anti-poverty Partnership emerged from this initial meeting with a remit to tackle social exclusion in the county of Worcestershire. A separate forum was established for Herefordshire and both partnerships were steered by their county's respective council. This chapter focuses on the Worcestershire Anti-poverty Partnership.

References

Alderson, P. (1995) *Listening to Children.* Barnardos, Essex.

Bailey, C., White, C., and Pain, R. (1999) Evaluating qualitative research: dealing with the tension between 'science' and 'creativity'. *Area* 31, 169-183.

Balloch, S. and Jones, B. (1990) *Poverty and Anti-poverty Strategy: the Local Government Response.* Association of Metropolitan Authorities, London.

Banks, M. and MacKian, S. (2000) Jump in! The water's warm: a comment on Peck's 'grey geography'. *Transactions of the Institute of British Geographers* 25, 249-254

Barret, H, Storey, D and Yarwood, R. (2002) Development through partnership: LEADER II in the Marches. In: Walford, N. and Evans, N. (eds) *Innovations in Rural Areas.* Ceramac, Presses Universitaires Blaise Pascel, Clermont, pp.311-328.

Berthoud, R., Brown, J., with Cooper, S. (1981*) Poverty and the Development of Anti-poverty Policy in the UK.* Heineman, London

Cahill, C. (2004) Defying gravity? Raising consciousness through collective research. *Children's Geographies* 2, 273-286.

Cloke, P. and Little, J. (1997) Introduction: other countrysides? In: Cloke, P. and Little, J. (eds) *Contested Countryside Cultures.* Routledge, London pp.1-18

Corcoran, C. (2003a) Partnership, exclusion and the young: the neglected voice of the rural young. In: Beesley B. (*et al)* (eds.) *The New Countryside.* Brandon University (Rural Development Institute) and Saint Mary's University, Canada.

Corcoran, C. (2003b) The effectiveness of partnership in the implementation of youth strategies: a case study of Bromyard and Wychavon. PhD Thesis. Coventry University in association with University College Worcester, Worcester.

Crang, M. (1998) *Cultural Geography.* Routledge, London.

Edwards, B., Goodwin, M., Pemberton, S. and Woods, M. (2000) *Partnership Working in Rural Regeneration.* The Policy Press, Bristol

England, K. (1994) Getting personal: reflexivity, positionality and feminist research *The Professional Geographer* 46.1, 80-89

Flowerdew, R. and Martin ,D. (1997) *Methods in Human Geography.* Longman, Essex.

Fuller, D. (1999) Part of the action, or 'going native'? Learning to cope with the 'politics of integration'. *Area* 31, 221-229.

Garonna, P. and Ryan, P. (1991) The problems facing youth. In: Ryan, P. *et al.* (eds) *The Problem of Youth.* Macmillan, London pp.1-8

Geddes, M. (1997) *Partnership Against Poverty and Exclusion?* The Policy Press, London

Hackett, C. (1997) Young people and political participation. In: Roche, J. and Tucker, S. (eds.) *Youth in Society.* Sage, London pp.81-88

Halfacree, K. (1999) 'Anarchy doesn't work unless you think about it': intellectual interpretation and DIY culture. *Area* 31, 209-221.

Halpern, D. (1998) Poverty, social exclusion and the policy-making process: the road from theory to practice. In: Oppenheim, C. (ed.) *An Inclusive Society.* IPPR, Southampton, pp.269-283.

HMSO (1995) People, Prosperity and Partnership. HMSO, Edinburgh.

Hoggart, K. Lees, L. and Davies, A. (2002) *Researching Human Geography.* Arnold, London.

Hubbard, P. (1999) Researching female sex work: reflections on geographical exclusion, critical methodologies and 'useful' knowledge *Area* 31, 220-239.

Imrie, R. and Raco, M. (1999) How new is the new local governance? Lessons from the UK. *Transactions of the Instituite of British Geographers* 24.1, 45-63

James, S. (1990) Is there a 'place' for children in geography? *Area* 22, 278-283.

Kitchin, R. and Hubbard, P. (1999) Research, action and 'critical' geographies. *Area* 31, 195-198.
Kitchin, R. and Tate, J. (2000) *Conducting Research into Human Geography* Pearson, Essex.
Lees, L. (1999) Critical geography and the opening up of the academy: lessons from 'real life' attempts. *Area* 31, 377-83.
Lefebvre H. (1991) *The Production of Space*. Blackwell, Oxford.
Lefebvre H. (1992) *Critique of Everyday Life*. Verso, London.
Lewis, J. (2002) *Cultural Studies - the Basics*. Sage, London.
Macmillan, R. and Scott, A. (2003) On the case? Dilemmas of collaborative research. *Area* 35, 101-109.
Martin, R. (2001) Geography and public policy: the case of the missing agenda. *Progress in Human Geography* 25, 189-210.
Matthews, H. (2001) Power games and oral territories: ethical dilemmas when working with children and young people. *Ethics, place and environment* 4.2, 117-118.
Maxey, I. (1999) Beyond boundaries? Activism, academia, reflexivity and research. *Area* 31, 199-208.
Mulgan, G. (1998) Social exclusion: joined up solutions to joined up problems. In: Oppenheim, C. (ed.) *An Inclusive Society*. IPPR, Southampton pp.259-268.
Murdoch, J. and Pratt, A. (1993) Rural studies: modernism, post-modernism and the 'post' rural *Journal of Rural Studies* 9.4, 411-427.
Neumann, W.L. (1994) *Social Research Methods Second edition*. Allyn and Bacon, Boston.
Pain, R. and Francis, P. (2003) Reflections on participatory research. *Area* 35, 46-54.
Peck, J. (1999) Editorial: grey geography? *Transactions of the Institution of British Geographers* 24, 131-135.
Peck, J. (2000) Jumping in, joining up and getting on. *Transactions of the Institution of British Geographers* 25, 255-258.
Philo, C. (1992) Neglected rural geographies: a review. *Journal of Rural Studies* 8, 193-207.
Pollard, J., Henry, N., Bryson, J. and Daniels, P. (2000) Shares of great? Geographers and policy. *Transactions of the Institution of British Geographers* 25, 243-248.
Robson, C. (2002) *Real World Research* Second edition. Blackwell, Oxford.
Rosenbaum, M. (1993) Independent mobility and children's rights. In: Hillman, M. (ed.) *Children, Transport and the Quality of Life*. Policy Studies Institute, London pp.19-27.
Snowdon, P. and Slee, B (1997) Partnerships in Rural Development in Scotland Paper presented at RESSG Annual Conference, University College Worcester.
Social Exclusion Unit, (1997) *Speech by the Prime Minister on Monday 2 June 1997, at the Aylesbury estate, Southwark*. Social exclusion Unit. [Accessed 12th December 1997] Available from http://www.open.government.uk/co/seu/more/html#title
Soja, E. (1989) Postmodern Geographies: the Reassertion of Space in Critical Social Theory.Verso, London.
Soja E. (1996) ThirdSpace: Journeys to Los Angeles and other Real-and-Imagined Places. Blackwell, Cambridge.
Stanley, L. and Wise S. (1993) Breaking Out Again: Feminist Ontology and Epistemology. Routledge, London.
Taylor, M. (2000) Communities in the lead: power, organizational capacity and social capital. *Urban Studies* 37.5-6, 1019-1035
Valentine, G. (1997) A safe place to grow up? Parenting perceptions of children's safety and the rural idyll. *Journal of Rural Studies*. 13, 137-148.
Valentine, G. (2001) *Social Geographies*. Pearson Education, Essex.

Chapter 24

Conclusions

John Smithers and Randall K. Wilson

Introduction

As suggested at the outset, this book has sought a middle ground between the eclecticism of academic symposia and the deliberately limiting structure of an academic text. As a result, what has been accomplished is an illustration of the multifaceted nature of some seemingly coherent *core dimensions* of contemporary rural change. The issues and examples presented in this text suggest that change seen in its broad form may be helpfully 'compartmentalized' in relation to organizing concepts such as farm adjustment, environment, community and governance. Yet, within these foci there is evidence that context, place and people matter, and that the potential for more nuanced interpretation lies in the empirical richness of how change is experienced and managed both within and between regions.

Taken together, the various contributions by the authors paint pictures of rural regions that are evolving, in different ways, in response to forces from without and within. The progressive refashioning of agricultural/rural policies and governance structures on both sides of the Atlantic is effectively 'changing the rules' within which land use and community systems operate. However, the manifestations and the impacts of such process are policy dependent, spatially variable and highly contingent. While state-sponsored support for, or regulation of, farming appears to fashion general system level responses that seem to characterize agricultural systems, they are not deterministic. In all three countries, there was evidence that diversity, change and 'hybridization' are increasingly evident (though not equally in degree) in land use and livelihood systems. There are hints that this diversity is important in relation to understanding how regions might pursue sustainability in different ways – perhaps in ways that do not involve land-based production.

On the community side, there is clear a indication that a myriad of interests and voices now compete for prominence and influence in the social construction of rural issues and the management of rural affairs. Here again, the authors have spoken to the struggle to sustain tradition (read agriculture) on the one hand and the desire to capitalize on assets and opportunity through deliberate action on the other. Simply put, the prominence of agriculture in economic, social and political terms is now in question: clearly alive and well in some areas and decidedly shaky in others. The increasing commodification, valuation and regulation of the countryside and nature by local people, elected officials and corporate interests has led to diverse agendas, uneven change and sporadic conflict in the realms of the 'old economy' and the new one.

In this concluding chapter, an attempt is made to sustain the balance between generalization and detail that has run though the book. Each of the four organizing themes is revisited with a view to (a) situating various empirical specifics within a wider frame of progressive changes in both land use systems and rural communities, and (b) to speculate on how selected specific forms and processes of change represent possible pathways of future development at both the local level and beyond.

Assessing agricultural response

Recent debates in rural geography have pondered the persistence and the evolving nature of selected 'trajectories' of farming. While some argue that production-focused forms of farming are giving way systematically to other land-based activities that differ in both form and function (Ilbery and Bowler, 1998; Walford, 1999), others have suggested that evidence for the demise of the industrial model is mixed and highly dependent on location (Wilson, 2001; Smithers and Johnson, 2004). A third interpretation is that elements of both productivist and non-productivist agriculture are present not only within single regions, but perhaps also on the same farm (Evans *et al.*, 2002). Based on the coverage provided in this volume, it appears that rural space is inhabited by several different agricultures both across and within regions.

One response, clearly evident in North America, is the consolidation of food production activities in the hands of fewer but larger actors and the integration of these production units with the corporate world of the agri-food system. As suggested in several of the initial chapters of the book, the proliferation of 'big agriculture' in North America is still clearly in evidence – and continues to be supported by the actions of corporate owners and government policy. Such forms of farming appear to be highly responsive to the shifting arrangements for commodity processing and distribution at the regional and global scale, and are attentive to consumer demand as expressed in the market place. Further, there is a suggestion that these practices (and ideologies) are being adopted not only by corporate farms, but by individual producers (family farms) such that the lines between these once distinct groups is increasingly blurred. Such responses are indicative of the abiding strength of the industrial model, however, they are also implicated in increasing instances of conflict with both farm and non-farm interests, and are thought to foster the development of economic linkages that redirect the farm away from its community ties. While the productive capacity of this pathway of farm development is not in dispute, the degree to which such responses have the capacity to support farm and community sustainability is unclear. However, coverage in this volume suggests that this trend may as often lead to division and alienation as to mutual progress.

In some respects, the response above is provoking responses of its own. In Marshall's account of the fishing communities of Grand Manan Island it was shown that industrialization, routinization and corporatization of fish farming (in the interest of both corporate gain and local economic development) has had an almost polarizing effect at the local level by creating actors and spectators or 'winners and losers' – especially at the community scale The principals in these new forms of development may reap rewards, but others in communities may experience alienation, a disruption in traditional patterns of activity, and ultimately a challenge to long-held values concerning independence and self reliance. Hence some members of communities (or rural regions) may embrace new change while others refrain – and perhaps actively resist. In this example, and many others, there is evidence of the differentiated nature of acceptance and rejection – and a preliminary glimpse of how class, ethnicity and values (rural, urban and corporate) mix in different ways in selected cases.

Two other forms of agricultural response to the 'evolving rural' were presented in the book and are important in the conceptualization of rural space. First, in both Britain and North America, food has become more than a material output from farming; it shows signs of becoming a means of communicating 'goodness and quality' to a wider world. In some regions it may even have transformative qualities where the reputation and distinctiveness of rural areas is associated with the food and drink that it produces. Ilbery and Maye's examination of specialist livestock production brought this new form of response into focus and illustrated how and why producers opt to achieve value via strategies that stand apart from the dominant production model and the mainstream market. This response, though shown to be somewhat ambiguous in its degree of 'localness', is indicative of another increasingly important trend in rural regions – the proliferation of local foods and the evident 'turn to quality' (Goodman, 2003) by both producers and consumers. Such choices by farmers stand as examples of entrepreneurial innovation and as lifestyle and value-based decisions to abandon what is often perceived to be a production-based treadmill.

Second, beyond the realms of agricultural commodities and local food lies a host of activities and responses that seem to take agriculture away from its productive traditions and towards wider forms of engagement with a rural-consuming public. The example of visitor-based farming provided in the chapter by Che, Veeck and Veeck is representative of the types of structural diversification that are increasingly common in all three countries – and especially in those regions that are more densely populated or are 'near urban'. In these instances, the product is a rural 'experience' and perhaps an opportunity to engage vicariously in perceived rural traditions (which may differ greatly from the daily practice of contemporary farming). In its treatment in this volume, there are glimpses of two important issues concerning this development trajectory. One is the appeal to many of a constructed, managed and romanticized image of agriculture and the countryside (as delivered in by entrepreneurs (Chapter 7) and portrayed in farm shows and other managed events (Chapter 8). Alternatively, this form of diversification represents a strategic business decision and perhaps a business plan where producers seek additional value from both farm commodities and the farm itself. Such business trajectories show signs of taking farming away from its productive roots and reorienting its purposes towards a new set of demands and opportunities in the countryside.

Environment

The environment has become a focal point for much interpretation and contestation over developments in rural areas. These include critical attention to the evolving nature of agriculture, the broadening public interest in the amenity value of rural land, the (re)regulation of rural land use and the desire (and increasing necessity) to engage in, and demonstrate, environmental compliance to others. Beyond these 'modern' concerns is an abiding reality in some less developed or diversified regions – the environment as a resource for living and working. Reflections on each of these themes were present in the contributions to this volume and some general observations are possible.

Attention to the environment as a means of subsistence and a basic resource for living and working (Chapters 11 and 12) are a reminder that issues of access and vulnerability are important at both the household and community scale. In the former, there is a need to recognize that increasing competition for rural land among the myriad of interests reflected in this text poses risks to those who depend on the environment for the provision of life's basic necessities. Sadly these voices are often less heard than others in public debate, but we are reminded that these traditional needs remain – even in modern times. Similarly, the traditional patterns of work and survival at the regional scale are subject to pressures from the natural environment. Here the focus shifts from the access to resources to the way in which resources are used – and the degree to which those patterns of use are viable in the face of environmental stress. In contrast to much recent attention to how environment is interpreted, commodified and defended, environmental resources (soils, climate, water) also underpin the means of primary production – and resource depletion or recurrent risks will challenge communities and governments to reflect upon what sorts of production they want, and what the environment will provide.

Though more analytical work remains to be done, the preliminary accounts in Mayda's chapter underscore the foundational dependence of local communities upon their local environments. Providing basic resources for human habitation – clean air, water and extractive natural resources – these environments play a vital role in maintaining the viability of rural communities. Their degradation due to intensified industrialization (Ohio), coastal erosion and climate change (Alaska) or the depletion of resources due to over-harvesting (Newfoundland), stress the vulnerability of marginalized rural populations and landscapes to global and regional forces of economic liberalization.

The environment, as reflected in the preservation of natural assets, the minimizing of negative externalities, and production of quality outputs has attracted the attention of a critical public and policy makers in both the UK and North America. Hence, it has become increasingly important for farmers and farming (see Chapter 4) to not only meet environmental standards, but to be seen (and certified) as meeting such goals. The example of Ontario's Environmental Farm Plan programme calls attention to several apparently strategic developments within the farm sector that can perhaps be generalized to other places. It was suggested that the specter of environmental regulation in some regions has heightened producer concerns over regulation and prompted desires to assume leadership in setting the environmental agenda and establishing the terms of engagement. Indeed, one of the defining characteristics of the OEFP, as described by Robinson, is the degree to which farmers in Ontario have achieved success in convincing public sector agencies that such initiatives (with the funds that support them) are best delivered by those within the farm sector itself. Hence this, and similar, initiatives are notable for their origins, their form, and the manner in which they are delivered. Perhaps most intriguing of all, such schemes are also noteworthy for what may yet transpire at the level of the market place. Indeed, the 'turn to quality', though most often invoked at the community or regional level, may have its equivalents at the international scale where the need to be environmentally compliant may become an increasingly important in order to differentiate products and gain access to selected markets. Hence, the rewards of environmentalism in (re)regulated rural spaces and food systems may well extend beyond the realms of intrinsic motivation and sustainability to include the wants of ability to conduct business in an increasingly demanding market place.

On both sides of the Atlantic, there is longstanding recognition of the impacts of urban expansion into agricultural regions. Gayler's account of the political, economic and social climate for urban expansion into the countryside is familiar to those with previous exposure to the farmland preservation debate in geography and related disciplines. Where this account differs, and becomes particularly salient for our desire to understand the various and competing trajectories for development is the marrying of the longstanding zeal for urban expansion with the arrival of new forms of agricultural development in the countryside (grape and wine production) that actually benefit from the earlier stages of urbanization of rural space, but ultimately suffer as this process reaches its fullest extent. Hence, in rural Niagara, as in other regions pursuing amenity-based opportunities, the partitioning of development versus preservationist interests defies easy categorization. Indeed, the goals of some fractions of the agricultural and food sector may indeed run contrary to prevailing ideologies for land preservation or the needs of productivist forms of farming. These will implicate different groups at the local level, not least of which are a host of state actors with development (development control) objectives of their own.

Community

The experiences of local communities are often the focal point for studies seeking to understand the dynamics of rural transformation. They provide windows into the diverse place specific implications of and response to the broader processes of rural restructuring. Though not intended as an exhaustive account of community-level impacts, the chapters in this section do paint an intriguing picture of the multifaceted experiences of rural change taking place in the North American context. Taken together, the works underscore three major themes.

First, the chapters draw out the diverse *types* of local impacts rendered by rural transformation. They bring to life important economic (Smithers), epistemological (Cadieux), historical (Algeo), socio-demographic (Barcus and Bruce), and socio-cultural (Furuseth) aspects of these processes as they are played out in real places.

Important social and economic implications of rural change at the community level are poignantly illustrated in Smither's chapter, which attempts to trace out the diverse ways family farmers are refashioning their relationships with local communities in southern Ontario. While some farmers continue to look to these towns as places of commercial agricultural exchange, other farmers have become increasingly reliant upon them as sources of supplemental employment or locations for retirement.

Cadieux's chapter speaks to the conceptual underpinnings driving much of the exurban migration to rural areas. The rural ideals envisioned by those engaging in 'rural tourism' can serve to alienate them from the productive processes that created the rural landscape in the first instance. Addressing and challenging these epistemologies is necessary for finding ways to engage new residents in their adopted communities, thereby achieving a better balance among the ways individuals reside, play and work in rural places.

The environment also provides opportunity for non-farm-based community development, as seen in relation to tourism-based initiatives. Indeed, the prospect of capitalizing on environmental assets in pursuit of local development is appealing at the community level, and the importance of local leadership is clear. Here the issues seem to revolve around opportunity and opportunity cost. As seen in both Kentucky and in Niagara, efforts to capitalize on nature and promote amenity can result from what are, in effect, local political projects designed to reinvent places around new interests and activities. To that end, some activities are more sought after than others. While the presence of distinctive environmental resources offers opportunity, it also mediates how other developments will be seen. Hence the question is not 'development or no development?' it's 'what kind of development?' 'By whom?' 'For Whom?' Even if the journey to an (economically) reoriented community is characterized by local successes and general advancement, the contributions to this book remind that the benefits of this transformation are distributed unevenly across and within communities, and some opportunities are let go in the pursuit of other forms of progress. Perhaps ultimately, the success of such initiatives will reside in their inclusiveness across the segments of local society.

Other significant indicators for mapping the social impacts of rural transformations are changes to the local housing markets. Chapters by Holly Barcus and David Bruce exemplify how rural economic restructuring is directly impacting marginal populations in rural areas by affecting the quality and availability of low cost housing. Striking in each of these chapters, is the eroded availability of suitable, appropriate and adequate housing for low income, elderly or minority populations not only in rural communities experiencing rapid socio-economic growth (e.g. as housing prices and property taxes soar, neighbourhoods and rural landscapes become gentrified), but also in the rural communities suffering overall economic decline. In the latter communities, the losses of traditional resource extraction-based industries have not been replaced with new post-productivist economic growth in sectors such as rural tourism, services or urbanization. In such places, economic and population decline create a surplus of housing, but in ever-worsening conditions.

The implications of rural change upon the socio-cultural and ethnic composition of local communities are illustrated in Owen Furuseth's case study of Mexican immigration to Montgomery County, North Carolina. In this instance, the transformation of traditional agriculture into increasingly mechanized, large-scale, integrated formulations has created a demand for low wage labor. The resultant influx of Latino migrants from Mexico to fill these jobs has enriched and diversified the rural socio-cultural landscape in ways that challenge stereotypical portrayals of the rural American South.

A second theme emerging from these chapters speaks to the diverse ways local residents have *responded* to the impacts of rural change, as reflected in the uneven trajectories of socio-economic change occurring within and between rural communities. Echoing many of the chapters in the agriculture section, John Smithers' study documents how some family farmers in Ontario are expanding and embracing new globalized industrial production regimes. Others, meanwhile, are looking to alternative and supplemental forms of income generation (by force or choice) in order to either stabilize or ease the decline of their agriculture-based livelihoods. Likewise, the chapters on housing issues focus upon some communities that have enjoyed service-sector socio-economic growth as a result of rural change and others that have suffered decline. However, the differing fates of rural communities are perhaps most severely illustrated in the contrast between Owen Furuseth's research on Montgomery County, North Carolina and Chris Maya's work on 'ecological ghost towns.' While Maya's case studies exemplify the final blow to rural communities unable to remain viable amidst changing social and ecological conditions, the socio-cultural change documented in Furuseth's study of Montgomery County is driven by the rapid growth and expansion of industrialized agriculture in the countryside.

Third and finally, the chapters underscore the *common global and regional forces* driving rural socio-economic change in rural North America. Despite place-based differences in the ways these forces have played out in rural communities – both in terms of local impacts and responses – the rural areas examined here all share the challenge of grappling with an increasingly globalized, dynamic and (neo-)liberalized socio-economic landscape. In all cases, the residents of rural communities – both old and new – are challenged to reconfigure local social, cultural and economic institutions, relationships and activities in the face of these broader processes of regional restructuring.

Coupled with chapters throughout this volume, the themes in this section underscore the complexities of rural transformation that strongly challenge oversimplified models of a linear rural transition from productivist to post-productivist landscapes (Murdoch *et al.*, 2003). Collectively, they are also emblematic of current trends in rural geography to better understand the experiences of 'hidden' cultures, identities and marginalized populations in rural communities that have often been ignored in studies of rural change (Cloke and Little, 1997).

Governance

One of the most dramatic changes associated with the dynamics of rural change are the newly emerging forms of rural governance, the decision making processes and institutions that integrate some measure of civil society with more formal governmental institutions at various spatial scales. The deregulatory tendencies of the neo-liberal economic policies of the past decades have been accompanied, in part, by shifting the regulatory burden (in terms of decision making, implementation, and/or enforcement) to lower tiers of government and by the creation of new collaborative 'partnerships' with a diverse range of actors, from private industry, to local residents, governmental and non-governmental agencies (Wondolleck and Yaffee, 2000).

The chapters in this section offer insights into this trend as it has played out in rural areas. Collectively, the studies stress the diversity of collaborative partnerships that are emerging as strategies for various actors seeking to take advantage of the new socio-economic conditions occasioned by rural restructuring. The chapters not only emphasize the complexity of these new collaborative approaches, but examine the ineffectiveness of more conventional decision making institutions in addressing the dynamics of rural decline (Deakin and Halseth). Several studies stress the obstacles to realizing egalitarian democratic ideals within the rubric of new collaborative partnerships, whether based upon conceptual and epistemological differences (Wilson) or incompatible social (Corcoran) or economic (Gill and Williams) interests. In short, while positive gains might be realized in terms of more effective decision making, greater inclusion in governance processes, and more socially and environmentally responsible forms of rural development, such outcomes are never guaranteed.

Ann Deakin's exploratory study of regional identities and governance begins to question the taken for granted county-based jurisdictional categories and boundaries that have traditionally served as the basis for public policy making and services provision in the USA. The historical, socio-economic and ecological investigation of Chataqua County, New York emphasizes the historical subjectivity and potential impermanence of these jurisdictional boundaries. The apparent incongruence between these boundaries and various regional identities opens the way for potential new governance forms based on these and other emergent consolidations of interests. Additional work on the dynamics of rural change in this region will help complete the picture of just how these new identities and interests emerge and come to assert themselves in new decision making forums.

Greg Halseth chapter further underscores the discordance between conventional governmental institutions – both local and provincial – and emerging grass roots interests and organizations in the face of declining resource dependent towns in British Columbia. Competing geographical imaginations of what rural British Columbia is and should be have stymied efforts to pursue alternative economic development initiatives. As diverse local interest groups debate the towns' future and local governments give lip service to economic diversification, the continued commitment of provincial governments to large-scale resource extraction enterprises dominates redevelopment action strategies. Such conditions are emblematic of those giving rise to collaborative community-state partnerships for resource management in the western USA. For advocates, these partnerships provide a means to resolve local conflicts and develop management strategies that can both improve the health of public lands and the socio-economic well-being of rural communities.

In practice, however, community-based resource management is not an obvious panacea to the challenges faced by resource dependent rural communities. Randall Wilson's chapter on the narratives of community-based collaboration in the governance of public lands in the American West highlights the challenges of bringing disparate rural residents together with federal land managers to resolve conflicts and improve resource decision making. As Wilson argues, merely bringing these interests together will not necessarily lead to the 'discovery' of some pre-defined common good, as some of the dominant narratives suggest. Nor does this always address questions of local participation, representation, or the political influence of local power relations. Rather, the success of community-based collaboration in public land management rests heavily upon the recognition of the political nature of constructing the common good via open dialogue, negotiation and debate among participants.

In addition to these conceptual concerns, challenges to new forms of governance include more pragmatic matters of organization, communication and bringing appropriate stakeholders to the table. Working in the context of UK, Christine Corcoran details the difficulties of conducting effective collaborative partnerships designed to target rural youth and poverty issues. Employing an action research paradigm, the researcher (Corcoran) acted as a bridge between local youth and a local government panel on social exclusion in the counties of Worcester and Hereford. While this arrangement was generally effective in transmitting the concerns of rural youths to government policy makers, the efforts were complicated by the lack of direct participation by the youth themselves in partnership deliberations.

But what happens when a large transnational corporation is one of two major stakeholders in a collaborative partnership? Alison Gill and Peter Williams explore this question in their chapter on corporate-community relations in British Columbia. Their case study of a ski resort town in Whistler BC, documents how a distinct local community interest emerged from this 'company town' created and managed by the Intrawest Corporation. While one might suspect corporate-community collaborations to be little more than a marketing tool that masks corporate efforts to generate maximum short-term profits at the expense of long-term local interests, Gill and Williams' nuanced examination of this partnership identifies several social and environmental initiatives that have benefited the local community. However, whether these outcomes – reached via agreements that also benefited Intrawest economically – are enough to characterize the partnership as something other than a shrewd business relationship remains an open question.

These chapters could not possibly represent the full range of geographic scholarship on rural governance. However, they do offer a sampling of the diverse forms of collaborative partnerships emerging in the rural areas of Canada, the UK and the USA. How to balance and accommodate the interests of existing state and local governmental agencies, large corporate actors, trade groups, non-governmental organizations, and a growing number of local residents, in a decision making framework that will encourage more sustainable and equitable socio-environmental relations in rural areas is a challenge faced by rural residents, planners and researchers in all three countries.

References

Cloke, P. and Little, J. (eds.) (1997) *Contested Countryside Cultures: Otherness, Marginalization and Rurality.* Routledge, London.

Evans, N., Morris, C. and Winter, M. (2002). Conceptualizing agriculture: a critique of post-productivism as the new orthodoxy. *Progress in Human Geography*, 26(3), 313-332.

Goodman, D. (2003). The quality 'turn' and alternative food practices: reflections and agenda. *Journal of Rural Studies*, 19(1): 1-7.

Ilbery, B., Bowler, I. (1998) From agricultural productivism to post-productivism. In: Ilbery, B. (ed.) *The Geography of Rural Change.* Longman, Essex, pp.57-84.

Murdoch, J., Lowe, P., Ward, N., and Marsden, T. (2003) *The Differentiated Countryside.* Routledge, London.

Smithers, J & Johnson, P. (2004). The dynamics of family farming in North Huron County, Ontario. Part I. Development Trajectories. *The Canadian Geographer*, Vol.48, No.2, pp.191-208.

Walford, N. (1999) Geographical transition from productivism to post-productivism: agricultural production in England and Wales 1950s to 1990s. In: Walford, N. Everitt, J.C., and Napton, D.E. (eds) *Reshaping the Countryside: Perceptions and Processes of Rural Change.* CAB International, Wallingford, 25-38.

Wilson, G.A. (2001) From productivism to post-productivism...and back again? Exploring the (un)changed natural and mental landscapes of European agriculture. *Transactions of the Institute of British Geographers*, 26, 77-102.

Wondolleck, J., and Yaffee, S. (2000) *Making Collaboration Work: Lessons from Innovation in Natural Resource Management.* Island Press, Washington DC.

Index